CBS Problems & Solutions Series

Problems & Solutions of
Electric Circuit
Analysis

Problems & Solutions of Electric Circuit Analysis

R.K. Mehta
Senior Faculty Member
Dept. of Electrical Engineering,
North Eastern Regional Institute of Science & Technology, Nirjuli,
Arunachal Pradesh

A.K. Mal
Senior Faculty Member
Dept. of Electronics & Communication Engineering,
North Eastern Regional Institute of Science & Technology, Nirjuli,
Arunachal Pradesh

C B S

CBS Publishers & Distributors Pvt. Ltd.

New Delhi • Bengaluru • Chennai • Kochi • Kolkata • Mumbai
Hyderabad • Uttarakhand • Nagpur • Patna • Pune • Jharkhand

ISBN: 81-239-0856-3

First Edition: 2002
Reprint: 2006, 2008, 2011, 2015, 2020

Published by **Satish Kumar Jain** and produced by **Varun Jain** for
CBS Publishers & Distributors Pvt. Ltd.,
4819/XI Prahlad Street, 24 Ansari Road, Daryaganj, New Delhi - 110002
delhi@cbspd.com, cbspubs@airtelmail.in • www.cbspd.com
Ph.: 23289259, 23266861, 23266867 • Fax: 011-23243014

Corporate Office: 204 FIE, Industrial Area, Patparganj, Delhi - 110 092
Ph: 49344934 • Fax: 011-49344935
E-mail: publishing@cbspd.com • publicity@cbspd.com

Branches:
• *Bengaluru:* 2975, 17th Cross, K.R. Road, Bansankari 2nd Stage,
  Bengaluru - 70 • Ph: +91-80-26771678/79 • Fax: +91-80-26771680
  E-mail: cbsbng@gmail.com, bangalore@cbspd.com
• *Chennai:* No. 7, Subbaraya Street, Shenoy Nagar, Chennai - 600030
  Ph: +91-44-26681266, 26680620 • Fax: +91-44-42032115
  E-mail: chennai@cbspd.com
• *Kochi:* Ashana House, 39/1904, A.M. Thomas Road, Valanjambalam,
  Ernakulum, Kochi • Ph: +91-484-4059061-65
  Fax: +91-484-4059065 • E-mail: cochin@cbspd.com
• *Kolkata:* 6-B, Ground Floor, Rameshwar Shaw Road, Kolkata - 700014
  Ph: +91-33-22891126/7/8 • E-mail: kolkata@cbspd.com
• *Mumbai:* 83-C, Dr. E. Moses Road, Worli, Mumbai - 400018
  Ph: +91-9833017933, 022-24902340/41 • E-mail: mumbai@cbspd.com

Representatives:

• Hyderabad: 0-9885175004   • Nagpur: 0-9021734563
• Patna: 0-9334159340   • Pune: 0-9623451994
• Jharkhand: 0-9811541605   • Uttarakhand: 0-9716462459

Printed at:
J.S. Offset Printers, Delhi (India)

Preface

"Electric Circuit Analysis" is one of the important subjects for students at the degree and diploma levels. The subject is taught in the degree course for Electrical and Electronics & Communication engineering for one semester course.

This book, "Problems & Solutions in Electric Circuit Analysis" is written on the basis of long experience of authors and feedbacks from students. They feel that the concept of subject becomes easy and can be better understood through examples. Exploring a new principle "less read and much understand" the book has been designed in that form. The book assumes that the students have prior knowledge of introductory course in electric circuits.

Each chapter of the book presents a systematic approach for understanding the subject. Attempts have been made to include wide variety of solved and drill problems. Drill problems have been given sufficient hints for understanding them quickly.

The chapter 1 to 3 explain the dc and ac transients in the first order and second order circuits. The chapter 4 to 5 explain the about the solution through the Laplace transform methods. The chapter 6 is dedicated to explaining the solutions to circuits with complex exponential inputs. The chapter 7 explains the steady-state solutions to circuits with sinusoidal inputs. Chapters 8 and 9 explain about the network functions and two-port parameters. Appendix A gives the concepts of coupled inductors. The Cramer's rule is explained in Appendix B. Appendix C contains objective and multiple choice questions with sufficient explanations.

The book in this shape will be quite helpful to students and teachers at various engineering colleges in India.

We feel pleasure to express our gratitude to the following who have helped and encouraged us in preparing the manuscript of the book: Professor P. D. Kashyap, HOD, Electrical Engg, NERIST, Nirjuli, Professor K. Kumar, Dean, P & D NERIST, Nirjuli, Professor S. K. Nagar, Dept. of Electrical Engg. IT, BHU, Varanasi, Mr. T. J. Singh, Mr. S. K. Bhagat, Mr. Sibal Chatterji, all of NERIST, Nirjuli.

Finally, we would like to appreciate to our many of former students who assisted us in checking the results for accuracy. We also want to express gratitude to Mr. H. S. Poplai, General Manager, CBS, Publishers & Distributor, and his teammate for their enthusiastic cooperation in publishing the book in a very short span of time.

Suggestions and comments for the further modification and improvement of the book will be highly appreciated.

Authors.

Contents

First-Order Circuits

SUMMARY OF THIS CHAPTER

1.1 CIRCUIT ELEMENTS

The following tables show quantities and their governing equations for the resistor, the inductor and the capacitor.

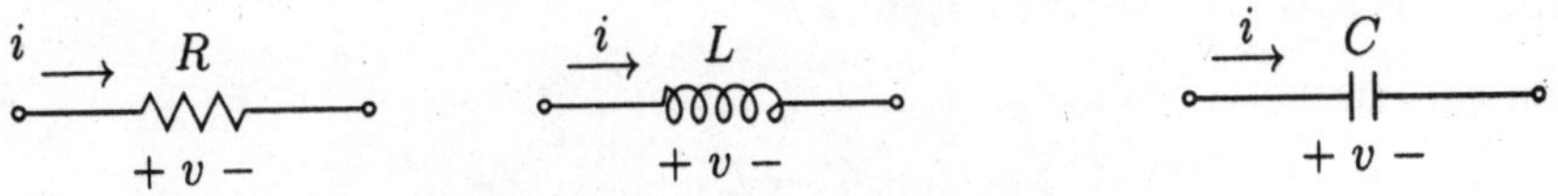

Figure 1.1: Circuit elements.

Resistor

Quantity	Unit	Equation
Voltage	volt (V)	$v = iR$ (Ohm's Law)
Current	ampere (A)	$i = v/R$
Power	watt (W)	$p_R = i^2 R$
Energy	joule (J)	$w_R = \int_0^t p_R dt = i^2 R t$

Inductor

Quantity	Equation
Voltage	$v = L\frac{di}{dt}$
Current	$i = \frac{1}{L}\int_{-\infty}^t v\,dt$
	or, $i = \frac{1}{L}\int_{-\infty}^0 v\,dt + \frac{1}{L}\int_0^t v\,dt$
	or, $i = i(0^-) + \frac{1}{L}\int_0^t v\,dt$
	where, $i(0^-)$ = initial current through the inductor
Power	$p_L = vi = L\frac{di}{dt}i = Li\frac{di}{dt}$
Energy	$w_L = \int_0^t p_L dt = \frac{1}{2}Li^2$

1

Capacitor

Quantity	Equation
Voltage	$v = \frac{1}{C} \int_{-\infty}^{t} i\, dt = v(0^-) + \frac{1}{C} \int_{0}^{t} i\, dt$ where, $v(0^-) =$ initial voltage across the capacitor
Current	$i = C \frac{dv}{dt}$
Power	$p_C = vi = Cv \frac{dv}{dt}$
Energy	$w_C = \int_{0}^{t} p_C \, dt = \frac{1}{2} C v^2 = \frac{1}{2} \frac{q^2}{C} = \frac{1}{2} qv$

Combination of Resistors

Resistors In Series: $R_{eq} = R_1 + R_2 + R_3 + \cdots\cdots + R_n$

Resistors In Parallel: $\frac{1}{R_{eq}} = \frac{1}{R_1} + \frac{1}{R_2} + \cdots\cdots + \frac{1}{R_n}$

Combination of Inductors

Inductors In Series: $L_{eq} = L_1 + L_2 + L_3 + \cdots\cdots + L_n$

Inductors In Parallel: $\frac{1}{L_{eq}} = \frac{1}{L_1} + \frac{1}{L_2} + \cdots\cdots + \frac{1}{L_n}$

Combination of Capacitors

Capacitors In Series: $\frac{1}{C_{eq}} = \frac{1}{C_1} + \frac{1}{C_2} + \cdots\cdots + \frac{1}{C_n}$

Capacitors In Parallel: $C_{eq} = C_1 + C_2 + C_3 + \cdots\cdots + C_n$

1.2   KIRCHHOFF'S VOLTAGE LAW (KVL)

At any point of time, the algebraic sum of voltages around any closed path in a specified direction is zero.

mathematically,

$$\sum v = 0$$

or

$$\sum v_{\text{rise}} = \sum v_{\text{drop}}$$

1.3   KIRCHHOFF'S CURRENT LAW (KCL)

At any point of time, the algebraic sum of all currents entering a node is zero.

mathematically,

$$\sum i = 0$$

or

$$\sum i_{\text{incoming}} = i_{\text{outgoing}}$$

1.4 STAR-DELTA TRANSFORMATION

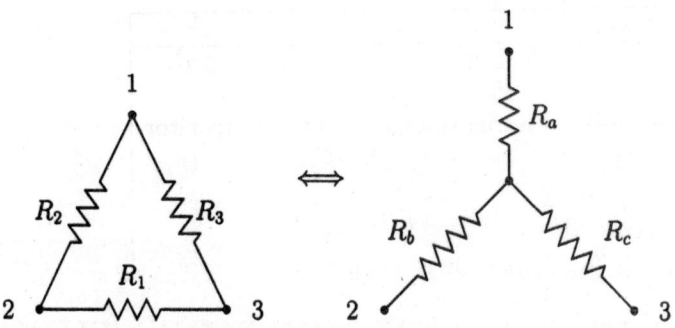

Figure 1.2: Showing star-delta transformation.

The star-delta transformation is also called wye-delta transformation. The element of the star connected network can be expressed in terms of the elements of delta connected network as:

$$R_a = \frac{R_2 R_3}{R_1 + R_2 + R_3}$$

$$R_b = \frac{R_1 R_3}{R_1 + R_2 + R_3}$$

$$R_c = \frac{R_1 R_2}{R_1 + R_2 + R_3}$$

The element of the delta connected network can be expressed in terms of the elements of the star connected network as:

$$R_1 = R_b + R_c + \frac{R_b R_c}{R_a}$$

$$R_2 = R_a + R_c + \frac{R_a R_c}{R_b}$$

$$R_3 = R_a + R_b + \frac{R_a R_b}{R_c}$$

1.5 VOLTAGE AND CURRENT DIVISION

With reference to Figure 1.3(a)

$$V_1 = \frac{R_1}{R_1 + R_2} V \qquad \text{and} \qquad V_2 = \frac{R_2}{R_1 + R_2} V$$

With reference to Figure 1.3(b)

$$I_1 = \frac{R_2}{R_1 + R_2} I \qquad \text{and} \qquad I_2 = \frac{R_1}{R_1 + R_2} I$$

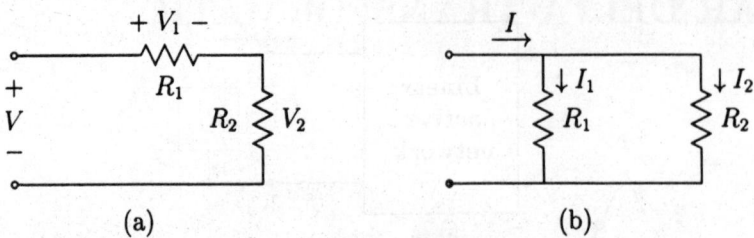

Figure 1.3: (a) Voltage division circuit; (b) Current division circuit.

1.6 THEVENIN'S AND NORTON'S THEOREMS

The figure 1.4(i) shows that the active linear network is supplying power to the load R_L. This network between terminal a and b may be equivalently replaced by a series circuit model or by a parallel circuit model. The figure 1.4(ii) shows the series circuit model containing one voltage source, V_{Th} in series with one resistance, R_{Th}. This model is called Thevenin's equivalent network. Similarly, the figure 1.4(iii) represents parallel circuit model containing one current source, I_N, in parallel with one resistance, R_N. This model is known as Norton's equivalent network. The Thevenin's equivalent circuit and the Norton's equivalent circuit parameters are defined as follows:

V_{Th} is the open circuit voltage across the terminals of interest (say a and b). It is given by $V_{Th} = V_{ab}$, when load is removed and terminals a and b are left open.

I_N is the short circuit current through the terminals of interest (say a and b). It is given by $I_N = I_{ab}$, when load is removed and terminals a and b are shorted.

(i) Network containing resistors and independent sources:

R_{Th} is the open circuit input resistance seen through the terminals of interest (say a and b). It is given by $R_{Th} = R_{in}$ when all the voltage sources are replaced by shorts and all the current sources are replaced by opens.

(ii) Network containing resistors, independent sources and dependent sources:

R_{Th} is the open circuit input resistance seen through the terminals of interest (say a and b) when all the voltage sources are replaced by shorts; all the current sources are replaced by opens and all the dependent sources are left unchanged.

R_N is equal to R_{Th}.

1.7 FIRST-ORDER DIFFERENTIAL EQUATION

The general form of first-order differential equation is

$$\frac{dy}{dt} + a_1 y = x(t) \tag{1.1}$$

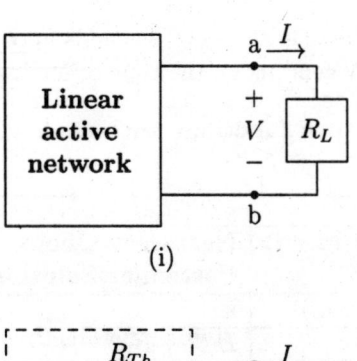

(i)

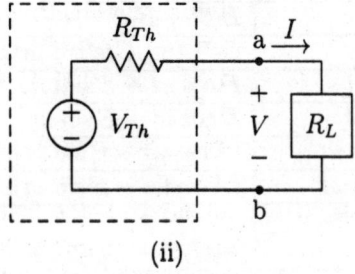

(ii)

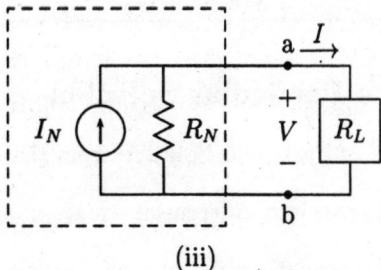

(iii)

Figure 1.4: (i) A linear network; (ii) Thevenin's equivalent circuit; (iii) Norton's equivalent circuit.

where
y = dependent variable (or response)
x = independent variable (or forcing function, or excitation)

The complete solution to above differential equation will be given by

$$y(t) = y_c(t) + y_p(t) \tag{1.2}$$

where

$y_c(t)$ = complementary solution (or natural solution, $y_n(t)$, or transient solution, y_{tr})
$y_p(t)$ = particular solution (or steady-state solution, y_{ss})

The form of the complementary solution to the differential eqn (1.1) is given by:

$$y_c(t) = Ke^{-a_1 t} \tag{1.3}$$

The form of the particular solution, $y_p(t)$ depends upon the forcing function, $x(t)$, and is selected from table 1.1 depending on the type of forcing function.

Table 1.1: Showing input functions with their choices for the particular solutions

	Factor in $x(t)$	Necessary Choice for Particular Solution $y_p(t)$
1	V (a const.)	k
2	$k_1 t^n$	$B_0 t^n + B_1 t^{n-1} + \cdots + B_{n-1}t + B_n$
3	$k_2 e^{rt}$	$C e^{rt}$ (or, $C t e^{rt}$ if first choice fails)
4	$k_3 \cos \omega t$	$D \cos \omega t + E \sin \omega t$
5	$k_4 \sin \omega t$	$D \cos \omega t + E \sin \omega t$
6	$k_5 e^{rt} \cos \omega t$	$(A \cos \omega t + B \sin \omega t)e^{rt}$
7	$k_6 e^{rt} \sin \omega t$	$(A \cos \omega t + B \sin \omega t)e^{rt}$
8	$k_7 t^n e^{rt} \cos \omega t$	$(F_1 t^n + F_2 t^{n-1} + \cdots + F_n)e^{rt} \cos \omega t$ $+(G_1 t^n + G_2 t^{n-1} + \cdots + G_n)e^{rt} \sin \omega t$
9	$k_8 t^n e^{rt} \sin \omega t$	$(F_1 t^n + F_2 t^{n-1} + \cdots + F_n)e^{rt} \cos \omega t$ $+(G_1 t^n + G_2 t^{n-1} + \cdots + G_n)e^{rt} \sin \omega t$

Steps to Determine Particular Solution

1. Write the trial form of Particular Solution from the table 1.1.

2. Substitute the trial form into differential equation.

3. Form a set of algebraic equations in unknown coefficients by equating the coefficients of like terms on both sides of the equation obtained from the step (2).

4. Solve for the unknown coefficient and so determine particular solution (PS).

The complete solution is obtained by:

$$y(t) = \text{CS} + \text{PS}$$

where, $\text{CS} = K e^{-a_1 t}$ (from eqn 1.3) and PS can be determined by the above steps. Now this constant K is determined from the knowledge of initial condition, i.e.; $y(0)$. As a precaution, the initial condition must always be applied to the total solution, not to the complementary solution alone.

- To determine $i_L(0)$ in the RL circuit, consider the steady state circuit for $t < 0$ and to determine $v_L(0)$, consider the circuit at $t = 0^+$.

- To determine $v_C(0)$ in RC circuit, consider the steady state circuit for $t < 0$ and to determine $i_C(0)$, consider the circuit at $t = 0^+$

- The $t < 0$, steady-state circuit is the same as the $t = 0^-$ circuit.

1.8 SOLUTION BY INSPECTION

For circuits containing one energy storage element, solutions may be obtained as follows:

1.8.1 RL circuit

The total solution for the inductor current for $t > 0$,

$$i_L(t) = [i_L(0^+) - i_{L,ss}]e^{-\frac{R_{Th}}{L}t} + i_{L,ss} \tag{1.4}$$

where $i_L(0^+)$ inductor current at $t = 0^+$,

          $i_{L,ss}$ steady state value of inductor current in the $t > 0$ circuit,

          R_{Th} Thevenin resistance seen by the inductor for the $t > 0$ circuit,

          $\frac{L}{R_{Th}}$ time constant of RL circuit.

The solution steps are:

1. Draw the $t > 0$ circuit.

2. Find the Thevenin resistance, R_{Th} seen by inductor terminals in the $t > 0$ circuit.

3. Find the steady state value of the inductor current, $i_{L,ss}$ in the $t > 0$ circuit.

4. Draw the $t < 0$ circuit and find the steady-state value of the inductor current, $i_L(0^-)$.

5. By continuity of inductor current, obtain $i_L(0^+) = i_L(0^-)$.

6. Write the solution for inductor current for $t > 0$

$$i_L(t) = [i_L(0^+) - i_{L,ss}]e^{-\frac{R_{Th}}{L}t} + i_{L,ss}$$

Note: Current through an inductor can not change instantaneously whereas voltage across it may change abruptly. Hence

$$i_L(0^-) = i_L(0^+)$$
$$v_L(0^-) \neq v_L(0^+)$$

1.8.2 RC-circuit

Total solution for the capacitor voltage for $t > 0$ is written as:

$$v_C(t) = [v_C(0^+) - v_{C,ss}]e^{-\frac{t}{R_{Th}C}} + v_{C,ss} \tag{1.5}$$

where $v_C(0_+)$ capacitor voltage at $t = 0^+$,

       v_{Css} steady state capacitor voltage in the $t > 0$ circuit,

       R_{Th} Thevenin resistance seen by the capacitor terminals in the $t > 0$ circuit,

       $R_{Th}C$ time constant of RC-circuit.

Note : *Voltage across capacitor can not change instantaneously although current through it may change abruptly.* Hence

$$v_C(0^+) = v_C(0^-)$$
$$i_C(0^-) \neq i_C(0^+)$$

1.9   SOLUTION BY INTEGRATION

Consider a first-order differential equation with a forcing function (or input function) written

$$\frac{dv}{dt} + Pv = Q \tag{1.6}$$

where P is a constant and Q may be a function of the independent variable t or a constant. The complete solution to this equation is given by

$$v = e^{-Pt} \int Qe^{Pt}dt + Ke^{-Pt} \tag{1.7}$$

The first term in eqn. 1.7 is called the particular integral and second term is called the complementary function. Note that the complementary function contains the arbitrary constant, K but does not depends on the forcing function, Q, and particular integral does not have arbitrary constant, but depends on forcing function, Q.

1.10   ZERO-INPUT AND ZERO-STATE RESPONSES

The zero-input response (ZIR) of any linear circuit is the response due to initial conditions with all the sources dead. It can be determined from the complementary (natural) solution using its initial conditions. The zero-state response (ZSR) is the complete response under zero initial conditions. Thus, we can write that

Complete Solution(CS) = ZIR + ZSR

1.11   BEHAVIOUR OF CIRCUIT ELEMENTS WITH TIME

The figure 1.5 shows the equivalent circuits of resistor, inductor, and capacitor at $t = 0$ and at $t = \infty$.

Circuit Model of Capacitor and Inductor with Their Initial Conditions

The capacitor with an initial condition (charge or voltage) may be treated as a series circuit model containing the capacitor in series with the voltage source. The inductor with an initial condition (magnetic flux or current) may be treated as a parallel circuit model containing the inductor in parallel with the current source. The figure 1.6 shows the circuit model for a capacitor and an inductor with their initial conditions.

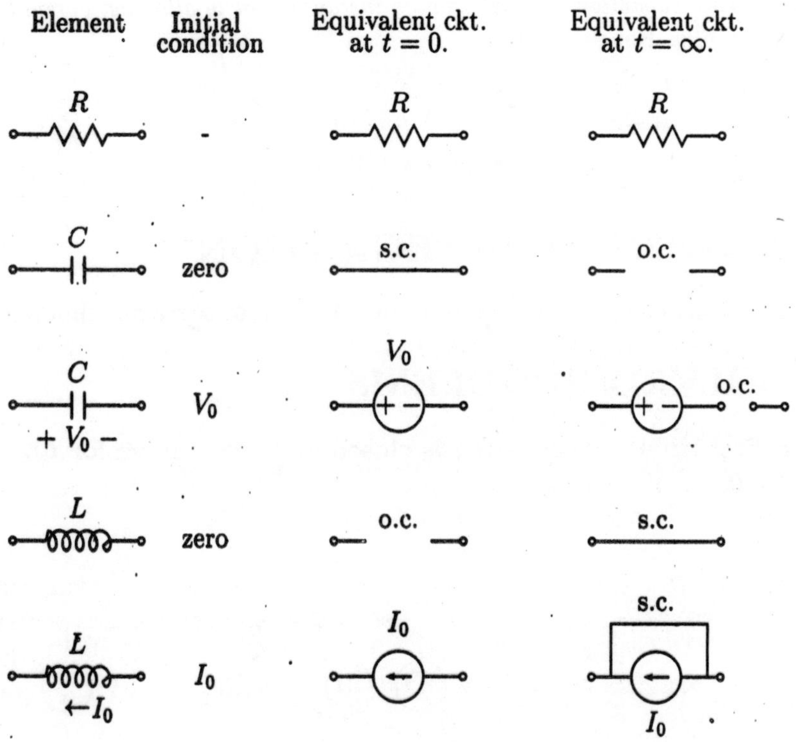

Element	Initial condition	Equivalent ckt. at $t = 0$.	Equivalent ckt. at $t = \infty$.
R (resistor)	-	R (resistor)	R (resistor)
C (capacitor)	zero	s.c.	o.c.
C (capacitor) $+ V_0 -$	V_0	V_0 (source)	V_0 (source) o.c.
L (inductor)	zero	o.c.	s.c.
L (inductor) $\leftarrow I_0$	I_0	I_0 (source)	s.c. I_0 (source)

Figure 1.5: Equivalent circuit at $t = 0$ and $t = \infty$.

Time-Constant

Consider a general response equation for a constant input

$$i(t) = I_{ss}(1 - e^{-t/\tau}) \tag{1.8}$$

and a general response equation for an unforced circuit

$$i(t) = I_0 e^{-t/\tau} \tag{1.9}$$

where τ is the time-constant of a circuit, I_{ss} is the steady-state value of $i(t)$ and I_0 is the initial condition of $i(t)$.

Some facts about the time-constant are:

- Time constant, τ is the time taken for the response to reach 63.2% of steady state value with reference to eqn 1.8.

- Time constant is the time taken for the response to decay to 36.8% of the initial change with reference to eqn 1.9.

- Time required for the response to reach steady state equals to 4τ sec. practically.

- Transient response stays approximately for 4τ sec.

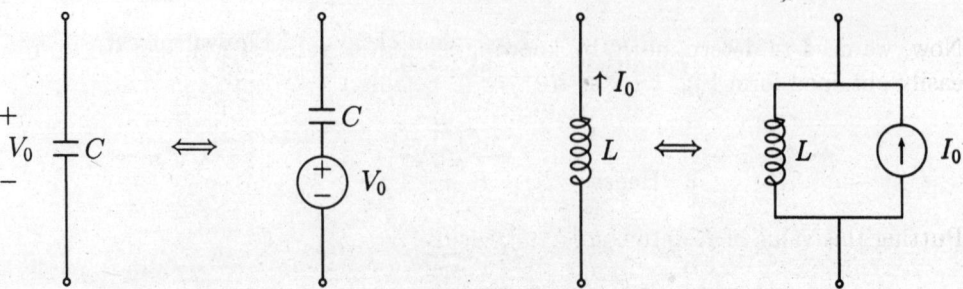

Figure 1.6: Circuit model for a capacitor and an inductor containing initial conditions.

1.12 SOLVED PROBLEMS

SP 1.1 In Fig. SP 1.1, the switch is closed at t = 0. Solve for $i(t)$, $v_L(t)$, and $v_R(t)$ for $t > 0$.

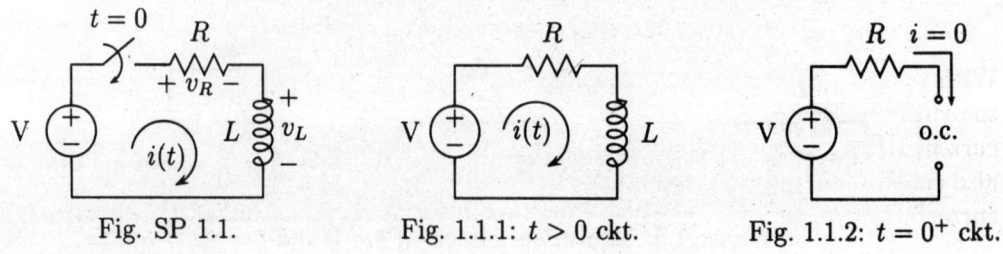

Fig. SP 1.1.   Fig. 1.1.1: $t > 0$ ckt.   Fig. 1.1.2: $t = 0^+$ ckt.

SOLUTION:

By KVL in the circuit of Fig. 1.1.1

$$Ri + L\frac{di}{dt} = V$$

Differentiating and rearranging

$$\frac{di}{dt} + \frac{R}{L}i = \frac{V}{L} \qquad [1.1.1]$$

Form of natural solution to eqn [1.1.1]

$$i_n = Ke^{-\frac{R}{L}t} \qquad [1.1.2]$$

Form of particular solution corresponding to constant input

$$i_p = k_1 \qquad [1.1.3]$$

Putting the value of $i = i_p = k_1$ into eqn [1.1.1] as

$$\frac{d}{dt}(k_1) + \frac{R}{L}(k_1) = \frac{V}{L} \quad \text{or,} \ 0 + \frac{R}{L}k_1 = \frac{V}{L} \quad \text{Thus,} \ k_1 = \frac{V}{R} = i_p \qquad [1.1.4]$$

The complete solution is

$$i(t) = i_n + i_p = Ke^{-\frac{R}{L}t} + V/R \qquad [1.1.5]$$

Now, we need to determine K by knowing the initial value of $i(t)$, i.e., $i(0^+)$. It can be easily obtained form Fig. 1.1.2 as $i(0^+) = 0$. Setting t = 0^+ into eqn [1.1.5],

$$i(0^+) = 0 = K + V/R$$
$$\text{Hence} \qquad K = -V/R$$

Putting this value of K into eqn [1.1.5], yields

$$i(t) = \frac{V}{R}(1 - e^{-\frac{R}{L}t}) \qquad \text{for t} > 0 \qquad [1.1.6]$$

The eqn [1.1.6] can be written in time constant form as:

$$i(t) = \frac{V}{R}(1 - e^{-t/\tau}) \qquad [1.1.7]$$

where τ is the time constant and will be given for the series RL-circuit by

$$\tau = \frac{L}{R} = \frac{\text{total inductance connected in series along with } R}{\text{total resistance connected in series along with } L} \qquad [1.1.8]$$

With reference to eqn [1.1.7], the initial value of the current is zero and final value of the current is $i(\infty) = V/R$. Thus, plot of the eqn [1.1.7] is an exponential curve. The current $i(t)$ rises exponentially and finally becomes constant, V/R at $t = \infty$. The V/R is also called the steady-state value of current $i(t)$. It is interesting to find the value of the current at $t = \tau$ as:

$$i(\tau) = \frac{V}{R}(1 - e^{-\tau/\tau}) = 0.63\frac{V}{R} \qquad [1.1.9]$$

Thus the value of the current at $t = \tau$ is 63% of the steady-state value. The speed with which $i(t)$ is increasing is called the speed of response and is given by the rate of change of current, i.e., di/dt. So from eqn [1.1.6]

$$\frac{di(t)}{dt} = \frac{V}{R}\frac{R}{L}e^{-\frac{R}{L}t} = \frac{V}{L}e^{-\frac{R}{L}t} \qquad [1.1.10]$$

The value of the $di(t)/dt$ at $t = 0$ is V/L, which is slope of the curve at $t = 0$. The slope is decreasing with the time. The plot of $i(t)$ is shown in Fig. 1.1.3. The time span during which current is changing toward its steady-state value is called the transient time.

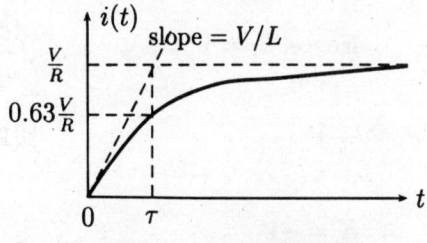

Fig. 1.1.3: Plot of $i(t)$.

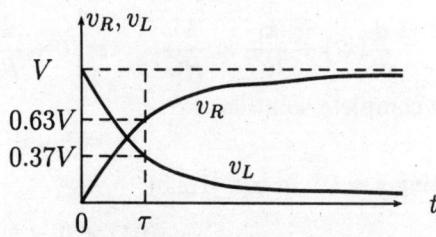

Fig. 1.1.4: Plot of v_L and v_R.

The voltage across the resistor is

$$v_R(t) = Ri(t) = V(1 - e^{-\frac{R}{L}t}) \qquad \text{for t} > 0 \qquad [1.1.11]$$

and the voltage across the inductor is

$$v_L(t) = V - v_R(t) = Ve^{-\frac{R}{L}t}$$ [1.1.12]

Thus, we find that the $v_R(t)$ increases with time while $v_L(t)$ decreases with time. At $t = \infty$, the v_R becomes V and the v_L becomes zero. The plots of $v_R(t)$ and $v_L(t)$ has been shown in Fig. 1.1.4.

SP 1.2 In Fig. SP 1.2, the switch is closed at t = 0. Solve for $v_C(t)$ for $t > 0$.

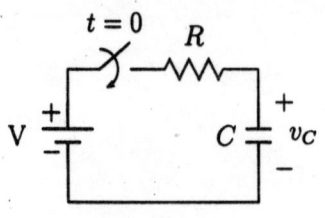 
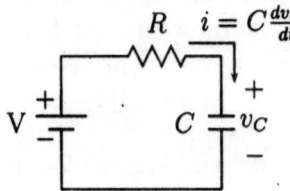
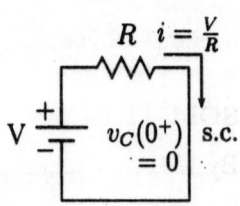

Fig. SP 1.2.   Fig. 1.2.1: $t > 0$ ckt.   Fig. 1.2.2: $t = 0^+$ ckt.

SOLUTION:

By KVL in the circuit of Fig. 1.2.1

$$Ri + v_C = V$$

Setting $i = C\frac{dv_C}{dt}$ in this equation,

$$RC\frac{dv_C}{dt} + v_C = V$$

Thus

$$\frac{dv_C}{dt} + \frac{1}{RC}v_C = \frac{V}{RC}$$ [1.2.1]

Natural solution

$$v_{Cn} = Ke^{-\frac{t}{RC}}$$ [1.2.2]

Particular solution corresponding to constant input :

$$v_{Cp} = k_1$$ [1.2.3]

Substituting eqn [1.2.3] into eqn [1.2.1]

$$\frac{d}{dt}(k_1) + \frac{k_1}{RC} = \frac{V}{RC} \quad \text{or, } 0 + \frac{k_1}{RC} = \frac{V}{RC} \quad \text{Hence, } k_1 = V = v_{Cp}$$ [1.2.4]

The complete solution is

$$v_C = v_{Cn} + v_{Cp} = Ke^{-\frac{t}{RC}} + V$$ [1.2.5]

Setting t $= 0^+$ in eqn [1.2.5]

$$v_C(0^+) = 0 = K + V \qquad \Rightarrow K = -V$$

(As from Fig. 1.2.2, $v_C(0^+) = 0$)

Therefore $v_C(t) = V(1 - e^{-\frac{t}{RC}})$ for $t > 0$

SP 1.3 In Fig. SP 1.3, the switch is closed at $t = 0$. Find the expression for $q(t)$ for $t > 0$ if initial charge across the capacitor is zero.

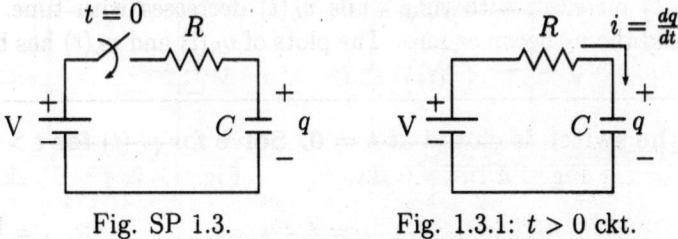

Fig. SP 1.3.   Fig. 1.3.1: $t > 0$ ckt.

SOLUTION:

By KVL in the circuit of Fig. 1.3.1

$$Ri + \frac{1}{C} \int_0^t i \, dt = V \qquad [1.3.1]$$

Using $i = \frac{dq}{dt}$

$$R\frac{dq}{dt} + \frac{q}{C} = V$$

Thus $\qquad \frac{dq}{dt} + \frac{1}{RC}q = \frac{V}{R} \qquad [1.3.2]$

Form of complementary solution

$$q_c = Ke^{-\frac{t}{RC}} \qquad [1.3.3]$$

Form of particular solution for constant excitation:

$$q_p = k_0 \qquad [1.3.4]$$

Substituting eqn [1.3.4] into eqn [1.3.2] gives

$$\frac{d}{dt}(k_0) + \frac{1}{RC}(k_0) = \frac{V}{R} \qquad \text{or, } 0 + \frac{k_0}{RC} = \frac{V}{R} \qquad \text{Thus, } k_0 = CV = q_p \qquad [1.3.5]$$

Complete solution

$$q(t) = q_c + q_p = Ke^{\frac{-t}{RC}} + CV \qquad [1.3.6]$$

Setting $t = 0$ in eqn [1.3.6]

$$q(0) = 0 = K + CV \qquad \Rightarrow K = -CV$$

Therefore, the complete solution is

$$q(t) = CV(1 - e^{\frac{-t}{RC}}) \qquad \text{for t} > 0$$

SP 1.4 In Fig. SP 1.4, the switch is closed at t = 0. Solve for $i(t)$ for $t > 0$.

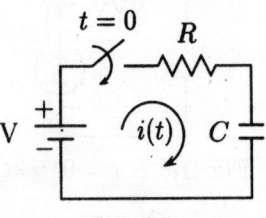

Fig. SP 1.4.   Fig. 1.4.1: $t > 0$ ckt.

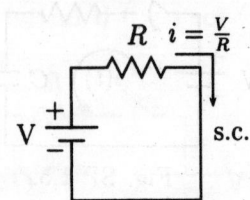

Fig. 1.4.2: $t = 0^+$ ckt.

SOLUTION:

By KVL in the circuit of Fig. 1.4.1

$$Ri + \frac{1}{C}\int_0^t idt = V \qquad\qquad [1.4.1]$$

Differentiating

$$R\frac{di}{dt} + \frac{i}{C} = 0$$

Thus $\qquad \dfrac{di}{dt} + \dfrac{i}{RC} = 0 \qquad\qquad [1.4.2]$

Form of natural solution

$$i_n = Ke^{-\frac{t}{RC}} \qquad\qquad [1.4.3]$$

PS is zero in absence of forcing function, hence the complete solution is

$$i(t) = i_n = Ke^{-\frac{t}{RC}} \qquad\qquad [1.4.4]$$

Setting t = 0^+ in eqn [1.4.4]

$$i(0^+) = \frac{V}{R} = K$$

Putting the value K into eqn [1.4.4]

$$i(t) = \frac{V}{R}e^{-\frac{t}{RC}} \qquad \text{for } t > 0$$

SP 1.5 In the Fig. SP 1.5, the switch is closed at $t = 0$. The initial voltage across the capacitor is V_0. Find the $i(t)$ for $t > 0$.

SOLUTION:

By KVL in the circuit of Fig.1.5.1

$$Ri + V_0 + \frac{1}{C}\int_0^t idt = V \qquad\qquad [1.5.1]$$

Differentiating

$$R\frac{di}{dt} + \frac{i}{C} = 0$$

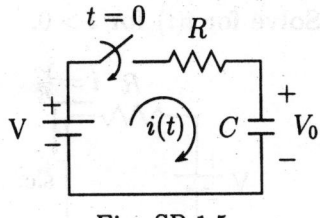

Fig. SP 1.5. Fig. 1.5.1: $t > 0$ ckt. Fig. 1.5.2: $t = 0^+$ ckt.

$$\text{Thus} \qquad \frac{di}{dt} + \frac{i}{RC} = 0 \qquad\qquad [1.5.2]$$

Form of solution to eqn [1.5.2]

$$i(t) = Ke^{-t/RC} \qquad\qquad [1.5.3]$$

From the Fig. 1.5.2

$$i(0^+) = \frac{V - V_0}{R} \qquad\qquad [1.5.4]$$

Setting $t = 0^+$ in eqn [1.5.3]

$$i(0^+) = \frac{V - V_0}{R} = K$$

$$\text{Hence} \qquad i(t) = \frac{V - V_0}{R} e^{-t/RC} \qquad \text{for } t > 0$$

SP 1.6 In the network shown in Fig. SP 1.6, the switch is closed at $t = 0$. Find $i(t)$ for $t > 0$, if (a) $v(t) = e^{-2t}$; (b) $v(t) = e^{-t}$; and (c) $v(t) = t$.

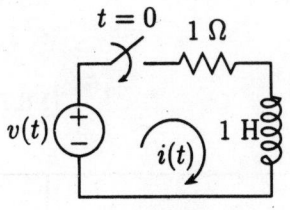
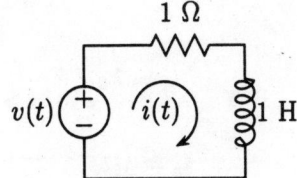
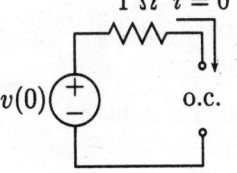

Fig. SP 1.6. Fig. 1.6.1: $t > 0$ ckt. Fig. 1.6.2: $t = 0^+$ ckt.

SOLUTION:

(a) By KVL in the circuit of Fig. 1.6.1

$$i + \frac{di}{dt} = v(t)$$

Thus

$$\frac{di}{dt} + i = e^{-2t} \qquad\qquad [1.6.1]$$

Form of natural solution to equation [1.6.1]

$$i_n = ke^{-t} \qquad [1.6.2]$$

Form of particular solution

$$i_p = k_0 e^{-2t} \qquad [1.6.3]$$

Substituting eqn [1.6.3] into eqn [1.6.1]

$$\frac{d}{dt}(k_0 e^{-2t}) + k_0 e^{-2t} = e^{-2t}$$

or $\qquad -2k_0 e^{-2t} + k_0 e^{-2t} = e^{-2t}$

or $\qquad -2k_0 + k_0 = 1 \qquad$ or $\qquad k_0 = -1$

Therefore

$$i_p = -e^{-2t} \qquad [1.6.4]$$

Complete solution

$$i(t) = i_n + i_p = ke^{-t} - e^{-2t} \qquad [1.6.5]$$

Setting $t = 0^+$ in eqn [1.6.5]

$$i(0^+) = 0 = k - 1 \qquad \Rightarrow k = 1$$

Therefore, the complete solution is

$$i(t) = e^{-t} - e^{-2t} \qquad \text{for} \qquad t > 0$$

(b) For the new value of $v(t) = e^{-t}$, the eqn [1.6.1] becomes

$$\frac{di}{dt} + i = e^{-t} \qquad [1.6.6]$$

Natural solution

$$i_n = ke^{-t} \qquad [1.6.7]$$

Particular solution

$$i_p = k_0 t e^{-t} \qquad [1.6.8]$$

Note that, when $\frac{R}{L} = \alpha$ where α is the coefficient of t in exponential input, the form of particular solution is given by equation [1.6.8], not by $i_p = k_0 e^{-t}$

Substituting eqn [1.6.8] in eqn [1.6.6], results

$$\frac{d}{dt}(k_0 t e^{-t}) + k_0 t e^{-t} = e^{-t}$$

or $\qquad k_0 e^{-t} - k_0 t e^{-t} + k_0 t e^{-t} = e^{-t}$

or $\qquad k_0 - k_0 t + k_0 t = 1 \qquad$ or $\qquad k_0 = 1$

Hence $\qquad i_p = t e^{-t} \qquad [1.6.9]$

The complete solution is

$$i(t) = i_n + i_p = ke^{-t} + te^{-t} \qquad [1.6.10]$$

Setting $t = 0^+$ in eqn [1.6.10]

$$i(0^+) = 0 = k + 0 \qquad \Rightarrow k = 0$$

Therefore complete solution is

$$i(t) = te^{-t} \qquad \text{for } t > 0$$

(c) For the new value of $v(t) = t$, eqn [1.6.1] becomes

$$\frac{di}{dt} + i = t \qquad [1.6.11]$$

Natural solution

$$i_n = ke^{-t} \qquad [1.6.12]$$

Particular solution

$$i_p = k_1 t + k_0 \qquad [1.6.13]$$

Substituting eqn [1.6.13] into eqn [1.6.11]

$$\frac{d}{dt}(k_1 t + k_0) + k_1 t + k_0 = t \qquad \text{or} \qquad k_1 t + (k_0 + k_1) = t$$

Equating coefficients of like terms of above equation, we have

$$k_1 = 1$$
$$\text{and} \qquad k_1 + k_0 = 0$$
$$\text{Hence} \qquad k_1 = 1 \qquad \text{and} \qquad k_0 = -1$$

Putting the values of k_1 and k_0 into eqn [1.6.13]

$$i_p = t - 1 \qquad [1.6.14]$$

The complete solution is

$$i(t) = i_n + i_p = ke^{-t} + t - 1 \qquad [1.6.15]$$

Setting $t = 0^+$ in eqn [1.6.15]

$$i(0^+) = 0 = k - 1 \qquad \Rightarrow k = 1$$

Therefore, the complete solution is

$$i(t) = e^{-t} + t - 1 \qquad \text{for } t > 0$$

SP 1.7 The switch is closed at $t = 0$ in the circuit of Fig. SP 1.7. Find the inductance L of coil if $i(0.2) = 0.5$ A.

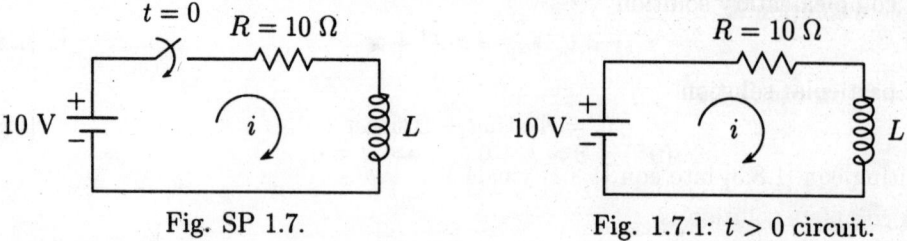

Fig. SP 1.7.   Fig. 1.7.1: $t > 0$ circuit.

SOLUTION:

The complete solution for $i(t)$ for the circuit of Fig. 1.7.1 will be written as

$$i(t) = \frac{V}{R}(1 - e^{-\frac{R}{L}t}) = (1 - e^{-\frac{10}{L}t})$$ [1.7.1]

Setting $t = 0.2$ in eqn [1.7.1]

$$i(0.2) = (1 - e^{-\frac{10}{L}0.2}) \qquad \Rightarrow 0.5 = (1 - e^{-\frac{2}{L}})$$

Therefore

$$e^{-\frac{2}{L}} = 0.5$$

Taking ln on both sides of the above equation

$$-\frac{2}{L} = \ln(0.5) = -0.693 \qquad \Rightarrow L = 2.89 \text{ H}$$

SP 1.8 In the Fig. SP 1.8 switch is closed at $t = 0$. Find $i(t)$ for $t > 0$.

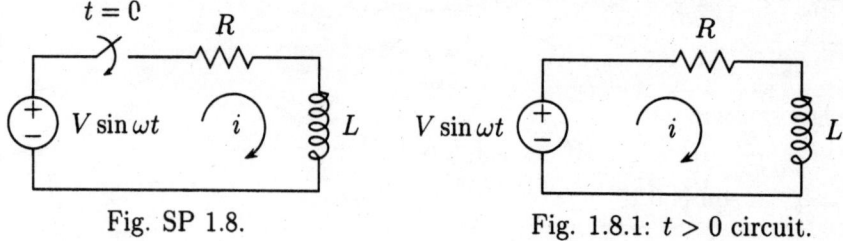

Fig. SP 1.8.   Fig. 1.8.1: $t > 0$ circuit.

SOLUTION:

By KVL in the circuit of Fig. 1.8.1

$$Ri + L\frac{di}{dt} = V \sin \omega t$$

or $\qquad \dfrac{di}{dt} + \dfrac{R}{L}i = \dfrac{V}{L} \sin \omega t$ [1.8.1]

Form of complementary solution

$$i_c = ke^{-\frac{R}{L}t} \qquad [1.8.2]$$

Form of particular solution

$$i_p = A\cos\omega t + B\sin\omega t \qquad [1.8.3]$$

Substituting eqn [1.8.3] into eqn [1.8.1] yields

$$\frac{d}{dt}(A\cos\omega t + B\sin\omega t) + \frac{R}{L}(A\cos\omega t + B\sin\omega t) = \frac{V}{L}\sin\omega t$$

or $\qquad -A\omega\sin\omega t + B\omega\cos\omega t + \frac{R}{L}A\cos\omega t + \frac{R}{L}B\sin\omega t = \frac{V}{L}\sin\omega t$

Thus

$$\left(\frac{RB}{L} - A\omega\right)\sin\omega t + \left(B\omega + \frac{RA}{L}\right)\cos\omega t = \frac{V}{L}\sin\omega t \qquad [1.8.4]$$

Equating the coefficients of like terms of above equation

$$\frac{RB}{L} - A\omega = \frac{V}{L} \qquad [1.8.5]$$

and $\qquad B\omega + \frac{RA}{L} = 0 \qquad \Rightarrow B = -\frac{AR}{\omega L} \qquad [1.8.6]$

Putting the value of B into eqn [1.8.5], we have

$$A = -\frac{\omega L}{R^2 + \omega^2 L^2}V \qquad [1.8.7]$$

Putting the value of A from eqn [1.8.7] into eqn [1.8.6] as

$$B = -\frac{R}{\omega L}\left(-\frac{\omega L}{R^2 + \omega^2 L^2}V\right) = \frac{RV}{R^2 + \omega^2 L^2} \qquad [1.8.8]$$

Hence i_p can be written as

$$
\begin{aligned}
i_p &= -\frac{\omega LV}{R^2 + \omega^2 L^2}\cos\omega t + \frac{RV}{R^2 + \omega^2 L^2}\sin\omega t \\
&= \frac{V}{R^2 + \omega^2 L^2}[R\sin\omega t - \omega L\cos\omega t] \\
&= \frac{V\sqrt{R^2 + \omega^2 L^2}}{R^2 + \omega^2 L^2}\left[\frac{R}{\sqrt{R^2 + \omega^2 L^2}}\sin\omega t - \frac{\omega L}{\sqrt{R^2 + \omega^2 L^2}}\cos\omega t\right] \\
&= \frac{V}{\sqrt{R^2 + \omega^2 L^2}}[\sin\omega t\cos\theta - \cos\omega t\sin\theta] \\
&= \frac{V}{\sqrt{R^2 + \omega^2 L^2}}\sin(\omega t - \theta) \\
&= \frac{V}{\sqrt{R^2 + \omega^2 L^2}}\sin\left(\omega t - \tan^{-1}\frac{\omega L}{R}\right)
\end{aligned}
$$

where $\tan\theta = \frac{\omega L}{R}$ from the impedance triangle of Fig. 1.8.3.

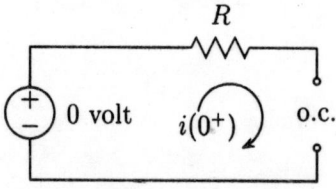

Fig. 1.8.2: $t = 0^+$ circuit.

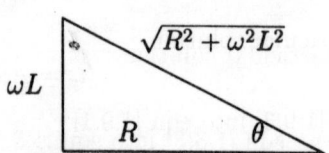

Fig. 1.8.3: Impedance triangle.

Hence complete solution is

$$i(t) = i_c + i_p = ke^{-\frac{R}{L}t} + \frac{V}{\sqrt{R^2 + \omega^2 L^2}} \sin\left(\omega t - \tan^{-1}\frac{\omega L}{R}\right) \qquad [1.8.9]$$

To determine k, put $t = 0^+$ in eqn [1.8.9] as

$$i(0^+) = k + \frac{V}{\sqrt{R^2 + \omega^2 L^2}} \sin\left(-\tan^{-1}\frac{\omega L}{R}\right)$$

or $\qquad 0 = k + \dfrac{V}{\sqrt{R^2 + \omega^2 L^2}} \sin\left(-\sin^{-1}\dfrac{\omega L}{\sqrt{R^2 + \omega^2 L^2}}\right)$

or $\qquad 0 = k + \dfrac{V}{\sqrt{R^2 + \omega^2 L^2}} \left(\dfrac{-\omega L}{\sqrt{R^2 + \omega^2 L^2}}\right)$

Hence $\qquad k = \dfrac{\omega L V}{R^2 + \omega^2 L^2}$

Putting the value of k into eqn [1.8.9], we have

$$i(t) = \frac{\omega L V}{R^2 + \omega^2 L^2} e^{-\frac{R}{L}t} + \frac{V}{\sqrt{R^2 + \omega^2 L^2}} \sin\left(\omega t - \tan^{-1}\frac{\omega L}{R}\right) \qquad \text{for } t > 0$$

SP 1.9 In circuit of Fig. SP 1.9, the switch is moved from a to b at $t = 0$. Find $i(t)$ for $t > 0$, assuming that the switch was at position a for a long time.

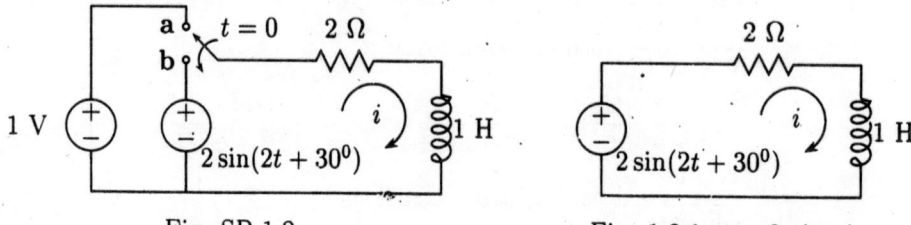

Fig. SP 1.9.

Fig. 1.9.1: $t > 0$ circuit.

SOLUTION:

By KVL in the circuit of Fig. 1.9.1

$$2i + \frac{di}{dt} = 2\sin(2t + 30^0) \qquad [1.9.1]$$

Form of complementary solution

$$i_c(t) = Ke^{-2t} \qquad\qquad [1.9.2]$$

Form of particular solution

$$i_p = A\sin(2t + 30^0 + \phi) \qquad\qquad [1.9.3]$$

Setting eqn [1.9.3] into eqn [1.9.1]

$$\frac{d}{dt}\left[A\sin(2t + 30^0 + \phi)\right] + 2A\sin(2t + 30^0 + \phi) = 2\sin(2t + 30^0)$$

or $\qquad 2A\cos(2t + 30^0 + \phi) + 2A\sin(2t + 30^0 + \phi) = 2\sin(2t + 30^0)$

or $\qquad A(\cos\phi + \sin\phi)\cos(2t + 30^0) + A(\cos\phi - \sin\phi)\sin(2t + 30^0) = \sin(2t + 30^0)$

Comparing the coefficients of like terms on both sides of last equation

$$A(\cos\phi + \sin\phi) = 0 \qquad\qquad [1.9.4]$$

and $\qquad A(\cos\phi - \sin\phi) = 1 \qquad\qquad [1.9.5]$

Squaring and then adding the eqns. [1.9.4] and [1.9.5] gives $\quad A = 0.707$
and eqn [1.9.4] gives $\qquad \phi = -45^0$ Hence

$$i_p = 0.707\sin(2t + 30^0 - 45^0) = 0.707\sin(2t - 15^0)$$

And complete solution

$$i(t) = i_c + i_p = Ke^{-2t} + 0.707\sin(2t - 15^0) \qquad\qquad [1.9.6]$$

The initial value of the current $i(t)$ is obtained from Fig. 1.9.2 as:

$$i(0^+) = i(0^-) = 1/2 \text{ A}$$

Setting t $= 0^+$ into eqn [1.9.6]

$$i(0^+) = 1/2 = K + 0.707\sin(-15^0)$$

Hence $\qquad K = 0.683$

Therefore the complete solution is

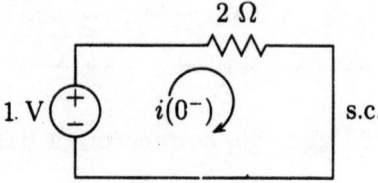

Fig. 1.9.2: $t = 0^-$ circuit.

$$i(t) = 0.683e^{-2t} + 0.707\sin(2t - 15^0)$$

SP 1.10 In the network shown in Fig. SP 1.10, the switch is closed at $t = 0$. Given that $i(0) = 0.1$ A . The transient appears to disappear in 0.3 sec. Find
    (a) value of R,
    (b) value of C, and
    (c) the equation of $i(t)$.

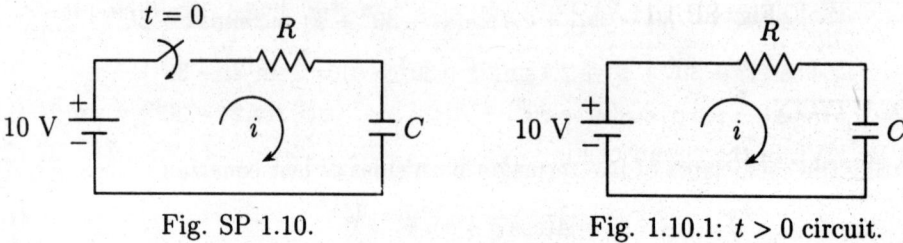

      Fig. SP 1.10.   Fig. 1.10.1: $t > 0$ circuit.

SOLUTION:

The expression for $i(t)$ for circuit shown in Fig. 1.10.1 is given by

$$i(t) = \frac{10}{R}e^{-t/RC} \qquad\qquad [1.10.1]$$

Here RC $= \tau$ (time constant of the circuit).
    (a) Setting t $= 0$ in eqn [1.10.1] as

$$i(0) = \frac{10}{R} = 0.1 \qquad \Rightarrow R = 100 \ \Omega$$

(b) We know that transient appears for four times of time constant in practice, i.e.,
    $4\tau = 4$ RC sec.
    Therefore
$$4 \ RC = 0.3 \qquad \Rightarrow C = 0.75 \text{ mF}$$

(c) Putting the value of R and C in the eqn [1.10.1] we have

$$i(t) = 0.1 \ e^{-13.33t} \qquad t > 0$$

SP 1.11 A switch is closed at $t = 0$, connecting a battery of voltage V with a series RC circuit.

    (a) Determine the ratio of energy delivered to the capacitor to the total energy supplied by the source as a function of time.

    (b) Show that the ratio approaches 0.50 as $t \to \infty$.

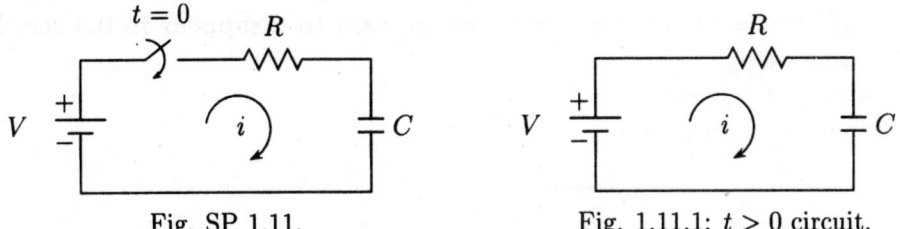

Fig. SP 1.11.   Fig. 1.11.1: $t > 0$ circuit.

SOLUTION:

(a) By KVL in the circuit of Fig. 1.11.1

$$Ri + \frac{1}{C}\int_0^t i\, dt = V \qquad [1.11.1]$$

Differentiating

$$R\frac{di}{dt} + \frac{1}{C}i = 0$$

Therefore

$$\frac{di}{dt} + \frac{1}{RC}i = 0 \qquad [1.11.2]$$

The solution for $i(t)$ to eqn [1.11.2] will be given by

$$i(t) = \frac{V}{R}\, e^{-\frac{t}{RC}} \qquad [1.11.3]$$

The energy stored in the capacitor is given by

$$w_C(t) = \frac{1}{2}Cv_C^2 = \frac{1}{2}C\,[V - iR]^2 = \frac{1}{2}C\left[V - R\frac{V}{R}e^{-\frac{t}{RC}}\right]^2$$

Therefore

$$w_C(t) = \frac{1}{2}\, CV^2 \left[1 - e^{-\frac{t}{RC}}\right]^2 \qquad [1.11.4]$$

Energy supplied by the source V is given by

$$\begin{aligned}
w_T(t) &= \int_0^t p\, dt = \int_0^t Vi\, dt = \frac{V^2}{R}\int_0^t e^{-\frac{t}{RC}}\, dt \\
&= \frac{V^2}{R}\left[\frac{e^{-\frac{t}{RC}}}{-1/RC}\right]_0^t = \frac{V^2}{R}\left[\frac{e^{-\frac{t}{RC}} - e^0}{-1/RC}\right]
\end{aligned}$$

Thus

$$w_T(t) = CV^2(1 - e^{-\frac{t}{RC}}) \qquad [1.11.5]$$

From eqn [1.11.4] and [1.11.5] we can obtain

$$\frac{w_C(t)}{w_T(t)} = \frac{1}{2}(1 - e^{-\frac{t}{RC}}) \qquad [1.11.6]$$

(b) Taking limits as $t \to \infty$ in eqn [1.11.6], we have

$$\lim_{t\to\infty}\frac{w_C}{w_T} = \lim_{t\to\infty}\frac{1}{2}(1 - e^{-\frac{t}{RC}}) = \frac{1}{2}(1) = 0.5 \qquad \text{Proved.}$$

SP 1.12 In the circuit of Fig. **SP 1.12**, switch is opened at $t = 0$. **Solve for** $v(t)$.

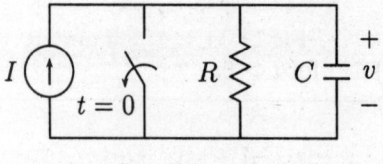   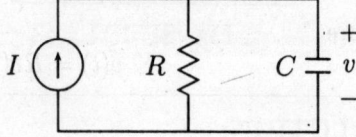

Fig. SP 1.12.   Fig. 1.12.1: $t > 0$ circuit.

SOLUTION:

By KCL in circuit of Fig.1.12.1

$$\frac{v}{R} + C\frac{dv}{dt} = I$$

Hence

$$\frac{dv}{dt} + \frac{v}{RC} = \frac{I}{C} \qquad [1.12.1]$$

Form of natural solution

$$v_n = Ke^{-t/RC} \qquad [1.12.2]$$

Form of particular solution for constant input

$$v_p = k_1 \qquad [1.12.3]$$

Substituting eqn [1.12.3] into eqn [1.12.1], results

$$\frac{d}{dt}(k_1) + \frac{k_1}{RC} = \frac{I}{C}$$

or $\qquad 0 + \frac{k_1}{RC} = \frac{I}{C}$

Hence $\qquad k_1 = RI = v_p$

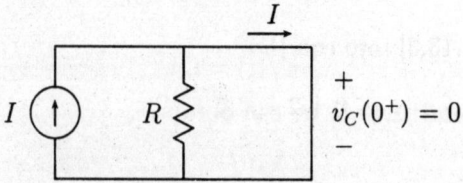

Fig. 1.12.2: $t = 0^+$ circuit.

Total solution

$$v(t) = v_n + v_p = Ke^{-t/RC} + RI \qquad [1.12.4]$$

From the circuit of Fig. 1.12.2

$$v(0^+) = 0$$

Setting $t = 0^+$ eqn [1.12.4] gives

$$v(0^+) = 0 = K + RI \qquad \Rightarrow K = -RI$$

Therefore

$$v(t) = RI(1 - e^{-t/RC}) \qquad \text{for } t > 0$$

SP 1.13 The switch in the Fig. SP 1.13 is opened at $t = 0$. Determine $v_2(t)$.

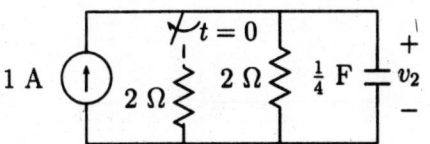

Fig. SP 1.13.

Fig. 1.13.1: $t > 0$ circuit.

SOLUTION:

By KCL in the circuit of Fig. 1.13.1

$$\frac{v_2}{2} + \frac{1}{4}\frac{dv_2}{dt} = 1$$

Therefore

$$\frac{dv_2}{dt} + 2v_2 = 4 \qquad\qquad [1.13.1]$$

Natural solution

$$v_n(t) = Ke^{-2t} \qquad\qquad [1.13.2]$$

Particular solution

$$v_p(t) = k \qquad\qquad [1.13.3]$$

Substituting eqn [1.13.3] into eqn [1.13.1]

$$0 + 2k = 4 \qquad \Rightarrow k = 2 = v_p$$

Complete solution

$$v_2 = v_n + v_p = Ke^{-2t} + 2 \qquad\qquad [1.13.4]$$

From the circuit of Fig.1.13.2, it is observed that

$$v_2(0^-) = (1)\left(\frac{1}{2}\right) = 1 \text{ V}$$

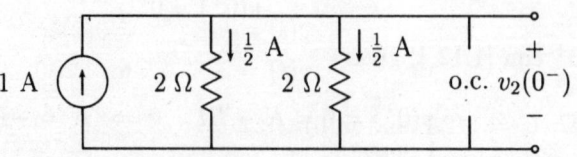

Fig. 1.13.2: $t = 0^-$ circuit.

Therefore

$$v_2(0^-) = v_2(0^+) = 1 \text{ V}$$

Setting t $= 0^+$ in eqn [1.13.4]

$$v_2(0^+) = 1 = K + 2 \qquad \Rightarrow K = -1$$

Therefore

$$v_2(t) = -e^{-2t} + 2 \qquad \text{for } t > 0$$

SP 1.14 Consider the parallel RC circuit of Fig. SP 1.14, where the capacitor is linear and time-invariant with $C = 1/2$ F and initial voltage at $t = 0$ is $v_C(0)$ = 2 volts. Find the zero input response, $v(t)$ for the following types of resistor
   (a) A linear time invariant resistor with $R = 2\ \Omega$.
   (b) A linear time varying resistor with $R(t) = \sec t\ \Omega$
   (c) A nonlinear time invariant resistor having a characteristic $i_R = v_R^2$

Fig. SP 1.14.

SOLUTION:

From Fig. SP 1.14, KVL gives

$$i_C + i_R = 0 \tag{1.14.1}$$

or

$$C\frac{dv_c}{dt} + \frac{v_R}{R} = 0 \tag{1.14.2}$$

and KVL gives

$$v_C = v_R = v \tag{1.14.3}$$

Using eqn [1.14.3], the eqn [1.14.2] can be written as

$$\frac{dv}{dt} + \frac{v}{RC} = 0 \tag{1.14.4}$$

Form of solution to eqn [1.14.4]

$$v(t) = ke^{-t/RC} = v(0)e^{-t/RC}$$

or $\quad v(t) = 2e^{-t} \quad$ for $t > 0$

where $R = 2\Omega$, $C = 1/2$ F and $v(0) = 2$ V.

(b) Given $R(t) = \sec t$. The eqn [1.14.4] becomes

$$\frac{dv}{dt} + (2\cos t)\, v = 0 \qquad\qquad [1.14.5]$$

or

$$\frac{dv}{v} + (2\cos t)dt = 0$$

Integrating above equation

$$\int \frac{dv}{v} + \int (2\cos t)dt + k = 0 \qquad \text{or} \qquad \ln v + 2\sin t + k = 0 . \qquad [1.14.6]$$

Setting t = 0 in the above equation

$$ln[v(0)] + 0 + k = 0 \qquad \Rightarrow k = -\ln 2 = -0.693$$

Hence, eqn [1.14.6] will become

$$\ln v + 2\sin t - 0.693 = 0$$

Therefore

$$v(t) = e^{-2\sin t + 0.693} \qquad \text{for } t > 0$$

(c) Setting $i_C = C\frac{dv_C}{dt} = \frac{dv}{dt}$ and $i_R = v_R^2 = v^2$ in eqn [1.14.1]

$$\frac{dv}{dt} + 2v^2 = 0 \qquad\qquad [1.14.6]$$

The eqn 1.14.6 is called a nonlinear differential equation and can be solved by separating the variables and integrating both sides as

$$\int\limits_{2}^{v(t)} \frac{dv}{v^2} = \int\limits_{0}^{t} -2dt \qquad \text{or,} \qquad -\frac{1}{v(t)} + \frac{1}{2} = -2t$$

Therefore $\qquad v(t) = \dfrac{2}{1 + 4t}u(t) \qquad$ for all t

---

SP 1.15 In the circuit shown in Fig. SP 1.15, the switch is closed at $t = 0$ .The initial voltage across the capacitor is $v_C(0) = 5$ V. Find

(a) $i(t)$ for $t > 0$,
(b) $v_C(t)$ for $t > 0$,
(c) $v_R(t)$ for $t > 0$,
(d) the initial energy stored in capacitor, W_0,
(e) the energy stored in capacitor, $w_C(t)$,
(f) the energy dissipated in resistor, $w_R(t)$.

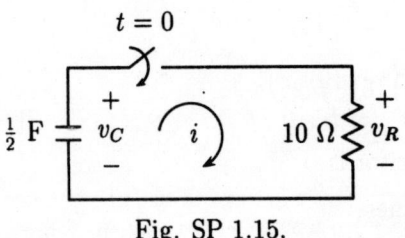

Fig. SP 1.15.   Fig. 1.15.1: $t > 0$ circuit.

SOLUTION:

By KVL in the circuit of Fig. 1.15.1

$$-5 + \frac{1}{1/2} \int_0^t i dt + 10i = 0$$

or

$$2 \int_0^t i dt + 10i = 5 \qquad\qquad [1.15.1]$$

Differentiating

$$2i + 10\frac{di}{dt} = 0$$

or

$$\frac{di}{dt} + \frac{1}{5}i = 0 \qquad\qquad [1.15.2]$$

Solution to eqn [1.15.2]

$$i(t) = i(0^+)e^{-t/5} = \frac{1}{2}e^{-t/5} \qquad \text{for} \quad t > 0$$

where $i(0^+) = 5/10 = 1/2$ A from Fig. 1.15.2

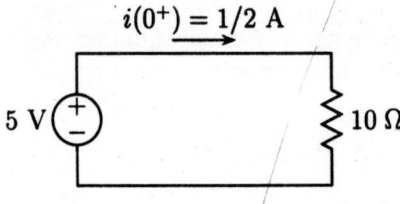

$i(0^+) = 1/2$ A

Fig. 1.15.2: $t = 0^+$ circuit.

(b) The voltage across the capacitor is

$$\begin{aligned} v_C(t) &= \frac{1}{1/2} \int_0^t i_C dt + v_C(0) = -2 \int_0^t i dt + v_C(0) \\ &= -2 \int_0^t \frac{1}{2}e^{-t/5} dt + 5 = 5e^{-t/5} \qquad \text{for} \quad t > 0 \end{aligned}$$

(c) The voltage across the resistor is

$$v_R(t) = 10i = 10\left(\frac{1}{2}e^{-t/5}\right) = 5e^{-t/5} \qquad \text{for} \quad t > 0$$

(d) The initial energy stored in capacitor is

$$W_0 = \frac{1}{2}Cv_C^2(0) = \frac{1}{2}\left(\frac{1}{2}\right)(5)^2 = 25/4 \text{ J}$$

(e) The instantaneous energy stored in the capacitor is

$$w_C(t) = \frac{1}{2}Cv_C^2 = \frac{1}{2}\left(\frac{1}{2}\right)\left(5e^{-t/5}\right)^2 = \frac{25}{4}e^{-2t/5} \qquad \text{for} \qquad t > 0$$

(f) The energy dissipated in the resistor is

$$w_R(t) = W_0 - w_C = \frac{25}{4} - \frac{25}{4}e^{-2t/5} = \frac{25}{4}\left(1 - e^{-2t/5}\right) \qquad \text{for } t > 0$$

SP 1.16 For the circuit shown in Fig. SP 1.16, determine $v_C(t)$, $i_C(t)$ and $v(t)$ for all time t.

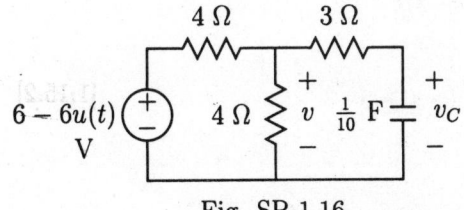

Fig. SP 1.16.

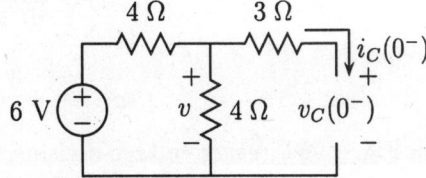

Fig. 1.16.1: $t = 0^-$ circuit.

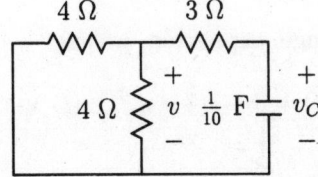

Fig. 1.16.2: $t > 0$ circuit.

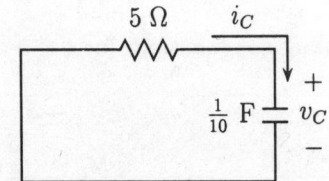

Fig. 1.16.3: Equivalent circuit of Fig. 1.16.2.

SOLUTION:

We know that

$$6 - 6u(t) = 6 \text{ V} \quad \text{for } t < 0$$
$$= 0 \text{ V} \quad \text{for } t \geq 0$$

From circuit of Fig. 1.16.1

$$v_C(0^-) = \frac{4}{4+4}(6) = 3 \text{ V} = v(0^-) \qquad [1.16.1]$$

and

$$i_C(0^-) = 0 \qquad [1.16.2]$$

Thus, by continuity of capacitor voltage

$$v_C(0^+) = v_C(0^-) = 3 \text{ V}$$

From circuit of Fig. Fig.1.9.3, using KVL

$$5i_C + v_C = 0$$

or

$$5\left(\frac{1}{10}\frac{dv_C}{dt}\right) + v_C = 0$$

Hence

$$\frac{dv_C}{dt} + 2v_C = 0 \qquad\qquad\qquad [1.16.3]$$

The solution of above equation is given by

$$v_C(t) = v_C(0^+)e^{-2t} = 3e^{-2t} \qquad \text{for } t \geq 0 \qquad\qquad [1.16.4]$$

Furthermore

$$i_C(t) = \frac{1}{10}\frac{dv_C}{dt} = \frac{1}{10}[3(-2)e^{-2t}]$$

Therefore

$$i_C(t) = -\frac{3}{5}e^{-2t} \text{ A} \qquad \text{for } t \geq 0 \qquad\qquad [1.16.5]$$

From Fig. 1.16.2, using voltage division formula

$$v(t) = \frac{(4\|4)}{(4\|4) + 3}v_C(t) = \frac{2}{5}\left(3e^{-2t}\right) \text{ V} = \frac{6}{5}e^{-2t} \text{ for } t \geq 0 \qquad [1.16.6]$$

Combining the functions for $t < 0$ and $t \geq 0$ into single expression, we get

$$v_C(t) = 3 - 3u(t) + 3e^{-2t}u(t) = 3 - 3(1 - e^{-2t})u(t) \text{ V}$$

$$i_C(t) = -\frac{3}{5}e^{-2t}u(t)$$

$$\text{and} \qquad v(t) = 3 - 3u(t) + \frac{6}{5}e^{-2t}u(t)$$

SP 1.17 In the circuit of Fig. SP 1.17, the switch is opened at $t = 0$. Solve for $v(t)$.

SOLUTION:

By KCL in Fig. 1.17.1

$$\frac{v}{R} + \frac{1}{L}\int_0^t v\,dt = I \qquad\qquad\qquad [1.17.1]$$

Differentiating

$$\frac{1}{R}\frac{dv}{dt} + \frac{v}{L} = 0$$

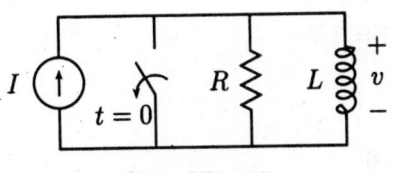

Fig. SP 1.17.

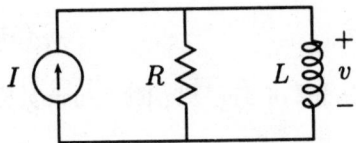

Fig. 1.17.1: $t > 0$ circuit.

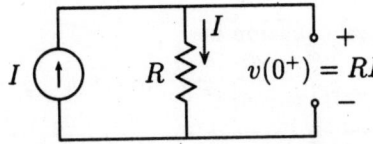

Fig. 1.17.2: $t = 0^+$ circuit.

Thus

$$\frac{dv}{dt} + \frac{R}{L}v = 0 \qquad [1.17.2]$$

Form of solution to eqn [1.17.2]

$$v(t) = v(0^+)e^{-Rt/L} = RIe^{-Rt/L}$$

where $v(0^+) = RI$ (from Fig. 1.17.2)

SP 1.18 The circuit shown in Fig. SP1.18, given that $i_L(0) = 3$ A . Find $i_L(t)$ for $t \geq 0$.

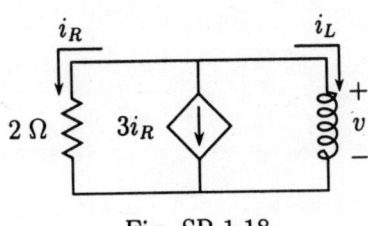

Fig. SP 1.18.

SOLUTION:

By KVL

$$i_R + 3i_R + i_L = 0$$

Thus

$$4i_R + i_L = 0 \qquad [1.18.1]$$

Since $i_R = v/2$, therefore eqn [1.18.1] can be written as

$$4v/2 + i_L = 0 \qquad [1.18.2]$$

However, $v = 1\frac{di_L}{dt}$. Thus eqn [1.18.2] will be

$$2\frac{di_L}{dt} + i_L = 0$$

and

$$\frac{di_L}{dt} + \frac{1}{2}i_L = 0 \qquad [1.18.3]$$

The solution of this differential equation is

$$i_L = i_L(0)e^{-t/2} = 3e^{-t/2} \qquad \text{for } t \geq 0$$

SP 1.19 The switch in Fig. SP 1.19 is thrown from a to b at $t = 0$. Find $i_L(t)$ for $t > 0$ if $i_s = e^{-t}$.

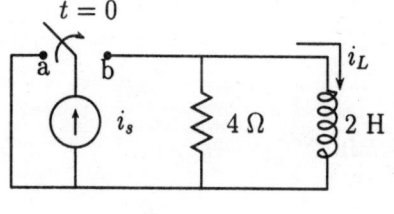

Fig. SP 1.19.

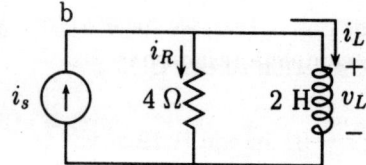

Fig. 1.19.1: $t > 0$ circuit.

SOLUTION:

By KCL at node **b** of circuit of Fig. 1.19.1

$$i_L + i_R = i_s$$

or

$$i_L + \frac{v_L}{4} = e^{-t}$$

or

$$i_L + \frac{1}{4}\left(2\frac{di_L}{dt}\right) = e^{-t}$$

Therefore

$$\frac{di_L}{dt} + 2i_L = 2e^{-t} \qquad [1.19.1]$$

Form of complementary solution

$$i_{L,c} = Ae^{-2t} \qquad [1.19.2]$$

Form of particular response to the exponential input

$$i_{L,p}(t) = k_0 e^{-t} \qquad [1.19.3]$$

Substituting eqn [1.19.3] into eqn [1.19.1]

$$-k_0 e^{-t} + 2k_0 e^{-t} = 2e^{-t}$$

or

$$-k_0 + 2k_0 = 2 \quad \text{or} \quad k_0 = 2$$

Thus

$$i_{L,p}(t) = 2e^{-t} \tag{1.19.4}$$

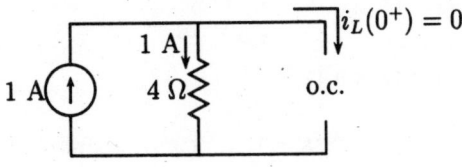

Fig. 1.19.2: $t = 0^+$ circuit.

From Fig. 1.19.2, we have $i_L(0^+) = 0$
The complete response is

$$i_L(t) = Ae^{-2t} + 2e^{-t} \tag{1.19.5}$$

Setting $t = 0^+$ in eqn [1.19.5] yields

$$i_L(0^+) = 0 = A + 2 \quad \Rightarrow A = -2$$

Finally

$$i_L(t) = 2(e^{-t} - e^{-2t}) \qquad t > 0$$

SP 1.20 In the network shown in Fig. SP 1.20, the switch is closed at $t = 0$ connecting a source $v(t) = e^{-2t}$ to the RC network. At $t = 0$, it is observed that the capacitor voltage has the value $v_2(0) = 2$ V. For the element values given, determine $v_2(t)$

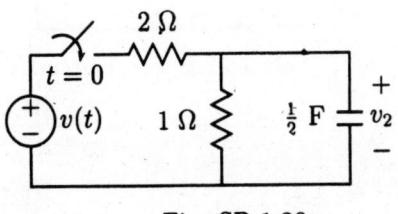

Fig. SP 1.20.

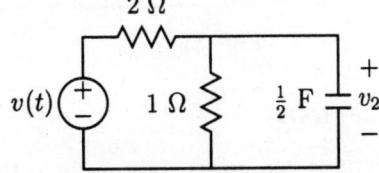

Fig. 1.20.1: $t > 0$ circuit.

SOLUTION:

By KCL at the right node in the circuit of Fig. 1.20.1

$$\frac{v - v_2}{2} = \frac{v_2}{1} + \frac{1}{2}\frac{dv_2}{dt}$$

Therefore

$$\frac{dv_2}{dt} + 3v_2 = e^{-2t} \qquad\qquad [1.20.1]$$

Natural response

$$v_{2n} = Ae^{-3t} \qquad\qquad [1.20.2]$$

Particular response

$$v_{2p} = ke^{-2t} \qquad\qquad [1.20.3]$$

Substituting eqn [1.20.3] into eqn [1.20.1]

$$\frac{d}{dt}(ke^{-2t}) + 3ke^{-2t} = e^{-2t}$$

or

$$-2k + 3k = 1 \qquad \text{or} \qquad k = 1$$

Therefore

$$v_{2p} = e^{-2t} \qquad\qquad [1.20.4]$$

The complete solution is

$$v_2(t) = Ae^{-3t} + e^{-2t} \qquad\qquad [1.20.5]$$

Setting t = 0 in eqn [1.20.5] yields

$$v_2(0) = 2 = A + 1 \qquad \Rightarrow A = 1$$

Therefore $\qquad v_2(t) = e^{-3t} + e^{-2t} \qquad t > 0$

SP 1.21 In the network shown in the Fig. SP 1.21, the switch is closed at $t = 0$. Find $v_2(t)$ for $t > 0$.

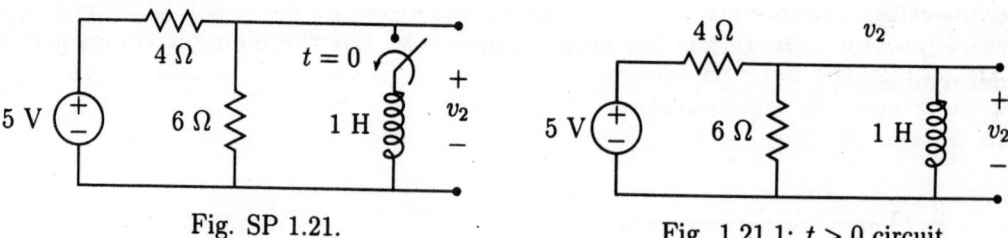

Fig. SP 1.21.   Fig. 1.21.1: $t > 0$ circuit.

SOLUTION:

By KCL at the upper right node in the circuit of Fig. 1.21.1

$$\frac{5 - v_2}{4} = \frac{v_2}{6} + \frac{1}{1}\int_0^t v_2 \, dt$$

Therefore
$$\frac{5}{4} = \frac{5}{12}v_2 + \int_0^t v_2 \, dt \qquad [1.21.1]$$

Differentiating
$$\frac{5}{12}\frac{dv_2}{dt} + v_2 = 0$$

Thus
$$\frac{dv_2}{dt} + \frac{12}{5}v_2 = 0 \qquad [1.21.2]$$

The form of solution to eqn [1.21.2] is
$$v_2(t) = Ae^{-12t/5} \qquad [1.21.3]$$

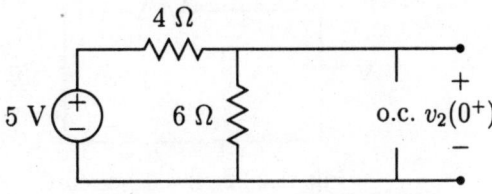

Fig. 1.21.2: $t = 0^+$ circuit.

From the Fig. 1.21.2,
$$v_2(0^+) = \frac{6}{6+4}(5) = 3 \text{ V}$$

Setting t = 0^+ in eqn [1.21.3], gives
$$v_2(0^+) = 3 = A \qquad \text{Therefore} \qquad v_2(t) = 3e^{-\frac{12}{5}t} \qquad t > 0$$

SP 1.22 In the network shown in Fig. SP 1.22, switch is moved from position 1 to 2 at $t = 0$. As a steady state current having previously been established in RL circuit, find $i(t)$ for $t > 0$.

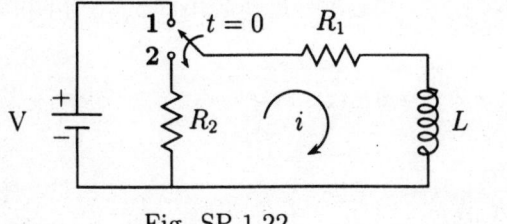

Fig. SP 1.22.

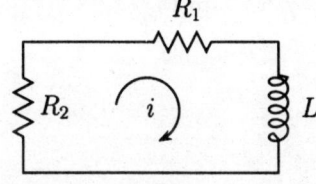

Fig. 1.22.1: $t > 0$ circuit.

SOLUTION:

By KVL in the circuit of Fig. 1.22.1

$$(R_1 + R_2)i + L\frac{di}{dt} = 0$$

or

$$\frac{di}{dt} + \frac{(R_1 + R_2)}{L}i = 0 \qquad [1.22.1]$$

The eqn 1.22.1 is first-oder differential equation without forcing function. Such a differential equation is called the homogeneous differential equation.

Form of solution

$$i(t) = i(0^+)e^{-(R_1+R_2)t/L} \qquad [1.22.2]$$

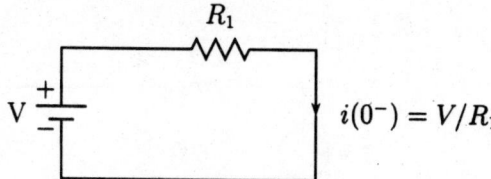

Fig. 1.22.2: $t = 0^-$ circuit.

From Fig. 1.22.2

$$i(0^-) = i(0^+) = \frac{V}{R_1}$$

Hence, the solution is

$$i(t) = \frac{V}{R_1}e^{-(R_1+R_2)t/L} \qquad t > 0$$

SP 1.23 In the circuit of Fig. **SP 1.23, the switch is moved from a to b at** $t = 0$.**Find :**

    (i) $i(t)$ for $t > 0$,

    (ii) $v_L(t)$ for $t > 0$,

    (iii) $v_R(t)$ for $t > 0$,

    (iv) **initial energy stored in inductor,** W_0,

    (v) **the energy stored in inductor,** $w_L(t)$ **for** $t > 0$,

    (vi) **the energy dissipated in resistor,** $w_R(t)$ **for** $t > 0$.

SOLUTION:

    (i) By KVL in Fig. 1.23.1

$$4i + 2\frac{di}{dt} = 0,$$

Hence

$$\frac{di}{dt} + 2i = 0 \qquad [1.23.1]$$

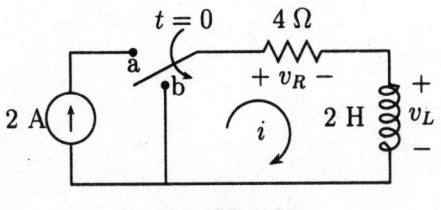

Fig. SP 1.23.

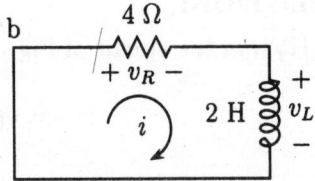

Fig. 1.23.1: $t > 0$ circuit.

Response to last equation

$$i(t) = i(0^+)e^{-2t} = 2e^{-2t} \qquad [1.23.2]$$

where $i(0^+) = i(0^-) = 2$ A

(ii)

$$v_L(t) = L\frac{di}{dt} = 2\frac{d}{dt}\left(2e^{-2t}\right) = -8e^{-2t} \qquad [1.23.2]$$

(iii)

$$v_R(t) = 4i = 8e^{-2t}$$

(iv)

$$W_0 = \frac{1}{2}Li^2(0^+) = 4 \text{ J}$$

(v)

$$w_L(t) = \frac{1}{2}Li^2 = 4e^{-4t}$$

(vi)

$$w_R(t) = W_0 - w_L(t) = 4\left(1 - e^{-4t}\right) \text{ J}$$

SP 1.24 The switch K in Fig. SP 1.24 is moved from position a to b at $t = 0$. The right capacitor is uncharged at $t = 0$.
   (a) Find the $i(t)$ for $t > 0$.
   (b) Find the $v_2(t)$ for $t > 0$.

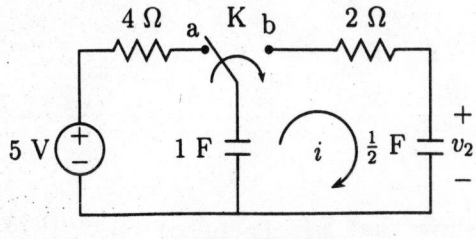

Fig. SP 1.24.

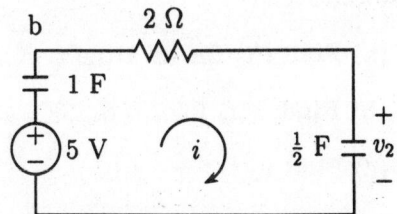

Fig. 1.24.1: $t > 0$ circuit.

SOLUTION:

Applying KVL in the circuit shown in Fig. 1.24.1

$$-5 + \frac{1}{1}\int_0^t i\,dt + 2i + \frac{1}{1/2}\int_0^t i\,dt = 0$$

or

$$3\int_0^t i\,dt + 2i = 5 \qquad\qquad [1.24.1]$$

Differentiating above equation and rearranging the terms, we have

$$\frac{di}{dt} + \frac{3}{2}i = 0 \qquad\qquad [1.24.2]$$

The form of solution to eqn [1.24.2] is

$$i(t) = i(0^+)e^{-\frac{3}{2}t} = \frac{5}{2}e^{-\frac{3}{2}t} \text{ for } t > 0$$

where $i(0^+) = \frac{5}{2}$ A   (from Fig. 1.24.2)

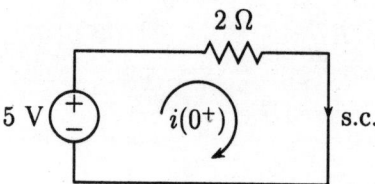

Fig. 1.24.2: $t = 0^+$ circuit.

$$v_2 = \frac{1}{1/2}\int_0^t i\,dt = 2\int_0^t \frac{5}{2}e^{-3t/2}\,dt$$

Therefore $$v_2 = 5\left.\frac{e^{-3t/2}}{-3/2}\right|_0^t = \frac{10}{3}(1 - e^{-3t/2})$$

SP 1.25 In the network given in Fig. SP 1.25, the initial voltage on C_1 is V_1 and on C_2 is V_2 such that $v_1(0) = V_1$ and $v_2(0) = V_2$. At t = 0, the switch is closed.

(a) Find $i(t)$ for all time t.

(b) Find $v_1(t)$ for $t > 0$.

(c) Find $v_2(t)$ for $t > 0$.

(d) From your results on (b) and (c), show that $v_1(\infty) = v_2(\infty)$

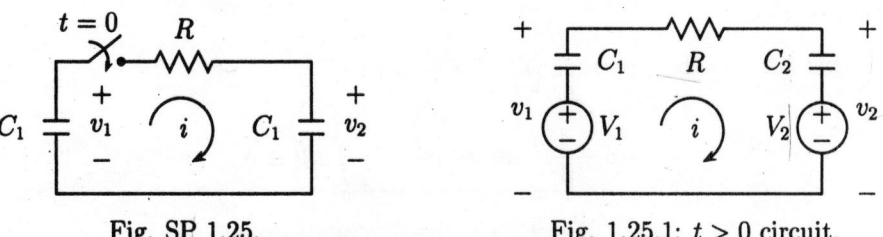

Fig. SP 1.25.   Fig. 1.25.1: $t > 0$ circuit.

(e) For the following values of the elements, R $= 1$ $\Omega, C_1 = 1$ F, $C_2 = 0.5$ F, $V_1 = 2$ V, $V_2 = 1$ V. Sketch $i(t)$ and $v_2(t)$ and identify the time constant.

SOLUTION:

By KVL in the circuit of Fig. 1.25.1

$$-V_1 + \frac{1}{C_1}\int_0^t idt + Ri + V_2 + \frac{1}{C_2}\int_0^t idt = 0$$

Differentiating

$$R\frac{di}{dt} + \left(\frac{1}{C_1} + \frac{1}{C_2}\right)i = 0$$

Thus

$$\frac{di}{dt} + \frac{C_1 + C_2}{C_1 C_2 R}i = 0 \qquad [1.25.1]$$

The form of solution is

$$i(t) = Ae^{-\frac{C_1+C_2}{C_1C_2R}t} \qquad [1.25.2]$$

It is noted from Fig. 1.25.2 that

$$i(0^+) = \frac{v_1(0) - v_2(0)}{R} = \frac{V_1 - V_2}{R} \qquad [1.25.3]$$

Setting t $= 0^+$ in the eqn [1.25.2]

$$i(0^+) = A = \frac{(V_1 - V_2)}{R}$$

Therefore

$$i(t) = \frac{V_1 - V_2}{R}e^{-\frac{C_1+C_2}{C_1C_2R}t} \qquad [1.25.4]$$

(b) From Fig. 1.25.1

$$
\begin{aligned}
v_1(t) &= -\frac{1}{C_1}\int idt + A \\
&= -\frac{1}{C_1}\int \frac{V_1 - V_2}{R}e^{-\frac{C_1+C_2}{C_1C_2R}t}dt + A
\end{aligned}
$$

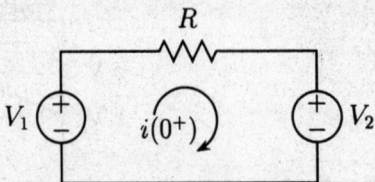

Fig. 1.25.2: $t = 0^+$ circuit.

Thus

$$v_1(t) = \frac{(V_1 - V_2)C_2}{C_1 + C_2} e^{-\frac{C_1+C_2}{C_1C_2R}t} + A \qquad [1.25.5]$$

Setting t $= 0^+$ in the last equation, we have

$$v_1(0^+) = V_1 = \frac{(V_1 - V_2)C_2}{C_1 + C_2} + A$$

Thus

$$A = \frac{V_1 C_1 + V_2 C_2}{C_1 + C_2}$$

Therefore

$$v_1(t) = \frac{(V_1 - V_2)C_2}{C_1 + C_2} e^{-\frac{C_1+C_2}{C_1C_2R}t} + \frac{V_1 C_1 + V_2 C_2}{C_1 + C_2} \qquad [1.25.6]$$

(c) Again from Fig. 1.25.1

$$v_2(t) = \frac{1}{C_2} \int i\, dt + B$$

$$= \frac{1}{C_2} \int \frac{V_1 - V_2}{R} e^{-\frac{C_1+C_2}{C_1C_2R}t} dt + B$$

Thus

$$v_2(t) = -\frac{C_1(V_1 - V_2)}{C_1 + C_2} e^{-\frac{C_1+C_2}{C_1C_2R}t} + B \qquad [1.25.7]$$

Setting t $= 0$ in the above equation

$$v_2(0) = V_2 = -\frac{C_1(V_1 - V_2)}{C_1 + C_2} + B$$

or

$$B = \frac{C_1 V_1 + C_2 V_2}{C_1 + C_2}$$

Thus

$$v_2(t) = \frac{(C_1 V_1 + C_2 V_2)}{C_1 + C_2} - \frac{C_1(V_1 - V_2)}{C_1 + C_2} e^{-\frac{C_1+C_2}{C_1C_2R}t} \qquad [1.25.8]$$

(d) From eqn [1.25.6] and eqn [1.25.8], we have

$$v_1(\infty) = v_2(\infty) = \frac{C_1 V_1 + C_2 V_2}{C_1 + C_2}$$

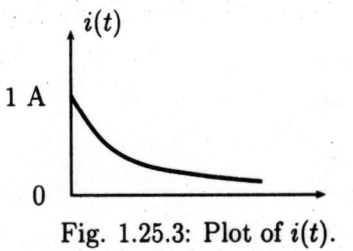

Fig. 1.25.3: Plot of $i(t)$.

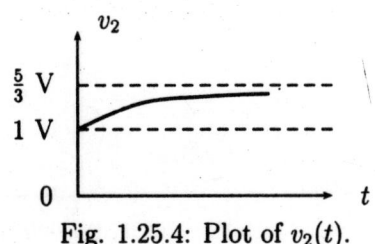

Fig. 1.25.4: Plot of $v_2(t)$.

(e) For the given values,

$$i(t) = e^{-3t} \quad \text{and} \quad v_2(t) = \frac{5}{3} - \frac{2}{3}e^{-3t} \quad t > 0$$

Time constant $\tau = \frac{1}{3}$(in each case)

SP 1.26 In the network shown in Fig. SP 1.26, the switch is in position a for a long period of time. At t = 0 switch is moved from a to b. Find $v_2(t)$. Assume that, the initial current in the 2 H inductor is zero.

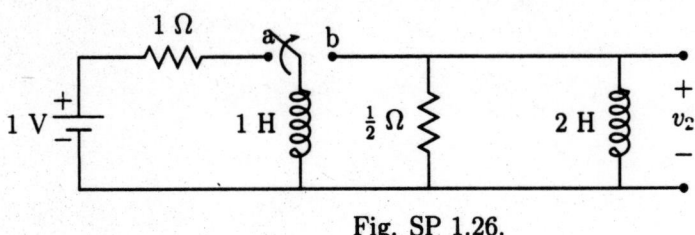

Fig. SP 1.26.

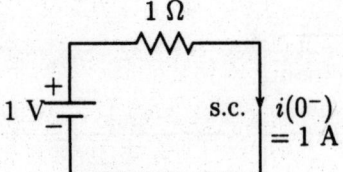
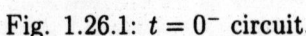

Fig. 1.26.1: $t = 0^-$ circuit.

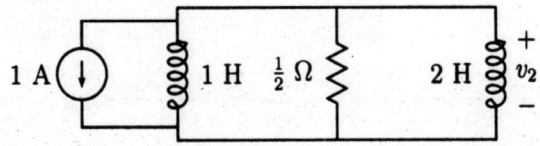

Fig. 1.26.2: $t > 0$ circuit.

SOLUTION:

It is obvious from the circuit of Fig. 1.26.1 that the current through the left inductor at $t = 0^-$ is 1 A, i.e., $i(0^-) = 1$ A. thus, by continuity of inductor current, we have

$$i(0^-) = i(0^+) = 1 \text{ A}$$

Hence, the left inductor of Fig. 1.26.2 has been replaced by the inductor with current source in parallel. The magnitude this current source is same as the $i(0^-)$.

Using KCL in Fig. 1.26.2

$$1 + \frac{1}{1} \int_0^t v_2 dt + \frac{v_2}{1/2} + \frac{1}{2} \int_0^t v_2 dt = 0$$

Differentiating

$$v_2 + 0 + 2\frac{dv_2}{dt} + \frac{1}{2}v_2 = 0$$

Thus

$$\frac{dv_2}{dt} + \frac{3}{4}v_2 = 0 \qquad\qquad [1.26.1]$$

The form of solution to eqn [1.26.1] is

$$v_2(t) = Ke^{-3t/4} \qquad\qquad [1.26.2]$$

From Fig. 1.26.2, $v_2(0^+)$ can be determined as:

$$v_2(0^+) = -\frac{1}{2} \times 1 = -1/2 \text{ V}$$

Setting $t = 0^+$ in eqn [1.26.2]

$$v_2(0) = -1/2 = K$$

Therefore

$$v_2(t) = -\frac{1}{2}e^{-3t/4} \text{ V} \qquad \text{for } t \geq 0$$

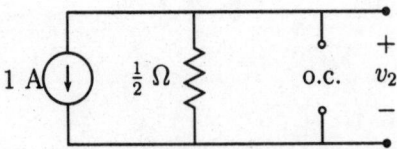

Fig. 1.26.3: $t = 0^+$ circuit.

SP 1.27 In the network of Fig. SP 1.27, the switch K is closed at t = 0, a steady state having previously been attained, solve for current, $i(t)$ in the circuit as a function of time.

    SOLUTION:

From the Fig. 1.27.1

$$i_L(0^-) = V/(R_1 + R_2)$$

Thus, by continuity of inductor current

$$i_L(0^+) = i_L(0^-) = V/(R_1 + R_2) \qquad\qquad [1.27.1]$$

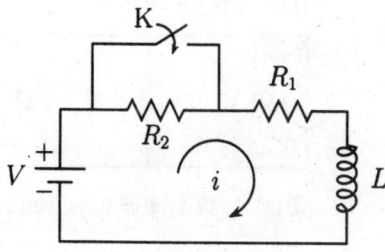

Fig. SP 1.27.

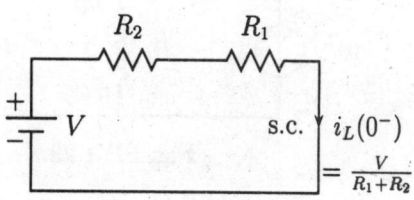

Fig. 1.27.1: $t = 0^-$ circuit.

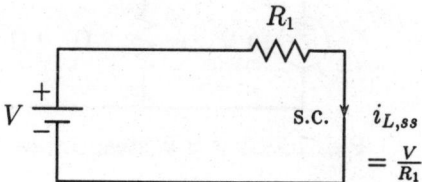

Fig. 1.27.2: $t > 0$, steady-state circuit.

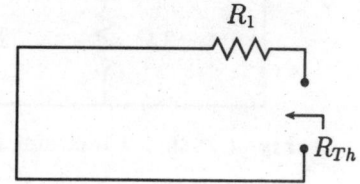

Fig. 1.27.3: $t > 0$ circuit for R_{Th}.

From Fig. 1.27.3

$$R_{Th} = R_1$$

Therefore

$$\tau = L/R_{Th} = L/R_1 \qquad [1.27.2]$$

From Fig. 1.27.2, the steady state current is given by

$$i_{L,ss} = V/R_1 \qquad [1.27.3]$$

Thus, inductor current for $t > 0$ may be written as

$$
\begin{aligned}
i_L(t) &= \left(i_L(0^+) - i_{L,ss}\right) e^{-\frac{R_{Th}}{L}t} + i_{L,ss} \\
&= \left(\frac{V}{R_1 + R_2} - \frac{V}{R_1}\right) e^{-\frac{R_1}{L}t} + \frac{V}{R_1} \\
&= \frac{R_1 - R_1 - R_2}{R_1(R_1 + R_2)} V e^{-\frac{R_1}{L}t} + \frac{V}{R_1}
\end{aligned}
$$

Therefore $\quad i(t) = i_L(t) = \dfrac{V}{R_1(R_1 + R_2)}\left(R_1 + R_2 - R_2 e^{-\frac{R_1}{L}t}\right)$ for $t > 0$

SP 1.28 For the circuit in Fig. SP 1.28, determine the solution of the current in the resistor R_1, $i_{R_1}(t)$ for $t > 0$. The switch opens at t = 0.

SOLUTION:

From the circuit of Fig. 1.28.1

$$i_L(0^-) = 10/(2\|2) = \frac{10}{1} = 10 \text{ A}$$

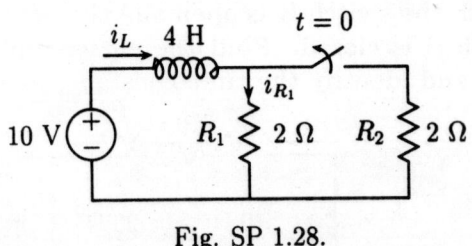

Fig. SP 1.28.

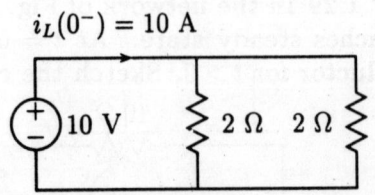

Fig. 1.28.1: $t = 0^-$ circuit.

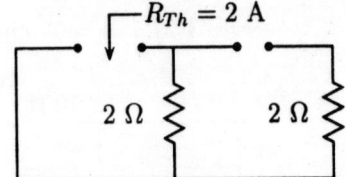

Fig. 1.28.2: $t > 0$ circuit for R_{Th}.

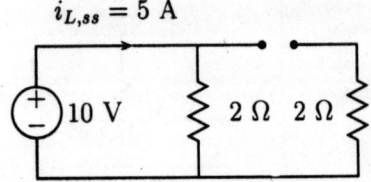

Fig. 1.28.3: $t > 0$, steady-state circuit.

Therefore

$$i_L(0^+) = i_L(0^-) = 10 \text{ A} \qquad [1.28.1]$$

Form circuit of Fig. 1.28.2

$$R_{Th} = 2 \ \Omega$$

Therefore

$$L/R_{Th} = \frac{4}{2} = 2 \text{ Sec.} \qquad [1.28.2]$$

From the circuit of Fig. 1.28.3

$$i_{L,ss} = \frac{10}{2} = 5 \ \text{A} \qquad [1.28.3]$$

So, the total solution is

$$
\begin{aligned}
i_L(t) &= \left[i_L(0^-) - i_{L,ss}\right] e^{-\frac{R_{Th}}{L}t} + i_{L,ss} \\
&= (10 - 5)e^{-t/2} + 5 \\
&= 5e^{-t/2} + 5
\end{aligned}
$$

Now, in order to find resistor current $i_{R_1}(t)$ for $t > 0$, we observe from Fig. 1.28.4 that

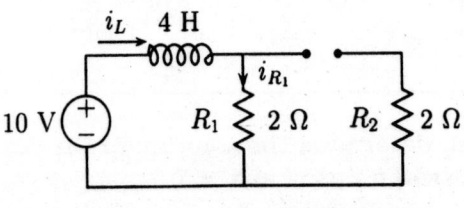

Fig. 1.28.4: $t > 0$ circuit.

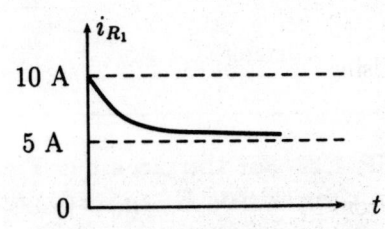

Fig. 1.28.5: Polt of $i_{R_1}(t)$.

$$i_{R_1}(t) = i_L(t) = 5e^{-t/2} + 5 \qquad t > 0$$

SP 1.29 In the network of Fig. **SP 1.29**, the switch K is open and the network reaches steady state. At $t = 0$, switch S is closed. Find the current in the inductor for $t > 0$. Sketch the current and identify the time constant.

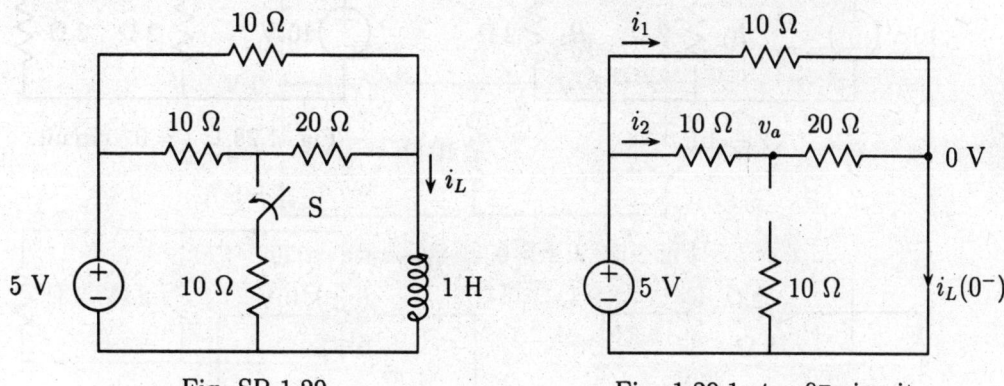

Fig. SP 1.29.   Fig. 1.29.1: $t = 0^-$ circuit.

SOLUTION:

From the circuit shown in Fig. 1.29.1, it is observed that

$$i_1(0^-) = \frac{5-0}{10} = \frac{5}{10} \text{ A}$$

and

$$i_2(0^-) = \frac{5-0}{10+20} = \frac{5}{30} \text{ A}$$

Therefore

$$
\begin{aligned}
i_L(0^-) &= i_1 + i_2 \\
&= \frac{5}{10} + \frac{5}{30} = \frac{20}{30} = \frac{2}{3} \text{ A.}
\end{aligned}
$$

From continuity of inductor current

$$i_L(0^+) = i_L(0^-) = 2/3 \text{ A} \qquad [1.29.1]$$

By KCL at the node v_a in Fig. 1.29.2

$$\frac{5 - v_a}{10} = \frac{v_a}{10} + \frac{v_a}{20} \qquad \Rightarrow v_a = 2 \text{ V.} \qquad [1.29.2]$$

Using circuit of Fig.1.29.2, steady state inductor current can be found as

$$L_{,ss} = i_1 + i_2 = \frac{5-0}{10} + \frac{2-0}{20} = \frac{5}{10} + \frac{2}{20} = \frac{3}{5} \text{ A} \qquad [1.29.2]$$

From the Fig. 1.29.3, the Thevenin resistance is

$$R_{Th} = \frac{50}{7}\Omega \qquad [1.29.4]$$

and

$$\tau(\text{time constant}) = L/R_{Th} = \frac{7}{50} \text{ sec}$$

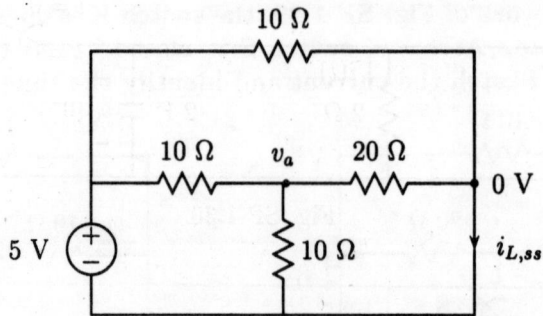

Fig. 1.29.2: $t > 0$, steady-state circuit.

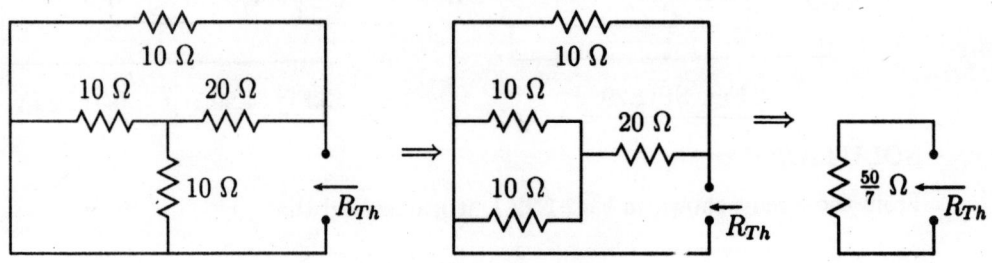

Fig. 1.29.3: $t > 0$ ckts. for R_{Th}.

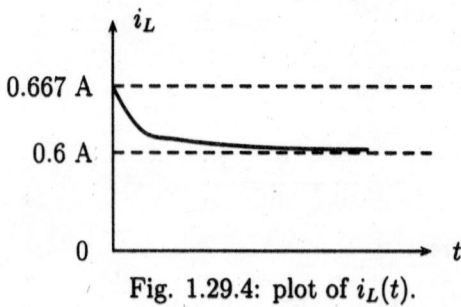

Fig. 1.29.4: plot of $i_L(t)$.

Thus inductor current for $t > 0$,

$$i_L(t) = \left[i_L(0^+) - i_{L,ss} \right] e^{\frac{R_{Th}}{L}t} + i_{L,ss}$$

$$= \left(\frac{2}{3} - \frac{3}{5} \right) e^{-\frac{50}{7}t} + \frac{3}{5}$$

Therefore $i_L(t) = 0.6 + 0.067 e^{-50t/7}$ for $t > 0$

SP 1.30 For the circuit shown in Fig.SP 1.30, determine $i(t)$ for $t > 0$. The switch opens at t = 0.

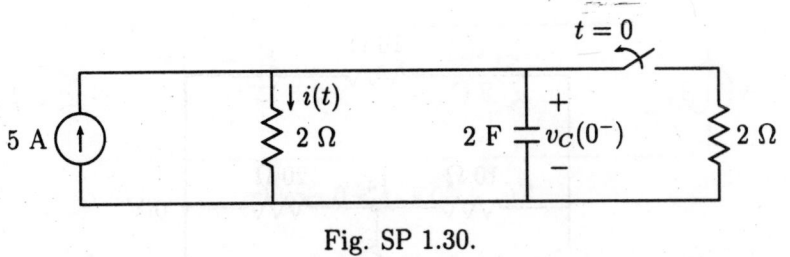

Fig. SP 1.30.

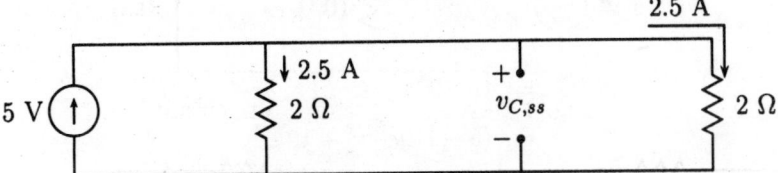

Fig. 1.30.1: $t = 0^-$ circuit.

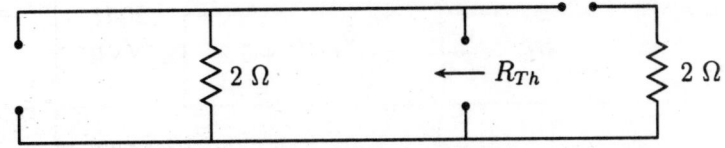

Fig. 1.30.2: $t > 0$ circuit for R_{Th}.

SOLUTION:

From Fig. 1.30.1 it is noted that

$$v_C(0^-) = 2(2.5) = 5 \text{ V}$$

Thus

$$v_C(0^+) = v_C(0^-) = 5 \text{ V}. \qquad [1.30.1]$$

From Fig. 1.30.2

$$R_{Th} = 2\Omega \qquad \text{Thus} \qquad \tau = R_{Th}C = 2 \times 2 = 4 \text{ sec}. \qquad [1.30.2]$$

From Fig. 1.30.3, steady-state capacitor voltage

$$v_{C,ss} = 2 \times 5 = 10 \text{ V}. \qquad [1.30.3]$$

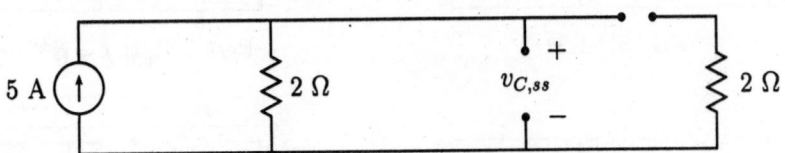

Fig. 1.30.3: $t > 0$, steady-state circuit.

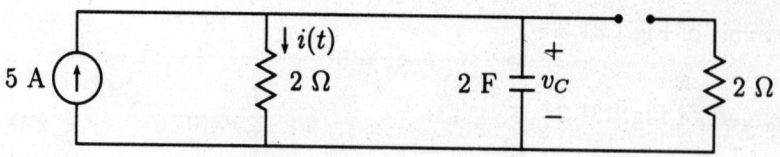

Fig. 1.30.4: $t > 0$ circuit.

Thus, for $t > 0$

$$
\begin{aligned}
v_C(t) &= \left[v_C(0^+) - v_{C,ss}\right] e^{-\frac{t}{R_{Th}C}} + v_{C,ss} \\
&= (5 - 10)e^{-t/4} + 10 \\
&= -5e^{-t/4} + 10 \qquad \text{for } t > 0
\end{aligned}
$$

Now it is observed from Fig. 1.30.4, that

$$
i(t) = \frac{v_C(t)}{2} = -\frac{5}{2}e^{-t/4} + 5 \qquad \text{for } t > 0
$$

SP 1.31 In the circuit shown in Fig. SP 1.31 the switch is opened at t = 0. Find $v_C(t)$ for $t > 0$.

SOLUTION:

From Fig.1.31.1

$$
v_C(0^-) = \frac{R_2 V}{R_1 + R_2}
$$

Thus

$$
v_C(0^+) = v_C(0^-) = \frac{R_2 V}{R_1 + R_2} \qquad\qquad [1.31.1]
$$

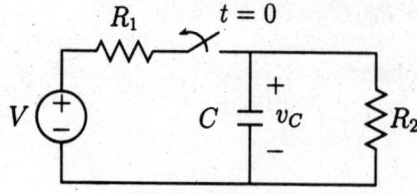

Fig. SP 1.31.

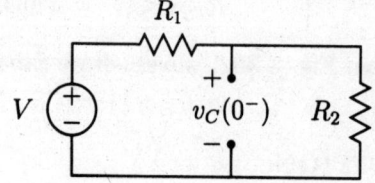

Fig. 1.31.1: $t = 0^+$ circuit.

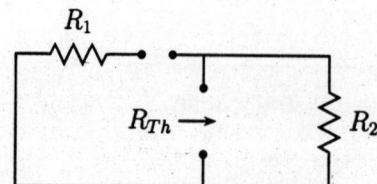

Fig. 1.31.2: $t > 0$ circuit for R_{Th}.

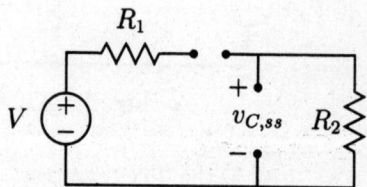

Fig. 1.31.3: $t > 0$, steady-state circuit.

From the circuit of Fig.1.31.3
$$v_{C,ss} = 0$$
[1.31.2]

From the circuit of Fig.1.31.2
$$R_{Th} = R_2$$
[1.31.3]

We know that for $t > 0$,

$$
\begin{aligned}
v_C(t) &= \left[v_C(0^+) - v_{C,ss} \right] e^{-\frac{t}{RC}} + v_{C,ss} \\
&= \left[\frac{R_2 V}{R_1 + R_2} - 0 \right] e^{-\frac{t}{R_2 C}} + 0
\end{aligned}
$$

Therefore
$$v_C(t) = \frac{R_2 V}{R_1 + R_2} e^{-\frac{t}{R_2 C}} \qquad t > 0$$

SP 1.32 In circuit of Fig. SP 1.32, the switch is closed at $t = 0$. The capacitor is uncharged. Find

(a) the voltage across the capacitor, $v_C(t)$ for $t > 0$,

(b) the current through capacitor, $i_C(t)$ for $t > 0$,

(c) the voltage, v_{AB} for $t > 0$.

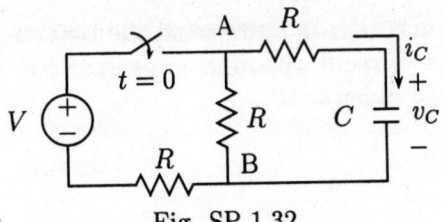

Fig. SP 1.32.

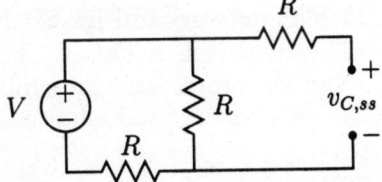

Fig. 1.32.1: $t > 0$ steady state circuit.

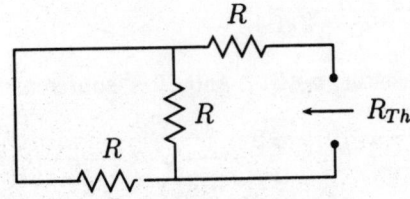

Fig. 1.32.2: $t > 0$ circuit for R_{Th}.

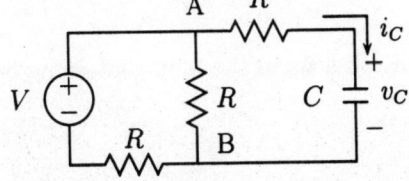

Fig. 1.32.3: $t > 0$ circuit.

SOLUTION:

The initial voltage across the capacitor is

$$v_C(0^+) = v_C(0^-) = 0 \text{(given)}$$

From Fig. 1.32.1, the steady state voltage is

$$v_{Css} = V/2$$

From Fig. 1.32.2, the Thevenin resistance is

$$R_{Th} = (R \| R) + R = 3R/2$$

(a) Hence for $t > 0$

$$
\begin{aligned}
v_C(t) &= \left[v_C(0^+) - V_{Css} \right] e^{-t/R_{Th}C} + v_{Css} \\
&= \left[0 - \frac{V}{2} \right] e^{-2t/3RC} + \frac{V}{2} \\
&= \frac{V}{2} \left[1 - e^{-2t/3RC} \right]
\end{aligned}
$$

(b) The capacitor current is

$$
i_C(t) = C \frac{dv_C}{dt} = \frac{V}{3R} e^{-2t/3RC}
$$

(c) The voltage across the terminals A and B is

$$
\begin{aligned}
v_{AB} &= Ri_C + v_C = \frac{V}{3} e^{-2t/3RC} + \frac{V}{2} \left[1 - e^{-2t/3RC} \right] \\
&= \frac{V}{2} \left[1 - \frac{1}{3} e^{-2t/3RC} \right]
\end{aligned}
$$

SP 1.33 The network of Fig. SP 1.33 is in a steady state with the switch open. At $t = 0$, the switch is closed. Find the current through capacitor for $t > 0$. Sketch the waveform and determine time constant.

SOLUTION:

From Fig. 1.33.1

$$
I = \frac{12}{3 + 2 + 1} = 2 \text{ A}
$$

By KVL around the loop containing 3-Ω resistor, $v_C(0^-)$, and 12 V source as :

$$
-12 + 3I + v_C(0^-) = 0
$$

Hence $v_C(0^-) = 12 - 3I = 12 - 3(2) = 6 \text{ V}$

Thus by continuity of capacitor voltage, we obtain

$$
v_C(0^+) = v_C(0^-) = 6 \text{ V}
$$

From the circuit of Fig. 1.33.2

$$
I = \frac{12}{[(3 + 2)\|4] + 1} = \frac{12}{29/9} = \frac{108}{29} \text{ A}
$$

Using current division formula to determine I_1 as :

$$
I_1 = \frac{4}{4 + (3 + 2)} I = \left(\frac{4}{9} \right) \left(\frac{108}{29} \right) = \frac{48}{29} \text{ A}
$$

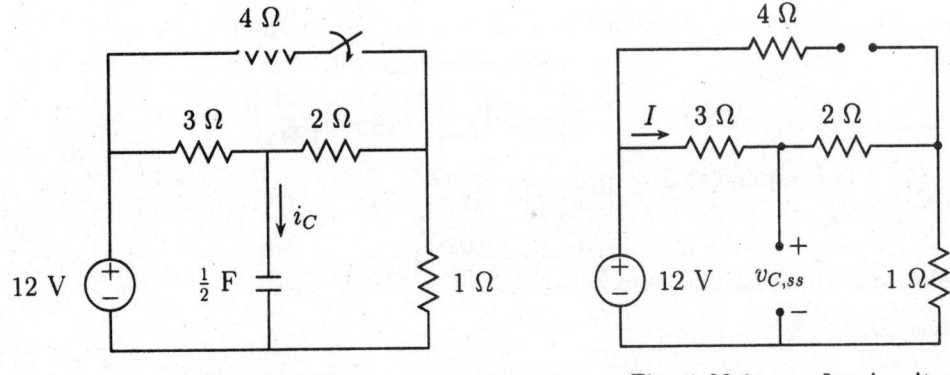

Fig. SP 1.33.   Fig. 1.33.1: t = 0⁻ circuit

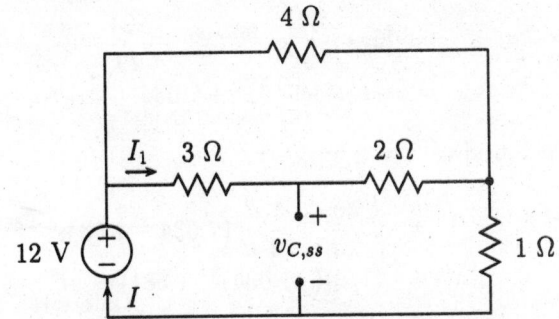

Fig. 1.33.2: $t > 0$, steady-state circuit

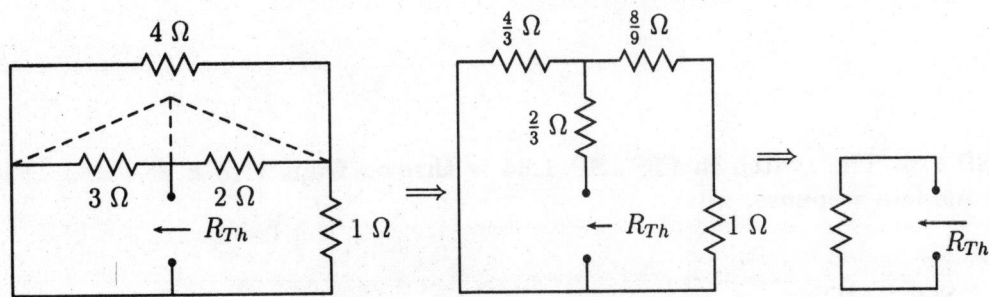

Fig. 1.33.3: $t > 0$ circuits for R_{Th}.

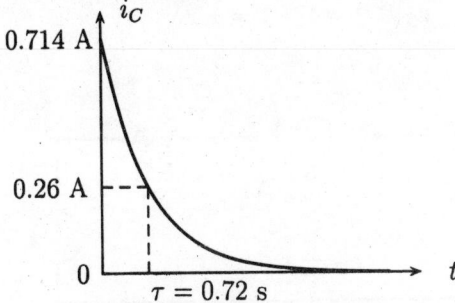

Fig. 1.33.4: plot of $i_C(t)$.

By KVL around the loop containing 12 V source, 3-Ω resistor, and $v_{C,ss}$ as :

$$-12 + 3I_1 + v_{C,ss} = 0$$

Hence $\qquad v_{C,ss} = 12 - 3\left(\frac{48}{29}\right) = \frac{204}{29} \text{ V}$

From Fig.1.33.3, we find that

$$R_{Th} = \frac{378}{261} \ \Omega$$

Therefore, the time constant is, $\tau = R_{Th}C = \frac{378}{261} \times \frac{1}{2} = \frac{189}{261}$ sec.

Thus, for $t > 0$

$$
\begin{aligned}
v_C(t) &= [v_C(0^+) - v_{C,ss}]e^{\frac{-t}{R_{Th}C}} + v_{C,ss} \\
&= [6 - \frac{204}{29}]e^{-261t/189} + \frac{204}{29} \\
&= -1.034e^{-1.381t} + 7.034
\end{aligned}
$$

Therefore, $i_C(t)$ can be found as

$$
\begin{aligned}
i_C(t) &= C\frac{dv_c}{dt} = \frac{1}{2}\frac{d}{dt}\left(7.034 - 1.034e^{-1.381t}\right) \\
&= (1/2)(-1.034)(-1.381)\ e^{-1.381t}
\end{aligned}
$$

Thus $\qquad i_C(t) = 0.714e^{-1.381t} \qquad t > 0$

SP 1.34 The switch in Fig. **SP 1.34** is thrown from a to b at $t = 0$. Find complete response, $v_C(t)$.

SOLUTION:

From circuit of Fig. 1.34.1

$$v_C(0^-) = 1 \times 1 = 1 \text{ V}.$$

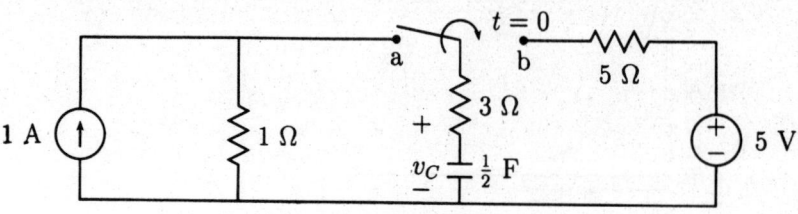

Fig. SP 1.34.

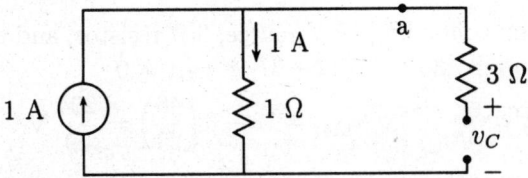

Fig. 1.34.1: $t = 0^-$ circuit.

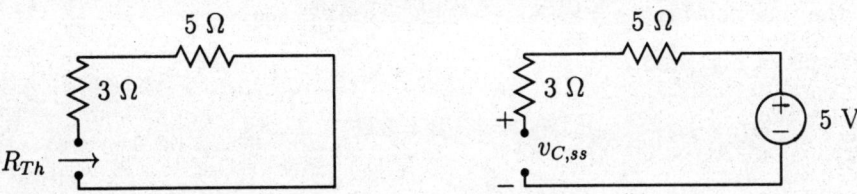

Fig. 1.34.2: $t > 0$ circuit for R_{TH}.   Fig. 1.34.3: $t > 0$, steady-state circuit.

Thus, by continuity of capacitor voltage, we obtain

$$v_C(0^+) = v_C(0^-) = 1 \text{ V}. \qquad [1.34.1]$$

From Fig. 1.34.3, $v_{C,ss}$ is

$$v_{c,ss} = 5 \text{ V}. \qquad [1.34.2]$$

From circuit of Fig. 1.34.2, Thevenin resistance is

$$R_{Th} = 8 \text{ }\Omega$$

Now time constant

$$\tau = R_{Th}C = 8 \times \frac{1}{2} = 4 \text{ sec}. \qquad [1.34.3]$$

Now, we know that for $t > 0$

$$
\begin{aligned}
v_C(t) &= [v_C(0^+) - v_{c,ss}]e^{\frac{-t}{R_{Th}C}} + v_{c,ss} \\
&= (1 - 5)e^{-t/4} + 5 \\
&= 5 - 4e^{-t/4} \qquad \text{for } t > 0
\end{aligned}
$$

SP 1.35 Determine the solution for the voltage, $v_R(t)$ across the 2Ω resistor in Fig. SP 1.35.

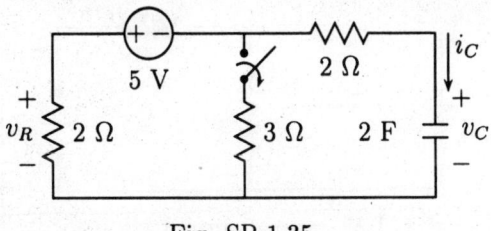

Fig. SP 1.35.

SOLUTION:

From the circuit of Fig. 1.35.1

$$v_C(0^-) = -(3\,\Omega)\left(\frac{5}{3+2}\ \text{A.}\right) = -3\ \text{V.}$$

Thus, by continuity of capacitor voltage, we obtain

$$v_C(0^+) = v_C(0^-) = -3\ \text{V.} \tag{1.35.1}$$

From the circuit of Fig. 1.35.2

$$R_{Th} = 2 + 2 = 3\ \Omega$$

Thus

$$\tau = R_{Th}C = 4 \times 2 = 8\ \text{sec.} \tag{1.35.2}$$

From the circuit of Fig. 1.35.3

$$v_{C,ss} = -5\ \text{V} \tag{1.35.3}$$

Thus, for $t > 0$,

$$
\begin{aligned}
v_C(t) &= [v_C(0^+) - v_{C,ss}]e^{\frac{-t}{R_{Th}C}} + v_{C,ss} \\
&= [-3 - (-5)]\,e^{-t/8} + (-5)
\end{aligned}
$$

$$v_C(t) = 2\,e^{-t/8} - 5 \tag{1.35.4}$$

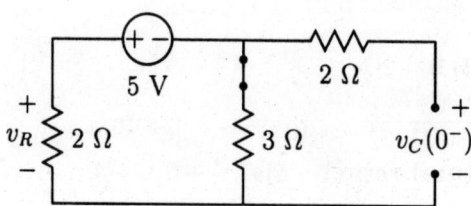

Fig. 1.35.1: $t = 0^-$ circuit.

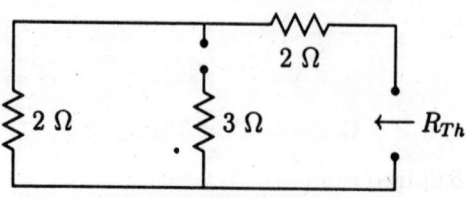

Fig. 1.35.2: $t > 0$ circuit for R_{Th}.

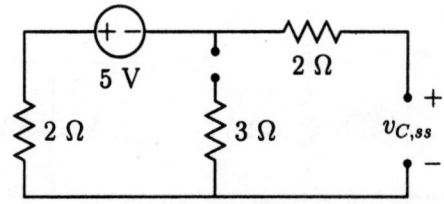

Fig. 1.35.3: $t > 0$, steady-state circuit.

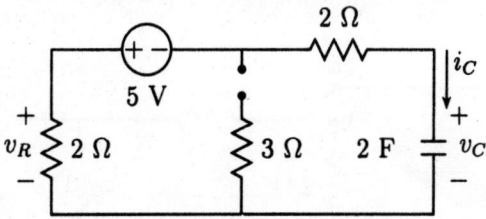

Fig. SP 1.35.4: $t > 0$ circuit.

Now $i_C(t)$ can be determined as :

$$i_C(t) = C\frac{dv_C}{dt} = 2\left[\frac{d}{dt}(2e^{-\frac{t}{8}} - 5)\right]$$
$$= 2\left[2(-1/8)e^{-\frac{t}{8}} + 0\right]$$

Thus

$$i_C(t) = -\frac{1}{2}e^{-\frac{t}{8}} \qquad [1.35.5]$$

Now, from Fig. 1.35.4, we observe that

$$v_R(t) = -2i_C(t) = -2\left[-\frac{1}{2}e^{-\frac{t}{8}}\right] = e^{-\frac{t}{8}} \qquad t > 0$$

SP 1.36 Find the natural response for $v_C(t)$ of the circuit in Fig. SP 1.36.

SOLUTION:

$v_C(0^-)$ may be determined from Fig. 1.36.1, by applying KVL around meshes as :
Left mesh:

$$4i + v_C(0^-) = 6 \qquad [1.36.1]$$

Right mesh

$$-v_C(0^-) - 2v_C(0^-) + 2i = 0$$

or

$$\frac{3}{2}v_C(0^-) = i \qquad [1.36.2]$$

Substituting eqn [1.36.2] into eqn [1.36.1], yields

$$4 \times \frac{3}{2}v_C(0^-) + v_C(0^-) = 6 \qquad \Rightarrow v_C(0^-) = \frac{6}{7}$$

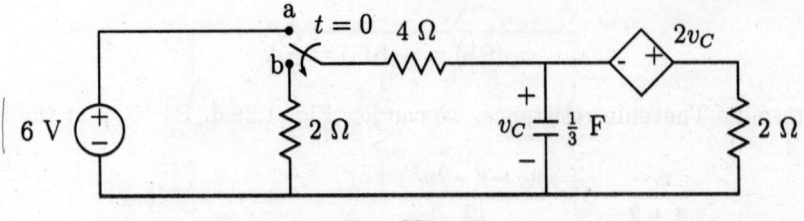

Fig. SP 1.36.

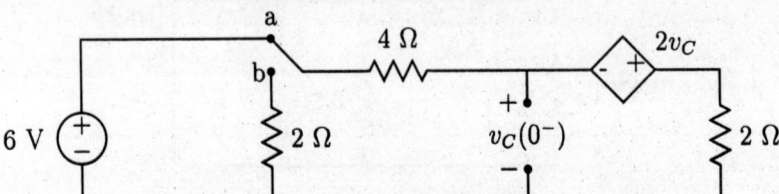

Fig. 1.36.1: $t = 0^-$ circuit.

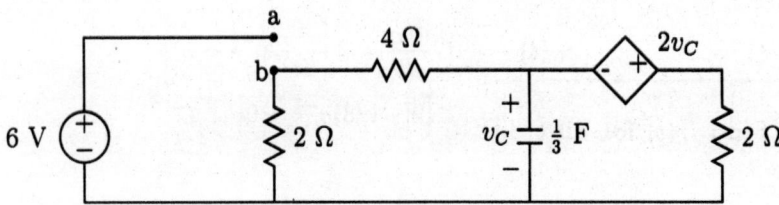

Fig. 1.36.2: $t > 0$ circuit.

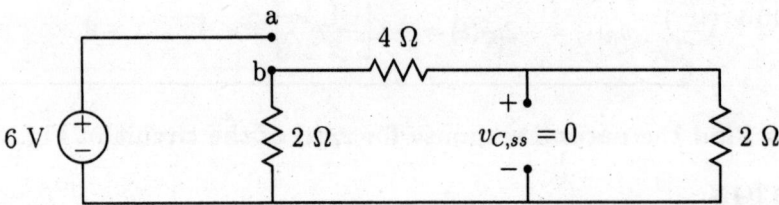

Fig. 1.36.3: $t > 0$, steady-state circuit.

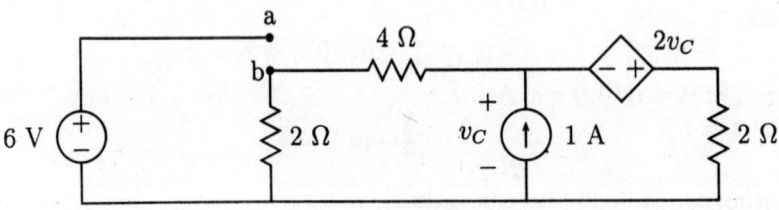

Fig. 1.36.4: $t > 0$ circuit for R_{Th}.

Thus

$$v_C(0^+) = v_C(0^-) = \frac{6}{7} \text{ V} \qquad\qquad [1.36.3]$$

To determine Thevenin resistance, we can use Fig. 1.36.4. By KCL at the node v_C in Fig. 1.36.4

$$\frac{v_C}{4+2} - 1 + \frac{v_C - (-2v_C)}{2} = 0 \qquad \Rightarrow v_C = 3/5 \text{ V}.$$

Thus, $R_{Th} = \frac{3}{5}$ Ω. The time constant τ is therefore,

$$\tau = R_{Th}C = \frac{3}{5} \times \frac{1}{3} = \frac{1}{5} \text{ sec.} \qquad\qquad [1.36.4]$$

From Fig. 1.36.3, we obtain

$$v_{C,ss} = 0 \qquad\qquad [1.36.5]$$

Thus for $t > 0$,

$$
\begin{aligned}
v_C(t) &= [v_C(0^+) - v_{C,ss}]e^{\frac{-t}{R_{Th}C}} + v_{C,ss} \\
&= \left(\frac{6}{7} - 0\right)e^{-5t} + 0 \\
&= \frac{6}{7}e^{-5t} \qquad \text{for } t > 0
\end{aligned}
$$

SP 1.37 Find $i_L(t)$ for all t, in the circuit shown in Fig. SP 1.37.

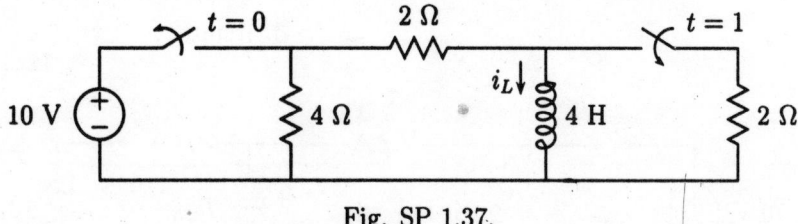

Fig. SP 1.37.

SOLUTION:

For $0 < t < 1$, $i_L(t)$ can be given as

$$i_L(t) = i_L(0^+)e^{-\frac{R_{Th}}{L}t} = 5e^{-\frac{3}{2}t} \text{ A} \qquad\qquad [1.37.1]$$

Note here $i_L(0^+) = i_L(0^-) = 5$ A from Fig. 1.37.1 and $R_{Th} = 6\Omega$ from Fig. 1.37.3.
    Now at $t = 1^-$

$$i_L(1^-) = 5e^{-1.5} = 1.1146 \text{ A}$$

Thus

$$i_L(1^+) = i_L(1^-) = 1.1146 \text{ A} \qquad\qquad [1.37.2]$$

For $t > 1$

$$i_L(t) = i_L(1^+)e^{-\frac{R_{Th}}{L}(t-1)} \text{ A} \qquad \text{for } t \geq 0 \qquad [1.37.3]$$

Putting the value of $i_L(1^+) = 1.1146$ A $R_{Th} = 3/2$ Ω and L = 4 H into eqn [1.37.3], we get

$$i_L(t) = 1.1146e^{-\frac{3}{8}(t-1)} \text{ A} \qquad \text{for } t > 1$$

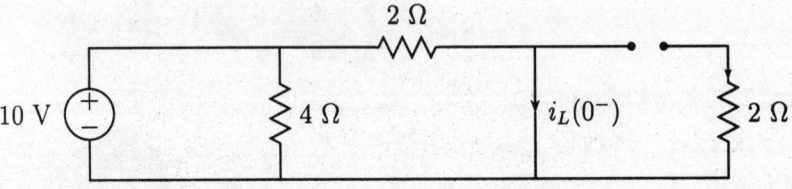

Fig. 1.37.1: $t = 0^-$ circuit.

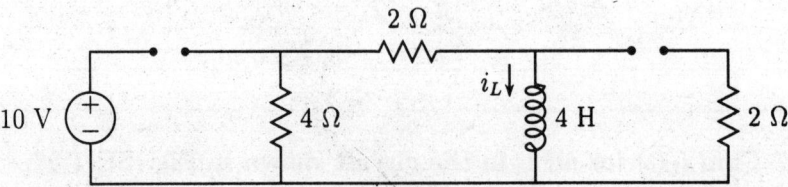

Fig. 1.37.2: $0 < t < 1$ circuit.

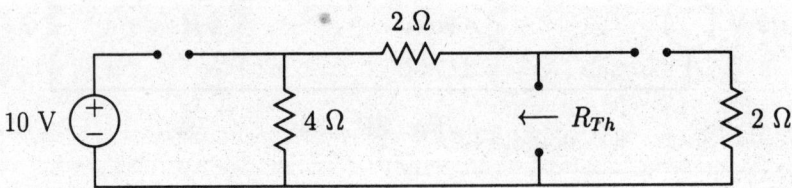

Fig. 1.37.3: $0 < t < 1$ circuit for R_{Th}.

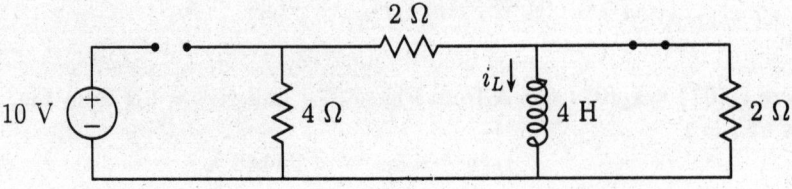

Fig. 1.37.4: $t > 1$ circuit.

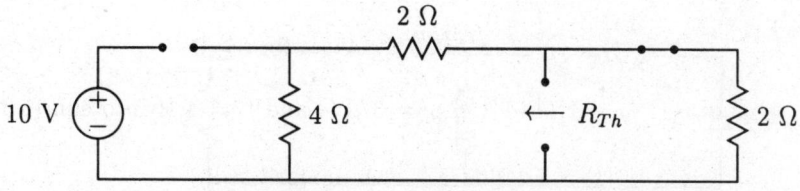

Fig. 1.37.5: $t > 1$ circuit for R_{Th}.

1.13   DRILL PROBLEMS

DP 1.1 Consider the differential equation

$$\frac{di}{dt} + i = v(t)$$

Find the general solutions of this equation for the following functions which exist only for $t \geq 0$:

(i) $v(t) = k_1 t$ [Hint: Select $i_p = a_0 t + a_1$]

(ii) $v(t) = te^{-2t}$ [Hint: Select $i_p = (a_1 t + a_0)e^{-2t}$]

(iii) $v(t) = \sin \omega t$ [Hint: Select $i_p = (A \cos \omega t + B \sin \omega t)$]

(iv) $v(t) = \cos \omega t$ [Hint: Select $i_p = (A \cos \omega t + B \sin \omega t)$]

(v) $v(t) = \sin^2 t$ [Hint: Use $\sin^2 t = \frac{1 - \cos 2t}{2}$ and select $i_p = k + (A \cos \omega t + B \sin \omega t)$]

(vi) $v(t) = \cos^2 t$ [Hint: Use $\cos^2 t = \frac{1 + \cos 2t}{2}$ and select $i_p = k + (A \cos \omega t + B \sin \omega t)$]

(vii) $v(t) = t \sin 2t$ [Hint: Select $i_p = (a_1 t + a_0) \cos 2t + (a_1 t + a_0) \sin 2t$]

(viii) $v(t) = e^{-t} \sin 2t$ [Hint: Select $i_p = (A \cos 2t + B \sin 2t)e^{-t}$]

DP 1.2 Write a differential equation in i_L valid for $t > 0$ in the circuit shown in Fig. DP 1.2.

Hints : By KCL at **A**

$$\frac{5 - v_A}{3} = i_L + \frac{v_A}{5}$$

Replace v_A by

$$v_A = \frac{1}{2}\frac{di_L}{dt}$$

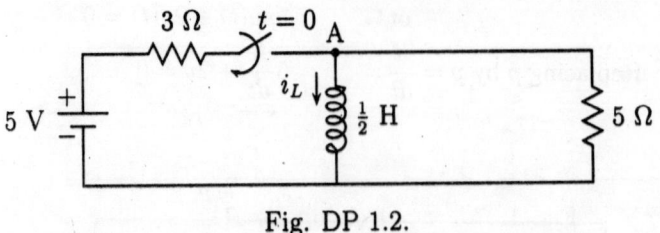

Fig. DP 1.2.

DP 1.3 Find the response for v_C for the circuit of Fig. DP 1.3. If
  (i) $v_s = 2$ V and
  (ii) $v_s = 2t^2$

Hints : By KCL at **a**

$$\frac{v_s - v_C}{4} = \frac{v_C}{2} + \frac{1}{2}\frac{dv_C}{dt}$$

(i) Select $v_{Cp} = k$
(ii) Select $v_{Cp} = k_1 t^2 + k_2 t + k_3$

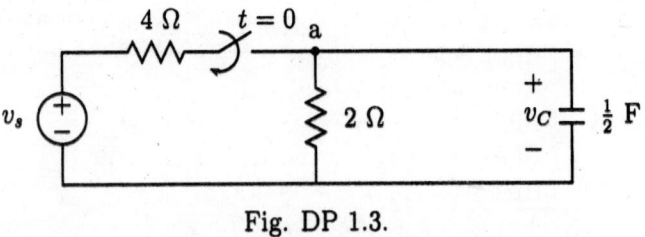

Fig. DP 1.3.

DP 1.4 Find a differential equation in $v(t)$ for $t > 0$

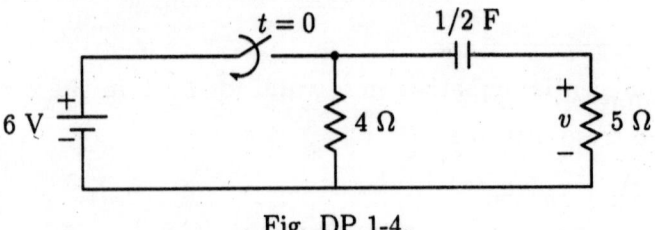

Fig. DP 1-4.

Hints : Use p-operator method to transform the circuit in Fig. DP 1.4.1 into p-domain as in Fig. DP 1.4.1
  By voltage division method

$$v(t) = \frac{5}{5 + 2/p}(6)$$

$$\text{or} \qquad 5pv(t) + 2v(t) = 0$$

$$\text{Replacing } p \text{ by } p = \frac{d}{dt} \qquad 5\frac{dv}{dt} + 2v = 0$$

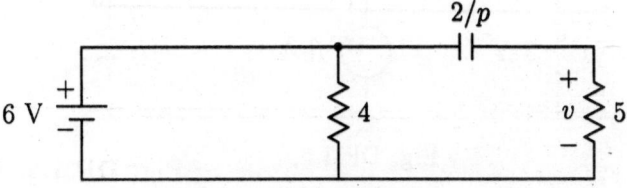

Fig. DP 1.4.1: *p*-domain circuit.

DP 1.5 Find v_C for $t > 0$ for the circuit of Fig. DP 1.5.

Hints : By inspection method

$$v_C(0^-) = v_C(0^+) = 0$$
$$v_{C,ss} = 6 \text{ V}$$
$$R_{Th} = 1 + 2 = 3 \ \Omega.$$

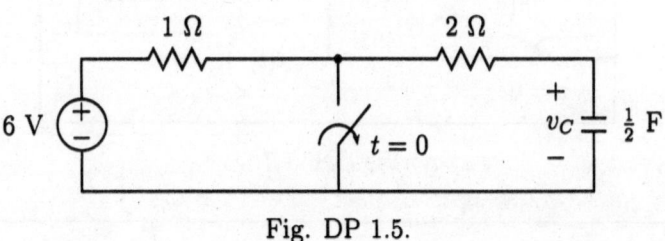

Fig. DP 1.5.

DP 1.6 Determine $v(t)$ and $i(t)$ for $t > 0$ for the circuit of Fig. DP 1.6. The switch is moved from a to b at $t = 0$.

Hints : By inspection method

$$v_C(0^-) = v_C(0^+) = 15 \text{ V}$$
$$v_{C,ss} = 11.33 \text{ V}$$
$$R_{Th} = 44/15 \ \Omega$$

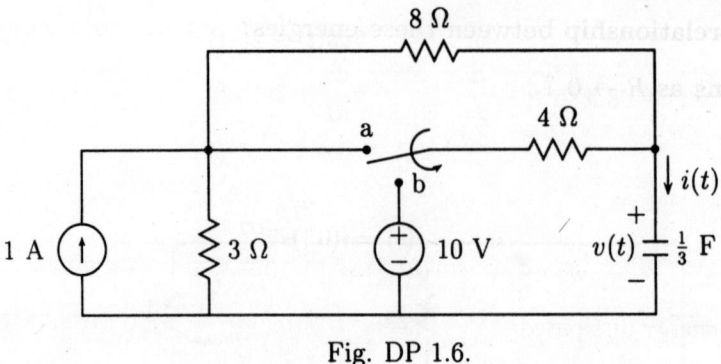

Fig. DP 1.6.

DP 1.7 Find $i(t)$ for $t > 0$ if switch is closed at t $= 0$ in the circuit of Fig. DP 1.7.

Hints : By inspection method

$$i_L(0^-) = 3/4 \text{ A} = i_L(0^+)$$
$$i_{L,ss} = 1/2 \text{ A}$$
$$R_{Th} = 10 \text{ }\Omega$$

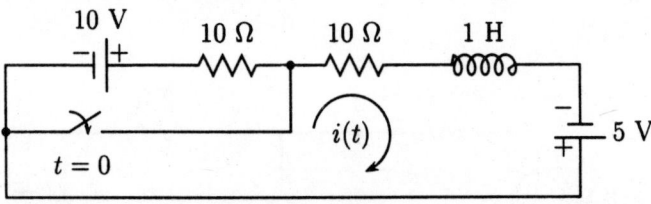

Fig. DP 1.7.

DP 1.8 The circuit shown in Fig. DP 1.8, is in steady state prior to time 0 when the left capacitor is charged to 10 V, and the right capacitor is uncharged. Switch is closed at $t = 0$. Calculate the following

1. **The current $i(t)$ for $t \geq 0$.**

2. **The energy dissipated during the interval 0 to T.**

3. **The limiting values for $t \to \infty$ of**

    (a) **the capacitor voltage v_1 and v_2,**

    (b) **the current,**

    (c) **the energy stored in the capacitor and the energy dissipated in the resistor.**

4. What is the relationship between these energies?

5. What happens as $R \to 0$?

Hints:

$$i(t) = i(0^+)e^{-t/2}$$

where $\qquad i(0^+) = 10/10 = 1 \text{ A}$

The energy dissipated $\qquad w_R = \int_0^T (10i^2)dt$

$$v_1 = 10 - \frac{1}{1/2}\int_0^t i\, dt$$

and $\qquad v_2 = \frac{1}{1/3}\int_0^t i\, dt$

$$v_1(\infty) = v_2(\infty) = 6 \text{ V}$$

$$w_C = \left(\frac{1}{2}\right)\left(\frac{1}{3}\right)v_2^2$$

$$w_R = (1/2)(1/2)(10)^2 - w_C$$

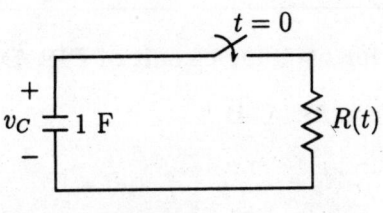

Fig.DP 1.8.

DP 1.9 Consider the circuit shown in Fig. **DP 1.9.** The resistor is time varying. Its R at time t is $R(t) = 2(t+1)$, and capacitor is time invariant with the capacitance 1/2 F. Calculate $v_C(t)$ for $t > 0$. The initial voltage is $v_C(0) = 5$ V.

Fig.DP 1.9.

Hints : By KCL

$$C\frac{dv_C}{dt} + \frac{v_C}{R(t)} = 0$$

Use variables -separable method and then integrate as

$$\int_5^{v_C} \frac{dv_C}{v_C} + \int_0^t \frac{dt}{1+t} = 0 \qquad \Rightarrow v_C = 5/(t+1)$$

DP 1.10 Find the time-constant of the circuit of Fig. DP 1.10.

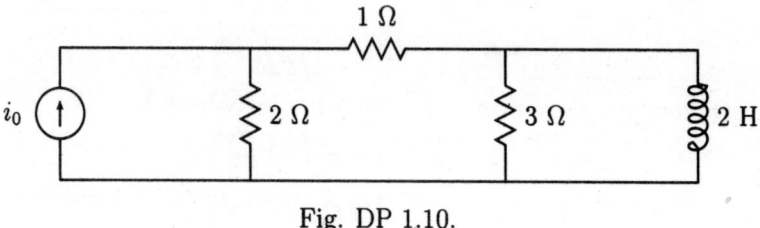

Fig. DP 1.10.

Hints : Find the Thevenin resistance R_{Th} at terminal of the inductor and use relationship, $\tau = L/R_{Th}$

DP 1.11 Find time- constant for the circuit of Fig. DP 1.11.

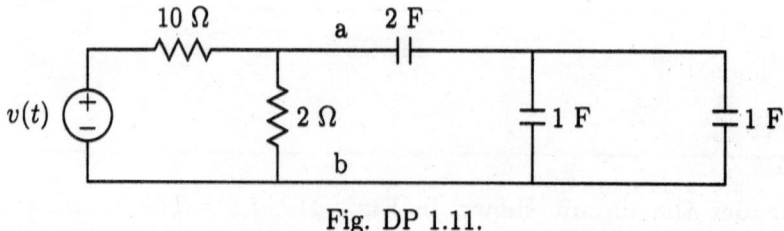

Fig. DP 1.11.

Hints : Find the equivalent capacitance, C_{eq} right to the terminal a-b and The Thevenin resistance, R_{Th} left to the terminal a-b. And use relationship $\tau = R_{Th}C_{eq}$

DP 1.12 Solve for $v_{AB}(t)$ for $t > 0$ for circuit of Fig. DP 1.12.

Hints : By KCL in the loop A-B-C-D-A

$$1i_1 + v_{AB} - \frac{1}{1}\int_0^t i_2 dt = 0$$

Put $i_1 = i_2 = i(0^+)e^{-t/RC} = (1)e^{-t}$ in the above equation and solve for v_{AB}.

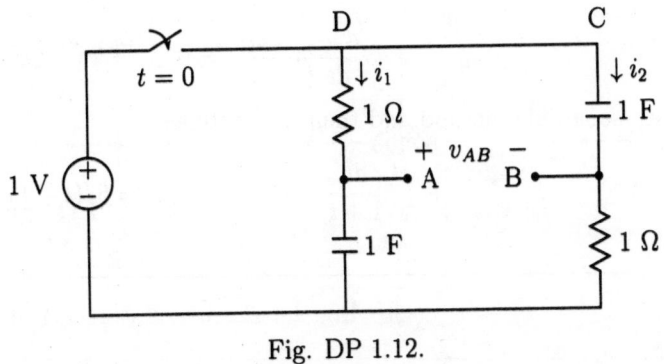

Fig. DP 1.12.

DP 1.13 In the circuit shown in Fig. DP 1.13, the capacitor is initially charged to **12 V**. Find the mathematical expression for the capacitor v_C after closing the switch at **t = 0.**   *GATE - 99(EE)*

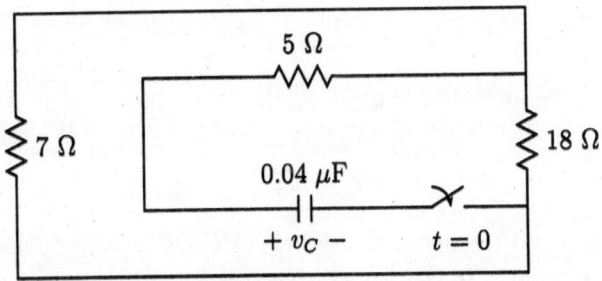

Fig. DP 1.13.

Hints : By inspection method

$$v_C(0^+) = v_C(0^-) = 12 \text{ V}$$
$$v_{C,ss} = 0$$
$$R_{Th} = [(2+7)||18] + 5 = 11 \text{ } \Omega$$

Hence $$v_C = 12e^{-2.27\times10^6 t}$$

DP 1.14 Find the time constant for circuit shown in Fig. DP 1.14.

Hints: Determine the Thevenin resistance R_{Th} seen through terminals of 2-H inductor with all the sources suppressed (current source replaced by open circuit and voltage source replaced by short circuit) and use formula $\tau = L/R_{Th}$.

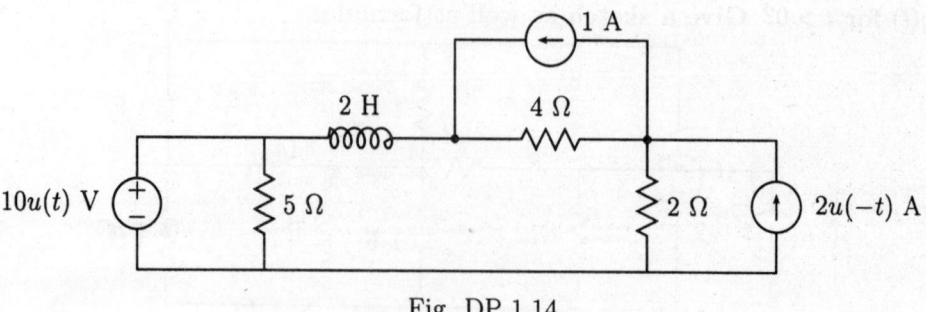

Fig. DP 1.14.

DP 1.15 Find the natural frequency for the circuit shown in Fig. DP 1.15.

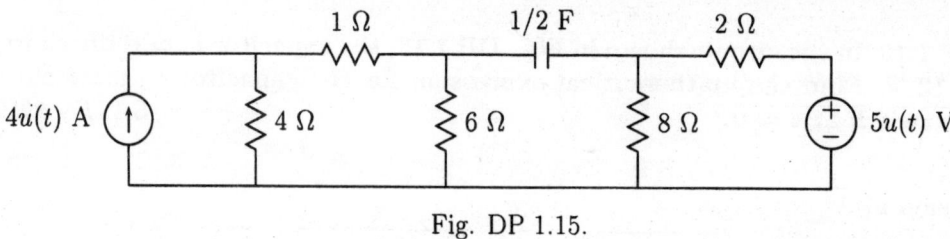

Fig. DP 1.15.

Hints: The natural frequency is given by

$$
\begin{aligned}
s_1 &= -R/L = -1/\tau \qquad \text{RL circuit} \\
    &= -1/RC = -1/\tau \qquad \text{RC circuit}
\end{aligned}
$$

To determine the natural frequency for the given circuit in Fig. DP 1.15, first find the Thevenin resistance seen through the terminals of 1/2-F capacitor, and then $-1/R_{eq}C$ which will be the natural frequency of this circuit.

DP 1.16 Find (a) the time constant (b) $v(\infty)$ for the circuit in Fig. DP 1.16. Also sketch $v(t)$ versus t. Given that $i(t) = 0.05u(t)$, $v(0) = 10$ V, $R = 1000$ Ω, and $C = 0.02$ μF.

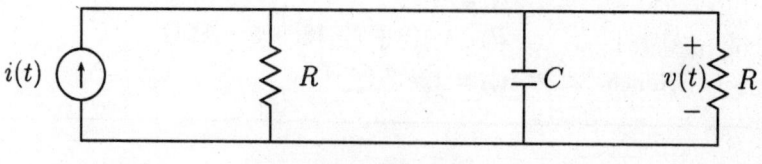

Fig. DP 1.16.

Hints: Time constant $\tau = R_{eq}C = (R/2)C$, $v(\infty) = R(i/2)$, and the expression for $v(t)$ is

$$
v(t) = (v(0) - v(\infty))e^{-t/\tau} + v(\infty)
$$

DP 1.17 In the circuit shown in Fig. DP 1.17, $i(t) = I_s u(t)$ and $v(0) = 0$. **What is $v(t)$ for $t > 0$? Give a sketch as well as formula.**

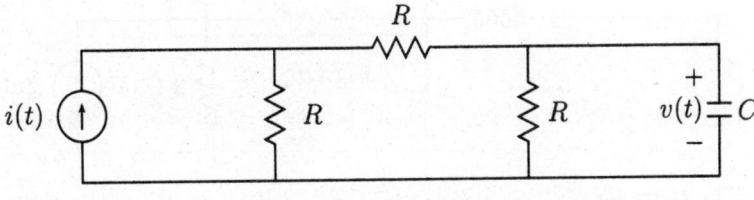

Fig. DP 1.17.

Hints: $v(t)$ is
$$v(t) = (v(0) - v(\infty))e^{-t/\tau} + v(\infty)$$
where $\tau = R_{eq}C$, $R_{eq} = R\|(R + R)$ and $v(\infty) = (RI_s/(R + R + R))R$

DP 1.18 The switch S in Fig. DP 1.18 is opened at $t = 0$. Obtain the expression for i_L for $t > 0$.

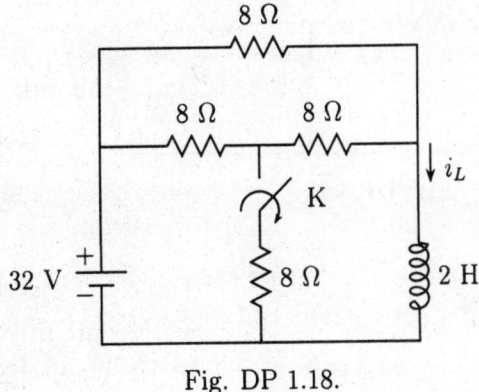

Fig. DP 1.18.

Hints: The inductor current is

$$i_L(t) = (i_L(0) - i_L(\infty))e^{-t/\tau} + i_L(\infty)$$

where $\tau = L/R_{eq}$ and $R_{eq} = 8\|(8 + 8)$

DP 1.19 In Fig. DP 1.19, the switch is closed at $t = 0$. Find
   (a) v_R, the voltage across the resistor at $t = 20$ μs
   (b) the time at which v_R is 1 V.

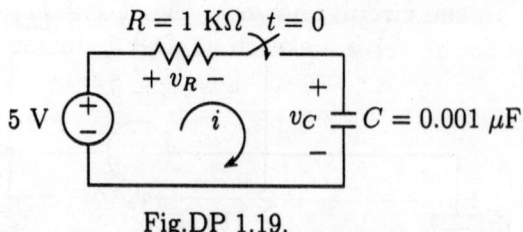

Fig.DP 1.19.

Hints: The current is

$$i = \frac{V}{R}e^{-t/RC}$$

and $\qquad v_R = Ri = Ve^{-t/RC}$

DP 1.20 Find the expression for $i_L(t)$ and $v_L(t)$ for $t > 0$ for the circuit of Fig. DP 1.20. The switch is closed at $t = 0$. The initial current $i_L(0) = 20$ mA.

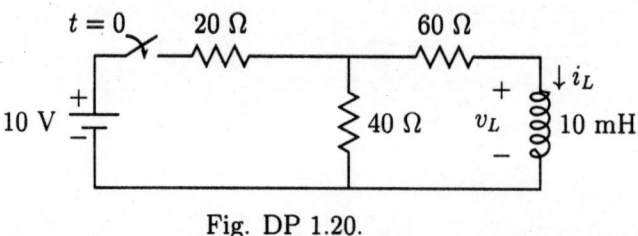

Fig. DP 1.20.

Hints: The inductor current is

$$i_L(t) = (i_L(0) - i_L(\infty))e^{-t/\tau} + i_L(\infty)$$

and voltage $\qquad v_L(t) = L\frac{di_L}{dt}$

DP 1.21 The switch shown in Fig. DP 1.21(a), closed at $t = 0$ and reopened at $t = 5$ ms. It is observed that the voltage across the across the switch gives the graphs shown in Fig. DP 1.21(b). Determine V_0 and L.

Hints: The sketch of inductor current is shown in Fig. DP 1.21.1. It is noted that when the switch is reopened for a long time, the i_L would have decayed to zero, that is

$$v_r = V_0 + 50i_L = V_0 = 20 \text{ V}$$

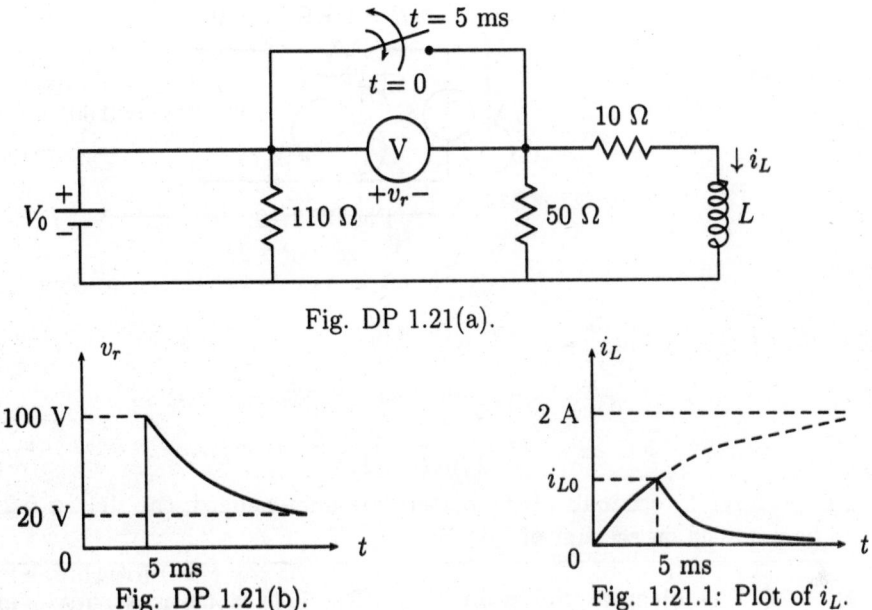

Fig. DP 1.21(a).

Fig. DP 1.21(b).

Fig. 1.21.1: Plot of i_L.

When the switch is reopened at $t = 5$ ms, i_L is at i_{L0} and the voltage across the switch is 100 V. Thus

$$v_r = V_0 + 50i_{0L}$$

or $$100 = 20 + 50i_L$$

Thus $$i_{L0} = 80/50 = 1.6 \text{ A}$$

Also the inductor current when switch was not reopened is $i_L(\infty) = V_0/10 = 20/10 = 2$ A. The inductor current for $0 < t < 5$ ms is

$$i_L(t) = [i_L(0^+) - i_L(\infty)]e^{-\frac{R_{eq}}{L}t} + i_L(\infty)$$
$$= [0 - 2]e^{-\frac{10}{L}t} + 2$$
$$= 2(1 - e^{-\frac{10}{L}t})$$

The L can be determined from the equation

$$i_L(5 \text{ ms}) = 1.6 = 2\left(1 - e^{-\frac{10}{L}(5\times10^{-3})}\right)$$

DP 1.22 In RL series circuit shown in Fig. DP 1.22, the switch S is thrown in position 1 at $t = 0$. After 2 ms, the switch is thrown to position 2. Find the expression for $i(t)$ for both position of the switch and sketch the variation of current. If, in position 2, terminals of the battery are interchanged, reevaluate $i(t)$.

Hints: The current $i(t)$ for $0 < t < 0.2$ ms is

$$i(t) = \frac{80}{40}(1 - e^{-\frac{40}{16}t})$$

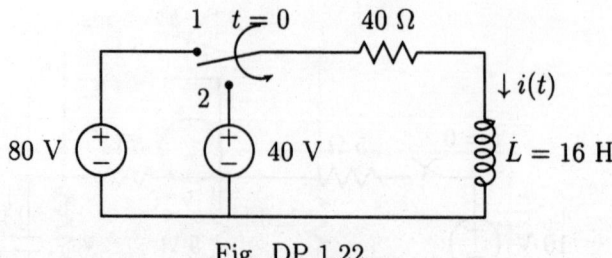

Fig. DP 1.22.

and $i(0.2 \text{ ms}) = 4/4001$ Thus, $i(t)$ for $0.2 \text{ ms} < t < \infty$ will be

$$
\begin{aligned}
i(t) &= [i(0.2 \text{ ms}) - i(\infty)] e^{-\frac{40}{16}t} + i(\infty) \\
&= \left[\frac{4}{4001} - 1\right] e^{-\frac{40}{40}t} + 1
\end{aligned}
$$

If, in position 2, terminals of the battery are interchanged, the $i(t)$ for $0.2 \text{ ms} < t < \infty$ will be modified on account of $i(\infty) = -1$.

DP 1.23 The circuit shown in Fig. DP 1.23 attains steady-state with the switch opened. The switch is closed at $t = 0$. Find the $v_C(t)$ and $i_C(t)$ for $t > 0$ and sketch the variations of $v_C(t)$ and $i_C(t)$ with time.

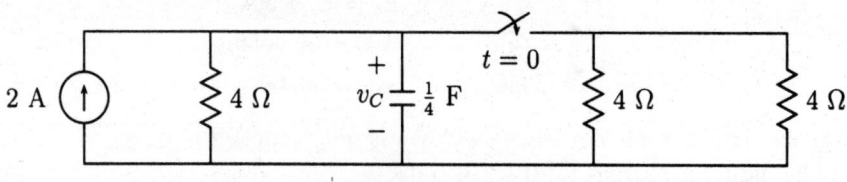

Fig. DP 1.23.

Hints:

$$
\begin{aligned}
v_C(0^-) &= 4 \times 2 = 8 \text{ V} = v_C(0^+) \\
R_{Th} &= \frac{1}{1/4 + 1/4 + 1/4} \\
v_C(\infty) &= 4\,(2/3) \text{ V}
\end{aligned}
$$

DP 1.24 In circuit shown in Fig. DP 1.24, the switch is closed at $t = 0$. Find the v_C and i_C for $t > 0$. The capacitor is initially uncharged.

Hints:

$$
\begin{aligned}
R_{Th} &= (5\|5) + (5\|5) = 5 \ \Omega \\
v_C(0^+) &= 0 \qquad \text{Given} \\
v_C(\infty) &= 5\left(\frac{10}{5+5}\right) = 5 \text{ V}
\end{aligned}
$$

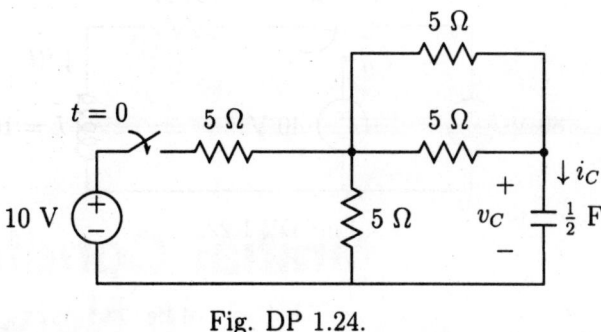

Fig. DP 1.24.

DP 1.25 In circuit shown in Fig. DP 1.25, at $t = 0$ the switch S_1 is closed and switch S_2 is opened. Find the expression for $i_L(t)$ for $t > 0$.

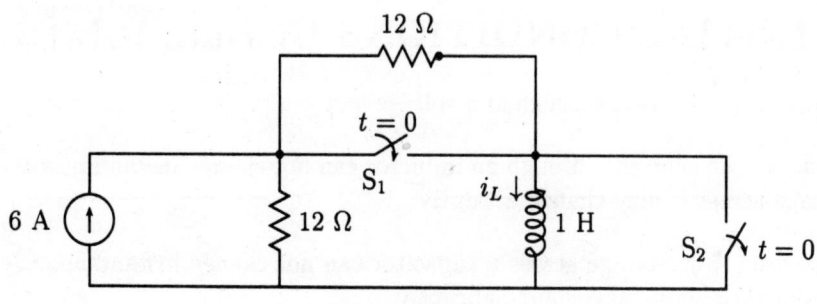

Fig. DP 1.25.

Hints:

$$i_L(0^+) = 0$$
$$R_{Th} = 12 \, \Omega$$
$$i_L(\infty) = 6 \, \text{A}$$

****

A(a) replacing each inductor with

(b) current source of value I_0 if current through it

(b) current source of value I_0 if current the inductor I_0.

(c) replacing each capacitor with

(d) short circuit if voltage across it say $v = 0$; with zero

CHAPTER

2

Initial Conditions
in Networks

SUMMARY OF THIS CHAPTER

2.1   INITIAL CONDITIONS IN ELEMENTS

The Resistor: The current through and voltage across a resistor changes instantaneously.

The Inductor: The current through an inductor can not change instantaneously whereas voltage across it may change abruptly.

The Capacitor: The voltage across a capacitor can not change instantaneously whereas current through it may change abruptly.

Meaning of the Following Switching Conditions

$t = 0$	indicates the time when switching action takes place.
$t = 0^+$	indicates the time immediately after switching.
$t = 0^-$	indicates the time immediately before switching.
$t < 0$, steady-state	indicates the time prior to switching in the steady-state condition.
$t > 0$, steady-state	indicates the time after switching in the steady-state condition.
$t = \infty$	indicates the time after switching in the steady-state condition.

Note that the circuit conditions under $t = 0^-$ is the same as the circuit conditions under $t < 0$, steady-state

2.2   A PROCEDURE FOR EVALUATING INITIAL CONDITIONS

Step 1: Draw an equivalent circuit at $t = 0^+$ from the given network by

(a) replacing each inductor with

   (i) open circuit if current through it at t = 0^- was zero,

   (ii) current source of value I_0 if current through it at t = 0^- was I_0.

(b) replacing each capacitor with

   (i) short circuit if voltage across it at t = 0^- was zero,

   (ii) voltage source of value V_0 if voltage across it at $t = 0^-$ was V_0.

(c) leaving the resistors in the network without change.

Step 2 : Determine initial values of the variables - currents, voltages, charges etc. from equivalent circuit obtained in step 1.

Step 3 : To determine the first derivative of the variables at $t = 0^+$, draw circuit for $t > 0$ for the given network.

Step 4 : Write corresponding differential equation(s) (or integro-differential equation(s)) in required variable(s) for $t > 0$ circuit.

Step 5 : Differentiate integro-differential equation(s) obtained in step (4) if required.

Step 6 : Set $t = 0^+$ in differential equation(s) obtained in step (5).

Step 7 : Obtain initial value(s) of first derivative by putting into them the known value(s) obtained in step (2).

Step 8 : To determine initial value(s) of second derivative, obtain second derivative of the variable(s) from equation(s) obtained in step (4) and repeat the steps (6) and (7).

Step 9 : Repeat the step (8) for calculating initial value(s) of any other higher order of derivatives.

2.3 SOLVED PROBLEMS

SP 2.1 In the network of Fig. SP 2.1, the switch is closed at t = 0 with the capacitor uncharged. Find the values for $i(0^+)$, $\frac{di}{dt}(0^+)$, and $\frac{d^2i}{dt^2}(0^+)$.

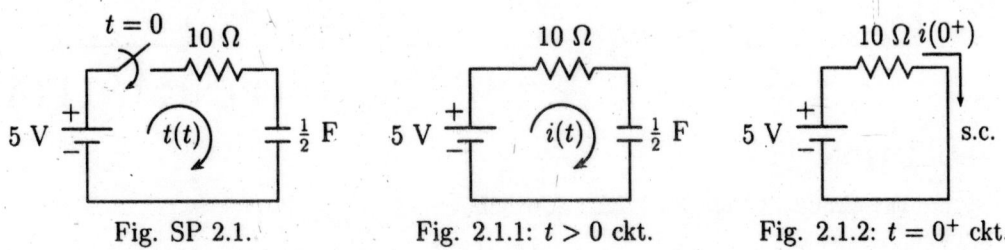

Fig. SP 2.1. Fig. 2.1.1: $t > 0$ ckt. Fig. 2.1.2: $t = 0^+$ ckt.

SOLUTION

From the circuit of Fig. 2.1.2, we obtain

$$i(0^+) = 5/10 = 0.5 \text{ A}$$

By KVL in circuit of Fig. 2.1.1,

$$10i + \frac{1}{1/2} \int_0^t i \, dt = 5 \qquad [2.1.1]$$

Differentiating eqn [2.1.1]

$$10\frac{di}{dt} + 2i = 0 \qquad [2.1.2]$$

Setting $t = 0^+$ in eqn [2.1.2]

$$10\frac{di}{dt}(0^+) + 2i(0^+) = 0$$

Hence

$$\frac{di}{dt}(0^+) = -\frac{2}{10}i(0^+) = -\frac{2}{10} \times 0.5 = -0.1 \text{ A/sec}$$

To find the 2nd derivative, eqn [2.1.2] must be differentiated as:

$$10\frac{d^2i}{dt^2} + 2\frac{di}{dt} = 0 \qquad [2.1.3]$$

Setting $t = 0^+$ in eqn[2.1.3]

$$10\frac{d^2i}{dt^2}(0^+) + 2\frac{di}{dt}(0^+) = 0$$

Therefore

$$\frac{d^2i}{dt^2}(0^+) = -\frac{2}{10}\frac{di}{dt}(0^+) = -\frac{2}{10} \times (-0.1) = 0.02 \text{ A/s.}^2$$

SP 2.2 In the network of Fig. SP 2.2, the switch is closed at $t = 0$ with the capacitor uncharged. Find the values for $v_C(0^+)$, $\frac{dv_C}{dt}(0^+)$ and $\frac{d^2v_C}{dt^2}(0^+)$.

Fig. SP 2.2.   Fig. 2.2.1: $t > 0$ ckt.   Fig. 2.2.2: $t = 0^+$ ckt.

SOLUTION:

From the circuit of Fig. 2.2.2, it is noted that

$$v_C(0^+) = 0$$

By KVL in Fig. 2.2.1,

$$10i + v_C = 10$$

or

$$10\left(10 \times 10^{-6}\frac{dv_C}{dt}\right) + v_C = 10$$

Hence

$$10^{-4}\frac{dv_C}{dt} + v_C = 10 \qquad\qquad [2.2.1]$$

Setting $t = 0^+$ in eqn [2.2.1]

$$10^{-4}\frac{dv_C}{dt}(0^+) + v_C(0^+) = 10$$

or

$$10^{-4}\frac{dv_C}{dt}(0^+) + 0 = 10$$

Hence

$$\frac{dv_C}{dt}(0^+) = \frac{10}{10^{-4}} = 10^5 \text{ V/sec.}$$

Differentiating eqn [2.2.1]

$$10^{-4}\frac{d^2v_C}{dt^2} + \frac{dv_C}{dt} = 0 \qquad\qquad [2.2.2]$$

Setting $t = 0^+$ in eqn [2.2.2]

$$10^{-4}\frac{d^2v_C}{dt^2}(0^+) + \frac{dv_C}{dt}(0^+) = 0$$

or

$$10^{-4}\frac{d^2v_C}{dt^2}(0^+) + 10^5 = 0$$

Hence

$$\frac{d^2v_C}{dt^2}(0^+) = -10^5/10^{-4} = -10^9 \text{ V/sec.}^2$$

SP 2.3 In the network of Fig. SP 2.3, switch is closed at t = 0 with zero current in the inductor. Find the values for $i(0^+)$, $\frac{di}{dt}(0^+)$ and $\frac{d^2i}{dt^2}(0^+)$.

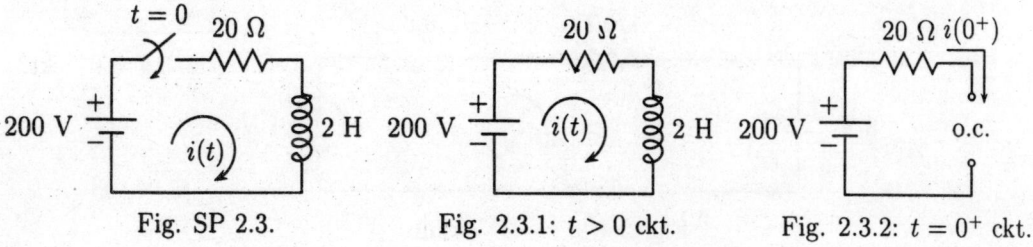

Fig. SP 2.3.   Fig. 2.3.1: $t > 0$ ckt.   Fig. 2.3.2: $t = 0^+$ ckt.

SOLUTION:

From the circuit of Fig. 2.3.2, we obtain

$$i(0^+) = 0$$

By KVL in the circuit of Fig. 2.3.1

$$20i(t) + 2\frac{di(t)}{dt} = 200 \qquad\qquad [2.3.1]$$

Setting $t = 0^+$

$$20i(0^+) + 2\frac{di}{dt}(0^+) = 200$$

Hence

$$\frac{di}{dt}(0^+) = \frac{200 - 20i(0^+)}{2} = 100 \text{ A/sec}$$

To find 2nd derivative, eqn [2.3.1] must be differentiated as :

$$20\frac{di(t)}{dt} + 2\frac{d^2i(t)}{dt^2} = 0 \qquad\qquad [2.3.2]$$

Setting $t = 0^+$

$$20\frac{di}{dt}(0^+) + 2\frac{d^2i}{dt^2}(0^+) = 0$$

Therefore

$$\frac{d^2i}{dt^2}(0^+) = -\frac{20}{2} \times 100 = -1000 \text{ A/sec.}^2$$

SP 2.4 In the circuit of Fig. SP 2.4, the switch is closed at $t = 0$. Find $i(0^+)$, $\frac{di}{dt}(0^+)$, and $\frac{d^2i}{dt^2}(0^+)$.

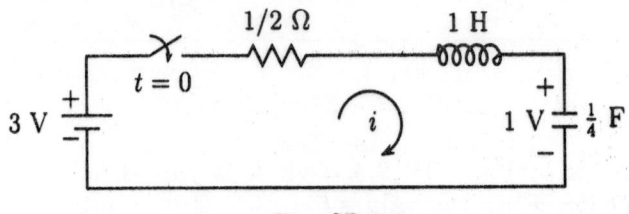

Fig. SP 2.4.

Fig. 2.4.1: $t = 0^+$ circuit.

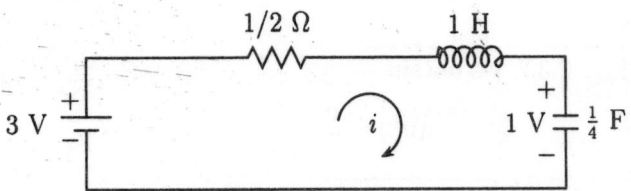

Fig. 2.4.2: $t > 0$ circuit.

SOLUTION:

From the circuit of Fig. 2.4.1, we obtain

$$i(0^+) = 0$$

By KVL in the circuit of Fig. 2.4.2

$$\frac{1}{2}i + 1\frac{di}{dt} + 1 + \frac{1}{1/4}\int_0^t i\,dt = 3 \qquad [2.4.1]$$

Setting $t = 0^+$

$$\frac{1}{2}i(0^+) + \frac{di}{dt}(0^+) + 4\int_0^{0^+} i\,dt = 2$$

Hence

$$\frac{di}{dt}(0^+) = 2 \text{ A/sec.}$$

To determine 2nd derivative, eqn [2.4.1] must be differentiated as :

$$\frac{1}{2}\frac{di(t)}{dt} + \frac{d^2i(t)}{dt^2} + 4i(t) = 0 \qquad [2.4.2]$$

Setting $t = 0^+$

$$\frac{1}{2}\frac{di}{dt}(0^+) + \frac{d^2i}{dt^2}(0^+) + 4i(0^+) = 0$$

Putting the values of known quantities in the last equation, we have

$$\frac{d^2i}{dt^2}(0^+) = -\frac{1}{2} \times 2 = -1 \text{ A/sec.}^2$$

SP 2.5 Find the values for $v(0^+)$, $\frac{dv}{dt}(0^+)$, and $\frac{d^2v}{dt^2}(0^+)$ if switch is moved from a to b at $t = 0$ in the network shown in **Fig. SP 2.5.**

SOLUTION:

From the Fig. 2.5.1, we observe that

$$v(0^+) = 0$$

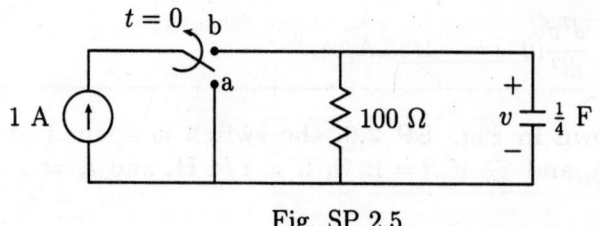

Fig. SP 2.5.

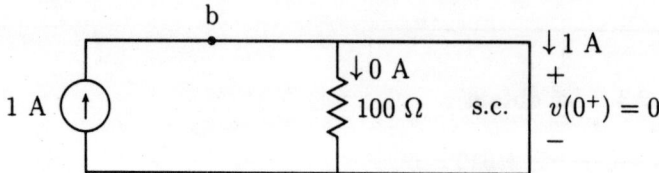

Fig. SP 2.5.1: $t = 0^+$ circuit.

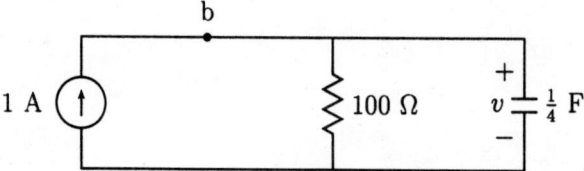

Fig. 2.5.2: $t > 0$ circuit.

By KCL in the circuit of Fig. 2.5.2

$$\frac{v(t)}{100} + \frac{1}{4}\frac{dv(t)}{dt} = 1 \qquad [2.5.1]$$

Setting $t = 0^+$

$$\frac{v(0^+)}{100} + \frac{1}{4}\frac{dv}{dt}(0^+) = 1$$

Therefore

$$\frac{dv}{dt}(0^+) = 4 \text{ V/sec.}$$

To determine 2nd derivative, equation [2.5.1] must be differentiated as :

$$\frac{1}{100}\frac{dv(t)}{dt} + \frac{1}{4}\frac{d^2v(t)}{dt^2} = 0 \qquad [2.5.2]$$

Setting $t = 0^+$

$$\frac{1}{100}\frac{dv}{dt}(0^+) + \frac{1}{4}\frac{d^2v}{dt^2}(0^+) = 0$$

or

$$\frac{1}{100} \times 4 + \frac{1}{4}\frac{d^2v}{dt^2}(0^+) = 0$$

Therefore

$$\frac{d^2v}{dt^2}(0^+) = -4/25 \text{ A/sec.}^2$$

SP 2.6 In the network shown in Fig. SP 2.6, the switch is opened at $t = 0$
Find values for $v(0^+)$, $\frac{dv}{dt}(0^+)$, and $\frac{d^2v}{dt^2}$ if $R = 10\ \Omega$, $L = 1/4$ H, and $I_s = 1$ A

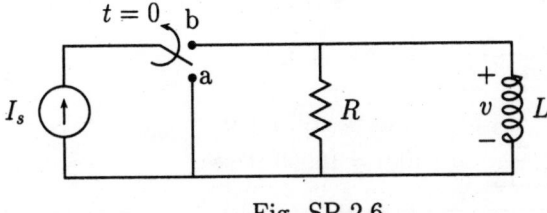

Fig. SP 2.6.

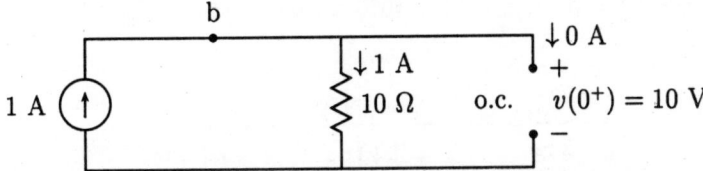

Fig. 2.6.1: $t = 0^+$ circuit.

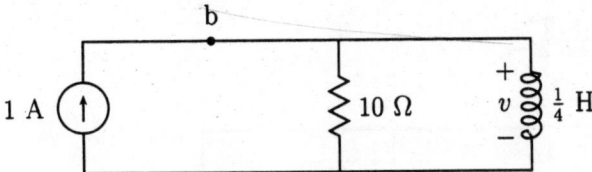

Fig. 2.6.2: $t > 0$ circuit.

SOLUTION:

From the circuit of Fig. 2.6.1, we obtain

$$v(0^+) = 1 \times 10 = 10 \text{ V}$$

By KCL in the circuit of Fig. 2.6.2

$$\frac{v(t)}{10} + \frac{1}{1/4} \int_0^t v(t)dt = 1 \qquad\qquad [2.6.1]$$

Differentiating

$$\frac{1}{10}\frac{dv(t)}{dt} + 4v(t) = 0 \qquad\qquad [2.6.2]$$

Setting $t = 0^+$

$$\frac{1}{10}\frac{dv}{dt}(0^+) + 4v(0^+) = 0$$

Therefore

$$\frac{dv}{dt}(0^+) = -\frac{4}{1/10} \times 10 = -400 \text{ V/sec.}$$

To find 2nd derivative, eqn [2.6.2] must be differentiated as :

$$\frac{1}{10}\frac{d^2v(t)}{dt^2} + 4\frac{dv(t)}{dt} = 0 \qquad\qquad [2.6.3]$$

Setting $t = 0^+$

$$\frac{1}{10}\frac{d^2v}{dt^2}(0^+) + 4\frac{dv}{dt}(0^+) = 0$$

Hence

$$\frac{d^2v}{dt^2}(0^+) = -\frac{4}{1/10} \times (-400) = 16000 \text{ V/sec.}^2$$

SP 2.7 In the network of Fig. SP 2.7, the switch is opened at $t = 0$. Find the values for $i_L(0^+), i_C(0^+), i_R(0^+), v_C(0^+), v_L(0^+), v_R(0^+), \frac{di_L}{dt}(0^+), \frac{d^2i_L}{dt^2}(0^+), \frac{dv_C}{dt}(0^+)$, and $\frac{d^2v_C}{dt^2}(0^+)$

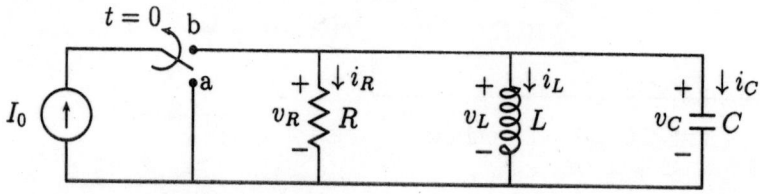

Fig. SP 2.7.

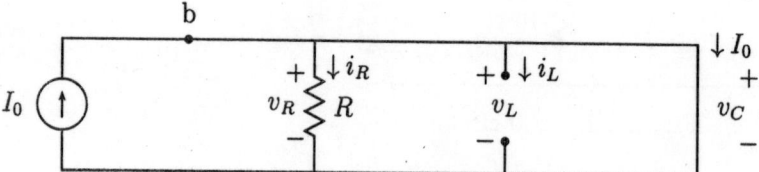

Fig. 2.7.1: $t = 0^+$ circuit.

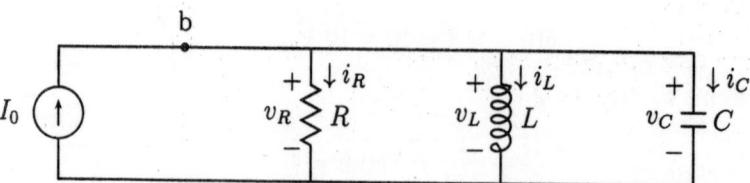

Fig. 2.7.2: $t > 0$ circuit.

SOLUTION:

From the circuit of Fig. 2.7.1, we obtain

$$i_L(0^+) = 0$$

$$i_C(0^+) = I_0$$
$$i_R(0^+) = 0$$
and $$v_C(0^+) = v_L(0^+) = v_R(0^+) = 0$$

The inductor voltage is written as:

$$v_L = L\frac{di_L}{dt} \qquad\qquad [2.7.1]$$

Setting $t = 0^+$ and solving for $\frac{di_L}{dt}(0^+)$ as :

$$\frac{di_L}{dt}(0^+) = \frac{v(0^+)}{L} = \frac{0}{L} = 0$$

The capacitor current is given by

$$i_C = C\frac{dv_C}{dt} \qquad\qquad [2.7.2]$$

Setting $t = 0^+$ in eqn [2.7.2] and solving for $\frac{dv_C}{dt}(0^+)$ as :

$$\frac{dv_C}{dt}(0^+) = \frac{i_C(0^+)}{C} = \frac{I_0}{C}$$

From the circuit of Fig. 2.7.2, KCL gives

$$i_R + i_C + i_L = I_0 \qquad\qquad [2.7.3]$$

and KVL gives

$$v_R = v_C = v_L \qquad\qquad [2.7.4]$$

Using eqn [2.7.3] and eqn [2.7.4], we can have differential equation in i_L as :

$$\frac{v_R}{R} + C\frac{dv_C}{dt} + i_L = I_0$$

or

$$\frac{v_L}{R} + C\frac{dv_L}{dt} + i_L = I_0$$

or

$$GL\frac{di_L}{dt} + CL\frac{d^2 i_L}{dt^2} + i_L = I_0 \qquad\qquad [2.7.5]$$

$$\text{(where } G = 1/R \text{ and } v_L = L\frac{di_L}{dt})$$

Setting $t = 0^+$ in eqn [2.7.5]

$$LC\frac{d^2 i_L}{dt^2}(0^+) + LG\frac{di_L}{dt}(0^+) + i_L(0^+) = I_0$$

or

$$LC\frac{d^2 i_L}{dt^2}(0^+) + LG \times 0 + 0 = I_0$$

Therefore

$$\frac{d^2 i_L}{dt^2}(0^+) = \frac{I_0}{LC}$$

Similarly, differential equation in v_C can be found using eqn [2.7.3] and eqn [2.7.4] as :

$$\frac{v_R}{R} + C\frac{dv_C}{dt} + \frac{1}{L}\int_0^t v_L dt = I_0$$

or

$$Gv_C + C\frac{dv_C}{dt} + \frac{1}{L}\int_0^t v_C dt = I_0$$

Differentiating last equation, we obtain

$$LC\frac{d^2 v_C}{dt^2} + LG\frac{dv_C}{dt} + v_C = 0 \qquad [2.7.6]$$

Setting t $= 0^+$ in eqn [2.7.7] and solving for $\frac{d^2 v_C}{dt^2}(0^+)$ as

$$LC\frac{d^2 v_C}{dt^2}(0^+) + LG\frac{dv_C}{dt}(0^+) + v_C(0^+) = 0$$

or

$$LC\frac{d^2 v_C}{dt^2}(0^+) + LG\left(\frac{I_0}{C}\right) + 0 = 0$$

Therefore

$$\frac{d^2 v_C}{dt^2}(0^+) = -\frac{GI_0}{C^2}$$

SP 2.8 In the given network of Fig. SP 2.8, switch is closed at $t = 0$. Find $i_1(0^+)$, $i_2(0^+)$, $\frac{di_1}{dt}(0^+)$, $\frac{di_2}{dt}(0^+)$, $\frac{d^2 i_1}{dt^2}(0^+)$, and $\frac{d^2 i_2}{dt^2}(0^+)$.

SOLUTION:

From the circuit of Fig. 2.8.1

$$i_1(0^+) = 10/10 = 10 \text{ A}$$
$$\text{and} \qquad i_2(0^+) = 0$$

By KVL in the circuit of Fig. 2.8.2

$$\frac{1}{1}\int i_1 dt + 10(i_1 - i_2) = 10 \qquad [2.8.1]$$

and

$$10(i_2 - i_1) + 20i_2 + 1\frac{di_2}{dt} = 0 \qquad [2.8.2]$$

Setting t $= 0^+$ in eqn [2.8.2], we have

$$10i_2(0^+) - 10i_1(0^+) + 20i_2(0^+) + \frac{di_2}{dt}(0^+) = 0$$

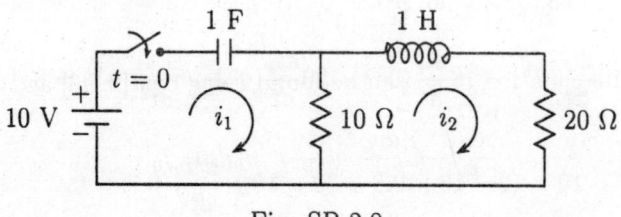

Fig. SP 2.8.

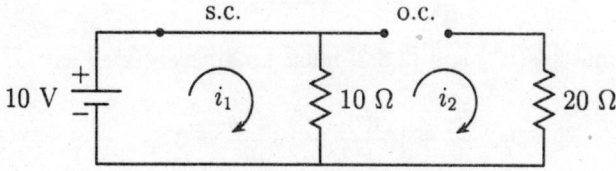

Fig. 2.8.1: $t = 0^+$ circuit.

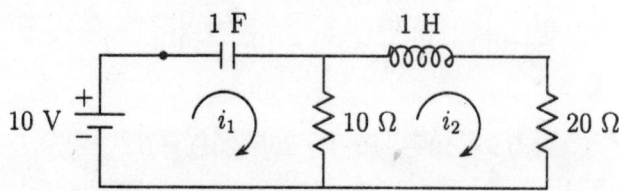

Fig. 2.8.2: $t > 0$ circuit.

or

$$10 \times 0 - 10 \times 1 + 20 \times 0 + \frac{di_2}{dt}(0^+) = 0$$

Therefore

$$\frac{di_2}{dt}(0^+) = 10 \text{ A/sec.}$$

Differentiating eqn [2.8.1]

$$i_1 + 10\frac{di_1}{dt} - 10\frac{di_2}{dt} = 0 \qquad [2.8.3]$$

Setting $t = 0^+$

$$i_1(0^+) + 10\frac{di_1}{dt}(0^+) - 10\frac{di_2}{dt}(0^+) = 0$$

or

$$1 + 10\frac{di_1}{dt}(0^+) - 10 \times 10 = 0$$

Therefore

$$\frac{di_1}{dt}(0^+) = \frac{10 \times 10 - 1}{10 \, .} = 9.9 \text{ A/sec}$$

To evaluate $\frac{d^2i_2}{dt^2}(0^+)$ the eqn [2.8.2] must be differentiated as :

$$10\frac{di_2}{dt} - 10\frac{di_1}{dt} + 20\frac{di_2}{dt} + \frac{d^2i_2}{dt^2} = 0 \qquad [2.8.4]$$

Now, setting $t = 0^+$ in eqn [2.8.4] and then putting the known values into it as :

$$10\frac{di_2}{dt}(0^+) - 10\frac{di_1}{dt}(0^+) + 20\frac{di_2}{dt}(0^+) + \frac{d^2i_2}{dt^2}(0^+) = 0$$

or

$$10 \times 10 - 10 \times 9.9 + 20 \times 10 + \frac{d^2i_2}{dt^2}(0^+) = 0$$

Hence

$$\frac{d^2i_2}{dt^2}(0^+) = -201 \text{ A/sec}^2.$$

Similarly, to determine $\frac{d^2i_1}{dt^2}(0^+)$ eqn [2.8.3] must be differentiated as :

$$\frac{di_1}{dt} + 10\frac{d^2i_1}{dt^2} - 10\frac{d^2i_2}{dt^2} = 0 \qquad\qquad [2.8.5]$$

Setting t = 0^+ in eqn [2.8.5] and then putting the known values into it as :

$$\frac{di_1}{dt}(0^+) + 10\frac{d^2i_1}{dt^2}(0^+) - 10\frac{d^2i_2}{dt^2}(0^+) = 0$$

or

$$9.9 + 10\frac{d^2i_1}{dt^2}(0^+) - 10(-201) = 0$$

Therefore

$$\frac{d^2i_1}{dt^2}(0^+) = -201.99 \text{ A/sec.}^2$$

SP 2.9 In the given network of Fig. SP 2.9, the left capacitor is charged to 10 V and the switch is closed at $t = 0$. Find $\frac{d^2i_2}{dt^2}(0^+)$.

SOLUTION:

By mesh current equations for the circuit of Fig. 2.9.1 are:
Mesh1:

$$20i_1(0^+) - 20i_2(0^+) = 10 \qquad\qquad [2.9.1]$$

Mesh2:

$$-20i_1(0^+) + 30i_2(0^+) = 0 \qquad\qquad [2.9.2]$$

Using Cramer's rule, $i_1(0^+)$ and $i_2(0^+)$ can be determined as

$$i_1(0^+) = \frac{\begin{vmatrix} 10 & -20 \\ 0 & 30 \end{vmatrix}}{\begin{vmatrix} 20 & -20 \\ -20 & 30 \end{vmatrix}} = 3/2 \text{ A} \quad \text{and} \quad i_2(0^+) = \frac{\begin{vmatrix} 20 & 10 \\ -20 & 0 \end{vmatrix}}{\begin{vmatrix} 20 & -20 \\ -20 & 30 \end{vmatrix}} = 1 \text{ A}$$

Writing mesh equations for the circuit shown in Fig 2.9.2 as :

$$\frac{1}{1}\int i_1 dt + 20i_1 - 20i_2 = 10 \qquad\qquad [2.9.3]$$

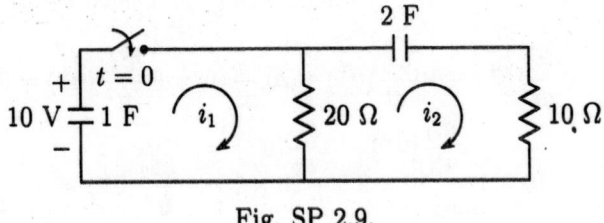

Fig. SP 2.9.

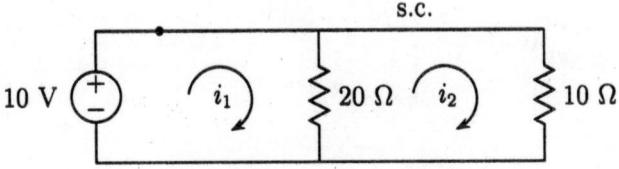

Fig. 2.9.1: $t = 0^+$ circuit.

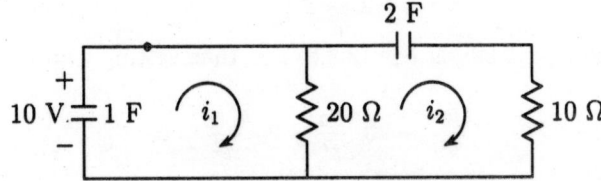

Fig. 2.9.2: $t > 0$ circuit.

and

$$30i_2 + \frac{1}{2}\int i_2 dt - 20i_1 = 0 \qquad\qquad [2.9.4]$$

Differentiating eqns [2.9.3] and [2.9.4] as :

$$i_1 + 20\frac{di_1}{dt} - 20\frac{di_2}{dt} = 0 \qquad\qquad [2.9.5]$$

and

$$30\frac{di_2}{dt} + \frac{i_2}{2} - 20\frac{di_1}{dt} = 0 \qquad\qquad [2.9.6]$$

Setting $t = 0^+$ in eqn [2.9.5] and [2.9.6] and then putting the known values into them as :

$$i_1(0^+) + 20\frac{di_1}{dt}(0^+) - 20\frac{di_2}{dt}(0^+) = 0$$

or

$$\frac{di_1}{dt}(0^+) - \frac{di_2}{dt}(0^+) = -\frac{i_1(0^+)}{20} = -\frac{3/2}{20}$$

Hence

$$\frac{di_1}{dt}(0^+) - \frac{di_2}{dt}(0^+) = -3/40 \qquad\qquad [2.9.7]$$

and

$$30\frac{di_2}{dt}(0^+) + \frac{i_2(0^+)}{2} - 20\frac{di_1}{dt}(0^+) = 0$$

or

$$-20\frac{di_1}{dt}(0^+) + 30\frac{di_2}{dt}(0^+) = -\frac{i_2(0^+)}{2} = -1/2$$

Hence

$$2\frac{di_1}{dt}(0^+) - 3\frac{di_2}{dt}(0^+) = 1/20 \qquad [2.9.8]$$

Using Cramer's rule for the equations [2.9.7] and [2.9.8] to determine $\frac{di_1}{dt}(0^+)$ and $\frac{di_2}{dt}(0^+)$ as :

$$\frac{di_1}{dt}(0^+) = \frac{\begin{vmatrix} -3/40 & -1 \\ 1/20 & -3 \end{vmatrix}}{\begin{vmatrix} 1 & -1 \\ 2 & -3 \end{vmatrix}} = -11/40 \text{ A/sec.}$$

$$\frac{di_2}{dt}(0^+) = \frac{\begin{vmatrix} 1 & -3/40 \\ 2 & 1/20 \end{vmatrix}}{\begin{vmatrix} 1 & -1 \\ 2 & -3 \end{vmatrix}} = -1/5 \text{ A/sec.}$$

Again differentiating eqn [2.9.5] & eqn [2.9.6] and then setting t = 0^+ as:

$$\frac{di_1}{dt} + 20\frac{d^2i_1}{dt^2} - 20\frac{d^2i_2}{dt^2} = 0$$

or

$$\frac{d^2i_1}{dt^2}(0^+) - \frac{d^2i_2}{dt^2}(0^+) = -\frac{1}{20}\frac{di_1}{dt}(0^+) = -\frac{1}{20}\left(\frac{-11}{40}\right)$$

Hence

$$\frac{d^2i_1}{dt^2}(0^+) - \frac{d^2i_2}{dt^2}(0^+) = 11/800 \qquad [2.9.9]$$

Next

$$30\frac{d^2i_2}{dt^2} + \frac{1}{2}\frac{di_2}{dt} - 20\frac{d^2i_1}{dt^2} = 0$$

or

$$2\frac{d^2i_1}{dt^2}(0^+) - 3\frac{d^2i_2}{dt^2}(0^+) = \frac{1}{20}\frac{di_2}{dt}(0^+) = \frac{1}{20}\left(\frac{-1}{5}\right)$$

Hence

$$2\frac{d^2i_1}{dt^2}(0^+) - 3\frac{d^2i_2}{dt^2}(0^+) = -1/100 \qquad [2.9.10]$$

Again using Cramer's rule, to eqn [2.9.9] and eqn [2.9.10], $\frac{d^2i_2}{dt^2}(0^+)$ can be determined as:

$$\frac{d^2i_2}{dt^2}(0^+) = \frac{\begin{vmatrix} 1 & 11/800 \\ 2 & -1/100 \end{vmatrix}}{\begin{vmatrix} 1 & -1 \\ 2 & -3 \end{vmatrix}} = 3/80 \text{ A/sec.}^2$$

SP 2.10 For the network of Fig. SP 2.10, determine $d^2i_1/dt^2(0^+)$ if switch is closed at $t = 0^+$.

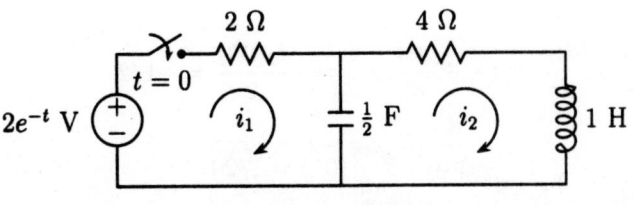

Fig. SP 2.10.

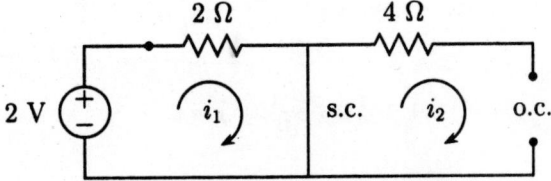

Fig. 2.10.1: $t = 0^+$ circuit.

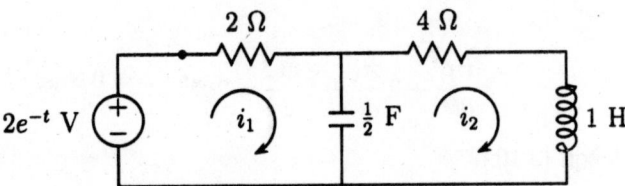

Fig. 2.10.2: $t > 0$ circuit.

SOLUTION:

From the circuit of Fig. 2.10.1

$$i_1(0^+) = 2/2 = 1 \text{ A}$$

and $\qquad i_2(0^+) = 0$

Writing mesh equations for the circuit of Fig. 2-10.2(a),

$$2i_1 + \frac{1}{1/2} \int_0^t (i_1 - i_2)dt = 2e^{-t} \qquad\qquad [2.10.1]$$

and

$$\frac{1}{1/2} \int_0^t (i_2 - i_1)dt + 4i_2 + 1\frac{di_2}{dt} = 0 \qquad\qquad [2.10.2]$$

Differentiating eqn [2.10.1]

$$2\frac{di_1}{dt} + 2(i_1 - i_2) = -2e^{-t} \qquad\qquad [2.10.3]$$

Setting $t = 0^+$ in eqn [2.10.3]

$$2\frac{di_1}{dt}(0^+) + 2i_1(0^+) - 2i_2(0^+) = -2$$

or

$$2\frac{di_1}{dt}(0^+) + 2 \times 1 - 2 \times 0 = -2$$

Hence

$$\frac{di_1}{dt}(0^+) = -2 \text{ A/sec.} \qquad [2.10.4]$$

Setting t = 0^+ in the eqn [2.10.2], we have

$$2\int_0^{0^+} (i_2 - i_1)dt + 4i_2(0^+) + \frac{di_2}{dt}(0^+) = 0$$

or

$$0 + 4 \times 0 + \frac{di_2}{dt}(0^+) = 0$$

Therefore

$$\frac{di_2}{dt}(0^+) = 0 \qquad [2.10.5]$$

Differentiating eqn [2.10.3]

$$2\frac{d^2i_1}{dt^2} + 2\frac{di_1}{dt} - 2\frac{di_2}{dt} = 2e^{-t} \qquad [2.10.6]$$

Setting $t = 0^+$ in the eqn [2.10.6]

$$2\frac{d^2i_1}{dt^2}(0^+) + 2\frac{di_1}{dt}(0^+) - 2\frac{di_2}{dt}(0^+) = 2$$

or

$$2\frac{d^2i_1}{dt^2}(0^+) + 2(-2) - 2 \times 0 = 2$$

Therefore

$$\frac{d^2i_1}{dt^2}(0^+) = 3$$

SP 2.11 In the circuit of Fig. SP 2.11, the switch is closed at $t = 0$. Determine $\frac{dv_0}{dt}(0^+)$ if $v(t) = \sin t$.

SOLUTION:

Writing mesh equations for the circuit of Fig. 2.11.1 as :

$$1\frac{di}{dt} + \frac{1}{1}\int_0^t idt = v(t) = \sin t \qquad [2.11.1]$$

and

$$\frac{1}{1}\int_0^t idt - 0.5\frac{di}{dt} = v_0(t) \qquad [2.11.2]$$

Setting $t = 0^+$ in eqn [2.11.1]

$$\frac{di}{dt}(0^+) + \int_0^{0^+} idt = 0$$

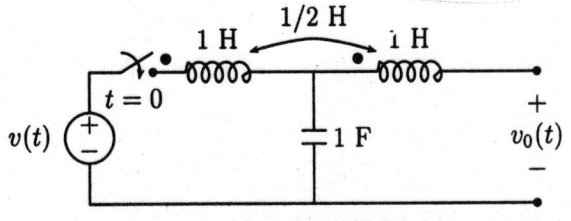

Fig. SP 2.11.

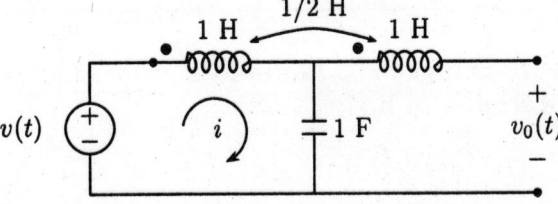

Fig. 2.11.1: $t > 0$ circuit.

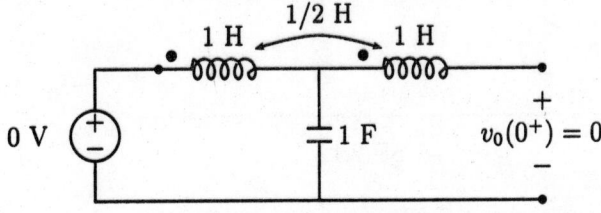

Fig. 2.11.2: $t = 0^+$ circuit.

Therefore

$$\frac{di}{dt}(0^+) = 0 \qquad [2.11.3]$$

Now setting $t = 0^+$ in eqn [2.11.2]

$$\int_0^{0^+} i\,dt - 0.5\frac{di}{dt}(0^+) = v_0(0^+)$$

or

$$0 - 0 = v_0(0^+) \qquad \Rightarrow v_0(0^+) = 0 \qquad [2.11.4]$$

Differentiating eqn [2.11.1] and eqn [2.11.2] as

$$\frac{d^2i}{dt^2} + i = \cos t \qquad [2.11.5]$$

and

$$i - 0.5\frac{d^2i}{dt^2} = \frac{dv_0}{dt} \qquad [2.11.6]$$

Setting $t = 0^+$ in eqn [2.11.5]

$$\frac{d^2i}{dt^2}(0^+) + i(0^+) = 1$$

Therefore

$$\frac{d^2i}{dt^2}(0^+) = 1 \qquad [2.11.7]$$

(Since $i(0^+) = 0$ from the Fig. 2.11.2)

Setting $t = 0^+$ in eqn [2.11.6]

$$i(0^+) - 0.5\frac{d^2i}{dt^2}(0^+) = \frac{dv_0}{dt}(0^+)$$

Therefore

$$\frac{dv_0}{dt}(0^+) = -1/2 \text{ V/sec.}$$

SP 2.12 For the network of Fig. SP 2.12, find the values for $i_L(0^+)$, and $\frac{di_L}{dt}(0^+)$ when switch is moved from a to b.

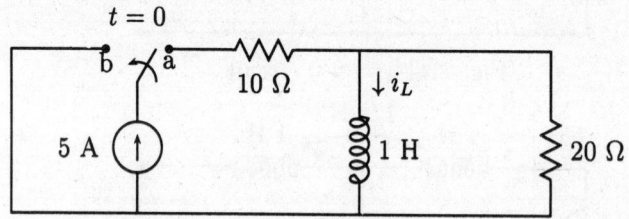

Fig. SP 2.12.

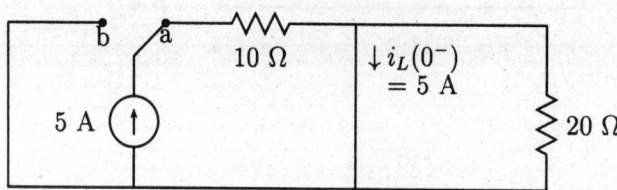

Fig. 2.12.1: $t = 0^-$ circuit.

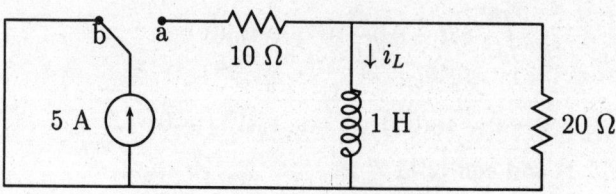

Fig. 2.12.2: $t > 0$ circuit.

SOLUTION:

From the circuit of Fig. 2.12.1, we obtain

$$i_L(0^-) = 5 \text{ A}$$

Thus by continuity of inductor current, we obtain

$$i_L(0^+) = i_L(0^-) = 5 \text{ A}$$

By KVL in the circuit of Fig. 2.12.2

$$1\frac{di_L}{dt} + 20i_L = 0 \qquad\qquad [2.12.1]$$

Setting $t = 0^+$ in eqn [2.12.1] and then putting known values, we get

$$\frac{di_L}{dt}(0^+) + 20i_L(0^+) = 0$$

Hence

$$\frac{di_L}{dt}(0^+) = -20 \times 5 = -100 \ \text{A/sec}.$$

SP 2.13 For the network of Fig. SP 2.13, the switch is moved from a to b at $t = 0$. Find the values for $i(0^+)$, $\frac{di}{dt}(0^+)$, and $\frac{d^2i}{dt^2}(0^+)$.

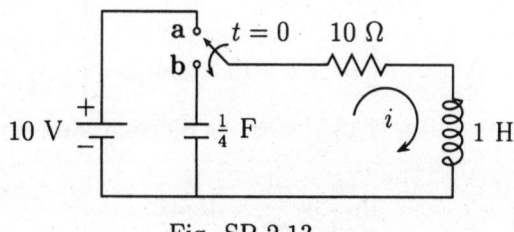

Fig. SP 2.13.

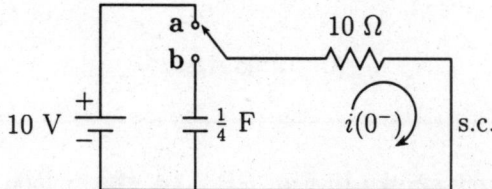

Fig. 2.13.1: $t = 0^-$ circuit.

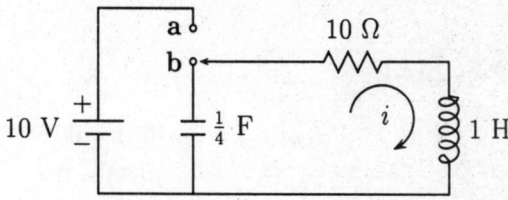

Fig. 2.13.2: $t > 0$ circuit.

SOLUTION:

From the circuit of Fig. 2.13.1, we obtain

$$i(0^-) = 10/10 = 1 \ \text{A}$$

Thus, by continuity of inductor current

$$i(0^+) = i(0^-) = 1 \text{ A}$$

Using KVL in the circuit of Fig. 2.13.2

$$10i + 1\frac{di}{dt} + \frac{1}{1/4}\int_0^t i\, dt = 0 \qquad [2.13.1]$$

Setting $t = 0^+$ in eqn [2.13.1]

$$10i(0^+) + \frac{di}{dt}(0^+) + 4\int_0^{0^+} i\, dt = 0$$

or

$$10 \times 1 + \frac{di}{dt}(0^+) + 0 = 0$$

Therefore

$$\frac{di}{dt}(0^+) = -10 \text{ A/sec.}$$

To determine 2nd derivative, eqn [2.13.1] must be differentiated as :

$$10\frac{di}{dt} + \frac{d^2i}{dt^2} + 4i = 0 \qquad [2.13.2]$$

Setting t $= 0^+$ in eqn [2.13.2] and solving for $\frac{d^2i}{dt^2}(0^+)$, we get

$$\frac{d^2i}{dt^2}(0^+) = 96 \text{ A/sec.}^2$$

SP 2.14 For the network shown in Fig. SP 2.14, the switch is moved from a to b at $t = 0$ after a steady state condition is established at position a. Calculate $i_1(0^+), i_2(0^+),$ and $i_3(0^+)$.

SOLUTION:

From the circuit of Fig. 2.14.1

$$i_{1H}(0^-) = i_{2H}(0^-) = 0$$
$$v_{\frac{1}{3}F}(0^-) = v_{\frac{1}{4}F}(0^-) = 0$$
and $\qquad v_{\frac{1}{2}F}(0^-) = 6 \text{ V}$

Thus by continuity of capacitor voltage and inductor current, we obtain

$$i_{1H}(0^+) = i_{2H}(0^+) = 0$$
$$v_{\frac{1}{3}F}(0^+) = v_{\frac{1}{4}F}(0^+) = 0$$
and $\qquad v_{\frac{1}{2}F}(0^+) = 6 \text{ V}$

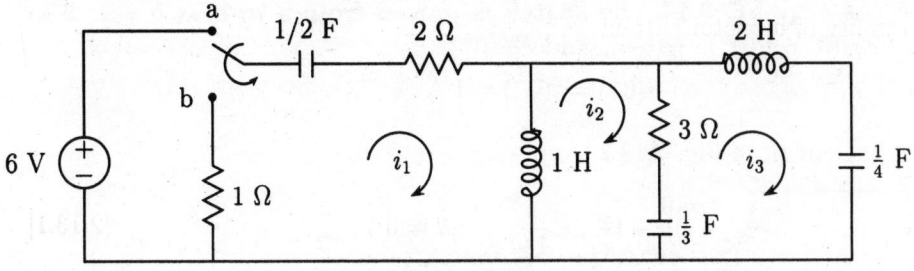

Fig. SP 2.14.

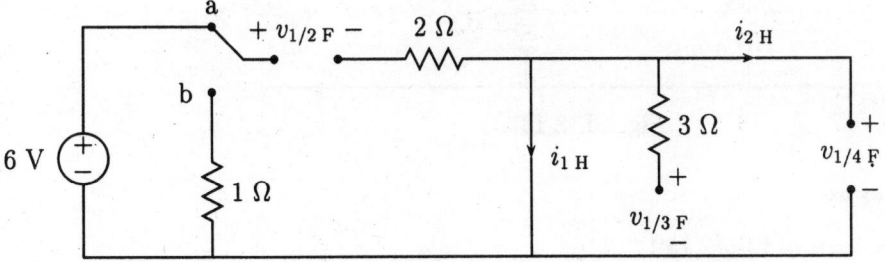

Fig. 2.14.1: $t = 0^-$ circuit.

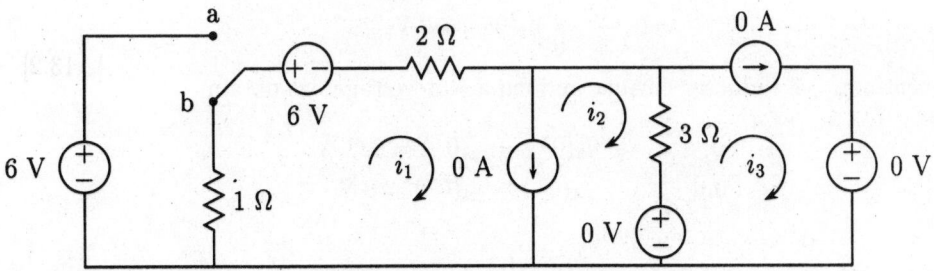

Fig. 2.14.2: $t = 0^+$ circuit.

From the circuit of Fig. 2.14.2

$$i_3(0^+) = 0$$

and

$$i_1(0^+) - i_2(0^+) = 0$$

Therefore

$$i_1(0^+) = i_2(0^+)$$

Taking KVL around the loop containing 1 Ω resistor, 6-V voltage source, 2 Ω resistor, 3 Ω resistor and 0-V source of Fig. 2.14.2

$$6 + 2i_1(0^+) + 3[i_2(0^+) - i_3(0^+)] + i_1(0^+) = 0$$

or

$$6 + 2i_1(0^+) + 3[i_1(0^+) - 0] + i_1(0^+) = 0$$

Thus

$$i_1(0^+) = -1$$

Then, we have

$$i_1(0^+) = i_2(0^+) = -1 \quad \text{and} \quad i_3(0^+) = 0$$

SP 2.15 In the Fig. SP 2.15, the switch is moved from a to b at $t = 0$. Find the values of $i_3(0^+)$, $v_4(0^+)$, $\frac{di_3}{dt}(0+)$, and $\frac{dv_4}{dt}(0^+)$.

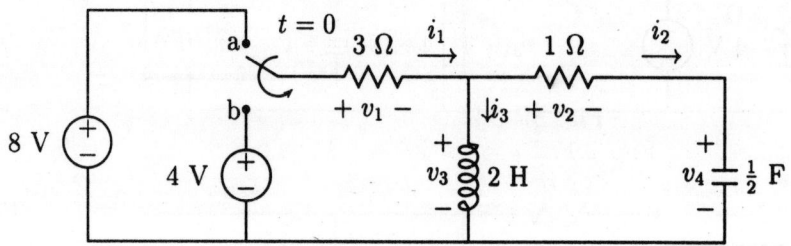

Fig. SP 2.15.

SOLUTION:

From the circuit of Fig. 2.15.1

$$i_3(0^-) = 8/3 \text{ A}$$

and $$v_4(0^-) = 0 \text{ V}$$

Thus by continuity of inductor current and capacitor voltage, we obtain

$$i_3(0^+) = i_3(0^-) = 8/3 \text{ A}$$

and $$v_4(0^+) = v_4(0^-) = 0 \text{ V}$$

By KCL at the node **c** of Fig. 2.15.2

$$i_1(0^+) = 8/3 + i_2(0^+) \tag{2.15.1}$$

Writing mesh equations for the both meshes of the circuit of Fig. 2.15.2 as :

$$3i_1(0^+) + v_3(0^+) = 4 \tag{2.15.2}$$

and

$$1i_2(0^+) - v_3(0^+) = 0 \tag{2.15.3}$$

Putting the value of $i_1(0^+)$ from eqn [2.15.1] into eqn [2.15.2] yields

$$3[8/3 + i_2(0^+)] + v_3(0^+) = 4$$

or

$$3i_2(0^+) + v_3(0^+) = -4 \tag{2.15.4}$$

Solving eqn [2.15.3] and eqn [2.15.4], we get

$$i_2(0^+) = -1 \text{ A} \quad \text{and} \quad v_3(0^+) = -1 \text{ V}$$

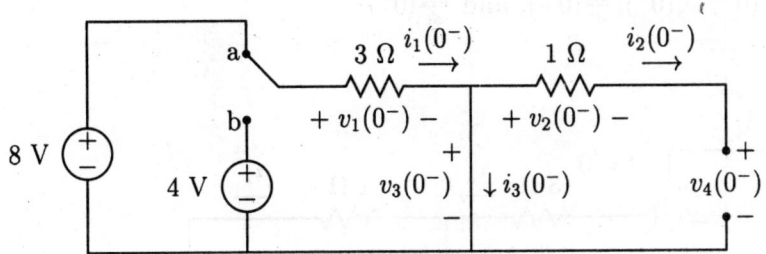

Fig. 2.15.1 $t = 0^-$ circuit.

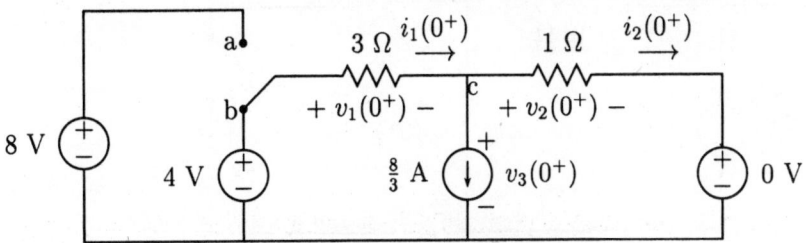

Fig. 2.15.2: $t = 0^+$ circuit.

For the capacitor:

$$\frac{dv_4}{dt}(0^+) = \frac{i_2(0^+)}{C} = \frac{-1}{1/2} = -2 \text{ V/sec.}$$

For the inductor:

$$\frac{di_3}{dt}(0^+) = \frac{v_3(0^+)}{L} = \frac{-1}{2} \text{ A/sec.}$$

SP 2.16 In the network of Fig. SP 2.16, the switch is closed at $t = 0$ with zero capacitor voltage and zero inductor current. Solve for

(a) $v_1(0^+)$ and $v_2(0^+)$,

(b) $v_1(\infty)$ and $v_2(\infty)$,

(c) $\frac{dv_1}{dt}(0^+)$ and $\frac{dv_2}{dt}(0^+)$.

SOLUTION:

(a) From the circuit of Fig. 2.16.1

$$i_C(0^+) = 10/10 = 1 \text{ A}$$
$$i_L(0^+) = 0$$
and $$v_1(0^+) + v_2(0^+) = 0$$
but $$v_2(0^+) = 5 \times 0 = 0$$
Hence $$v_1(0^+) = 0$$

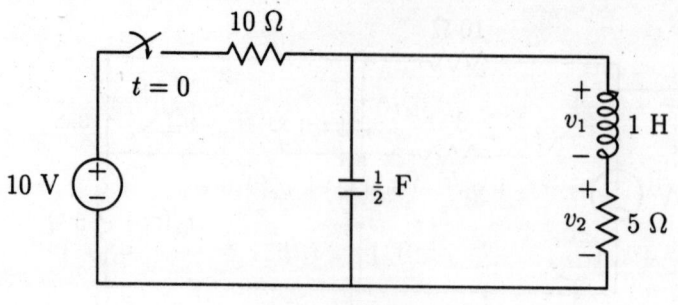

Fig. SP 2.16.

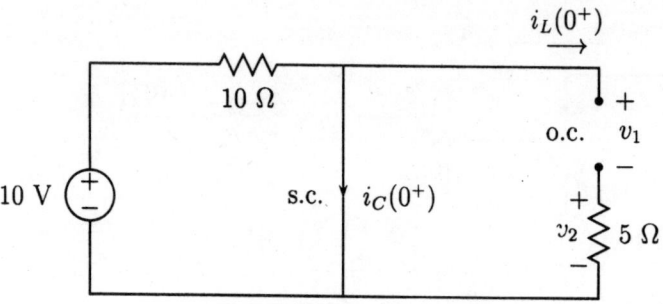

Fig. 2.16.1: $t = 0^+$ circuit.

(b) From the circuit of Fig. 2.16.2

$$i_C(\infty) = 0$$
$$i_L(\infty) = 10/(10 + 5) = 2/3 \text{ A}$$
$$v_1(\infty) = 0$$

and $\qquad v_2(\infty) = 5i_L(\infty) = 5(2/3) = 10/3 \text{ V}$

(c) By KVL around the right mesh in the circuit of Fig. 2.16.3

$$-\frac{1}{1/2} \int_0^t i_C dt + 1\frac{di_L}{dt} + 5i_L = 0 \qquad\qquad [2.16.1]$$

Setting $t = 0^+$ in eqn [2.16.1]

$$-2 \int_0^{0^+} i_C dt + \frac{di_L}{dt}(0^+) + 5i_L(0^+) = 0$$

or

$$0 + \frac{di_L}{dt}(0^+) + 5 \times 0 = 0 \qquad \Rightarrow \qquad \frac{di_L}{dt}(0^+) = 0$$

Differentiating eqn [2.16.1]

$$-2i_C + \frac{d^2 i_L}{dt^2} + 5\frac{di_L}{dt} = 0 \qquad\qquad [2.16.2]$$

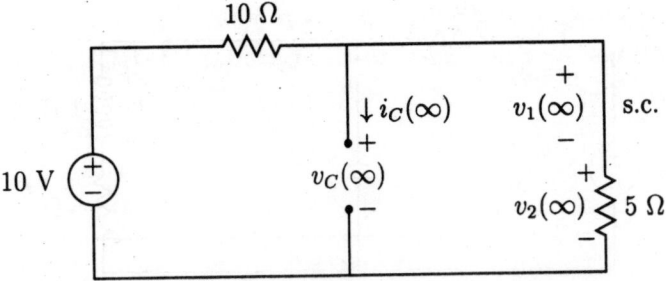

Fig. 2.16.2: $t = \infty$ circuit.

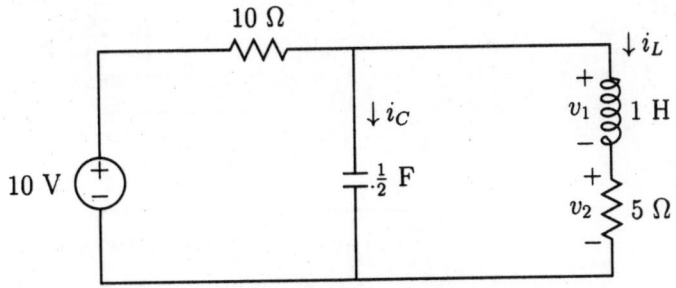

Fig. 2.16.3: $t > 0$ circuit.

Setting $t = 0^+$ in eqn [2.16.2]

$$-2i_C(0^+) + \frac{d^2 i_L}{dt^2}(0^+) + 5\frac{di_L}{dt}(0^+) = 0$$

or

$$-2 \times 1 + \frac{d^2 i_L}{dt^2}(0^+) + 5 \times 0 = 0$$

Therefore

$$\frac{d^2 i_L}{dt^2}(0^+) = 2 \text{ A/sec}^2.$$

We know that

$$v_1 = L\frac{di_L}{dt}$$

Differentiating last equation and then setting $t = 0^+$, we have

$$\frac{dv_1}{dt}(0^+) = L\frac{d^2 i_L}{dt^2}(0^+) = 1 \times 2 = 2 \text{ V/sec.}$$

And we have also

$$v_2 = 5i_L$$

Differentiating last equation and then setting $t = 0^+$

$$\frac{dv_2}{dt}(0^+) = 5\frac{di_L}{dt}(0^+) = 5 \times 0 = 0$$

SP 2.17 In the circuit shown in the Fig. SP 2.17, the switch is closed at $t = 0$ connecting a voltage source, $5\sin t$, to the parallel RL-RC circuit.Find :

(a) $\frac{di_1}{dt}(0^+)$.

(b) $\frac{di_2}{dt}(0^+)$.

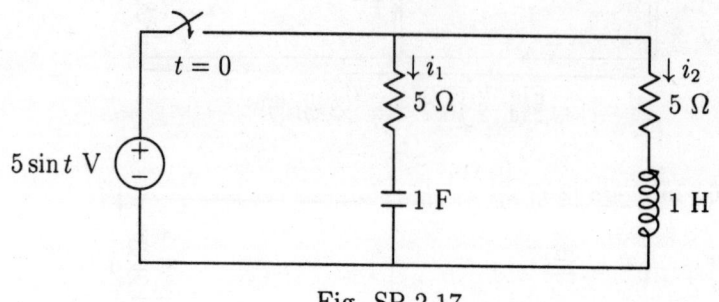

Fig. SP 2.17.

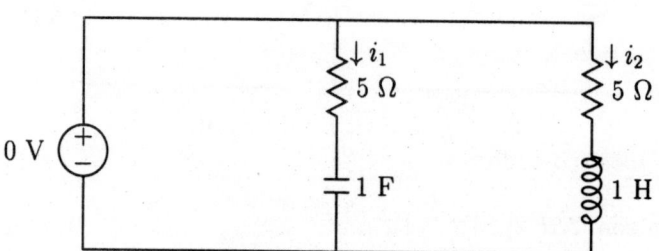

Fig. 2.17.1: $t = 0^+$ circuit.

SOLUTION:

It is noted from the circuit of Fig. 2.17.1 that

$$i_1(0^+) = i_2(0^+) = 0$$

Writing loop equations for the circuit of Fig. 2.17.2 as :

$$5i_1 + \frac{1}{1}\int_0^t i_1 dt = 5\sin t \qquad [2.17.1]$$

and

$$5i_2 + 1\frac{di_2}{dt} = 5\sin t \qquad [2.17.2]$$

Differentiating eqn [2.17.1]

$$5\frac{di_1}{dt} + i_1 = 5\cos t \qquad [2.17.3]$$

Setting $t = 0^+$ in eqn [2.17.3]

$$5\frac{di_1}{dt}(0^+) + i_1(0^+) = 5 \qquad \Rightarrow \qquad \frac{di_1}{dt}(0^+) = 5/5 = 1 \text{ A/sec.}$$

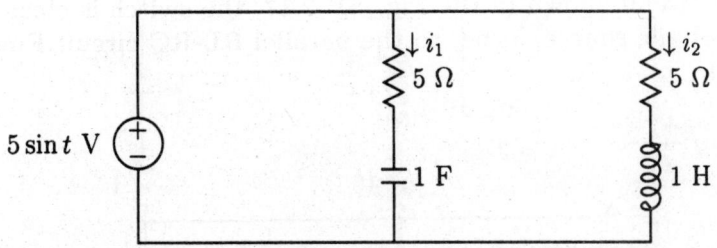

Fig. 2.17.2: $t > 0$ circuit.

Now setting $t = 0^+$ into eqn [2.17.2]

$$5i_2(0^+) + \frac{di_2}{dt}(0^+) = 0 \qquad \Rightarrow \qquad \frac{di_2}{dt}(0^+) = 5 \times 0 = 0$$

SP 2.18 In the network shown in Fig. SP 2.18, a steady state is reached with the switch open. At time $t = 0$, the switch is closed.

(a) **Write the integro-differential equations for the network after switch is closed.**

(b) **What is the voltage V_0 across capacitor before the switch is closed?**

(c) **Solve for $i_1(0^+)$ and $i_2(0^+)$.**

(d) **Solve for $\frac{di_1}{dt}(0^+)$ and $\frac{di_2}{dt}(0^+)$.**

(e) **What is the value of $\frac{di_1}{dt}(\infty)$.**

SOLUTION:

(a) By KVL in the circuit of Fig. 2.18.1

$$10i_1 + \frac{di_1}{dt} = 10 \tag{2.18.1}$$

and

$$10i_2 + V_0 + 4\int_0^t i_2 dt = 10 \tag{2.18.2}$$

(b) It is observed from the circuit of Fig. 2.18.2 that

$$i_1(0^-) = 10/(5 + 10) = 2/3 \text{ A}$$

and $\qquad v_C(0^-) = V_0 = i_1(0^-) \times 10 = \frac{2}{3} \times 10 = 20/3 \text{ V}$

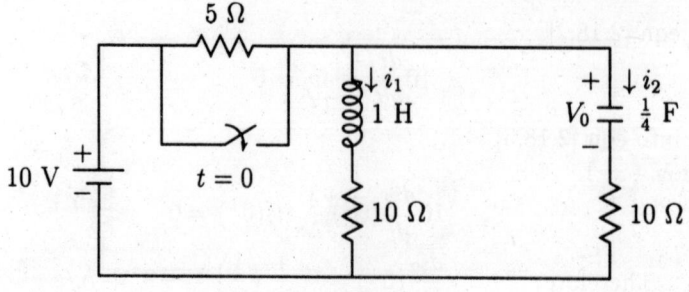

Fig. SP 2.18.

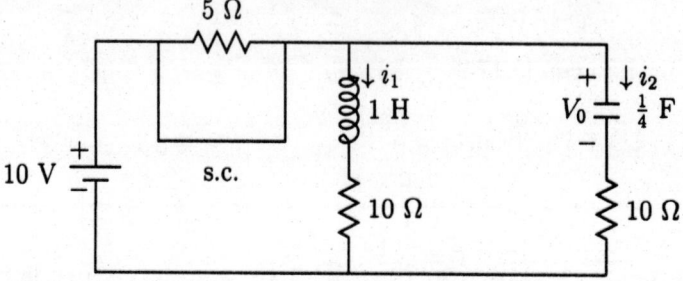

Fig. 2.18.1: $t > 0$ circuit.

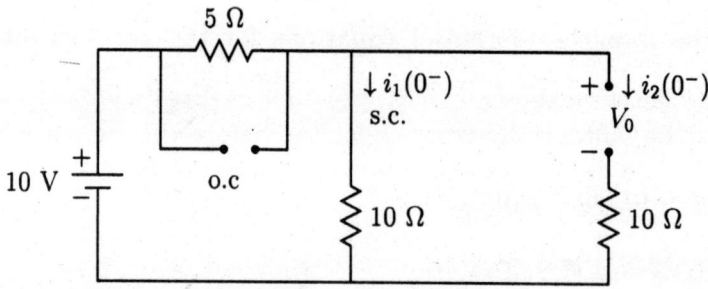

Fig. 2.18.2: $t = 0^-$ circuit.

(c) Setting $t = 0^+$ in eqn [2.18.2]

$$10i_2(0^+) + 20/3 + 4 \int_0^{0^+} i_2 dt = 10$$

or $10i_2(0^+) + 20/3 + 0 = 10$

Therefore $i_2(0^+) = 1/3$ A

By continuity of inductor current, we obtain

$$i_1(0^+) = i_1(0^-) = 2/3 \text{ A}$$

(d) Setting $t = 0^+$ in eqn [2.18.1]

$$10i_1(0^+) + \frac{di_1}{dt}(0^+) = 10$$

Therefore $\dfrac{di_1}{dt}(0^+) = 10 - 10(2/3) = 10/3$ A/sec

Differentiating eqn [2.18.2]

$$10\frac{di_2}{dt} + 4i_2 = 0 \qquad [2.18.3]$$

Setting $t = 0^+$ into eqn [2.18.3]

$$10\frac{di_2}{dt}(0^+) + 4i_2(0^+) = 0$$

Therefore $\qquad \dfrac{di_2}{dt}(0^+) = -\dfrac{4}{10}\left(\dfrac{1}{3}\right) = -2/15 \text{ A/sec}$

(e) Setting $t = \infty$ in eqn [2.18.1]

$$10i_1(\infty) + \frac{di_1}{dt}(\infty) = 10 \qquad [2.18.4]$$

or

$$10(1) + \frac{di_1}{dt}(\infty) = 10$$

$$\text{(Since } i_1(\infty) = 10/10 = 1 \text{ A)}$$

Therefore $\qquad \dfrac{di_1}{dt}(\infty) = 0$

SP 2.19 The network shown in Fig. SP 2.19, reaches a steady state. At $t = 0$, the switch is opened. Solve for
    (a) $i(0^+)$, $\frac{di}{dt}(0^+)$, and $\frac{d^2i}{dt^2}(0^+)$,
    (b) $v_2(0^+)$, $\frac{dv_2}{dt}(0^+)$, and $\frac{d^2v_2}{dt^2}(0^+)$.

SOLUTION:

It is observed from the circuit of Fig. 2.19.1 that

$$i_L(0^-) = 5/10 = 0.5 \text{ A} \qquad [2.19.1]$$

$$v_C(0^-) = 5 \text{ V} \qquad [2.19.2]$$

Thus by continuity of capacitor voltage and inductor current, we obtain

$$i_L(0^+) = i_L(0^-) = 0.5 \text{ A} \qquad [2.19.3]$$

$$v_C(0^+) = v_C(0^-) = 5 \text{ V} \qquad [2.19.4]$$

From the circuit of Fig. 2.19.2, it is obvious that the circulating current will be equal to the current source, i.e.,

$$i(0^+) = 1/2 \text{ A} \qquad [2.19.5]$$

By KVL around right mesh in the circuit of Fig. 2.19.3

$$15i + 1\frac{di}{dt} + \frac{1}{1/2}\int_0^t i\,dt - 5 = 0 \qquad [2.19.6]$$

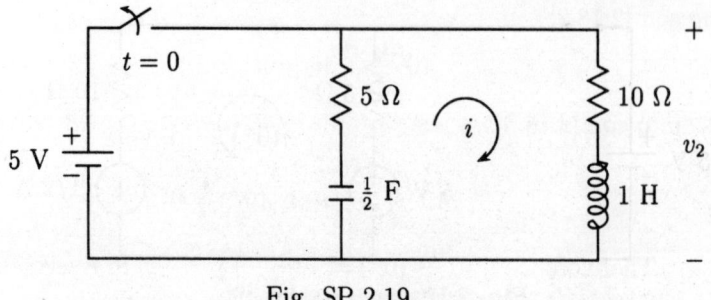

Fig. SP 2.19.

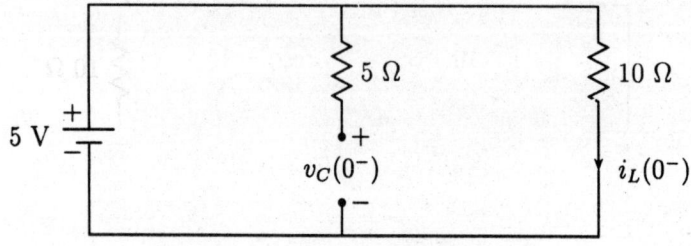

Fig. 2.19.1: $t = 0^-$ circuit.

Setting $t = 0^+$ in eqn [2.19.6]

$$15i(0^+) + \frac{di}{dt}(0^+) + 2\int_0^{0^+} i\,dt = 5$$

or

$$15(1/2) + \frac{di}{dt}(0^+) + 0 = 5$$

Therefore

$$\frac{di}{dt}(0^+) = -2.5 \text{ A/sec.} \qquad [2.19.7]$$

Differentiating eqn [2.19.6]

$$15\frac{di}{dt} + \frac{d^2i}{dt^2} + 2i = 0 \qquad [2.19.8]$$

Setting $t = 0^+$ in eqn [2.19.8]

$$15\frac{di}{dt}(0^+) + \frac{d^2i}{dt^2}(0^+) + 2i(0^+) = 0$$

or $\qquad 15(-2.5) + \dfrac{d^2i}{dt^2}(0^+) + 2(1/2) = 0$

Therefore $\qquad \dfrac{d^2i}{dt^2}(0^+) = 36.5 \text{ A/sec.}^2$

(b) The expression for $v_2(t)$ is deduced from the circuit of Fig. 2.19.3 as :

$$v_2(t) = \frac{di(t)}{dt} + 10i(t) \qquad [2.19.9]$$

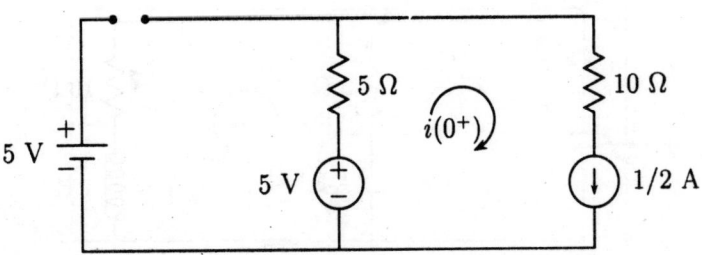

Fig. 2.19.2: $t = 0^+$ circuit.

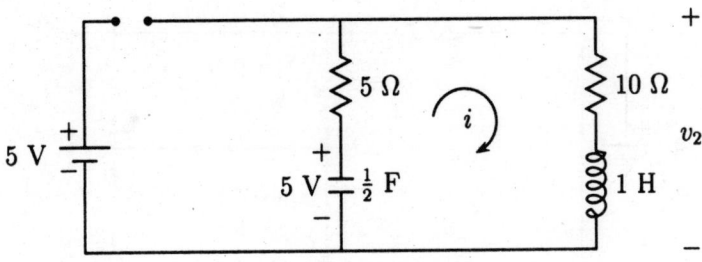

Fig. 2.19.3: $t > 0$ circuit.

Hence

$$v_2(0^+) = \frac{di}{dt}(0^+) + 10i(0^+) = -2.5 + 10(1/2) = 2.5 \text{ V}$$

Differentiating eqn [2.19.9]

$$\frac{dv_2}{dt} = \frac{d^2i}{dt^2} + 10\frac{di}{dt} \qquad [2.19.10]$$

Thus

$$\frac{dv_2}{dt}(0^+) = \frac{d^2i}{dt^2}(0^+) + 10\frac{di}{dt}(0^+) \qquad [2.19.11]$$

Putting the values for $\frac{d^2i}{dt^2}(0^+)$ and $\frac{di}{dt}(0^+)$ into eqn [2.19.11]

$$\frac{dv_2}{dt}(0^+) = 36.5 + 10(-2.5) = 11.5 \text{ V/sec.}$$

In the similar way $\frac{d^2v_2}{dt^2}(0^+)$ can be found out.

SP 2.20 Determine the initial conditions on i_1, i_2, and i_3 of the circuit of Fig. SP 2.20, if switch is closed at $t = 0$.

SOLUTION:

We find from the circuit of Fig. 2.20.1 that

$$i_{4\Omega}(0^-) = i_{1H}(0^-) = \frac{6}{2+4} = 1 \text{ A} \qquad [2.20.1]$$

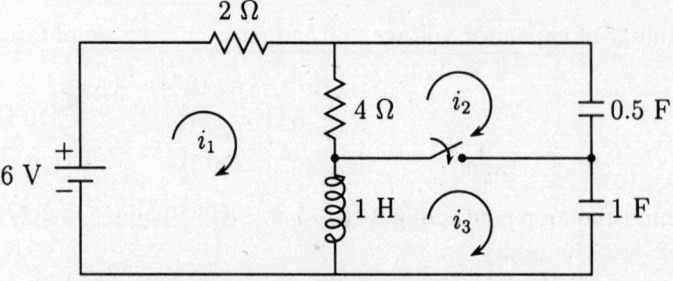

Fig. SP 2.20.

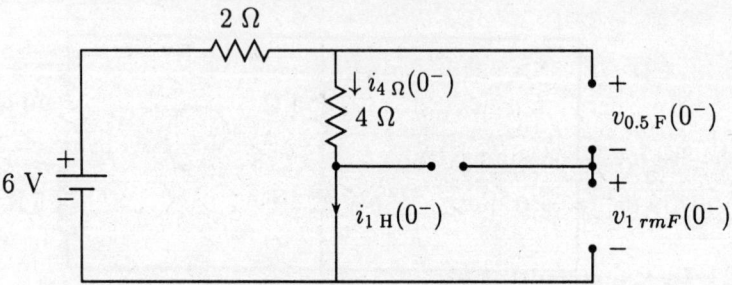

Fig. 2.20.1: $t = 0^-$ circuit.

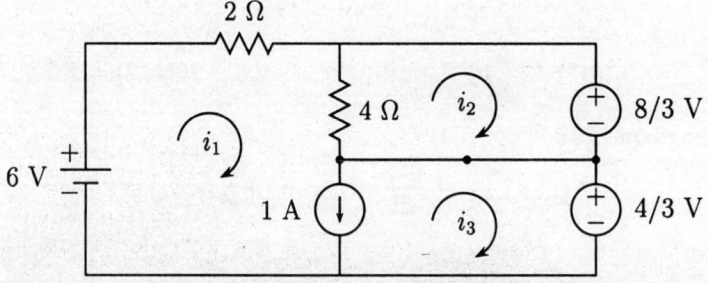

Fig. 2.20.2: $t = 0^+$ circuit.

and

$$v_{0.5F}(0^-) + v_{1.0F}(0^-) = 4i_{4\Omega}(0^-) = 4 \text{ V} \qquad [2.20.2]$$

Since two capacitors are connected in series, charge flow will be the same. Hence,

$$v_{0.5F}(0^-) = q/0.5 \qquad \text{and} \qquad v_{1.0F}(0^-) = q/1$$

Ratio:

$$\frac{v_{0.5F}(0^-)}{v_{1.0F}(0^-)} = 1/0.5 = 2$$

Therefore

$$v_{0.5F}(0^-) = 2v_{1.0F}(0^-) \qquad [2.20.3]$$

Solving eqn [2.20.2] and eqn [2.20.3], we get

$$v_{0.5F}(0^-) = 4/3 \text{ V} \quad \text{and} \quad v_{1.0F}(0^-) = 8/3 \text{ V}$$

By the continuity of capacitor voltage and inductor current we obtain

$$i_{1H}(0^+) = i_{1H}(0^-) = 1A$$
$$v_{0.5F}(0^+) = 8/3V$$
and $\quad v_{1.0F}(0^+) = 4/3V$

By KVL around the loop containing 2-Ω resistor, 8/3 V source, 4/3 V source, 6 V source of Fig. 2.20.2

$$2i_1(0^+) + 8/3 + 4/3 = 6 \qquad \Rightarrow \ i_1(0^+) = 1A$$

By KVL around the mesh containing 4-Ω resistor and 8/3 V source of circuit of Fig. 2.20.2

$$4[i_2(0^+) - i_1(0^+)] + 8/3 = 0$$
or $\qquad 4[i_2(0^+) - 1] + 8/3 = 0 \qquad \Rightarrow \ i_2(0^+) = 1/3A$

From the 3rd mesh containing 1 A source and 4/3 V source of Fig. 2.20.2, we have

$$i_1(0^+) - i_3(0^+) = 1 \qquad \Rightarrow \ i_3(0^+) = -1 + i_1(0^+) = -1 + 1 = 0$$

SP 2.21 In the circuit of Fig. SP 2.21, the switch opens at $t = 0$. Find :
(a) $i_L(0^-)$, (b) $i_L(0^+)$, (c) $i_L(\infty)$, (d) $\frac{di_L}{dt}(0^-)$, (e) $\frac{di_L}{dt}(0^+)$.

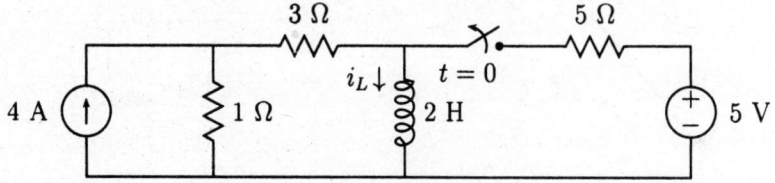

Fig. SP 2.21.

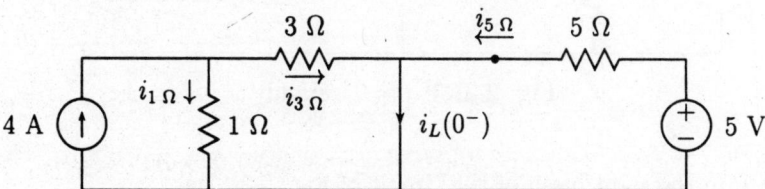

Fig. 2.21.1: $t = 0^-$ circuit.

SOLUTION:

(a) From the circuit of Fig. 2.21.1, current division method gives

$$i_{1\Omega} = \frac{3}{1+3}(4) = 3 \text{ A}$$

$$i_{3\Omega} = \frac{1}{1+3}(4) = 1 \text{ A}$$

and $\qquad i_{5\Omega} = 5/5 = 1 \text{ A}$
Hence $\qquad i_L(0^-) = i_{3\Omega} + i_{5\Omega} = 1 + 1 = 2 \text{ A}$

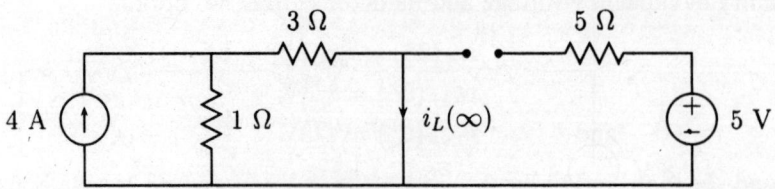

Fig. 2.21.2: $t = \infty$ circuit.

(b) Thus, by continuity of inductor current, we obtain

$$i_L(0^+) = i_L(0^-) = 2 \text{ A}$$

(c) From the circuit of Fig. 2.21.2

$$i(\infty) = i_{3\Omega} = \frac{1}{1+3}(4) = 1 \text{ A}$$

(d) Since $i(0^-) = 6$ A is the steady state value, the $\frac{di_L(0^-)}{dt}$ must be zero, i.e.,

$$\frac{di_L}{dt}(0^-) = 0$$

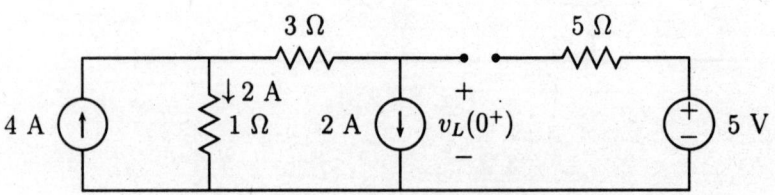

Fig. 2.21.3: $t = 0^+$ circuit.

(e) By KVL in the right mesh of the circuit of Fig. 2.21.3 as :

$$3 \times 1 + v_L(0^+) - 1 \times 2 = 0$$

Hence $$v_L(0^+) = 2 - 3 = -1 \text{ V}$$

For the inductor $$\frac{di_L}{dt}(0^+) = \frac{v_L(0^+)}{L} = -1/2 \text{ A/sec}.$$

SP 2.22 The switch in Fig. SP 2.22 opens at $t = 0$. Find the values for
  (a) $i_L(0^+)$, (b) $v_C(0^+)$, (c) $\frac{di_L}{dt}(0^+)$, and (d) $\frac{dv_C}{dt}(0^+)$.

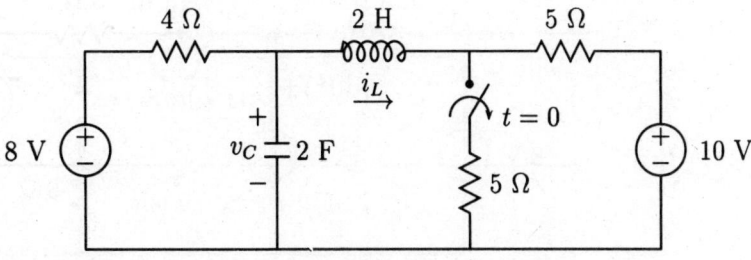

Fig. SP 2.22.

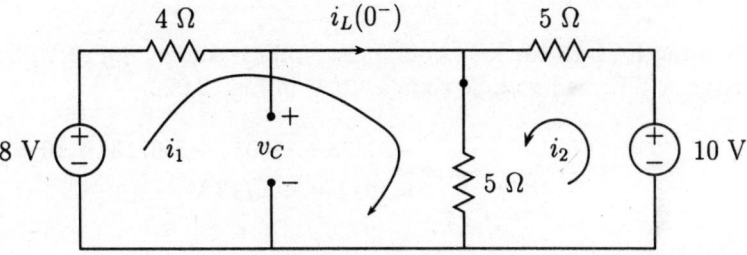

Fig. 2.22.1: $t = 0^-$ circuit.

SOLUTION:

(a) For the circuit of Fig. 2.22.1, the mesh current equations are :

$$9i_1 + 5i_2 = 8 \qquad\qquad [2.22.1]$$

$$5i_1 + 10i_2 = 10 \qquad\qquad [2.22.2]$$

from which

$$i_1 = 6/13 \text{ A} \qquad \text{and} \qquad i_2 = 10/13 \text{ A}$$

Therefore

$$i_L(0^-) = i_1 = 6/13 \text{ A}$$

(b) $v_C(0^-)$ can be determined as :

$$
\begin{aligned}
v_C(0^-) &= \text{ voltage across 5 }\Omega\text{ resistor} \\
&= 5\,(i_1 + i_2) \\
&= 5(6/13 + 10/13) = 80/13 \text{ V}
\end{aligned}
$$

Thus by continuity of capacitor voltage and inductor current, we obtain

$$v_C(0^+) = v_C(0^-) = 80/13 \text{ V}$$
$$i_L(0^+) = i_L(0^-) = 6/13 \text{ A}$$

(c) Taking KCL at the node **a** in Fig. 2.22.2

$$\frac{8 - 80/13}{4} = i_C(0^+) + 6/13 \qquad \Rightarrow i_C(0^+) = 0$$

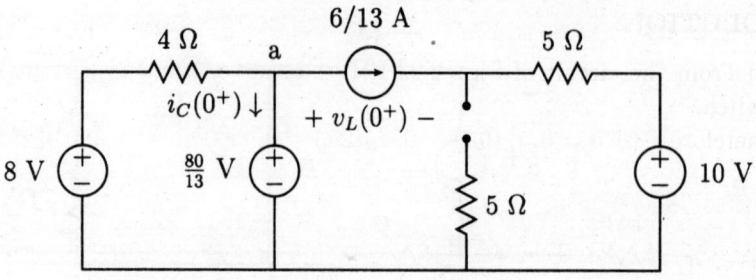

Fig. 2.22.2: $t = 0^+$ circuit.

Now using KVL in the loop containing voltage source, 80/13 V, current source, 6/13 A, resistance, 5 Ω, and voltage source, 10 V in Fig. 2.22.2

$$-80/13 + v_L(0^+) + 5(6/13) + 10 = 0$$

or $v_L(0^+) = -80/13$ V

For the inductor

$$v_L = L\frac{di_L}{dt}$$

Hence

$$\frac{di_L}{dt}(0^+) = \frac{v_L(0^+)}{L} = \frac{-80/13}{2} = -40/13 \text{ A/sec}.$$

For the capacitor

$$i_C = C\frac{dv_C}{dt}$$

Therefore

$$\frac{dv_C}{dt}(0^+) = \frac{i_C(0^+)}{C} = \frac{0}{2} = 0$$

SP 2.23 In the circuit of Fig. SP 2.23, find the values of
   (a) $i_1(0^-)$, $i_L(0^-)$, $i_R(0^-)$, $i_C(0^-)$, $v_L(0^-)$, **and** $v_C(0^-)$,
   (b) $i_1(0^+)$, $i_L(0^+)$, $i_R(0^+)$, $i_C(0^+)$, $v_L(0^+)$, **and** $v_C(0^+)$.

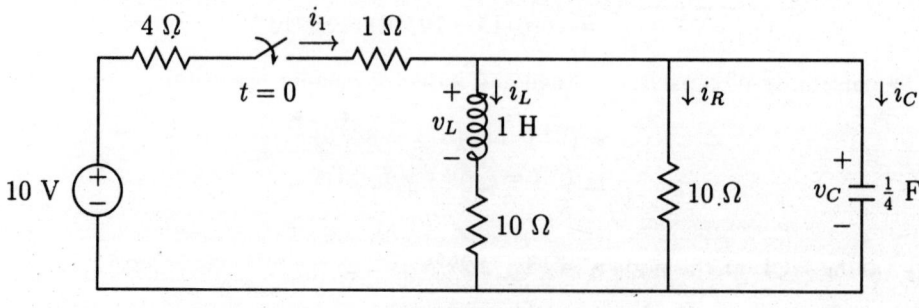

Fig. SP 2.23.

SOLUTION:

(a) From the circuit of Fig. 2.23.1, it is observed that no current is flowing right to the switch.

Therefore $i_1(0^-) = 0$, $i_L(0^-) = 0$, $i_R(0^-) = 0$, $v_C(0^-) = 0$, and $i_C(0^-) = 0$.

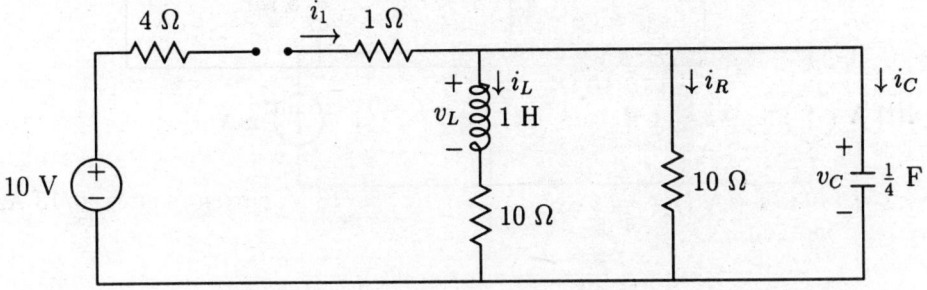

Fig. 2.23.1: $t = 0^-$ circuit.

(b) By continuity of capacitor voltage and inductor current, we obtain

$$v_C(0^+) = v_C(0^-) = 0$$
$$i_L(0^+) = i_L(0^-) = 0$$

From the circuit of Fig. 2.23.2

$$i_R(0^+) = 0/10 = 0$$
$$i_1(0^+) = \frac{10}{4+1} = 2 \text{ A}$$

and
$$i_C(0^+) = i_1(0^+) = 2 \text{ A}$$

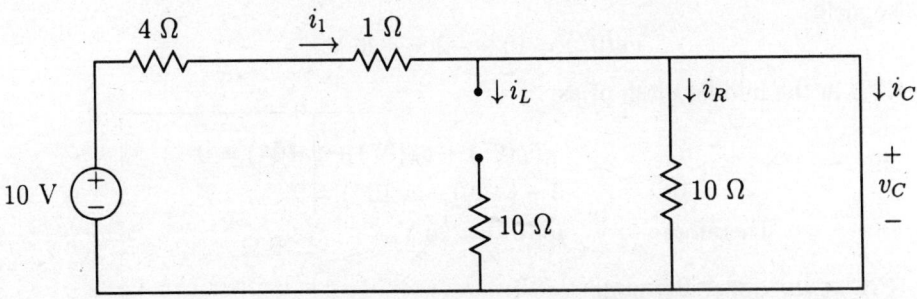

Fig. SP 2.23.2: $t = 0^+$ circuit.

SP 2.24 For circuit shown in Fig. SP 2.24, find the values for

(a) $v_L(0^-)$, $i_C(0^-)$, $i_R(0^-)$, $v_R(0^-)$, $v_C(0^-)$, and $i_L(0^-)$.

(b) $v_L(0^+)$, $i_C(0^+)$, $i_R(0^+)$, $v_R(0^+)$, $v_C(0^+)$, and $i_L(0^+)$.

(c) $\frac{di_L}{dt}(0^+)$, $\frac{dv_C}{dt}(0^+)$, and $\frac{di_R}{dt}(0^+)$.

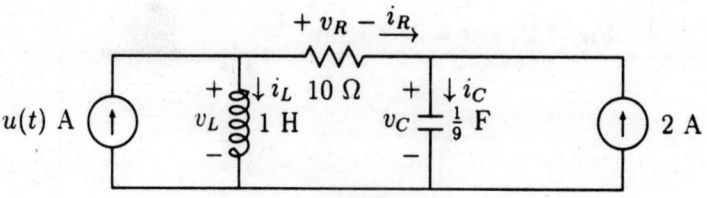

Fig. SP 2.24.

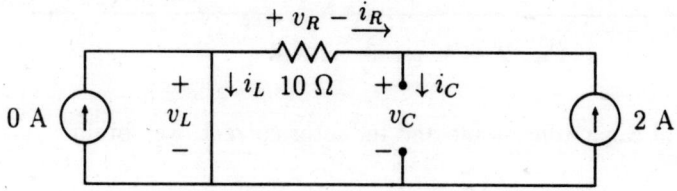

Fig. 2.24.1: $t = 0^-$ circuit.

SOLUTION:

From the circuit of Fig. 2.24.1, it is noted that a constant current through the inductor requires zero voltage across it,

$$v_L(0^-) = 0$$

and a constant voltage across the capacitor requires zero current through it,

$$i_C(0^-) = 0$$

Using KCL to the upper right node

$$i_R(0^-) = i_C(0^-) - 2 \qquad \Rightarrow i_R(0^-) = 0 - 2 = -2 \text{ A}$$

which also yields

$$v_R(0^-) = 10 \times -2 = -20 \text{ V}$$

Taking KVL in the middle mesh of as:

$$-v_L(0^-) + v_R(0^-) + v_C(0^-) = 0$$
$$\text{or} \qquad 0 + (-20) + v_C(0^-) = 0$$
$$\text{Therefore} \qquad v_C(0^-) = 20 \text{ V}$$

Taking KCL at the upper left node

$$i_R(0^-) + i_L(0^-) = 0$$
$$\text{or} \qquad i_L(0^-) = -i_R(0^-) = 2 \text{ A}$$

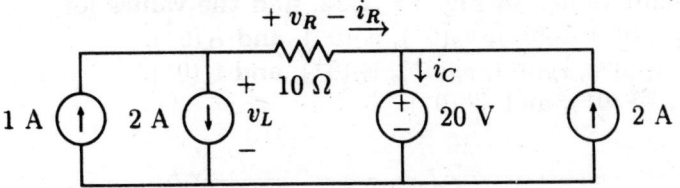

Fig. 2.24.2: $t = 0^+$ circuit.

(b) By continuity of inductor current and capacitor voltage, we have

$$i_L(0^+) = 2 \text{ A} = i_L(0^-)$$

and $\qquad v_C(0^+) = 20 \text{ V} = v_C(0^-)$

From Fig. 2.24.2, by KCL at upper left node

$$-1 + 2 + i_R(0^+) = 0 \qquad \Rightarrow i_R(0^+) = -1 \text{ A}$$

and $\qquad v_R(0^+) = 10 i_R(0^+) = -10 \text{ V}$

By KCL at the upper right node

$$i_R(0^+) + 2 - i_C(0^+) = 0 \qquad \Rightarrow i_C(0^+) = i_R(0^+) + 2 = -1 + 2 = 1 \text{ A}$$

By KVL in the middle mesh

$$-v_L(0^+) + v_R(0^+) + 20 = 0$$

or $\qquad -v_L(0^+) + (-10) + 20 = 0$

and $\qquad v_L(0^+) = 10 \text{ V}$

(c) For inductor:

$$v_L = L \frac{di_L}{dt}$$

Therefore

$$\frac{di_L}{dt}(0^+) = \frac{v_L(0^+)}{L} = \frac{10}{1} = 10 \text{ A/sec.}$$

Similarly, for capacitor

$$\frac{dv_C}{dt}(0^+) = \frac{i_C(0^+)}{C} = \frac{1}{1/9} = 9 \text{ V/sec.}$$

Writing KCL at the upper left node of circuit of Fig. SP 2.24

$$1 - i_L - i_R = 0 \qquad t > 0$$

and $\qquad 0 - \dfrac{di_L}{dt} - \dfrac{di_R}{dt} = 0 \qquad t > 0$

Therefore $\qquad \dfrac{di_R}{dt}(0^+) = -\dfrac{di_L}{dt}(0^+) = -10 \text{ A/sec.}$

SP 2.25 Find $v_C(0^+)$ **and** $i_L(0^+)$ **in the circuit of Fig. SP 2.25. Note the switches are opened at** $t = 0$.

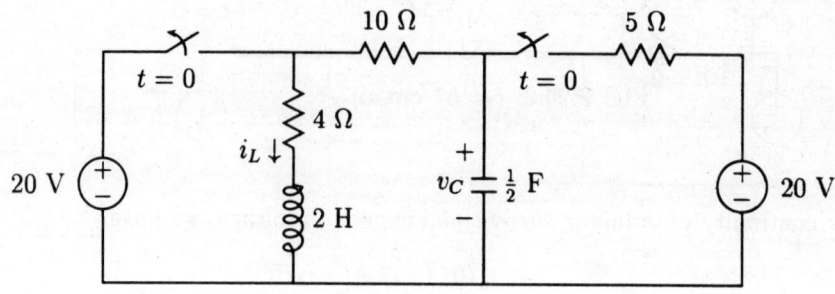

Fig. SP 2.25.

SOLUTION

Writing mesh equations in the circuit of Fig. 2.25.1 as

$$4(i_1 + i_2) = 20$$

$$4i_1 + 19i_2 = 20$$

from which $i_1 = 5$ A and $i_2 = 0$. Therefore $i_L(0^-) = i_1 + i_2 = 5$ A Thus by continuity of inductor current, we obtain

$$i_L(0^+) = i_L(0^-) = 5 \text{ A}$$

and $v_C(0^-) = 20 - 5i_2 = 20 \text{ V}$

Thus, by continuity of capacitor voltage, we obtain

$$v_C(0^+) = v_C(0^-) = 20 \text{ V}$$

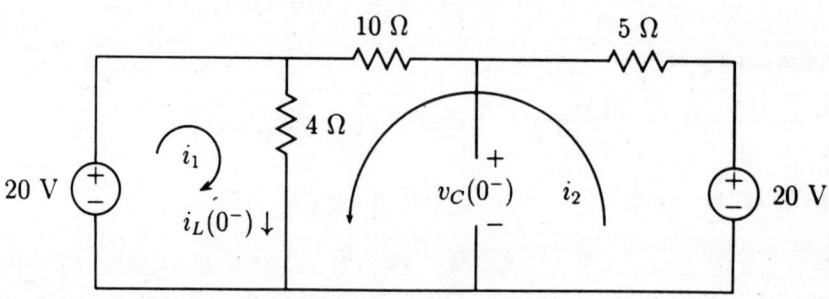

Fig. 2.25.1: $t = 0^-$ circuit.

SP 2.26 In the circuit of Fig. SP 2.26, find $v_C(0^+)$, $\frac{dv_C}{dt}(0^+)$, $v_C(\infty)$ and the energy stored in the capacitor $W_C(0^+)$. Both switches are thrown in the direction shown at time zero.

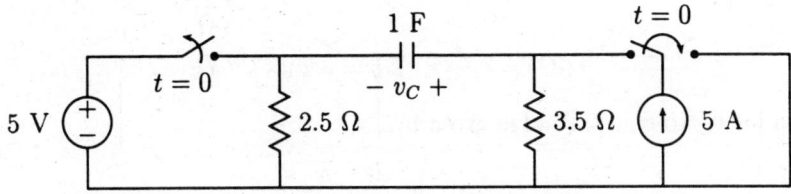

Fig. SP 2.26.

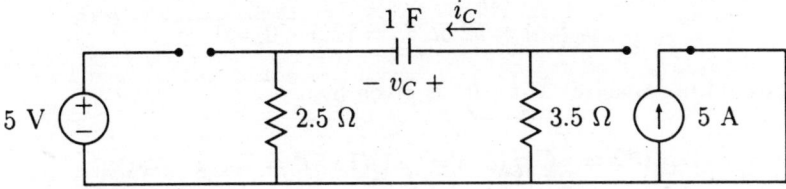

Fig. 2.26.1: $t > 0$ circuit.

SOLUTION:

From the circuit of Fig. 2.26.2

$$v_C(0^-) = 3.5 \times 5 - 5 = 12.5 \text{ V}$$

Thus by continuity of capacitor voltage

$$v_C(0^+) = v_C(0^-) = 12.5 \text{ V}$$

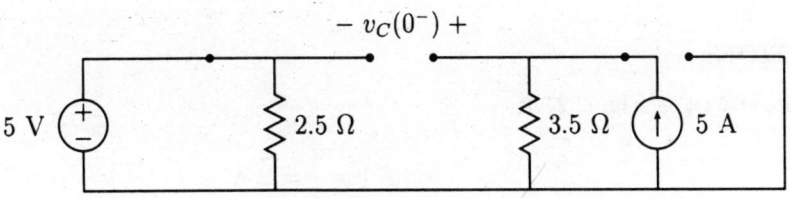

Fig. SP 2.26.2: $t = 0^-$ circuit.

By KVL in the circuit of Fig. 2.26.1

$$v_C + 6i_C = 0 \qquad \Rightarrow v_C + 6\left(1\frac{dv_C}{dt}\right) = 0$$

Therefore

$$\frac{dv_C}{dt} + \frac{1}{6}v_C = 0 \qquad\qquad [2.26.1]$$

Setting $t = 0^+$ in eqn [2.26.1]

$$\frac{dv_C}{dt}(0^+) + \frac{1}{6}v_C(0^+) = 0$$

Therefore

$$\frac{dv_C}{dt}(0^+) = -\frac{1}{6} \times 12.5 = -2.08 \text{ V/sec}$$

The solution for v_C to eqn [2.26.1] is given by

$$v_C(t) = v_C(0^+)e^{-t/8} = 12.5e^{-t/6} \qquad\qquad [2.26.2]$$

Setting $t = \infty$ in the eqn [2.26.2], we have

$$v_C(\infty) = 12.5e^{-\infty} = 12.5 \times 0 = 0$$

The energy stored in capacitor at t $=0^+$ is given by

$$W_C(0^+) = \frac{1}{2}Cv_C^2(0^+) = \frac{1}{2}(1)(12.5)^2 = 78.125 \text{ Watts}$$

SP 2.27 In the circuit shown in Fig. SP 2.27, find $i_L(0^+), v_C(0^+),$ **and** $\frac{di_L}{dt}(0^+),$
$\frac{dv_C}{dt}(0^+)$ **and** $\frac{di_C}{dt}(0^+).$

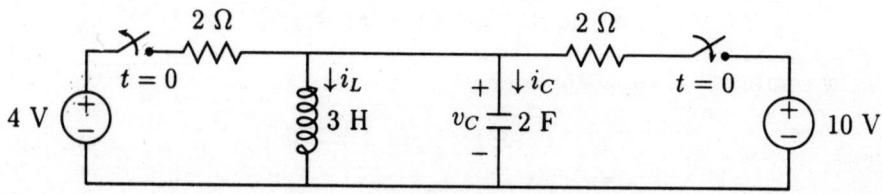

Fig. SP 2.27.

SOLUTION:

From the circuit of Fig. 2.27.1,

$$i_L(0^-) = \frac{4}{2} = 2 \text{ A}$$
$$v_L(0^-) = 0 \text{ V}$$
$$v_C(0^-) = 0 \text{ V}$$
$$\text{and} \qquad i_C(0^-) = 0$$

Thus by continuity of inductor current and capacitor voltage, we obtain

$$v_C(0^+) = v_C(0^-) = 0 \text{ V}$$
$$i_L(0^+) = i_L(0^-) = 2 \text{ A}$$

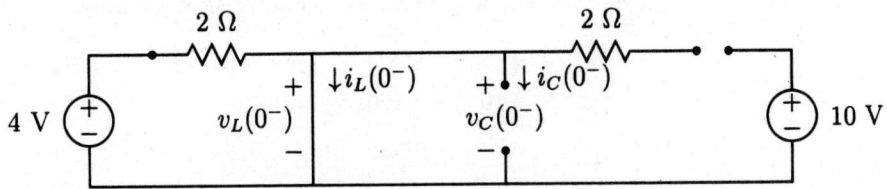

Fig. 2.27.1: $t = 0^-$ circuit.

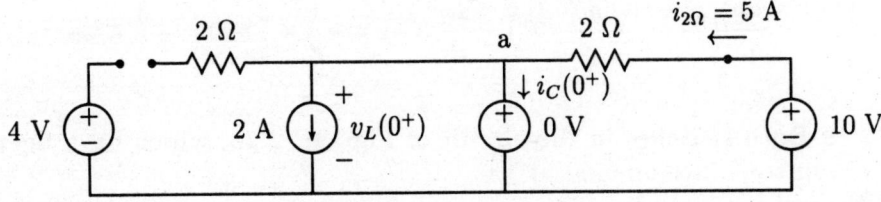

Fig. 2.27.2: $t = 0^+$ circuit.

From the circuit of Fig. 2.27.2

$$v_L(0^+) = 0$$

and $\qquad i_{2\Omega} = 10/2 = 5 \text{ A}$

By KCL at upper left node a

$$5 = i_C(0^+) + 2 \qquad \Rightarrow \quad i_C(0^+) = 3 \text{ A}$$

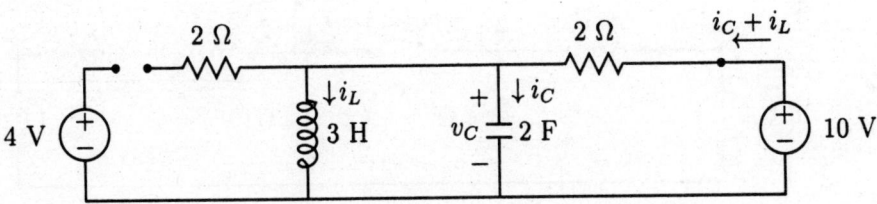

Fig. 2.27.3: $t > 0$ circuit.

For the capacitor:

$$i_C = C \frac{dv_C}{dt}$$

Hence

$$\frac{dv_C}{dt}(0^+) = \frac{i_C(0^+)}{C} = \frac{3}{2} \text{ V/sec.}$$

Similarly for the inductor:

$$\frac{di_L}{dt}(0^+) = \frac{v_L(0^+)}{L} = \frac{0}{L} = 0$$

By KVL around the rightmost mesh in the circuit of Fig. 2.27.3, we have

$$2(i_C + i_L) + v_C = 10 \qquad \text{for } t > 0$$

or

$$2\frac{di_C}{dt} + 2\frac{di_L}{dt} + \frac{dv_C}{dt} = 0 \qquad t > 0$$

Therefore

$$\frac{di_C}{dt}(0^+) = -\frac{1}{2}\frac{dv_C}{dt}(0^+) - \frac{di_L}{dt}(0^+) = -\frac{1}{2} \times \frac{3}{2} - 0 = -\frac{3}{4} \text{ A/sec.}$$

SP 2.28 Both switches in the circuit of Fig. SP 2.28, which have been closed for a long time, are opened at $t = 0$.
   (a) Find $i_L(0^+)$ and $v_C(0^+)$.
   (b) Find $\frac{dv_C}{dt}(0^+)$ and $\frac{di_L}{dt}(0^+)$

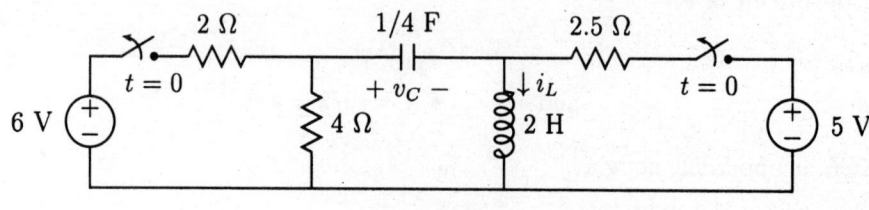

Fig. SP 2.28.

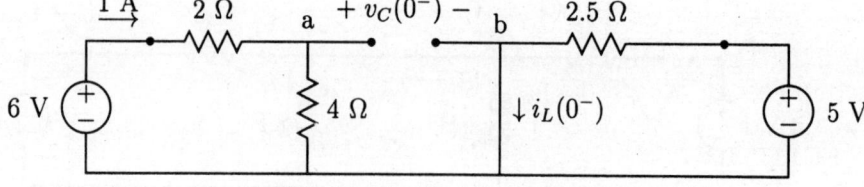

Fig. 2.28.1 : t = 0⁻ circuit.

SOLUTION:

(a) It is observed from circuit of Fig. 2.28.1 that

$$i_L(0^-) = 5/2.5 = 2 \text{ A}$$
$$\text{and} \qquad v_C(0^-) = V_a - V_b = (4 \times 1) - 0 = 4 \text{ V}$$

Thus by continuity of inductor current and capacitor voltage, we obtain

$$v_C(0^+) = v_C(0^-) = 4 \text{ V}$$
$$i_L(0^+) = i_L(0^-) = 2 \text{ A}$$

(b) It is noted from the Fig. 2.28.2 that only 2 A current is flowing through it. So $v_L(0^+)$ can be determined by KVL as:

$$4 + v_L(0^+) + 4 \times 2 = 0$$

Therefore

$$v_L(0^+) = -4 - 8 = -12 \text{ V}$$

It is also evident that $i_C(0^+) = 2$ A

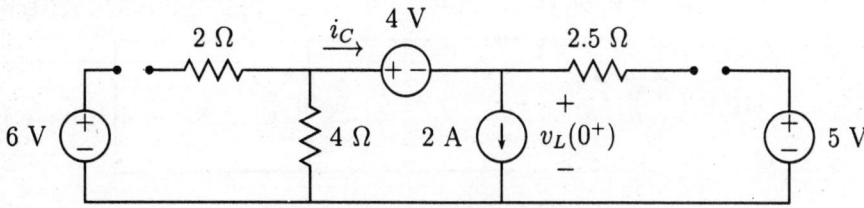

Fig. 2.28.2: $t = 0^+$ circuit.

For the inductor:

$$v_L = L\frac{di_L}{dt}$$

Therefore

$$\frac{di_L}{dt}(0^+) = \frac{v_L(0^+)}{L} = \frac{-12}{2} = -6 \text{ A/sec.}$$

Similarly for the inductor:

$$\frac{dv_C}{dt}(0^+) = \frac{i_C(0^+)}{C} = \frac{2}{1/4} = 8 \text{ V/sec.}$$

SP 2.29 The network shown in Fig. SP 2.29, is in the steady state with the switch closed. At $t = 0$, the switch is opened. Determine $v_K(0^+)$ and $\frac{dv_K}{dt}(0^+)$.

SOLUTION:

From the circuit of Fig. 2.29.1, we obtain

$$i(0^-) = 10/20 = 1/2 \text{ A}$$
$$v_C(0^-) = 0$$

From continuity of inductor current and capacitor voltage, we find

$$i(0^+) = i(0^-) = 1/2 \text{ A}$$
$$v_C(0^+) = v_C(0^-) = 0$$

We can obtain $v_K(0^+)$ from the of Fig. 2.29.2 as:

$$v_K(0^+) = 0$$

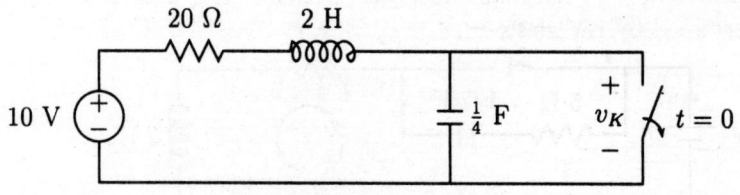

Fig. SP 2.29.

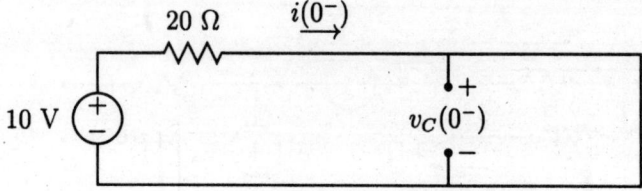

Fig. 2.29.1: $t = 0^-$ circuit.

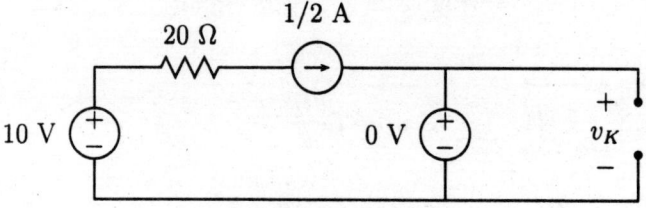

Fig. 2.29.2: $t = 0^+$ circuit.

and

$$\frac{dv_K}{dt} = \frac{dv_C}{dt} = \frac{i(t)}{C}$$

Therefore

$$\frac{dv_K}{dt}(0^+) = \frac{i(0^+)}{C} = \frac{1/2}{1/4} = 2 \text{ V/sec.}$$

SP 2.30 In the network of Fig. SP 2.30, the switch is opened at $t = 0$ after the network has attained a steady state with the switch closed.

Find $v_K(0^+)$ and $\frac{dv_K}{dt}(0^+)$.

SOLUTION:

From the circuit of Fig. 2.30.1,

$$i(0^-) = 10/5 = 2 \text{ A}$$

Thus by continuity of inductor current we obtain

$$i(0^+) = i(0^-) = 2 \text{ A}$$

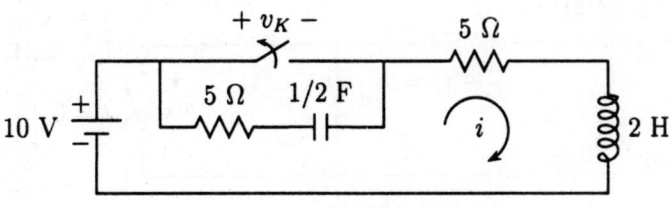

Fig. SP 2.30.

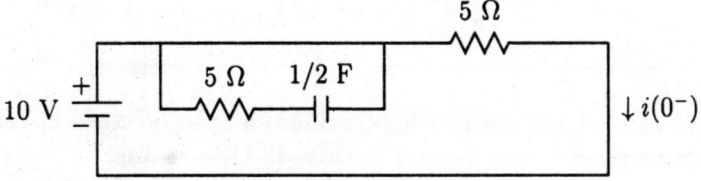

Fig. 2.30.1: $t = 0^-$ circuit.

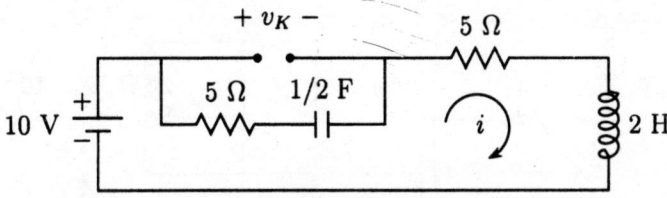

Fig. 2.30.2: $t > 0$ circuit.

Using KVL in the loop excluding switch of Fig. 2.30.2, we have

$$10i + \frac{1}{1/2}\int_0^t +2\frac{di}{dt} = 10 \qquad\qquad [2.30.1]$$

Setting t = 0^+ in eqn [2.30.1]

$$10i(0^+) + 2\int_0^{0^+} idt + 2\frac{di}{dt}(0^+) = 10$$

Hence $\qquad 10 \times 2 + 0 + 2\frac{di}{dt}(0^+) = 10 \qquad \Rightarrow \frac{di}{dt}(0^+) = -5 \text{ A/sec.}$

Now taking KVL in the mesh containing switch, we have

$$v_K(t) = 5_1 i(t) + \frac{1}{1/2}\int_0^t i(t)dt \qquad\qquad [2.30.2]$$

Setting $t = 0^+$ in the eqn [2.30.2], we have

$$v_K(0^+) = 5_1 i(0^+) + 2\int_0^{0^+} i(t)dt$$

or $\qquad v_K(0^+) = 5 \times 2 + 0 \qquad \Rightarrow v_K(0^+) = 10 \text{ V}$

(b) Differentiating eqn [2.30.2]

$$\frac{dv_K}{dt}(t) = 5\frac{di}{dt}(t) + 2i(t) \qquad [2.30.3]$$

Setting $t = 0^+$ in eqn [2.30.3]

$$\frac{dv_K}{dt}(0^+) = 5\frac{di}{dt}(0^+) + 2i(0^+)$$

Therefore
$$\frac{dv_K}{dt}(0^+) = 5(-5) + 2 \times 2 = -21 \text{ V/sec.}$$

SP 2.31 In the network shown in Fig. SP 2.31, a steady state is reached with the switch open. At $t = 0$, the switch is closed. Determine:
(a) $i_L(0^-)$, (b) $i_L(0^+)$, (c) $v_b(0^-)$, (d) $v_b(0^+)$, (e) $v_c(0^+)$, and (f) $i_L(\infty)$.

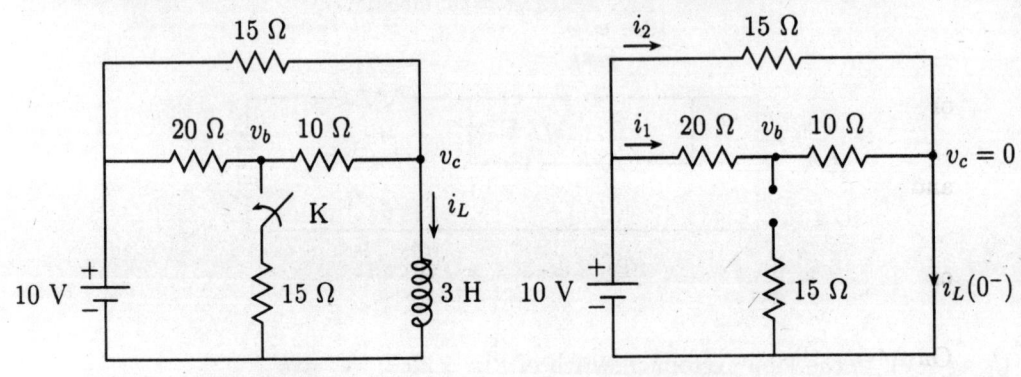

Fig. SP 2.31.   Fig. 2.31.1: $t = 0^-$ circuit.

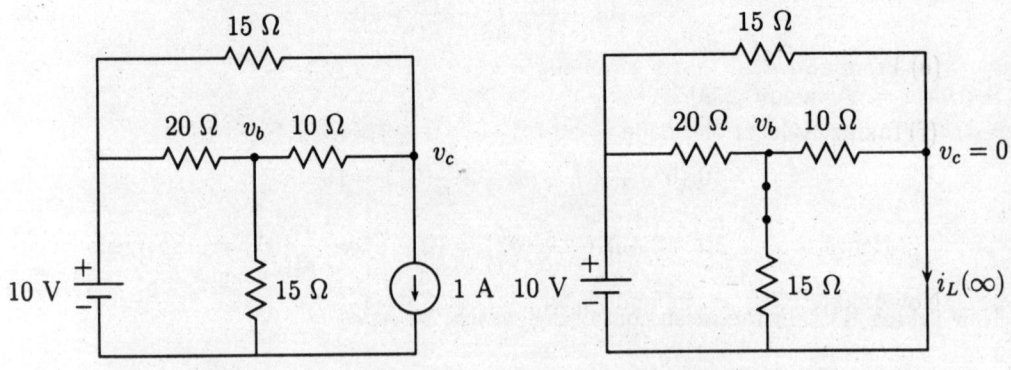

Fig. 2.31.2: $t = 0^+$ circuit.   Fig. 2.31.3: $t = \infty$ circuit.

SOLUTION:

(a) From the circuit of Fig. 2.31.1,

$$i_1 = \frac{10}{10 + 20} = \frac{1}{3} \text{ A}$$

$$i_2 = \frac{10}{15} = \frac{2}{3} \text{ A}$$

Therefore $\qquad i_L(0^-) = i_1 + i_2 = \frac{1}{3} + \frac{2}{3} = 1 \text{ A}$

(b) From continuity of inductor current, we obtain

$$i_L(0^+) = i_L(0^-) = 1 \text{ A}$$

(c) By KVL around the lower left mesh containing 10 V-source, 20 Ω-resistance, and $v_b(0^-)$ in the circuit of Fig. 2.31.1.

$$20i_1 + v_b(0^-) = 10$$

Hence $\qquad v_b(0^-) = 10 - 20(1/3) = 10/3 \text{ V}$

(d) Writing KCL at the node v_b and v_c for the circuit of Fig. 2.31.2 as :

$$\frac{10 - v_b(0^+)}{20} = \frac{v_b(0^+)}{15} + \frac{v_b(0^+) - v_c(0^+)}{10}$$

or

$$13v_b(0^+) - 6v_c(0^+) = 30 \qquad\qquad\qquad [2.31.1]$$

and

$$\frac{v_b(0^+) - v_c(0^+)}{10} + \frac{10 - v_c(0^+)}{15} = 1$$

or

$$3v_b(0^+) - 5v_b(0^+) = 10 \qquad\qquad\qquad [2.31.2]$$

On solving the eqn [2.31.1] and eqn [2.31.2], we get

$$v_b(0^+) = \frac{90}{47} \text{ V} \qquad \text{and} \qquad v_c(0^+) = -\frac{40}{47} \text{ V}$$

(e) From above $\qquad v_c(0^+) = -40/47 \text{ V}$

(f) Taking KCL at the node v_b of the circuit of Fig. 2.31.3

$$\frac{10 - v_b}{20} = \frac{v_b}{15} + \frac{v_b}{10} \qquad \Rightarrow v_b = 30/13 \text{ V}$$

Now taking KCL at the node v_c of the Fig. 2.31.3

$$i_L(\infty) = \frac{10 - 0}{15} + \frac{30/13 - 0}{10} = 35/39 \text{ A}$$

SP 2.32 In the accompanying network shown in Fig. SP 2.32, a steady state is reached with switch K open. At $t = 0$, the switch is closed. Find:
(a) $v_b(0^-)$, (b) $v_b(0^+)$, (c) $v_c(0^+)$, and (d) $v_c(\infty)$.

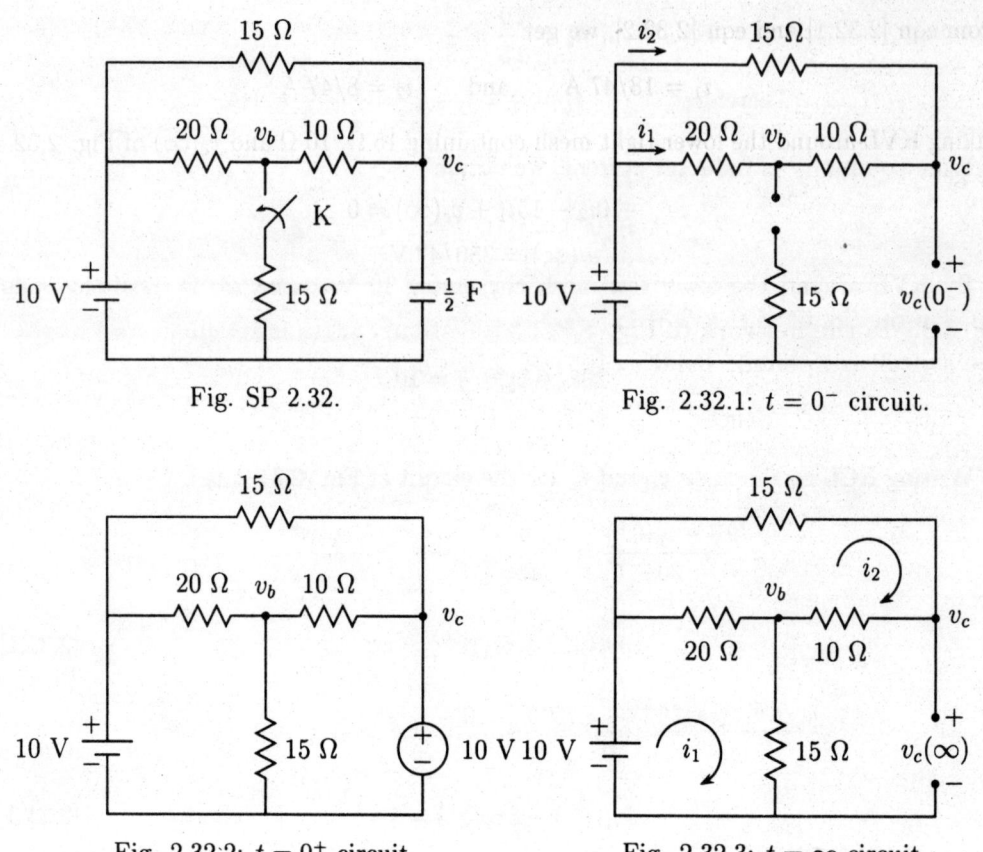

Fig. SP 2.32.

Fig. 2.32.1: $t = 0^-$ circuit.

Fig. 2.32:2: $t = 0^+$ circuit.

Fig. 2.32.3: $t = \infty$ circuit.

SOLUTION:

(a) From observation of the circuit of Fig. 2.32.1, we find that

$$i_1 = i_2 = 0$$

and $\qquad v_b(0^-) = v_c(0^-) = 10 \text{ V}$

(b) Using KCL at the node v_b of the circuit of Fig. 2.32.2

$$\frac{10 - v_b(0^+)}{20} = \frac{v_b(0^+)}{15} + \frac{v_b(0^+) - 10}{10} \qquad \Rightarrow v_b(0^+) = 90/13 \text{ V}$$

(c) Here v_c is the voltage across capacitor, therefore

$$v_c(0^+) = v_c(0^-) = 10 \text{ V}$$

(d) Writing mesh equations for the circuit of Fig. 2.32.3 as :

$$35i_1 - 20i_2 = 10 \qquad\qquad\qquad [2.32.1]$$

and

$$-20i_1 + 45i_2 = 0 \qquad\qquad\qquad [2.32.2]$$

From eqn [2.32.1] and eqn [2.32.2], we get

$$i_1 = 18/47 \text{ A} \quad \text{and} \quad i_2 = 8/47 \text{ A}$$

Taking KVL around the lower right mesh containing 15 Ω, 10 Ω and $v_c(\infty)$ of Fig. 2.32.3, we get

$$-10i_2 - 15i_1 + v_c(\infty) = 0$$
$$v_c(\infty) = 350/47 \text{ V}$$

SP 2.33 In the network of Fig. SP 2.33, a steady state is reached, and at $t = 0$, the switch is opened. Find $v_{sw}(0^+)$ and $\frac{dv_{sw}}{dt}(0^+)$.

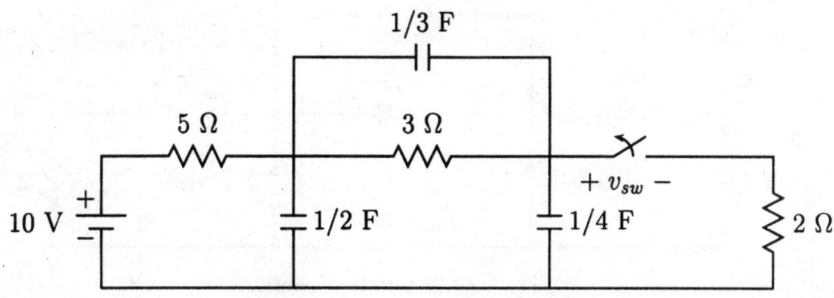

Fig. SP 2.33.

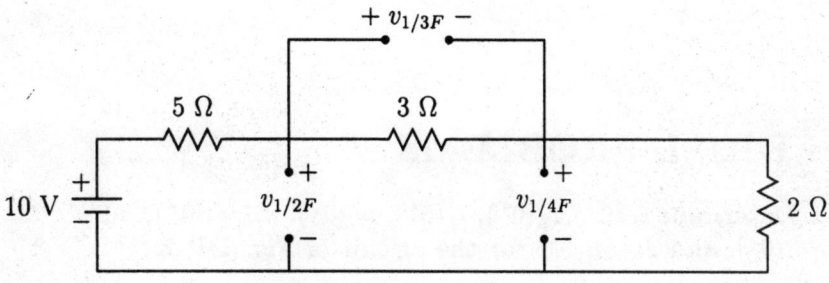

Fig. 2.33.1: $t = 0^-$ circuit.

SOLUTION:

It is observed from the circuit of Fig. 2.33.1 that

$$v_{\frac{1}{4}F}(0^-) = \frac{10}{5 + 3 + 2}(2) = 2 \text{ V}$$

$$v_{\frac{1}{3}F}(0^-) = \frac{10}{5 + 3 + 2}(3) = 3 \text{ V}$$

$$\text{and} \qquad v_{\frac{1}{2}F}(0^-) = 10 - \frac{10}{5 + 3 + 2}(5) = 5 \text{ V}$$

It obvious from the circuit of Fig. 2.33.2 that the voltage across the switch $v_K(0^+)$ will be equal to the initial voltage across the capacitor $C_{1/4F}$.

Therefore $$v_K(0^+) = v_{\frac{1}{4}F}(0^-) = 2 \text{ V}$$

With reference to Fig. 2.33.2, the current through 2-V source is

$$i_{\frac{1}{4}F}(0^+) = 0$$

Thus $$\frac{dv_{sw}}{dt}(0^+) = \frac{dv_{\frac{1}{4}}F}{dt}(0^+) = \frac{i_{\frac{1}{4}F}(0^+)}{1/4} = 0$$

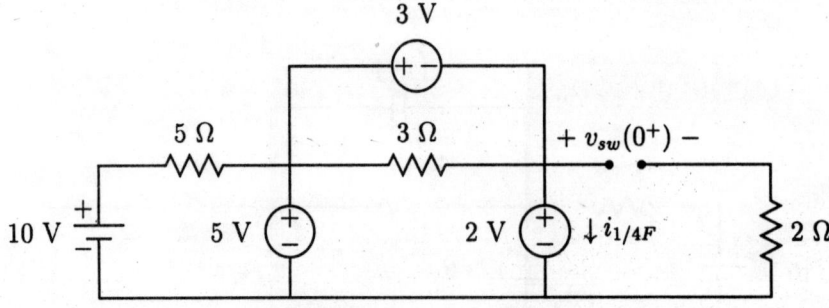

Fig. 2.33.2: $t = 0^+$ circuit.

2.4   DRILL PROBLEMS

DP 2.1 Determine $i_C(0^+), v_L(0^+), i_L(0^+), v_C(0^+), di_L/dt(0^+),$
$dv_C/dt(0^+),$ **and** $di_C/dt(0^+)$ **for the circuit of Fig. DP 2.1.**

Fig. DP 2-1.

Hints:

$$
\begin{aligned}
v(t) &= 5 - 5u(t) = 0 \quad \text{for } t \geq 0 \\
&= 5 \text{ V} \qquad \text{for } t < 0
\end{aligned}
$$

DP 2.2 Determine $i_C(0^+)$, $v_L(0^+)$, $i_L(0^+)$, $v_C(0^+)$, $di_L/dt(0^+)$, $dv_C/dt(0^+)$, **and** $di_C/dt(0^+)$ **for the circuit of Fig. DP 2.2.**

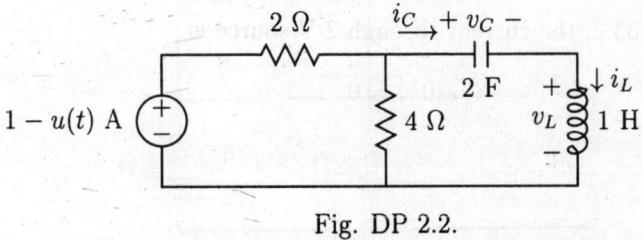

Fig. DP 2.2.

DP 2.3 Find $v_C(0^+)$, $i_C(0^+)$, $i_L(0^+)$, **and** $v_L(0^+)$ **for the circuit of Fig. DP 2.3.**

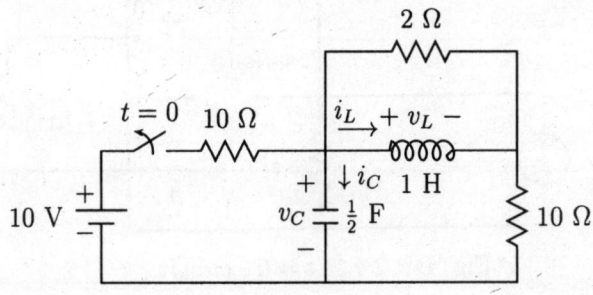

Fig. DP 2.3.

Hints : The circuit at $t = 0^-$ shows that
$$_L(0^-) = 10/(10 + 10) = 1/2 \text{ A} = i_L(0^+)$$
$$v_C(0^-) = 10(1/2) = 5 \text{ V} = v_C(0^+)$$

Draw circuit at $t = 0^+$ and then find the answers to the rest parts of the problem.

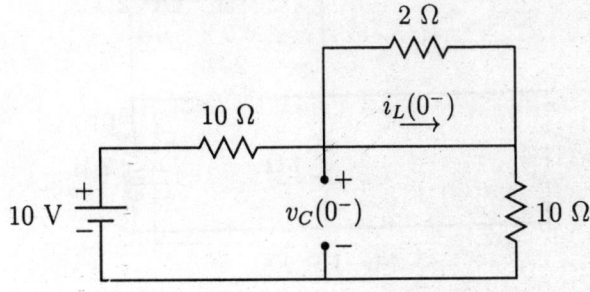

Fig. DP 2.3.1: $t = 0^-$ circuit.

Hints : The circuit at t $= 0^-$ shows that

$$i_L(0^-) = 10/6 = 5/3 \text{ A} \quad \text{and} \quad v_C(0^-) = 10 \text{ V}$$

DP 2.4 For the circuit of Fig. DP 2.4, determine $i_L(0^+)$, $v_L(0^+)$, $i_C(0^+)$, $v_C(0^+)$, $di_L/dt(0^+)$, and $dv_C/dt(0^+)$.

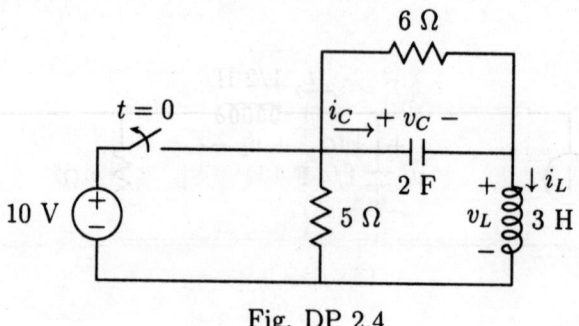

Fig. DP 2.4.

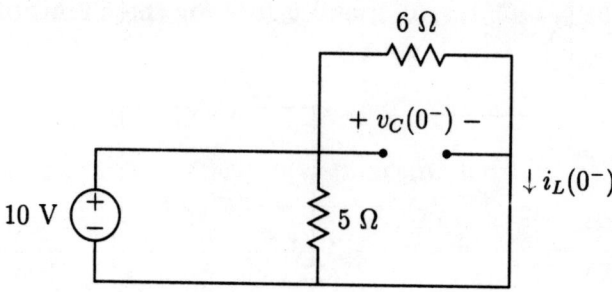

Fig. DP 2.4.1: $t = 0^-$ circuit.

DP 2.5 For the circuit of Fig. DP 2.5, determine $i_L(0^+)$, $v_L(0^+)$, $i_C(0^+)$, $v_C(0^+)$, $di_L/dt(0^+)$, and $dv_C/dt(0^+)$.

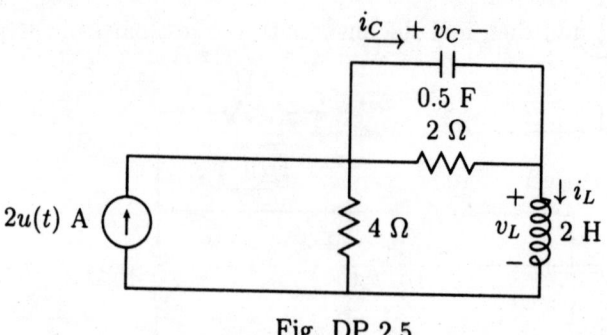

Fig. DP 2.5.

Hints :

$$2u(-t) = 2\text{ A} \qquad \text{for } t \leq 0$$
$$= 0 \qquad \text{for } t > 0$$

DP 2.6 Determine $i_L(0^+), v_L(0^+), i_C(0^+), v_C(0^+), dv_C/dt(0^+)$, and $di_L/dt(0^+)$ for the circuit of Fig. DP 2.6.

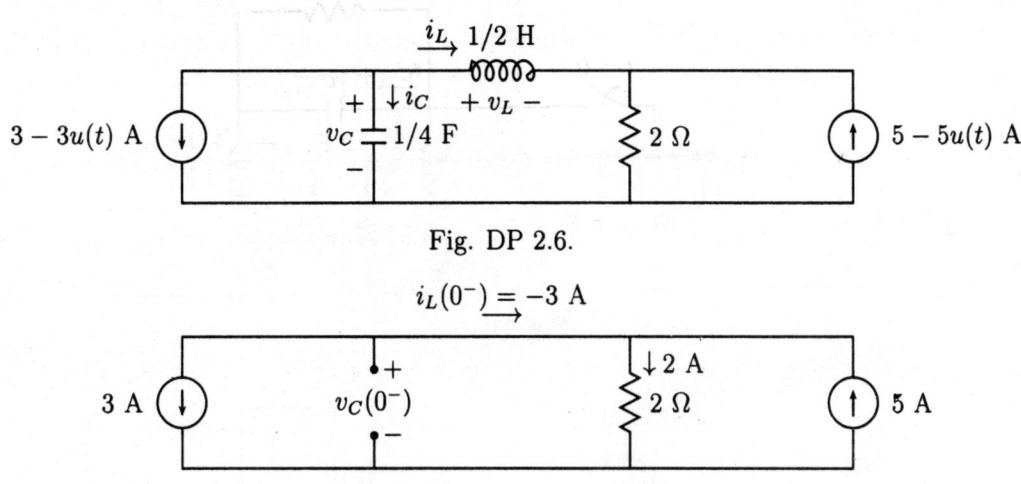

Fig. DP 2.6.

$$i_L(0^-) = -3 \text{ A}$$

Fig. DP 2.6.1: $t = 0^-$ circuit.

Hints : Fig. DP 2.6.1 explains that $i_L(0^-) = -3$ A and $v_C(0^-) = 2 \times 2 = 4$ V

DP 2.7 Determine $i_L(0^+), v_L(0^+), i_C(0^+), v_C(0^+), dv_C/dt(0^+)$, and $di_L/dt(0^+)$ for the circuit of Fig. DP 2.7.

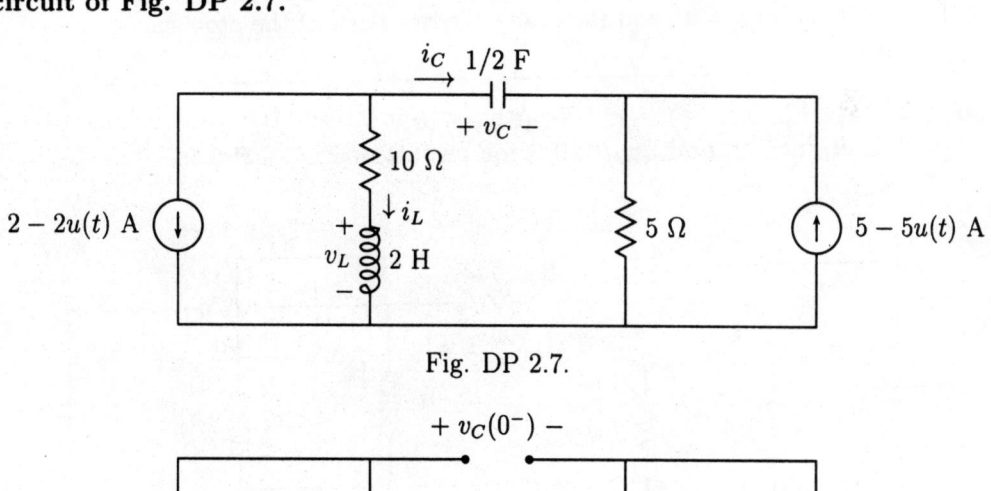

Fig. DP 2.7.

$$+ \, v_C(0^-) \, -$$

Fig. DP 2.7.1: $t = 0^-$ circuit.

Hints : Fig. DP 2.7.1 shows that $i_L(0^-) = -2$ A and by KVL in the middle mesh gives

$$v_C(0^-) + 25 + 20 = 0 \qquad \Rightarrow v_C(0^-) = -45 \text{ V}$$

DP 2.8 Determine $i_L(0^+), v_C(0^+), dv_C/dt(0^+)$, **and** $di_L/dt(0^+)$ **for the circuit of Fig. DP 2.8.**

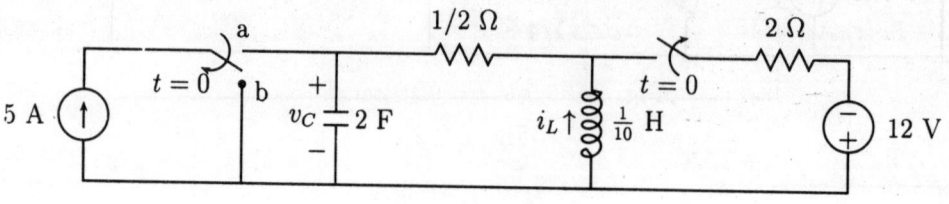

Fig. DP 2.8.

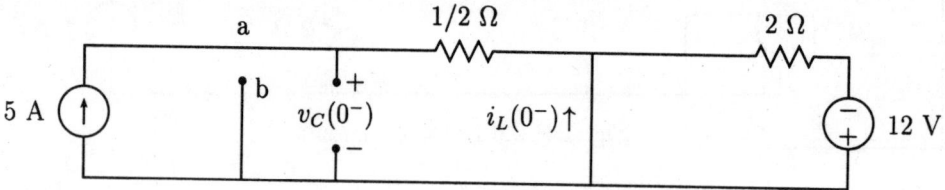

Fig. DP 2.8.1: $t = 0^-$ circuit.

Hints : It is noted from circuit of Fig. DP 2.8.1 that $i_L(0^-) = 1$ A and $v_C(0^-) = 2.5$ V. Draw a circuit at t = 0^+ and then solve the rest parts of the problem.

DP 2.9 Determine $i_1(0^-), i_L(0^+), i_C(0^+), v_L(0^+),$
$v_C(0^-), di_L/dt(0^+)$, **and** $dv_C/dt(0^+)$ **for circuit of Fig. DP 2.9.**

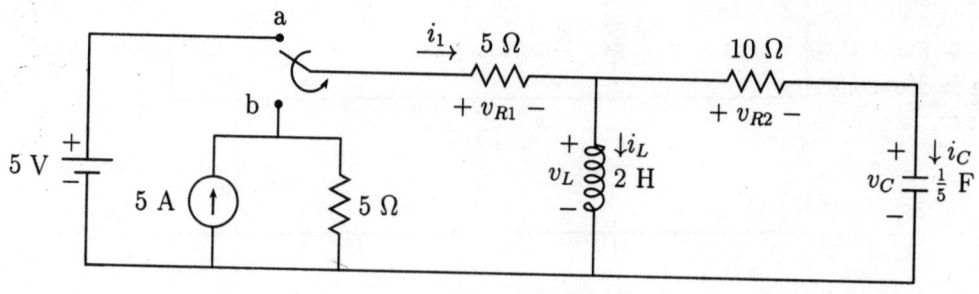

Fig. DP 2.9.

Hints : With reference to Fig. DP 2.9.1, we have
$i_1(0^-) = 5/5 = 1$ A ; $i_C(0^-) = 0$; $i_L(0^-) = i_1(0^-) = 1$A; $v_L(0^-) = 0$, and
$v_C(0^-) = 0$

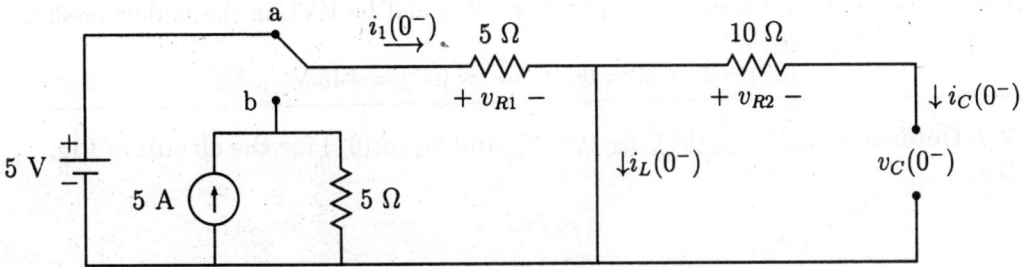

Fig. DP 2.9.1: $t = 0^-$ circuit.

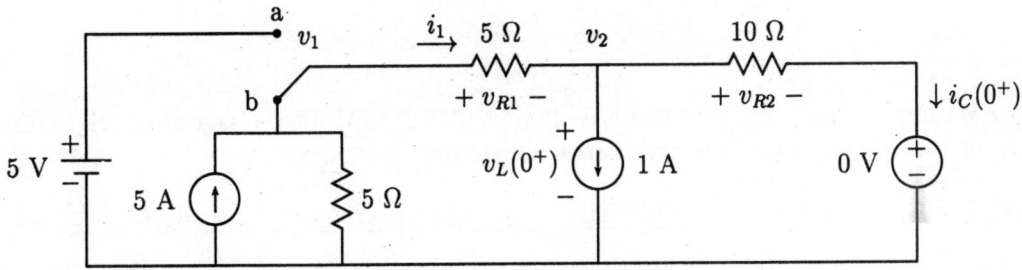

Fig. DP 2.9.2: $t = 0^+$ circuit.

KCL at v_1 and v_2 in circuit of Fig. DP 2.9.2, give

$$v_1/5 + (v_1 - v_2)/5 = 5$$

and

$$(v_1 - v_2)/5 = 1 + (v_2 - 0)/10$$

from which $v_2 = v_L(0^+) = 15/2$ V and $i_C(0^+) = v_2/10 = 3/4$ A. Use $di_L/dt(0^+) = v_L(0^+)/L$ and $dv_C/dt(0^+) = i_C(0^+)/C$ to solve the rest parts of the problem.

DP 2.10 Find $v_L(0^+)$ and $v_C(\infty)$ for the circuit shown in Fig. DP 2.10. The switch is closed at $t = 0$.

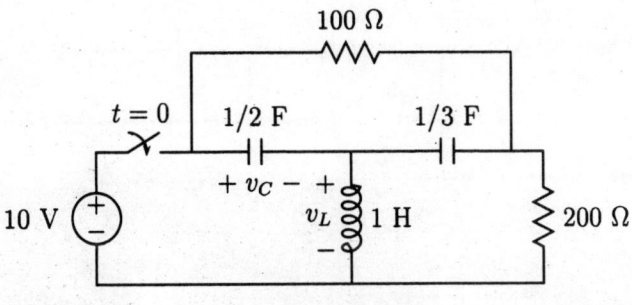

Fig. DP 2.10.

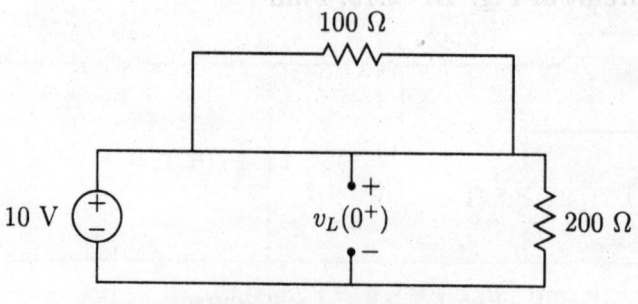

Fig. DP 2.10.1 $t = 0^+$ circuit.

Hints :

With reference to Fig. DP 2.10.1, we have $v_L(0^+) = 10$ V. The voltage across capacitor, v_C at t = ∞ will be equal to the source voltage that is 10 V.

DP 2.11 Explain following statements below what are normally useful in finding the responses of the circuits to step input.

(a) At $t = 0^+$ an uncharged capacitor acts as a shunt circuit.

(b) At $t = 0^+$ an uncharged inductor acts as an open circuit.

(c) At $t \to \infty$ a capacitor becomes an open circuit.

(d) At $t \to \infty$ an inductor becomes a short circuit.

DP 2.12 In the circuit of Fig. DP 2.12, the switch is closed at $t = 0$. Find the initial current in each capacitor at $t = 0^+$. Also Find the current in each capacitor at $t \to \infty$.

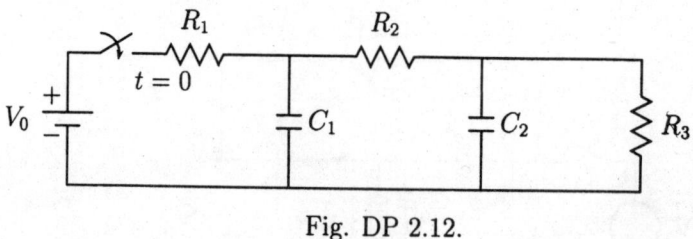

Fig. DP 2.12.

Hints: $i_{C1}(0^-) = V_0/R_1$, $i_{C2}(0^+) = 0$, and $i_{C1}(\infty) = i_{C2}(\infty) = 0$.

DP 2.13 In the circuit of Fig. DP 2.13, Find

(a) v_P at $t = 0^-$.

(b) v_P at $t = 0^+$.

(c) i_1 at $t = 0^-$.

(b) i_1 at $t = 0^+$.

if $v_s(t) = 1$ **V** for $t < 0$ and $v_s(t) = -1$ **V** for $t > 0$.

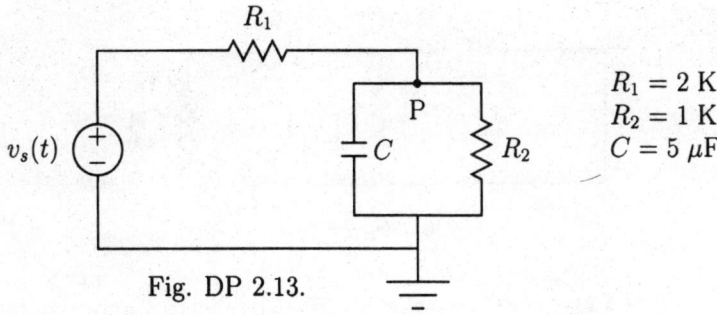

$$R_1 = 2 \text{ K}$$
$$R_2 = 1 \text{ K}$$
$$C = 5 \ \mu\text{F}$$

Fig. DP 2.13.

Hints:

$$v_P(0^-) = \frac{v_s(0^-)}{R_1 + R_2} R_2 \quad \text{and} \quad i_1(0^-) = \frac{v_s(0^-)}{R_1 + R_2}$$

DP 2.14 In the circuit of Fig. DP 2.14, find:
(a) $i_1(0^-)$ (b) $i_1(0^+)$ (c) $i_2(0^-)$ (d) $i_2(0^+)$ (e) $v_A(0^-)$ (f) $v_A(0^+)$ if
$i_s(t) = -1$ **mA** for $t < 0$ and $i_s(t) = 1$ **mA** for $t > 0$.

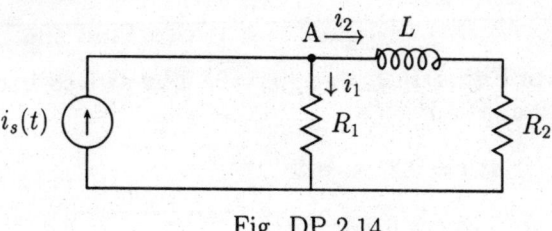

Fig. DP 2.14.

Hints:

$$i_1(0^-) = \frac{R_2}{R_1 + R_2} i_s(0^-)$$
$$i_2(0^-) = \frac{R_1}{R_1 + R_2} i_s(0^-)$$
$$i_2(0^-) = i_2(0^+)$$

DP 2.15 In the circuit of Fig. DP 2.15, the switch is closed at $t = 0$ and re-opened at $t = 1$ s. If $i_L(0) = 0$, find:

(a) v_L at $t = 0.8$ s.

(b) di_L/dt at $t = 0.8$ s.

(c) The current through 100-V source at $t = 0.8$ s.

(d) di_L/dt at $t = 1.25$ s.

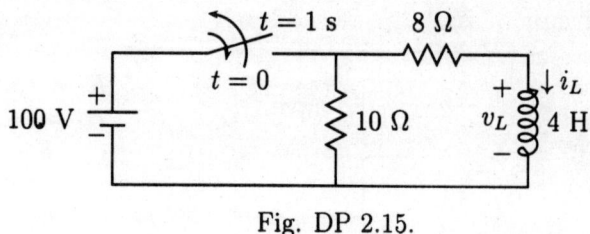

Fig. DP 2.15.

Hints: For $0 < t \leq 1$ s

$$i_L(t) = (i_L(0) - i_L(\infty))e^{-t/\tau_1} + i_L(\infty)$$

where $\tau_1 = L/R_{eq} = 4/8 = 1/2$ s and $i_L(\infty) = 100/8$ A.

For $1 < t < \infty$, we have

$$i_L(t) = (i_L(1^+) - i_L(\infty))e^{-t/\tau_2} + i_L(\infty)$$

where $\tau_2 = L/R_{eq} = 4/(8 + 10)$ s and $i_L(\infty) = 0$

To determine v_L, use formula $v_L = L(di_L/dt)$ and to determine source current at $t = 0.8$ s, find the $[\,100/10 - i_L(0.8)\,]$.

DP 2.16 In the circuit of Fig. DP 2.16, $i_L(0) = 0$. The switch is closed at $t = 0$ and re-opened at $t = 0.2$ s. Find:

(a) $i_L(t)$, $v_L(t)$, and $v_{sw}(t)$ at $t = 0.2^+$ s.

(b) $i_L(t)$, $v_L(t)$, and $di/dt(t)$ at $t = 0.3$ s.

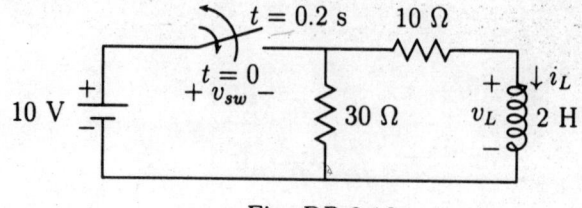

Fig. DP 2.16.

Hints: For $0 < t \leq 0.2$ s, the inductor current is

$$i_L(t) = (i_L(0) - i_L(\infty))e^{-t/\tau_1} + i_L(\infty)$$

where $\tau_1 = L/R_{eq} = 2/10 = 1/5$ s and $i_L(\infty) = 10/10 = 1$ A.
 For $0.2 < t < \infty$,

$$i_L(t) = (i_L(0.2^+) - i_L(\infty))e^{-t/\tau_2} + i_L(\infty)$$

where $\tau_2 = L/R_{eq} = 2/(30 + 10)$ s and $i_L(\infty) = 0$
 And v_{sw} may be determined from the equation $v_{sw} = 10 + 30i_L$ when switch is opened.

DP 2.17 In the circuit of Fig. DP 2.17, find:

(a) i_L at $t = 0.5$ s.

(b) t at which $v_L = 50$ V.

(c) t at which $v_R = v_L$.

There is no initial current flowing in the inductor.

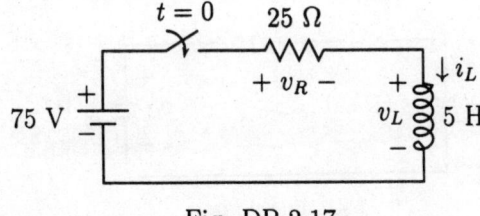

Fig. DP 2.17.

DP 2.18 Steady-state is reached with switch at position a. At $t = 0$, the switch is moved to position b. Find:

(a) $v_C(0^+)$ and $dv_C/dt(0^+)$.
(b) $i_L(0^+)$ and $di_L/dt(0^+)$

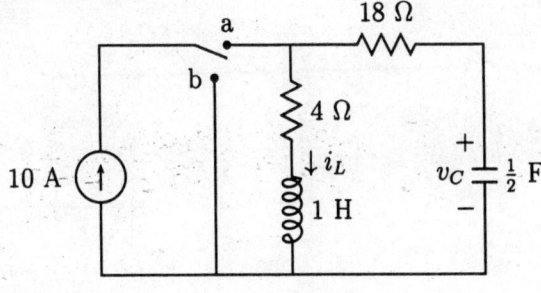

Fig. DP 2.18.

Hints: $v_C(0^-) = 4 \times v_C(0^+)$ and $i_L(0^-) = 10 = i_L(0^+)$

DP 2.19 In the circuit of Fig. DP 2.19, the switch is moved from position a to position b at $t = 0$. Find:

    (a) $v_P(0^-)$ **(b)** $v_Q(0^-)$ **(c)** $v_P - v_Q$ at $t = 0^+$ **(d)** $v_P(0^+)$ **(e)** $v_Q(0^+)$.

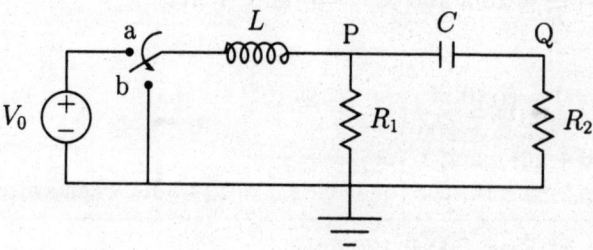

Fig. DP 2.19.

Hints: $v_P(0^-) = V_0$, $v_P - v_Q$ at $t = 0^- = V_0$, and $v_Q(0^-) = 0$.

DP 2.20 Find the $v_C(0.1)$ and $dv_C/dt(0.1)$ for the circuit in Fig. DP 2.20 if $v_C(0) = -2$ **V**.

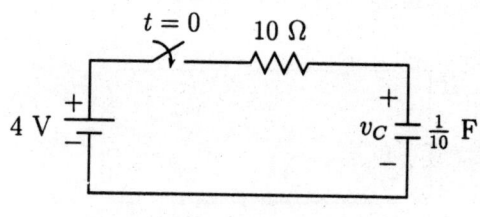

Fig. DP 2.20.

Hints: $v_C(t) = 4(1 - e^{-t/\tau})$, where $\tau = RC = 1$ s.

DP 2.21 In the circuit shown in Fig. DP 2.21, the capacitor and the inductor do not have initial stored charge. On closing the switch at $t = 0$, it is found that $i(0^+) = 15$ **mA** and that $v_{ab}(t) = 0$ for all $t \geq 0$. Evaluate R and L.

IES-92(EC)

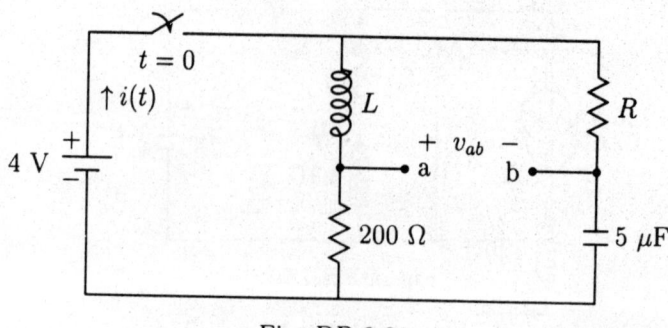

Fig. DP 2.21.

Hints: At $t = 0^+$, the inductor is replaced by open circuit and capacitor by short circuit. Thus

$$i(0^+) = \frac{4}{R} = 15 \times 10^{-3} \qquad \Rightarrow R = \frac{4}{15} \; K\Omega$$

Current in inductor branch

$$i_L(t) \doteq \frac{4}{200} \left(1 - e^{-\frac{200}{L}t} \right)$$

Current in capacitor branch

$$i_C(t) = \frac{4}{R} e^{-t/5 \times 10^{-6} R}$$

Taking KVL in the upper right loop as

$$R i_C - v_{ab} - L \frac{di_L}{dt} = 0$$

or

$$v_{ab} = R i_C - L \frac{di_L}{dt} = 0$$

or

$$R \left(\frac{4}{R} e^{-t/5 \times 10^{-6} R} \right) = L \frac{4}{200} \frac{200}{L} e^{-\frac{200}{L}t}$$

or

$$\frac{1}{5 \times 10^{-6} R} = \frac{200}{L}$$

DP 2.22 After keeping it open for a long time, the switch 'S' in the circuit shown in Fig. DP 2.11 is closed at $t = 0$. Find capacitor voltage $v_C(0^+)$ and inductor current $i_L(0^+)$.

IES-96(EC)

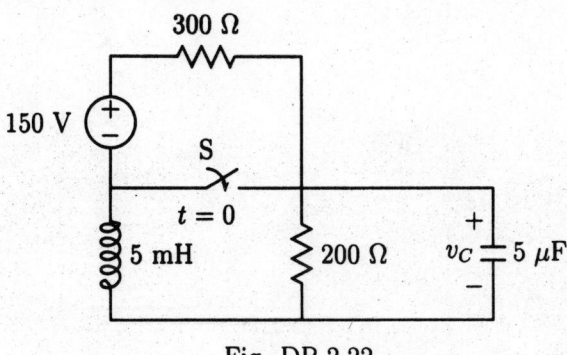

Fig. DP 2.22.

Hints: For $t < 0$ (steady-state), inductor is replaced with short circuit and capacitor with open circuit. Thus

$$i_L(0^-) = -\frac{150}{300 + 200} = -0.3 \text{ A} = i_L(0^+)$$
$$v_C(0^-) = 200 \times 0.3 = 60 \text{ V}$$

****

Second-Order Circuits

SUMMARY OF THIS CHAPTER

3.1   SECOND-ORDER DIFFERENTIAL EQUATION

The general form of second-order constant coefficient linear ordinary differential equation is

$$\frac{d^2x(t)}{dt^2} + a\frac{dx(t)}{dt} + bx(t) = f(t)$$

The complete solution to above equation for $x(t)$ is $x(t) = x_n(t) + x_{ss}(t)$

where $x_n(t)$ is the natural solution (or transient solution) and $x_{ss}(t)$ is the steady-state solution (or particular solution).

3.2   SERIES RLC CIRCUIT

By KVL in the circuit of Figure 3.1

$$Ri + L\frac{di}{dt} + \frac{1}{C}\int_{-\infty}^{t} i\,dt = 0$$

Differentiating $$L\frac{d^2i}{dt^2} + R\frac{di}{dt} + \frac{i}{C} = 0$$

Hence, we get the homogeneous differential eqution (without input function)

$$\frac{d^2i}{dt^2} + \frac{R}{L}\frac{di}{dt} + \frac{i}{LC} = 0$$

Characteristic equation

$$s^2 + \frac{R}{L}s + \frac{1}{LC} = 0$$

137

Roots are

$$s_{1,2} = -\frac{R}{2L} \pm \sqrt{\left(\frac{R}{2L}\right)^2 - \frac{1}{LC}}$$

The roots of the characteristic equation are called the eigenvalues. We define

$\alpha = \frac{R}{2L}$ = damping factor

$\omega_n = \frac{1}{\sqrt{LC}}$ = natural frequency (or undamped natural frequency)

$\omega_d = \sqrt{\omega_n^2 - \alpha^2}$ = damped natural frequency

$R_c = 2\sqrt{\frac{L}{C}}$ = critical resistance

$\zeta = \frac{R}{2}\sqrt{\frac{C}{L}}$ = damping ratio

$Q = \omega_n/2\alpha = \omega_n L/R$ = quality factor

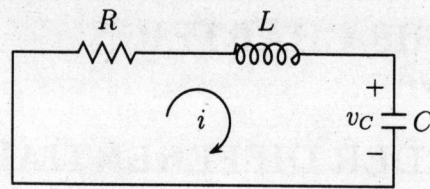

Figure 3.1: Series RLC circuit with initial conditions.

3.3   PARALLEL RLC CIRCUIT

By KCL in the circuit of Figure 3.2

$$C\frac{dv}{dt} + \frac{v}{R} + \frac{1}{L}\int_{-\infty}^{t} v\,dt = 0$$

Differentiating

$$C\frac{d^2v}{dt^2} + \frac{1}{R}\frac{dv}{dt} + \frac{v}{L} = 0$$

Hence

$$\frac{d^2v}{dt^2} + \frac{1}{RC}\frac{dv}{dt} + \frac{v}{LC} = 0$$

Characteristic equation

$$s^2 + \frac{1}{RC}s + \frac{1}{LC} = 0$$

Roots are

$$s_{1,2} = -\frac{1}{2RC} \pm \sqrt{\left(\frac{1}{2RC}\right)^2 - \frac{1}{LC}}$$

We define

$\alpha = \frac{1}{2RC}$ = damping factor

$\omega_n = \frac{1}{\sqrt{LC}}$ = natural frequency

$$\omega_d = \sqrt{\omega_n^2 - \alpha^2} = \text{damped natural frequency}$$

$$G_c = 2\sqrt{\frac{C}{L}} = \text{critical conductance}$$

$$\zeta = \frac{1}{2R}\sqrt{\frac{L}{C}} = \text{damping ratio}$$

$$Q = \omega_n/2\alpha = \omega_n RC = \text{quality factor}$$

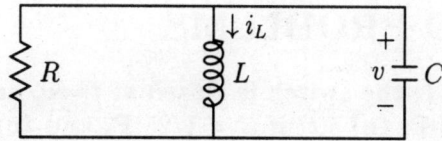

Figure 3.2: Parallel RLC circuit with initial conditions.

It is observed that the natural frequency for series RLC and parallel RLC circuits is the same, but damping factors are different.

$$\alpha = \frac{1}{2RC} \quad \text{(parallel RLC)}$$

$$\alpha = \frac{R}{2L} \quad \text{(series RLC)}$$

Case I: If $\alpha > \omega_n$, the circuit is overdamped and the roots of the characteristic equation are real and distinct. The natural response has the form:

$$f_n(t) = A_1 e^{s_1 t} + A_2 e^{s_2 t} \tag{3.1}$$

where

$$s_{1,2} = -\alpha \pm \sqrt{\alpha^2 - \omega_n^2}$$

Case II: If $\alpha = \omega_n$, the circuit is critically damped and the roots of the characteristic equation are real and equal (repeated roots). The natural response has the form:

$$f_n(t) = (A_1 + A_2 t)e^{-\alpha t} \tag{3.2}$$

where

$$s_{1,2} = -\alpha, -\alpha$$

Case III: If $\alpha < \omega_n$, then the circuit is underdamped and the roots of the characteristic equation are complex conjugate as:

$$s_{1,2} = -\alpha \pm j\sqrt{\omega_n^2 - \alpha^2} = -\alpha \pm j\omega_d$$

where $\omega_d = \sqrt{\omega_n^2 - \alpha^2}$. The natural response has the form:

$$f_n(t) = A_1 e^{(-\alpha + j\omega_d)t} + A_2 e^{(-\alpha - j\omega_d)t}$$

Using the Euler's identity $e^{\pm j\theta} = \cos\theta \pm j\sin\theta$, finally we can achieve

$$f_n(t) = e^{-\alpha t}(A_1 \cos\omega_d t + A_2 \sin\omega_d t) \qquad (3.3)$$

Method of determination of steady-state solution (or, particular solution) to second order differential equation is the same as stated in chapter-1 for the first order differential equation.

3.4   SOLVED PROBLEMS

SP 3.1 In Fig. SP 3.1, the switch is closed at $t = 0$, find:
(a) $i(t)$ if $C = 1/9$ F; (b) $i(t)$ if $C = 1/25$ F; and (c) $i(t)$ if $C = 1/50$ F. The initial voltage across the capacitor, $v_C(0) = 10$ V and the initial current through the inductor is zero.

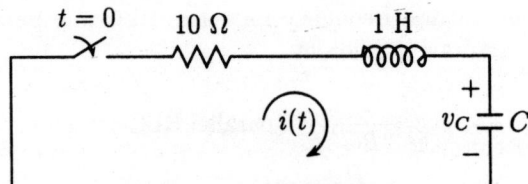

Fig. SP 3.1.

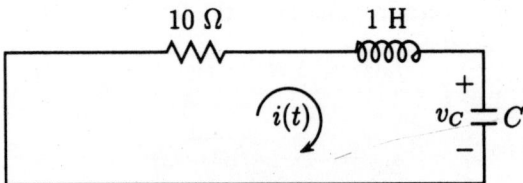

Fig. 3.1.1: $t > 0$ circuit.

SOLUTION:

(a) Given, R = 10 Ω, L= 1 H, and C = 1/9 F. We thus find

$$\alpha = \frac{R}{2L} = 5$$

$$\omega_n = \frac{1}{\sqrt{LC}} = 3$$

and $\qquad \sqrt{\alpha^2 - \omega_n^2} = 4$

Roots are $\qquad s_{1,2} = -\alpha \pm \sqrt{\alpha^2 - \omega_n^2} = -1, -9.$

Since $\alpha > \omega_n$, the response is overdamped and is given by

$$i(t) = Ae^{s_1 t} + Be^{s_2 t} = Ae^{-t} + Be^{-9t} \qquad [3.1.1]$$

It's derivative is

$$\frac{di}{dt} = -Ae^{-t} - 9Be^{-9t} \qquad [3.1.2]$$

The initial condition $i(0^+)$ and $\frac{di}{dt}(0^+)$ are determined as:

[in series RLC] $\qquad\qquad i(0^+) = 0$

By KVL in the circuit of Fig. 3.1.1

$$10i + \frac{di}{dt} + v_C = 0$$

or $\qquad 10i(0^+) + \frac{di}{dt}(0^+) + v_C(0^+) = 0$

Thus $\qquad \frac{di}{dt}(0^+) = -v_C(0^+) = -10 \text{ A/sec}$

From eqn [3.1.1]

$$i(0^+) = 0 = A + B \qquad [3.1.3]$$

From eqn [3.1.2]

$$\frac{di}{dt}(0^+) = -10 = -A - 9B \qquad [3.1.4]$$

Solving eqn [3.1.3] and eqn [3.1.4] gives

$$A = -10/8 \qquad \text{and} \qquad B = 10/8$$

Hence

$$i(t) = \frac{10}{8}\left(e^{-9t} - e^{-t}\right) \qquad \text{for } t > 0 \qquad [3.1.5]$$

(b) Given, R = 10 Ω, L= 1 H, and C = 1/25 F. We find

$$\alpha = \frac{R}{2L} = 5$$

and $\qquad \omega_n = \frac{1}{\sqrt{LC}} = 5$

Roots are : $\qquad s_{1,2} = -\alpha \pm \sqrt{\alpha^2 - \omega_n^2} = -5, -5$

Since $\alpha = \omega_n$, the response is critically damped and is given by

$$i(t) = (A_1 + A_2 t)e^{-5t} \qquad [3.1.6]$$

And it's derivative is

$$\frac{di}{dt} = A_2 e^{-5t} - 5(A_1 + A_2 t)e^{-5t} \qquad [3.1.7]$$

From eqn [3.1.6] $\qquad i(0^+) = 0 = A_1$

From eqn [3.1.7] $\qquad \frac{di}{dt}(0^+) = -10 = A_2$

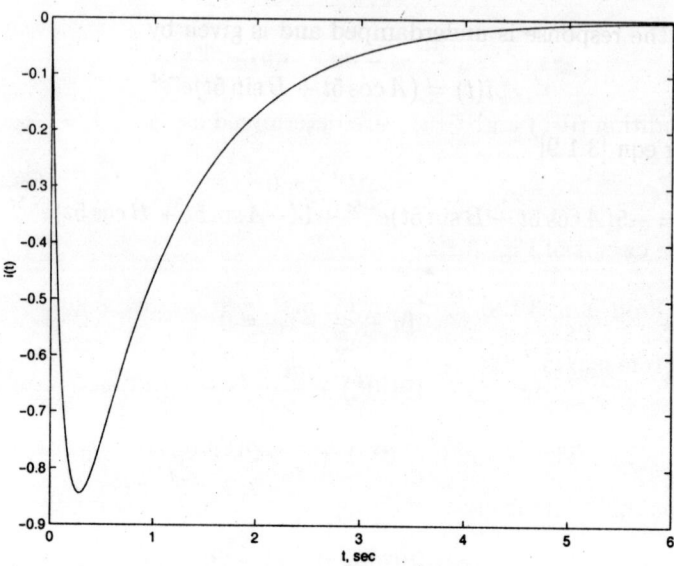

Fig. 3.1.2: $i(t)$ plot for eqn [3.1.5].

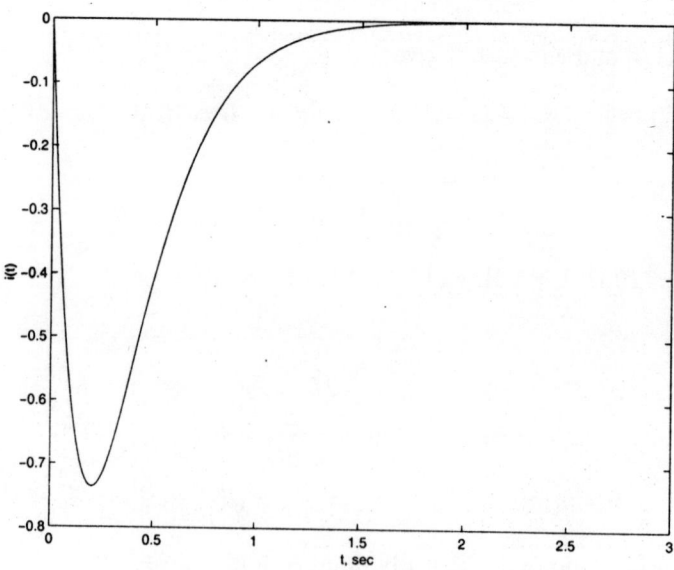

Fig. 3.1.3: $i(t)$ plot for eqn [3.1.8].

Therefore

$$i(t) = -10te^{-5t} \qquad \text{for } t > 0 \tag{3.1.8}$$

(c) Given, R = 10 Ω, L = 1 H, and C = 1/50 F. We thus find

$$\alpha = \frac{R}{2L} = 5$$

$$\omega_n = \frac{1}{\sqrt{LC}} = \sqrt{50} \qquad \text{and} \qquad \omega_d = \sqrt{\omega_n^2 - \alpha^2} = 5$$

Since $\alpha < \omega_n$, the response is underdamped and is given by

$$i(t) = (A\cos 5t + B\sin 5t)e^{-5t} \qquad\qquad [3.1.9]$$

Differentiating eqn [3.1.9]

$$\frac{di}{dt} = -5(A\cos 5t + B\sin 5t)e^{-5t} + 5(-A\sin 5t + B\cos 5t)e^{-5t} \qquad [3.1.10]$$

From eqn [3.1.9] $\qquad i(0^+) = 0 = A$

From eqn [3.1.10] $\qquad \dfrac{di}{dt}(0^+) = -10 = 5B \qquad \Rightarrow\ B = -2$

Therefore

$$i(t) = -2e^{-5t}\sin 5t \qquad t > 0 \qquad\qquad [3.1.11]$$

The plot of this eqn 3.1.11 is given in Figure 3.1.4. Such a plot is called damped sinusoidal plot.

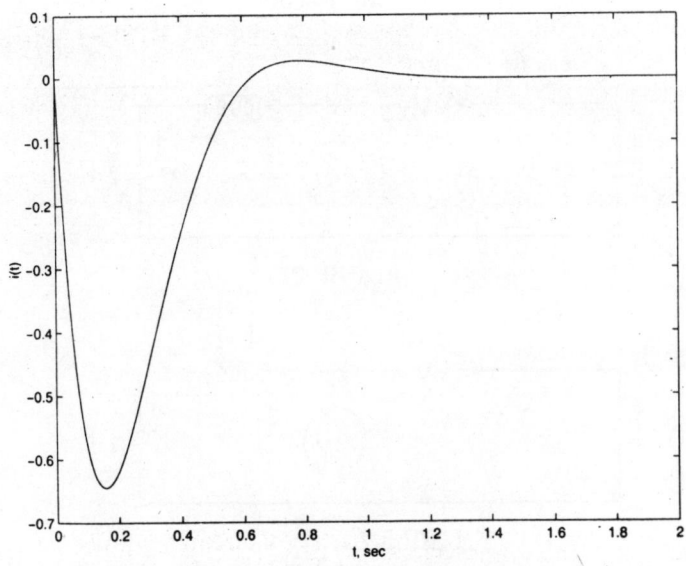

Fig. 3.1.4: $i(t)$ plot for eqn [3.1.11].

SP 3.2 In the circuit of Fig. SP 3.2, the switch is closed at $t = 0$. Find:
   (a) $i(t)$ for $t > 0$,
   (b) the time and amplitude of the first maximum,
   (c) the first zero crossover and second zero crossover,
   (d) the first negative extremum.
The initial voltage across the capacitor, $v_C(0)$ is 1 V.

SOLUTION:

(a) Given   R = 2 Ω, L = 1 H, and C = 1/4 F. We find

$$\alpha = \frac{R}{2L} = 1$$

$$\omega_n = \frac{1}{\sqrt{LC}} = 2$$

and $$\omega_d = \sqrt{\omega_n^2 - \alpha^2} = \sqrt{3}$$

Since $\alpha < \omega_n$, the uderdamped response is given by

$$i(t) = (A \cos \sqrt{3}t + B \sin \sqrt{3}t)e^{-t} \qquad [3.2.1]$$

and

$$\frac{di}{dt} = -(A \cos \sqrt{3}t + B \sin \sqrt{3}t)e^{-t} + \sqrt{3}(-A \sin \sqrt{3}t + B \cos \sqrt{3}t)e^{-t} \qquad [3.2.2]$$

Initial conditions are determined as follows:

$$i(0^+) = 0$$

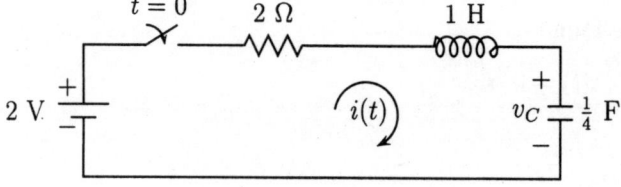

Fig. SP 3.2.

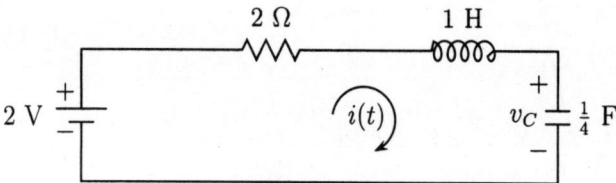

Fig. SP 3.2.1: $t > 0$ circuit.

By KVL in Fig. 3.2.1

$$2i + \frac{di}{dt} + v_C = 2$$

or

$$2i(0^+) + \frac{di}{dt}(0^+) + v_C(0^+) = 2$$

Therefore

$$\frac{di}{dt}(0^+) = 2 - 1 = 1 \text{ A/sec.}$$

From eqn [3.2.1] $i(0^+) = 0 = A$

From eqn [3.2.2] $\dfrac{di}{dt}(0^+) = 1 = \sqrt{3}B$

Hence $A = 0$ and $B = 0.58$

Therefore

$$i(t) = 0.58e^{-t} \sin \sqrt{3}t \quad \text{for } t > 0 \qquad [3.2.3]$$

(b) The maximum value is found by setting $\dfrac{di}{dt} = 0$, i.e.,

$$\frac{di}{dt} = 0.58[-e^{-t} \sin \sqrt{3}t + \sqrt{3}e^{-t} \cos \sqrt{3}t] = 0$$

or $\sin \sqrt{3}t = \sqrt{3} \cos \sqrt{3}t$

or $\tan \sqrt{3}t = \sqrt{3}$

or $\sqrt{3}t = \pi/3$

Hence $t = \dfrac{\pi/3}{\sqrt{3}} = 0.605$ sec.

which is $t_p = 0.605$ sec. = time for first maximum. Now setting t = t_p = 0.605 sec. into eqn [3.2.3], we have

$$i(t_p) = I_m = 0.58e^{-0.605} \sin 1.732 \times 0.605 = 0.274 \text{ A}$$

(c) Zero crossings are found as

$$i(t) = 0 = 0.58e^{-t} \sin \sqrt{3}t \qquad \Rightarrow \quad \sqrt{3}t = n\pi$$

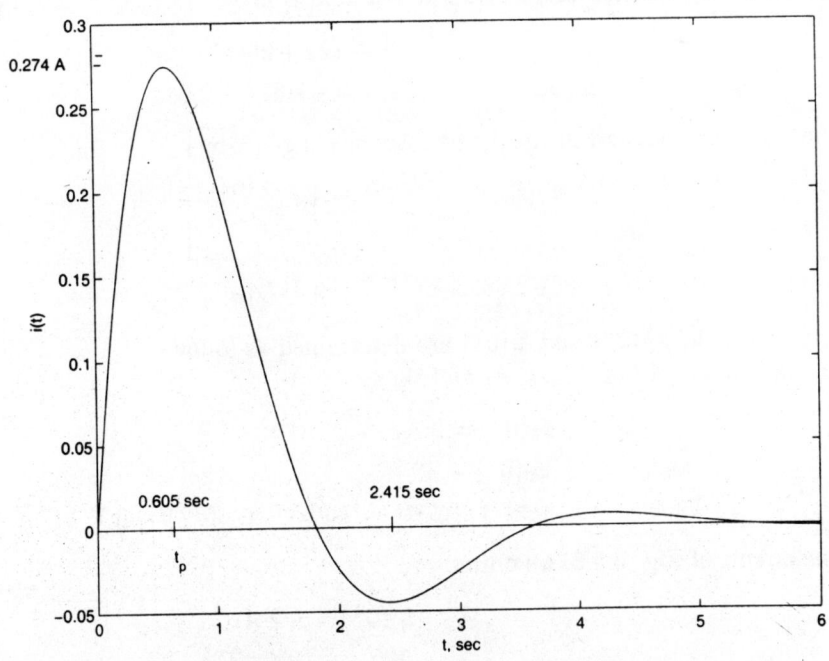

Fig. 3·2.2: $i(t)$ plot of eqn 3.2.3

When n = 0; t = $0 \times \pi/\sqrt{3} = 0$ = time for first zero crossing.
When n = 1; t = $\pi/\sqrt{3} = 1.81$ sec. = time for 2nd zero crossing and so on.

(d) The time for first negative extremum

$$t = 1.81 + 0.605 = 2.415 \text{ sec.}$$

Hence, the first negative maximum is $i(2.415) = -0.045$ A

SP 3.3 Find $v_C(t)$ for the circuit shown in Fig. SP 3.3, if the switch is moved from a to b at $t = 0$.

SOLUTION:

By KVL in the circuit of Fig. 3.3.1

$$2\frac{di_C}{dt} + 24i_C + v_C = 0$$

or

$$2\frac{d}{dt}\left(\frac{1}{50}\frac{dv_C}{dt}\right) + 24\left(\frac{1}{50}\frac{dv_C}{dt}\right) + v_C = 0$$

or

$$\frac{d^2v_C}{dt^2} + 12\frac{dv_C}{dt} + 25v_C = 0 \qquad\qquad [3.3.1]$$

The characteristic equation corresponds to eqn [3.3.1] is

$$s^2 + 12s + 25 = 0$$

Roots are : $s_{1,2} = -9.317, -2.683$

since the roots are real and unequal, the response is given by

$$v_C(t) = A_1 e^{-2.683t} + A_2 e^{-9.317t} \qquad\qquad [3.3.2]$$

and

$$\frac{dv_C}{dt} = -2.683 A_1 e^{-2.683t} - 9.317 A_2 e^{-9.317t} \qquad\qquad [3.3.3]$$

The initial conditions $i(0^+)$ and $\frac{di}{dt}(0^+)$ are determined as follows:
From the circuit of Fig. 3.3.2, we find that

$$i_L(0^-) = 2 \text{ A}$$

and $v_C(0^-) = 48$ V

Thus $i_L(0^+) = 2$ A   and $v_C(0^+) = 48$ V

From the circuit of Fig. 3.3.3, we obtain

$$i_C(0^+) = -2 \text{ A}$$

Hence for the capacitor $\dfrac{dv_C}{dt}(0^+) = \dfrac{i_C(0^+)}{C} = -100$ V/sec.

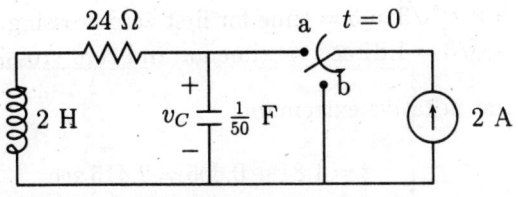

Fig. SP 3.3.

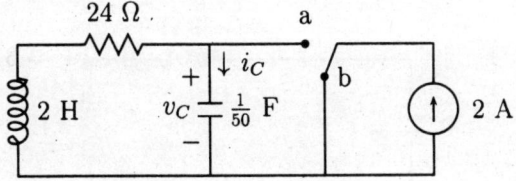

Fig. 3.3.1: $t > 0$ circuit.

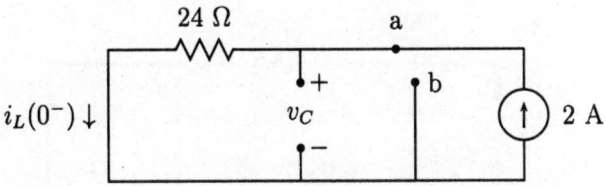

Fig. 3.3.2: $t = 0^-$ circuit.

From eqn [3.3.2] $v_C(0^+) = 48 = A_1 + A_2$ [3.3.4]

From eqn [3.3.3] $\dfrac{dv_C}{dt}(0^+) = -100 = -2.683A_1 - 9.317A_2$ [3.3.5]

Solving eqn [3.3.4] and eqn [3.3.5], we have

$$A_1 = 52.339 \quad \text{and} \quad A_2 = -4.339$$

Hence $v_C(t) = 52.339e^{-2.683t} - 4.339e^{-9.317t} \qquad \text{for } t > 0$

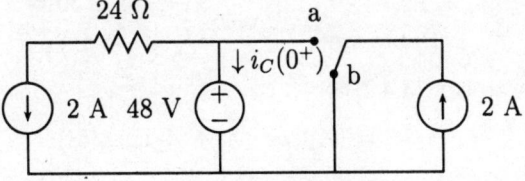

Fig. 3.3.3: $t = 0^+$ circuit.

SP 3.4 For the circuit of Fig. SP 3.4, (a) Find $i_L(t)$ for all t . (b) Find $v_C(t)$ for $t > 0$.

SOLUTION:

KVL in the circuit of Fig. 3.4.1 gives

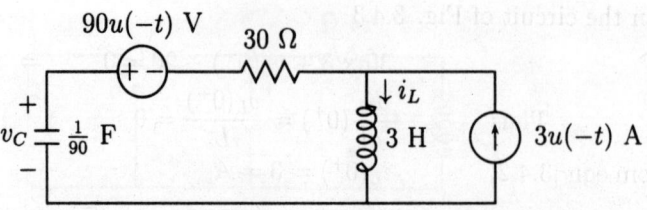

Fig. SP 3.4.

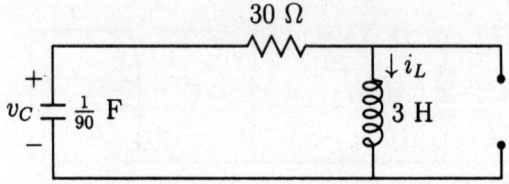

Fig. 3.4.1: $t > 0$ circuit.

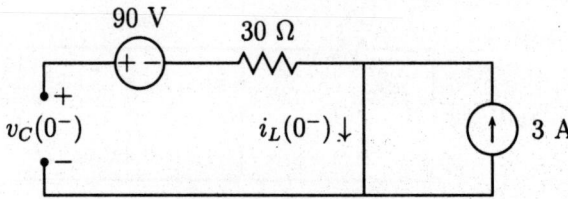

Fig. 3.4.2: $t = 0^-$ circuit.

$$30 i_L + 3\frac{di_L}{dt} + 90 \int_{-\infty}^{t} i_L dt = 0$$

Differentiating and rearranging

$$\frac{d^2 i_L}{dt^2} + 10\frac{di_L}{dt} + 30 i_L = 0 \qquad \text{[3.4.1]}$$

The characteristic equation corresponds to eqn [3.4.1] is

$$s^2 + 10s + 30 = 0$$

Roots are : $s_{1,2} = -5 \pm j\sqrt{5}$

Hence the response to eqn [3.4.1] is

$$i_L(t) = (A_1 \cos \sqrt{5}t + A_2 \sin \sqrt{5}t)e^{-5t} \qquad \text{[3.4.2]}$$

And it's derivative is

$$\frac{di_L}{dt} = -5(A_1 \cos \sqrt{5}t + A_2 \sin \sqrt{5}t)e^{-5t} + \sqrt{5}(-A_1 \sin \sqrt{5}t + A_2 \cos \sqrt{5}t)e^{-5t} \qquad \text{[3.4.3]}$$

Determination of initial conditions:

It is noted from the circuit of Fig. 3.4.2 that

$$i_L(0^-) = 3\ A = i_L(0^+)$$
$$v_C(0^-) = 90\ V = v_C(0^+)$$

By KVL in the circuit of Fig. 3.4.3

$$30 \times 3 + v_L(0^+) - 90 = 0 \qquad \Rightarrow v_L(0^+) = 0$$

Thus
$$\frac{di_L}{dt}(0^+) = \frac{v_L(0^+)}{L} = 0$$

From eqn [3.4.2]
$$i_L(0^+) = 3 = A_1$$

From eqn [3.4.3]
$$\frac{di_L}{dt}(0^+) = 0 = -5A_1 + \sqrt{5}A_2 \qquad \Rightarrow A_2 = 3\sqrt{5}$$

Therefore
$$i_L(t) = (3\cos\sqrt{5}t + 3\sqrt{5}\sin\sqrt{5}t)e^{-5t} \qquad \text{for } t > 0$$

Since $i_L(t) = 3$ A for $t < 0$, the inductor current for all t can given by

$$i_L(t) = 3u(-t) + (3\cos\sqrt{5}t + 3\sqrt{5}\sin\sqrt{5}t)e^{-5t}u(t) \qquad \text{for all } t$$

(b) The capacitor voltage, $v_C(t)$ for $t > 0$ with reference to Fig. 3.4.1, may be given by

$$
\begin{aligned}
v_C(t) &= 30i_L + 3\frac{di_L}{dt} \\
&= 30\left[(3\cos\sqrt{5}t + 3\sqrt{5}\sin\sqrt{5}t)e^{-5t}\right] \\
&\quad -15(3\cos\sqrt{5}t + 3\sqrt{5}\sin\sqrt{5}t)e^{-5t} \\
&\quad +3\sqrt{5}(-3\sin\sqrt{5}t + 3\sqrt{5}\cos\sqrt{5}t)e^{-5t} \\
&= 9e^{-5t}\left[10\cos\sqrt{5}t + 4\sqrt{5}\sin\sqrt{5}t\right] \qquad \text{for } t > 0
\end{aligned}
$$

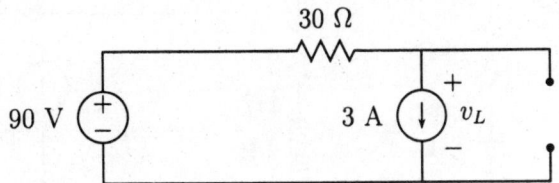

Fig. 3.4.3: $t = 0^+$ circuit.

SP 3.5 Find $i_L(t)$ for $t > 0$ in the circuit of Fig. SP 3.5. If the switch is moved from a to b at $t = 0$.

SOLUTION:

By KVL in the left mesh of Fig. 3.5.1

$$6i_L + 2\frac{di_L}{dt} - \frac{1}{1/14}\int_{-\infty}^{t}(2 - i_L)dt - 10(2 - i_L) = 0$$

or

$$2\frac{di_L}{dt} + 16i_L - 14\int_{-\infty}^{t}(2 - i_L)dt = 20$$

Differentiating and rearranging

$$\frac{d^2 i_L}{dt^2} + 8\frac{di_L}{dt} + 7i_L = 14 \qquad\qquad [3.5.1]$$

The characteristic equation of eqn [3.5.1] is

$$s^2 + 8s + 7 = 0$$

Roots are : $s_{1,2} = -1, -7$

Form of complimentary solution

$$i_{Lc} = k_1 e^{-t} + k_2 e^{-7t} \qquad\qquad [3.5.2]$$

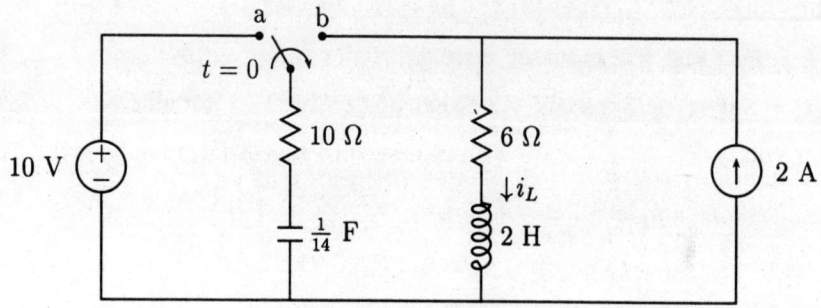

Fig. SP 3.5.

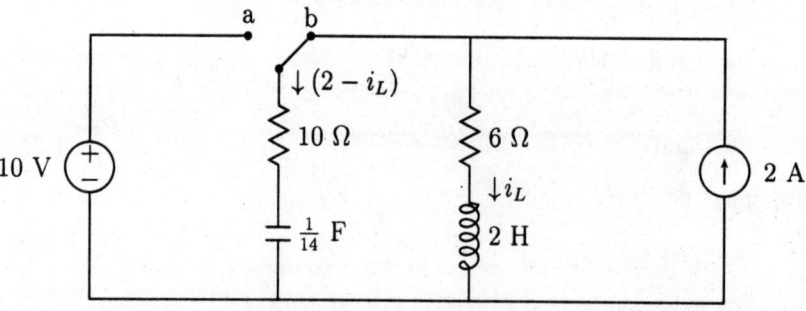

Fig. 3.5.1: $t > 0$ circuit.

The particular solution corresponding to constant excitation is

$$i_{Lp} = K \qquad\qquad [3.5.3]$$

Substituting $i_L = i_{Lp} = K$ into eqn [3.5.1] yields

$$0 + 0 + 7K = 14 \qquad \Rightarrow K = 2 = i_{Lp}$$

Complete solution

$$i_L(t) = k_1 e^{-t} + k_2 e^{-7t} + 2 \qquad\qquad [3.5.4]$$

and

$$\frac{di_L}{dt} = -k_1 e^{-t} - 7k_2 e^{-7t} \qquad\qquad [3.5.5]$$

The initial conditions are determined as follows: It is obvious from the circuit of Fig. 3.5.2 that

$$i_L(0^-) = 2\ A = i_L(0^+)$$

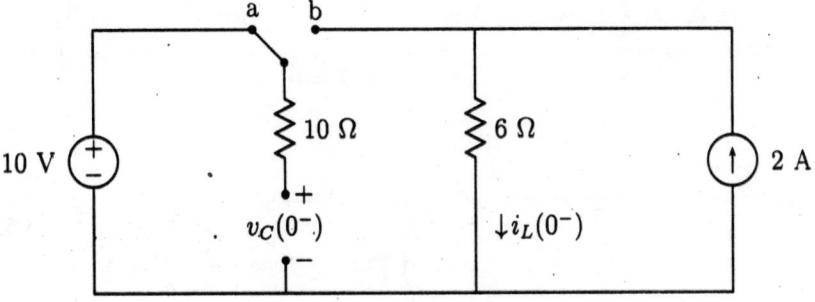

Fig. 3.5.2: $t = 0^-$ circuit.

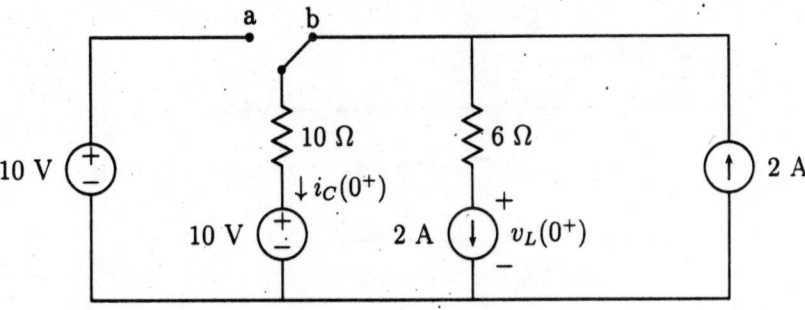

Fig. 3.5.3: $= 0^+$ circuit.

and $\qquad v_C(0^-) = 10\ V = v_C(0^+)$

By KCL at the upper node of Fig. 3.5.3

$$i_C(0^+) + 2 - 2 = 0 \qquad \Rightarrow i_C(0^+) = 0$$

Now by KVL in the left mesh of Fig. 3.5.3

$$6 \times 2 + v_L(0^+) - 10 = 0 \qquad \Rightarrow v_L(0^+) = -2\ V$$

Therefore, for inductor:

$$\frac{di_L}{dt}(0^+) = \frac{v_L(0^+)}{L} = -2/2 = -1\ \text{A/sec.}$$

From eqn [3.5.4]

$$i_L(0^+) = 2 = k_1 + k_2 + 2 \qquad \Rightarrow k_1 + k_2 = 0 \qquad\qquad [3.5.6]$$

From eqn [3.5.5]

$$\frac{di_L}{dt}(0^+) = -1 = -k_1 - 7k_2 \qquad \Rightarrow k_1 + 7k_2 = 1 \qquad\qquad [3.5.7]$$

Solving last two equations, we get

$$k_1 = -1/6 \qquad \text{and} \quad k_2 = 1/6$$

Therefore $\qquad i_L(t) = \dfrac{1}{6}(e^{-7t} - e^{-t}) + 2 \quad$ for $t > 0$

SP 3.6 Consider the circuit shown in Fig. SP 3.6. **(a)** Solve for $i(t)$ for $t > 0$, if both switches are thrown at $t = 0$. **(b)** Find the negative extremum value of the $i(t)$, if any.

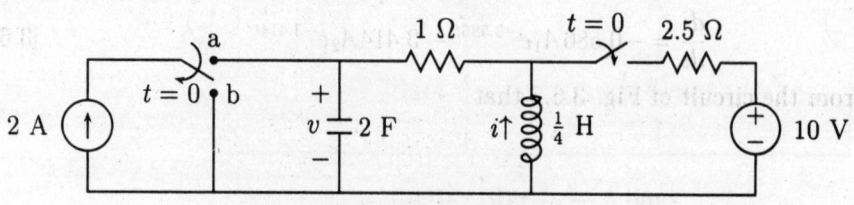

Fig. SP 3.6.

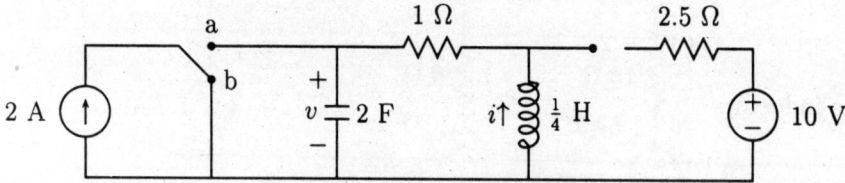

Fig. 3.6.1: $t > 0$ circuit.

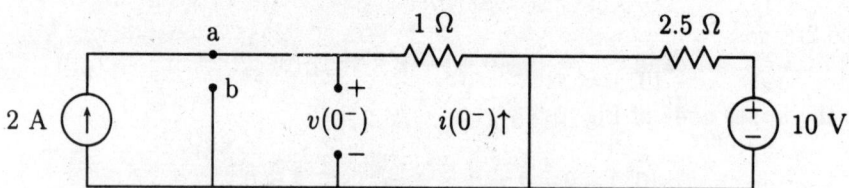

Fig. 3.6.2: $t = 0^-$ circuit.

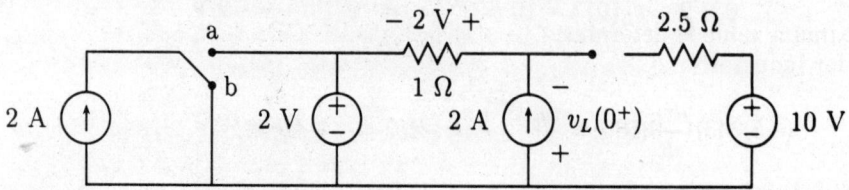

Fig. 3.6.3: $t = 0^+$ circuit.

SOLUTION:

(a) For time $t > 0$, the circuit is series RLC shown in Fig. 3.6.1, from which $R = 1\ \Omega$, $L = 1/4$ H, and $C = 2$ F. Thus we find

$$\alpha = \frac{R}{2L} = 2$$

$$\omega_n = \frac{1}{\sqrt{LC}} = \sqrt{2}$$

Roots are : $s_{1,2} = -0.586, -3.414$

Since, $\alpha > \omega_n$, the circuit is overdamped and response is given by

$$i(t) = A_1 e^{-0.586t} + A_2 e^{-3.414t} \tag{3.6.1}$$

and

$$\frac{di}{dt} = -0.586 A_1 e^{-0.586t} - 3.414 A_2 e^{-3.414t} \tag{3.6.2}$$

It is noted from the circuit of Fig. 3.6.2 that

$$i(0^-) = i(0^+) = 2 \text{ A}$$

and $\qquad v(0^-) = v(0^+) = 2 \text{ V}$

By KVL in the circuit of Fig. 3.6.3

$$v_L(0^+) + 2 + 2 = 0 \qquad \Rightarrow v_L(0^+) = -4 \text{ V}$$

Hence for the inductor:

$$\frac{di}{dt}(0^+) = \frac{v_L(0^+)}{L} = \frac{-4}{1/4} = -16 \text{ A/sec.}$$

From eqn [3.6.1]

$$i(0^+) = 2 = A_1 + A_2 \tag{3.6.3}$$

From eqn [3.6.2]

$$\frac{di}{dt}(0^+) = -16 = -0.586 A_1 - 3.414 A_2 \tag{3.6.4}$$

where $\qquad A_1 = -3.243 \qquad$ and $\qquad A_2 = -5.243$

So that

$$i(t) = -3.243 e^{-0.586t} + 5.243 e^{-3.414t} \qquad \text{for } t > 0$$

(b) The maximum value is determined by setting $di/dt = 0$, i.e.,

$$(-3.243)(-0.586)e^{-0.586t_m} + (5.243)(-3.414)e^{-3.414t_m}) = 0$$

or $\qquad 1.9 e^{-0.586t_m} - 17.9 e^{-3.414t_m} = 0$

or $\qquad e^{2.828t_m} = 17.9/1.9 = 9.421$

or $\qquad 2.828 t_m = \ln(9.421) = 2.243$

Hence $\qquad t_m = 0.793$ sec. $\qquad$ [where t_m is the time for the max. value of current]

Therefore, the negative extremum value of current will be

$$i_m = i(0.793) = -3.243 e^{(-0.586)(0.793)} + 5.243 e^{(-3.414)(0.793)} = -1.687 \text{ A}$$

SP 3.7 Consider parallel RLC circuit shown in Fig. SP 3.7. Find $v_C(t)$ for $t > 0$ for (a) $R = 5\ \Omega$ (b) $R = 6.71\ \Omega$ (c) $R = 7.5\ \Omega$. Initial conditions are $v_C(0) = 0$ and $i_L(0) = 5$ A.

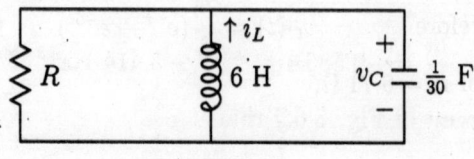

Fig. SP 3.7.

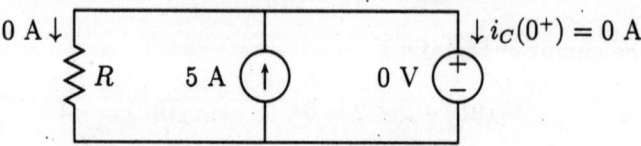

Fig. SP 3.7.1: $t = 0^+$ circuit.

SOLUTION:

(a) Given $R = 5\ \Omega$, $L = 6$ H, and $C = 1/30$ F. Thus we obtain

$$\alpha = \frac{1}{2RC} = 3$$

and $$\omega_n = \frac{1}{\sqrt{LC}} = \sqrt{5}$$

Roots are : $$s_{1,2} = -\alpha \pm \sqrt{\alpha^2 - \omega_n^2} = -1, -5$$

Since $\alpha > \omega_n$, the response is overdamped and is given by

$$v_C(t) = k_1 e^{-t} + k_2 e^{-5t} \qquad [3.7.1]$$

and

$$\frac{dv_C}{dt}(t) = -k_1 e^{-t} - 5k_2 e^{-5t} \qquad [3.7.2]$$

From the circuit of Fig. 3.7.1, we have

$$v_C(0^+) = 0 \qquad \text{Given}$$

$$i_C(0^+) = 5 \text{ A}$$

Hence $$\frac{dv_C}{dt}(0^+) = \frac{i_C(0^+)}{C} = 150 \text{ V/s}$$

From eqn [3.7.1]

$$v_C(0^+) = 0 = k_1 + k_2 \qquad [3.7.3]$$

From eqn [3.7.2]

$$\frac{dv_C}{dt}(0^+) = 150 = -k_1 - 5k_2 \qquad [3.7.4]$$

Solving last two equations, we get

$$k_1 = -k_2 = 75/2$$

Therefore $\qquad v_C(t) = \dfrac{75}{2}(e^{-t} - e^{-5t}) \qquad$ for $t > 0$

(b) For the new value of $R = 6.71\ \Omega$,

$$\alpha = \frac{1}{2RC} = 2.236$$

and $\qquad \omega_n = \dfrac{1}{\sqrt{LC}} = 2.236$

Roots are : $\qquad s_{1,2} = -2.236, -2.236$

Since $\quad \alpha = \omega_n$, the response is critically damped and is given by

$$v_C(t) = (A_1 + A_2 t)e^{-2.236t} \qquad\qquad [3.7.5]$$

and

$$\frac{dv_C}{dt} = -2.236(A_1 + A_2 t)e^{-2.236t} + A_2 e^{-2.236t} \qquad\qquad [3.7.6]$$

From eqn [3.6.5] : $\qquad v_C(0) = A_1 = 0 \qquad \Rightarrow A_1 = 0$

From eqn [3.6.6] : $\qquad \dfrac{dv_C}{dt}(0^+) = A_2 = 150 \qquad \Rightarrow A_2 = 150$

Therefore $\qquad v_C(t) = 150t e^{-2.236t} \qquad$ for $t > 0$

(c) For the new value of $R = 7.5\ \Omega$, we find

$$\alpha = \frac{1}{2RC} = 2$$

$$\omega_n = \frac{1}{\sqrt{LC}} = \sqrt{5}$$

and $\qquad \omega_d = \sqrt{\omega_n^2 - \alpha^2} = 1$

Since $\alpha < \omega_n$, the response is underdamped and is given by

$$v_C(t) = (A\cos t + B\sin t)e^{-2t} \qquad\qquad [3.7.7]$$

and

$$\frac{dv_C}{dt} = -2(A\cos t + B\sin t)e^{-2t} + (-A\sin t + B\cos t)e^{-2t} \qquad\qquad [3.7.8]$$

From eqn [3.7.7] $\qquad v_C(0) = A + 0 = 0 \qquad \Rightarrow A = 0$

From eqn [3.7.8] $\qquad \dfrac{dv_C}{dt}(0^+) = 150 = B$

Hence $\qquad v_C(t) = 150 e^{-2t}\sin t \qquad$ for $t > 0$

SP 3.8 For the RLC circuit of Fig. SP 3.8: find (a) $v(t)$; (b) $i(t)$ for $t > 0$.

SOLUTION:

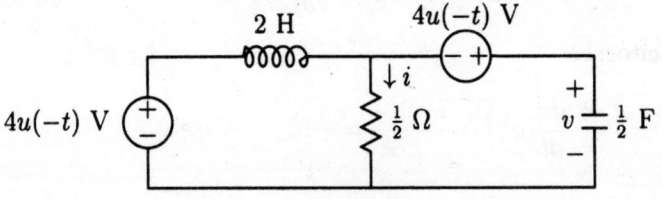

Fig. SP 3.8.

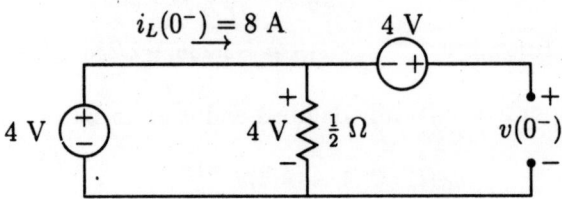

Fig. 3.8.1: $t = 0^-$ circuit.

(a) At $t > 0$, the circuit is parallel with $R = 1/2\ \Omega$, $L = 2$ H, and $C = 1/2$ F. Thus we obtain

$$\alpha = \frac{1}{2RC} = 2$$

and
$$\omega_n = \frac{1}{\sqrt{LC}} = 1$$

Roots are :
$$s_{1,2} = -\alpha \pm \sqrt{\alpha^2 - \omega_n^2} = -0.268, -3.732$$

Since $\alpha > \omega_n$, the response is overdamped and is given by

$$v(t) = k_1 e^{-0.268t} + k_2 e^{-3.732t} \qquad [3.8.1]$$

Its derivative is
$$\frac{dv}{dt} = -0.268 k_1 e^{-0.268t} - 3.732 k_2 e^{-3.732t} \qquad [3.8.2]$$

Determination of initial conditions:

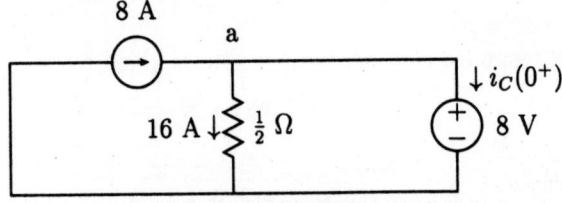

Fig. 3.8.2: $t = 0^+$ circuit.

It is observed from the circuit of Fig. 3.8.1, that

$$i_L(0^-) = \frac{4}{1/2} = 8\ \text{A} = i_L(0^+)$$

and
$$v_C(0^-) = 4 + 4 = 8\ \text{V} = v_C(0^+)$$

By KCL at the node a in the circuit of Fig. 3.8.2,

$$8 - 16 - i_C(0^+) = 0 \qquad \Rightarrow i_C(0^+) = -8 \text{ A}$$

Hence for the capacitor:

$$\frac{dv}{dt}(0^+) = \frac{i_C(0^+)}{C} = -16 \qquad \text{V/sec.}$$

From eqn [3.8.1]:

$$v(0^+) = k_1 + k_2 = 8 \qquad\qquad\qquad [3.8.3]$$

From eqn [3.8.2]:

$$\frac{dv}{dt}(0^+) = -16 = -0.268k_1 - 3.732k_2 \qquad\qquad [3.8.4]$$

Solving last two equations, we obtain

$$k_1 = 4 \qquad \text{and} \qquad k_2 = 4$$

Therefore $\qquad v(t) = 4\left(e^{-0.268t} + e^{-3.732t}\right) \qquad$ for $t > 0$

(b) We thus find

$$i(t) = \frac{v(t)}{1/2} = 8\left(e^{-0.268t} + e^{-3.732t}\right) \qquad \text{for } t > 0$$

SP 3.9 After being opened for a long time, the switch in the circuit of Fig. SP 3.9 closes at $t = 0$.
 (a) Find $v_C(t)$ for $t > 0$.
 (b) Find the minimum and maximum value of $v_C(t)$.

 SOLUTION:
 (a) At $t > 0$, the circuit is parallel RLC with $R = 1/12 \ \Omega$, $L = 1/50$ H, and $C = 2$ F. We thus find

$$\alpha = \frac{1}{2RC} = 3$$

$$\omega_n = \frac{1}{\sqrt{LC}} = 5$$

and $\qquad \omega_d = \sqrt{\omega_n^2 - \alpha^2} = 4$

The response is underdamped and is given by

$$v_C(t) = e^{-3t}(A_1 \cos 4t + A_2 \sin 4t) \qquad\qquad [3.9.1]$$

and

$$\frac{dv_C}{dt}(t) = -3e^{-3t}(A_1 \cos 4t + A_2 \sin 4t) + 4e^{-3t}(-A_1 \sin 4t + A_2 \cos 4t) \qquad [3.9.2]$$

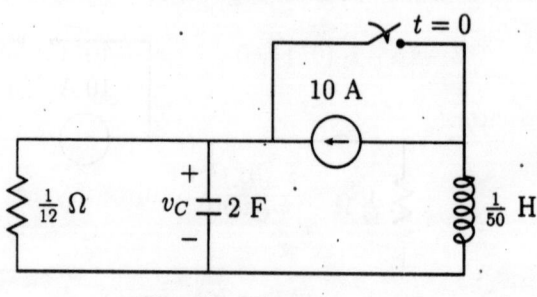

Fig. SP 3.9.

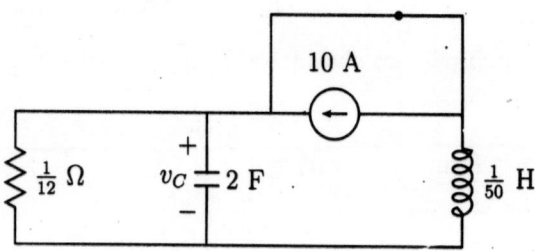

Fig. 3.9.1: $t > 0$ circuit.

It is noted from the circuit of Fig. 3.9.2 that

$$i_L(0^-) = -10 \text{ A} = i_L(0^+)$$

and $\qquad v_C(0^-) = \frac{1}{12} \times 10 = 5/6 \text{ V} = v_C(0^+)$

Using KCL at the node **a** of Fig. 3.9.3,

$$\frac{5/6}{1/12} + i_C(0^+) + (-10) = 0 \qquad \Rightarrow i_C(0^+) = 0$$

Therefore $\qquad \dfrac{dv_C}{dt}(0^+) = \dfrac{i_C(0^+)}{C} = 0$

From eqn [3.9.1] : $\qquad v_C(0^+) = 5/6 = A_1$

From eqn [3.9.2] : $\qquad \dfrac{dv_c}{dt}(0^+) = 0 = -3A_1 + 4A_2 \qquad \Rightarrow A_2 = 5/8$

And $\qquad v_C(t) = \left(\dfrac{5}{6}\cos 4t + \dfrac{5}{8}\sin 4t\right)e^{-3t} \qquad$ for $t > 0$

(b) The extremum values of $v_C(t)$ is determined from the equation

$$\frac{dv_C}{dt}(t) = 0$$

or $\qquad -3\left(\dfrac{5}{6}\cos 4t + \dfrac{5}{8}\sin 4t\right)e^{-3t} + 4\left(-\dfrac{5}{6}\sin 4t + \dfrac{5}{8}\cos 4t\right)e^{-3t} = 0$

or $\qquad \sin 4t = 0$

or $\qquad 4t = n\pi$

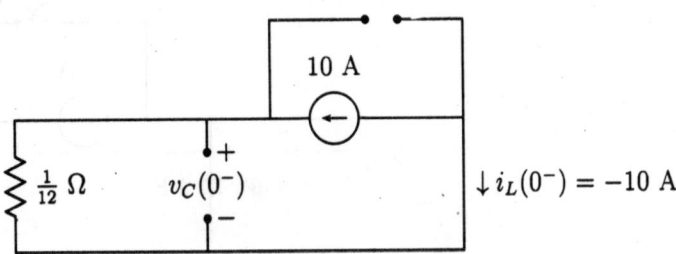

Fig. 3.9.2: $t = 0^-$ circuit.

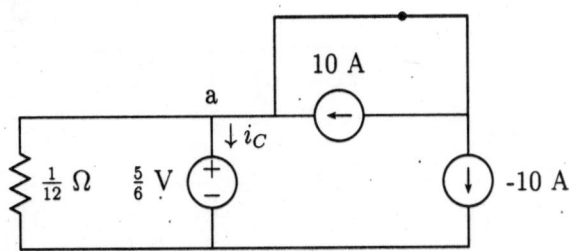

Fig. 3.9.3: $t = 0^+$ circuit.

Therefore, $t = \frac{n\pi}{4}$ (where $n = 0, 1, 2, \ldots$)
   When $n = 0$, $t = 0$, and $v_C(t = 0) = 5/6 = 0.833$ V.
   When $n = 1$, $t = \frac{\pi}{4}$, and $v_C(t = \frac{\pi}{4}) = -0.079$ V.
   When $n = 2$, $t = \frac{\pi}{2}$, and $v_C(t = \frac{2\pi}{4}) = 7.46 \times 10^{-3}$ V.
   When $n = 3$, $t = \frac{3\pi}{4}$, and $v_C(t = \frac{3\pi}{4}) = -7.06 \times 10^{-4}$ V and so on.
   We conclude from above calculations that the minimum value is -0.079 V and the maximum value is 0.833 V.

SP 3.10 In the network of Fig. SP 3.10, the switch is closed and a steady state is reached in the network. At $t = 0$, the switch is opened. Find the expression for the current in the inductor, $i_L(t)$.

SOLUTION:

By KVL in the circuit of fig. 3.10.1,

$$2\frac{di_L}{dt} + \frac{1}{1/10} \int_{-\infty}^{t} i_L dt = 0 \qquad [3.10.1]$$

Differentiating

$$\frac{d^2 i_L}{dt^2} + 5i_L = 0 \qquad [3.10.2]$$

Characteristic equation

$$s^2 + 5 = 0$$

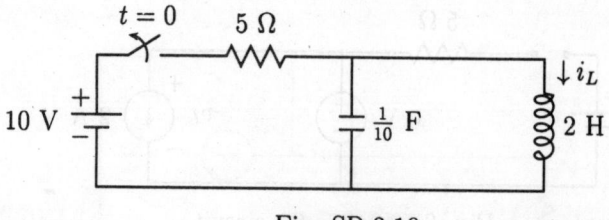

Fig. SP 3.10.

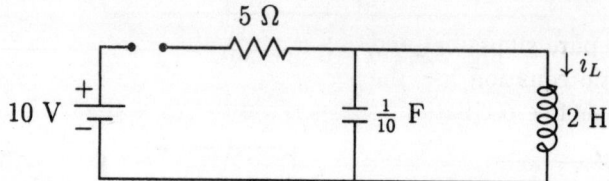

Fig. 3.10.1: $t > 0$ circuit.

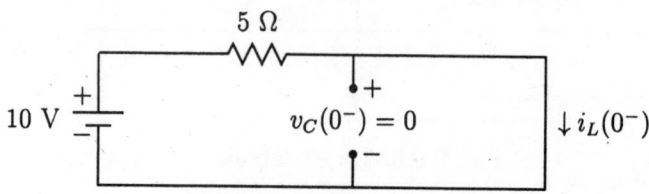

Fig. 3.10.2: $t = 0^-$ circuit.

Roots are:

$$s_{1,2} = \pm j\sqrt{5}$$

The natural solution is

$$i_L(t) = A_1 \cos \sqrt{5}\, t + A_2 \sin \sqrt{5}\, t \qquad\qquad [3.10.3]$$

and

$$\frac{di_L}{dt} = -\sqrt{5}A_1 \sin \sqrt{5}\, t + \sqrt{5}A_2 \cos \sqrt{5}\, t \qquad\qquad [3.10.4]$$

We find from the circuit of Fig. 3.10.2 that

$$i_L(0^-) = \frac{10}{5} = 2 \text{ A} = i_L(0^+)$$

and $\qquad v_C(0^-) = 0 = v_C(0^+)$

It is noted in the circuit of Fig. 3.10.3 that

$$v_L(0^+) = 0$$

Hence for inductor $\qquad \dfrac{di_L}{dt}(0^+) = \dfrac{v_L(0^+)}{L} = 0$

From eqn [3.10.3] $\qquad i_L(0^+) = 2 = A_1$

From eqn [3.10.4] $\qquad \dfrac{di_L}{dt}(0^+) = 0 = \sqrt{5}A_2 = 0 \qquad \Rightarrow A_2 = 0$

Finally we have $\qquad i_L(t) = 2 \cos \sqrt{5}\, t \qquad$ for $t \geq 0$

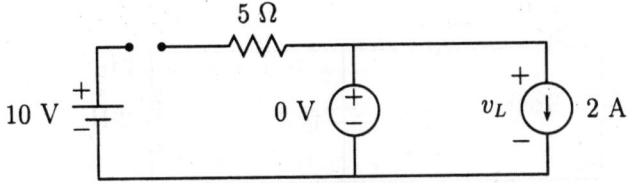

Fig. 3.10.3: $t = 0^+$ circuit.

The response $i_L(t)$ is pure sinusoidal and called undamped case. This situation occurs when the characteristic equation has the form: $s_{1,2} = \pm j\omega_n$. The response will have sustained oscillations or free oscillations ($\omega_n = \sqrt{5}$ rad/sec) with constant amplitude (5 A).

SP 3.11 **In the circuit of SP 3.11, the voltage source is $v_s = (t^2 + 2)u(t)$.**
   Also the initial conditions are given as $i_L(0^-) = 1/10$ A and $v_C(0^-) = 1$ V.
   Find $i_L(t)$ for $t > 0$.

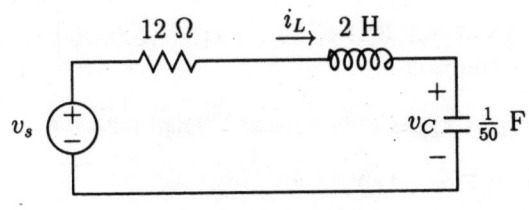

Fig. SP 3.11.

Fig. 3.11.1: $t = 0^+$ circuit.

SOLUTION:

Writing KVL for $t > 0$ in the circuit of Fig. SP 3.11

$$12i_L + 2\frac{di_L}{dt} + 50 \int_{-\infty}^{t} i_L dt = v_s = t^2 + 2 \qquad [3.11.1]$$

Differentiating above eqn [3.11.1]

$$\frac{d^2 i_L}{dt^2} + 6\frac{di_L}{dt} + 25i_L = t \qquad [3.11.2]$$

Characteristic equation

$$s^2 + 6s + 25 = 0$$

Roots are : $s_{1,2} = -3 \pm j4$

Form of natural response

$$i_{Ln} = e^{-3t}(A\cos 4t + B\sin 4t) \qquad [3.11.3]$$

Particular response corresponding to the input

$$i_{Lp} = k_1 + k_2 t \qquad [3.11.4]$$

Substituting eqn [3.11.4] into eqn [3.11.2] yields

$$0 + 6k_2 + 25k_1 + 25k_2 t = t$$

Equating the coefficients of like terms on the both sides of this equation

$$6k_2 + 25k_1 = 0 \qquad [3.11.5]$$

$$25k_2 = 1 \qquad [3.11.6]$$

On solving eqn [3.11.5] and eqn [3.11.6] we get $k_1 = -0.0096$ and $k_2 = 0.04$
Hence the complete solution is

$$i_L(t) = i_{Ln} + i_{Lp} = e^{-3t}(A\cos 4t + B\sin 4t) + 0.04t - 0.0096 \qquad [3.11.7]$$

By KVL in the circuit of Fig. 3.11.1,

$$v_L(0^+) - 2 + 12/10 + 1 = 0 \qquad \Rightarrow v_L(0^+) = -1/5 \text{ V}$$

Therefore $\dfrac{di_L}{dt}(0^+) = \dfrac{v_L(0^+)}{L} = \dfrac{-1/5}{2} = -1/10 \text{ A/sec.}$

From eqn [3.11.7] $i_L(0^+) = 1/10 = A - 0.0096 \qquad \Rightarrow A = 0.1096$

The derivative of eqn [3.11.7] is

$$\frac{di_L}{dt} = -3e^{-3t}(A\cos 4t + B\sin 4t) + e^{-3t}(-4A\sin 4t + 4B\cos 4t) + 0.04 \qquad [3.11.8]$$

From eqn [3.11.8]

$$\frac{di_L}{dt}(0^+) = 0.04 - 3A + 4B = -1/10$$

$$\Rightarrow B = 0.0472$$

Thus $i_L(t) = 0.04t - 0.0096 + e^{-3t}(0.1096\cos 4t + 0.0472\sin 4t)$ for $t > 0$

SP 3.12 The switch in the circuit of Fig. SP 3.12, which has been closed for a long time, is opened at $t = 0$.
   (a) Find the initial conditions
   (b) Find $v_C(t)$ for $t > 0$.

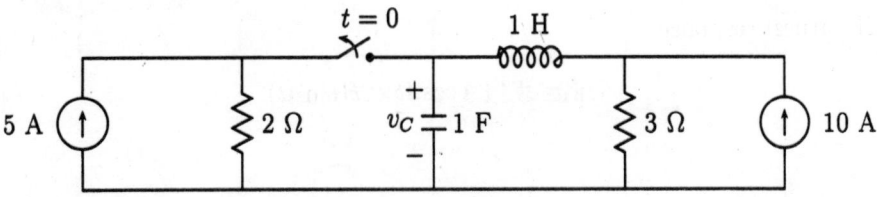

Fig. SP 3.12.

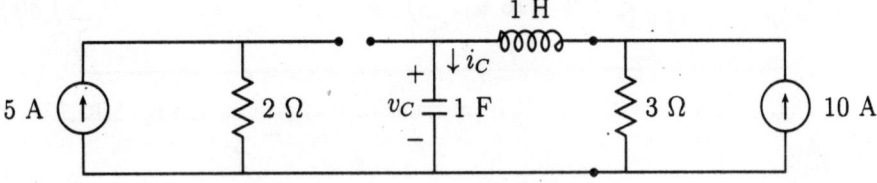

Fig. 3.12.1: $t > 0$ circuit.

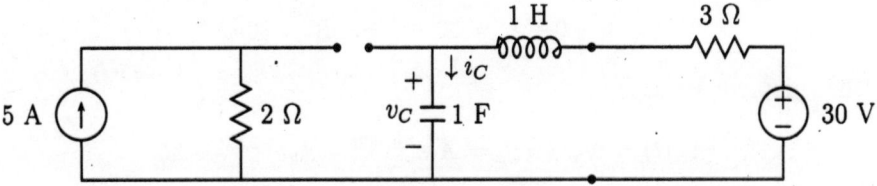

Fig. 3.12.2: $t > 0$ transformed circuit of Fig. 3.12.1.

SOLUTION:

By KVL in the circuit of Fig. 3.12.2

$$3i_C + 1\frac{di_C}{dt} + v_C = 30$$

Putting $i_C = C\frac{dv_C}{dt} = \frac{dv_C}{dt}$ in this equation

$$\frac{d^2v_C}{dt^2} + 3\frac{dv_C}{dt} + v_C = 30 \qquad [3.12.1]$$

The characteristic equation is

$$s^2 + 3s + 1 = 0$$

Roots are : $s_{1,2} = -0.382, -2.62$

and the natural response is

$$v_{Cn}(t) = A_1 e^{-0.382t} + A_2 e^{-2.62t} \qquad [3.12.2]$$

The particular response corresponding to constant input is

$$v_{Cp}(t) = K \qquad [3.12.3]$$

$$i_L(0^-) = 4 \text{ A}$$

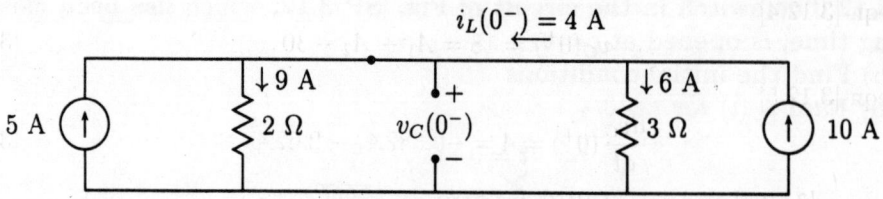

Fig. 3.12.3: $t = 0^-$ circuit.

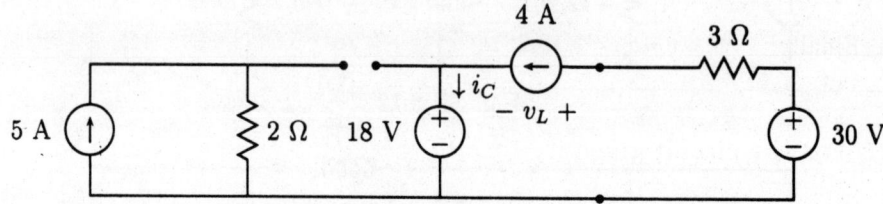

Fig. 3.12.4: $t = 0^+$ circuit from transformed circuit of Fig. 3.12.1.

Substituting eqn [3.12.3] into eqn [3.12.1] yields

$$0 + 0 + K = 30 \qquad \Rightarrow K = 30$$

Hence the complete solution is

$$v_C(t) = v_{Cn} + v_{Cp} = A_1 e^{-0.382t} + A_2 e^{-2.62t} + 30 \qquad [3.12.4]$$

Its derivative is

$$\frac{dv_C}{dt} = -0.382 A_1 e^{-0.382t} - 2.62 A_2 e^{-2.62t} \qquad [3.12.5]$$

Determination of initial conditions:

From the circuit of Fig. 3.12.3, using current division formula as

$$i_{2\Omega} = \frac{3}{3+2}(10+5) = 9 \text{ A}$$

and $\qquad i_{3\Omega} = \dfrac{2}{3+2}(10+5) = 6 \text{ A}$

It is also observed in the circuit of Fig. 3.12.3 that

$$i_L(0^-) = 4 \text{ A} = i_L(0^+)$$
$$\text{and} \qquad v_C(0^-) = 18 \text{ V} = v_C(0^+)$$

It is obvious from the circuit of Fig. 3.12.4 that

$$i_C(0^+) = 4 \text{ A}$$

Therefore, for the capacitor

$$\frac{dv_C}{dt}(0^+) = \frac{i_C(0^+)}{C} = 4 \text{ A/sec.}$$

From eqn [3.12.4]

$$v_C(0^+) = 18 = A_1 + A_2 + 30 \qquad \text{[3.12.6]}$$

From eqn [3.12.5]

$$\frac{dv_C}{dt}(0^+) = 4 = -0.382A_1 - 2.62A_2 \qquad \text{[3.12.7]}$$

Solving eqn [3.12.6] and eqn [3.12.7], we have

$$A_1 = -12.261 \qquad \text{and} \qquad A_2 = 0.261$$

Finally $\qquad v_C(t) = -12.261e^{-0.382t} + 0.261e^{-2.62t} + 30 \qquad$ for $t > 0$

SP 3.13 For the circuit given in Fig. SP 3.13 determine $v(t)$ and $i(t)$.

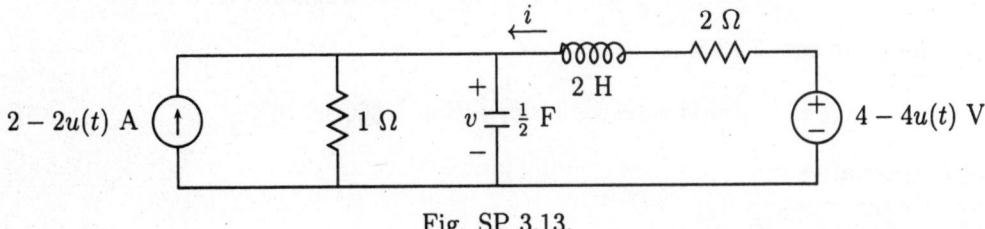

Fig. SP 3.13.

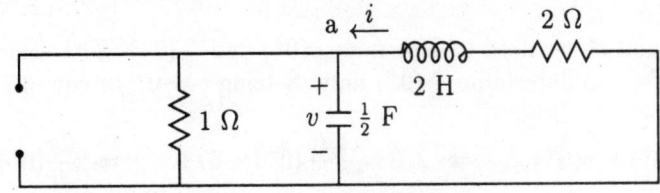

Fig. 3.13.1: $t > 0$ circuit.

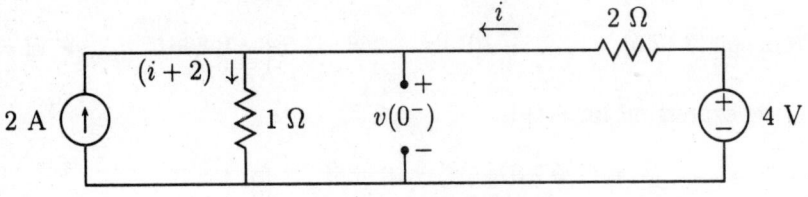

Fig: 3.13.2: $t = 0^-$ circuit.

SOLUTION:

By KVL in the loop containing 1-Ω resistor, 2-Ω resistor and 4-V source in circuit of Fig. 3.13.2,

$$2i(0^-) + 1\left[i(0^-) + 2\right] - 4 = 0 \qquad \Rightarrow i(0^-) = 2/3 \text{ A} = i(0^+)$$

and $\qquad v(0^-) = 1 \times \left[i(0^-) + 2\right] = 2/3 + 2 = 8/3 \text{ V} = v(0^+)$

From Fig. 3.13.1, by KVL to the mesh on the right in the circuit, we get

$$2i + 2\frac{di}{dt} + v = 0 \qquad [3.13.1]$$

and applying KCL at the node **a**

$$i = \frac{1}{2}\frac{dv}{dt} + \frac{v}{1} \qquad [3.13.2]$$

Substituting eqn [3.13.2] into eqn [3.13.1] and then simplifying, we get

$$\frac{d^2v}{dt^2} + 3\frac{dv}{dt} + 3v = 0 \qquad [3.13.3]$$

The characteristic equation is

$$s^2 + 3s + 3 = 0$$

$$\text{Roots are :} \qquad s_{1,2} = -1.5 \pm j0.866$$

The solution to eqn [3.13.3] is

$$v(t) = (A\cos 0.866t + B\sin 0.866t)e^{-1.5t} \qquad [3.13.4]$$

and it's derivative is

$$\frac{dv}{dt} = -1.5(A\cos 0.866t + B\sin 0.866t)e^{-1.5t} + 0.866(-A\sin 0.866t + B\cos 0.866t)e^{-1.5t}$$
$$[3.13.5]$$

To determine A and B, we need to determine $v(0^+)$ and $\frac{dv}{dt}(0^+)$. The value of $v(0^+) = 8/3$ V is known, so need to determine $\frac{dv}{dt}(0^+)$ only. Setting t $= 0^+$ in eqn [3.13.2] as

$$i(0^+) = \frac{1}{2}\frac{dv}{dt}(0^+) + v(0^+) \quad \Rightarrow \quad 2/3 = \frac{1}{2}\frac{dv}{dt}(0^+) + 8/3 \quad \Rightarrow \quad \frac{dv}{dt}(0^+) = -4 \text{ V/sec.}$$

From eqn [3.12.4] $\qquad v(0^+) = 8/3 = A$

From eqn [3.12.5] $\qquad \frac{dv}{dt}(0^+) = -4 = -1.5A + 0.866B \qquad \Rightarrow B = 0$

Therefore, the expression for $v(t)$ is

$$v(t) = 2.67e^{-1.5t}\cos 0.866t \qquad \text{for } t > 0$$

In addition, from eqn [3.13.2]

$$i(t) = \frac{1}{2}\frac{d}{dt}\left(2.67e^{-1.5t}\cos 0.866t\right) + 2.67e^{-1.5t}\cos 0.866t$$

On simplifying

$$i(t) = (0.67\cos 0.866t - 1.156\sin 0.866t)\,e^{-1.5t} \qquad \text{for } t > 0$$

SP 3.14 For the circuit given in Fig. SP 3.14, determine $v_2(t)$ for $t > 0$.

SOLUTION:

Writing KCL for $t > 0$ at the node v_2 of Fig. SP 3.14

$$\frac{1}{1} \int_0^t (v_1 - v_2)dt = 1\frac{dv_2}{dt} + \frac{v_2}{1/2} \qquad [3.14.1]$$

Differentiating

$$v_1 - v_2 = \frac{d^2v_2}{dt^2} + 2\frac{dv_2}{dt}$$

Hence

$$\frac{d^2v_2}{dt^2} + 2\frac{dv_2}{dt} + v_2 = v_1 = 4\sin(2t + 30^0) \qquad [3.14.2]$$

The characteristic equation is

$$s^2 + 2s + 1 = 0$$

Roots are : $s_{1,2} = -1, -1$

Hence the natural solution is

$$v_{2n} = (A_1 + A_2 t)e^{-t} \qquad [3.14.3]$$

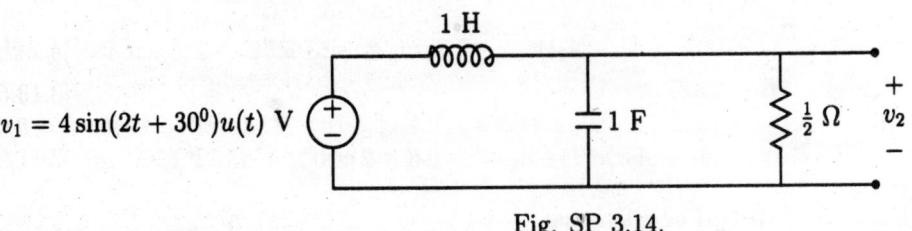

$v_1 = 4\sin(2t + 30^0)u(t)$ V

1 H

1 F

$\frac{1}{2}\,\Omega$

$+$ v_2 $-$

Fig. SP 3.14.

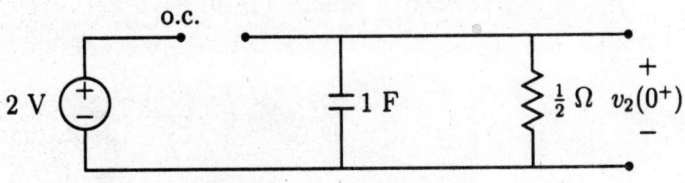

2 V

o.c.

1 F

$\frac{1}{2}\,\Omega$ $v_2(0^+)$

$+$ $-$

Fig. 3.14.1: $t = 0^+$ circuit.

The form of particular solution may be

$$v_{2p} = A\sin(2t + 30^0 + \phi) \qquad [3.14.4]$$

Substituting eqn [3.14.4] into eqn [3.14.2], we have

$$\frac{d^2}{dt^2}[A\sin(2t + 30^0 + \phi)] + 2\frac{d}{dt}[A\sin(2t + 30^0 + \phi)] + A\sin(2t + 30^0 + \phi) = 4\sin(2t + 30^0)$$

On simplifying, we get

$$(-3A\cos\phi - 4A\sin\phi)\sin(2t + 30^0) + (4A\cos\phi - 3A\sin\phi)\cos(2t + 30^0) = 4\sin(2t + 30^0)$$
$$[3.14.5]$$

Equating coefficients of like terms on both sides of eqn [3.14.5], we have

$$-3A\cos\phi - 4A\sin\phi = 4 \qquad [3.14.6]$$

and

$$4A\cos\phi - 3A\sin\phi = 0 \qquad [3.14.7]$$

From eqn [3.14.7]

$$\tan\phi = 4/3 \qquad \Rightarrow \phi = 53.13^0$$

Hence $\qquad \sin\phi = 4/5 \qquad$ and $\qquad \cos\phi = 3/5$

From the eqn [3.14.6]

$$-3A \times (3/5) - 4A(4/5) = 4 \qquad \Rightarrow A = -0.8$$

Hence $\qquad v_{2p} = -0.8\sin(2t + 30^0 + 53.13^0) = -0.8\sin(2t + 83.13^0)$

Complete solution

$$v_2(t) = (A_1 + A_2 t)e^{-t} - 0.8\sin(2t + 83.13^0) \qquad [3.14.8]$$

and

$$\frac{dv_2}{dt} = -(A_1 + A_2 t)e^{-t} + A_2 e^{-t} - 0.8 \times 2\cos(2t + 83.13^0) \qquad [3.14.9]$$

Determination of initial conditions:
It is observed from the circuit of Fig. 3.14.1 that

$$v_2(0^+) = 0 \qquad \text{and} \qquad i_C(0^+) = 0$$

Hence for capacitor $\qquad\qquad \dfrac{dv_2}{dt}(0^+) = \dfrac{i_C(0^+)}{C} = 0$

From eqn [3.14.8] $\qquad v_2(0^+) = 0 = A_1 - 0.8\sin 83.13^0 \qquad \Rightarrow A_1 = 0.79$

from eqn; [3.14.9] $\qquad \dfrac{dv_2}{dt}(0^+) = 0 = -A_1 + A_2 - 0.8 \times 2\cos(83.13^0) \qquad \Rightarrow A_2 = 0.98$

And finally $\qquad v_2(t) = (0.79 + 0.98t)e^{-t} - 0.8\sin(2t + 83.13^0) \qquad$ for $t > 0$

3.5   DRILL PROBLEMS

DP 3.1 For the circuit of Fig. DP 3.1, find:
  (a) $i(t)$ for $t > 0$,
  (b) damping factor,
  (c) natural frequency,
  (d) damped natural frequency,
  (e) damping ratio,
  (f) time constant,
  (g) settling time.

  Hints :

$$\text{Damping ratio} \qquad \zeta = \frac{R}{2}\sqrt{\frac{C}{L}} = \frac{1}{2}\sqrt{1/1} = 0.5$$

$$\text{Time constant} \qquad \tau = \frac{1}{\alpha} = \frac{2L}{R} = 2\,\text{S}$$

$$\text{Settling time} \qquad t_s = 4\tau = 4 \times 2 = 8\,\text{S}$$

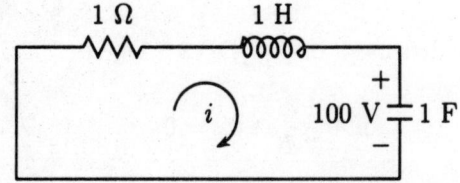

Fig. DP 3.1.

DP 3.2 Find $i(t)$ for all t for the circuit of Fig. DP 3.2. The initial conditions are $v_C(0^-) = 1$ V and $i(0^-) = 2$ A.

  Hints :

$$\alpha = 1, \qquad \omega_n = 1, \qquad s_{1,2} = -1, -1.$$

  Hence $\qquad i(t) = (A_1 + A_2 t)e^{-t}$

  and $\qquad \dfrac{di}{dt} = -(A_1 + A_2 t)e^{-t} + A_2 e^{-t}$

By continuity of inductor current and capacitor voltage, we have

$$i(0^-) = 2\,\text{A} = i(0^+)$$
$$v_C(0^-) = 1\,\text{V} = v_C(0^+)$$

To determine $\frac{di}{dt}(0^+)$, consider circuit at $t = 0^+$ shown in Fig. DP 3.2.1.

  KVL gives $\qquad 4 \times 2 + v_L(0^+) + 1 = 0 \qquad \Rightarrow v_L(0^+) = -9$ V

  Hence $\qquad \dfrac{di}{dt}(0^+) = \dfrac{v_L(0^+)}{L} = -4.5$

Use these initial conditions to determine A_1 and A_2.

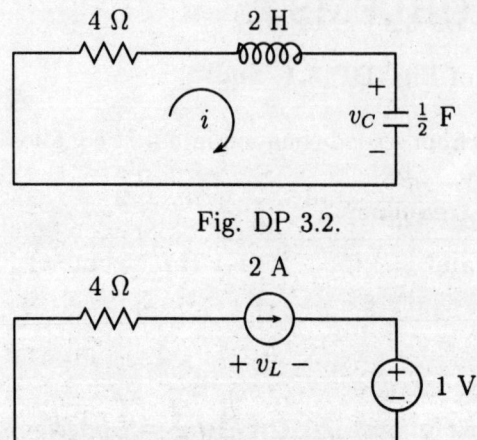

Fig. DP 3.2.

Fig. DP 3.2.1: $t = 0^+$ circuit.

DP 3.3 Determine $v_C(t)$ and $i(t)$ for $t > 0$ for the circuit of Fig. DP 3.3.

Hints : KVL around the left mesh of Fig. DP 3.3.1 gives

$$5i + \frac{1}{2}\frac{di}{dt} + v_C = 0$$

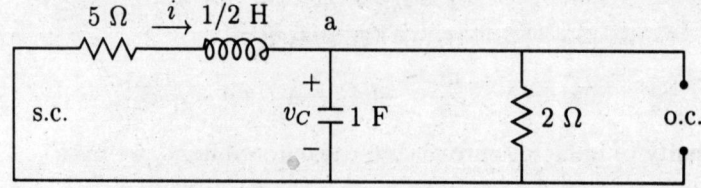

Fig. DP 3.3.

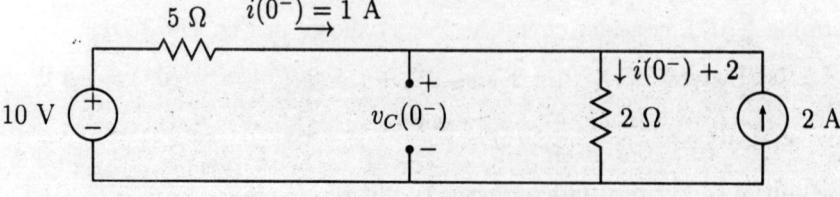

Fig. DP 3.3.1: $t > 0$ circuit.

Fig. DP 3.3.2: $t = 0^-$ circuit.

KCL at the node **a** of Fig. DP 3.3.1 gives

$$i = 1\frac{dv_C}{dt} + \frac{v_C}{2}$$

Substituting the value of i from second equation into first equation, we have

$$\frac{d^2v_C}{dt^2} + 10.5\frac{dv_C}{dt} + 7v_C = 0$$

DP 3.4 Determine $i(t)$ and $v(t)$ for $t > 0$ for the circuit of Fig. DP 3.4.

Hints : For $t > 0$, the circuit is parallel RLC with $\alpha = 1$, $\omega_n = 1$ and $s_{1,2} = -1, -1$. Hence

$$v(t) = (k_1 + k_2t)e^{-t}$$

Use Fig. DP 3.4.1 and Fig. DP 3.4.2 to determine $v(0^+)$ and $\frac{dv}{dt}(0^+)$. And for the current use

$$i = \frac{1}{2}\frac{dv}{dt} + v$$

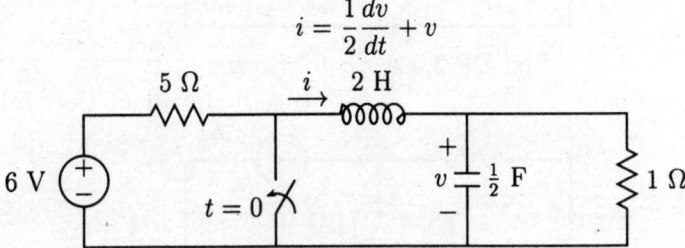

Fig. DP 3.4.

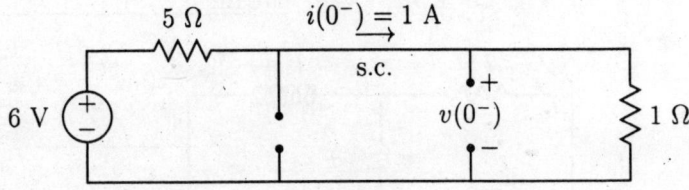

Fig. DP 3.4.1: $t = 0^-$ circuit.

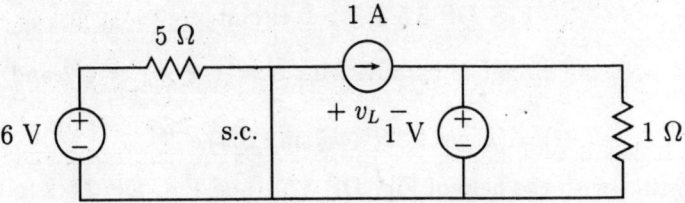

Fig. DP 3.4.2: $t = 0^+$ circuit.

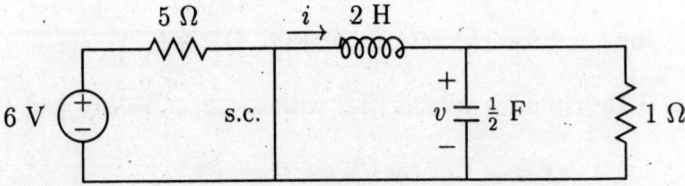

Fig. DP 3.4.3: $t > 0$ circuit.

DP 3.5 Determine $v(t)$ for $t > 0$ for the circuit of Fig. DP 3.5.

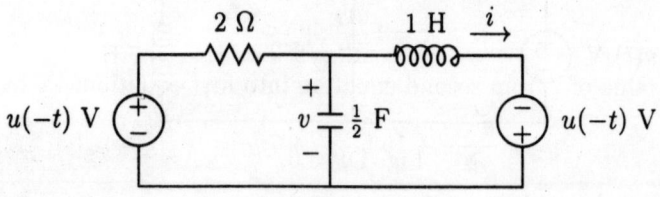

Fig. DP 3.5.

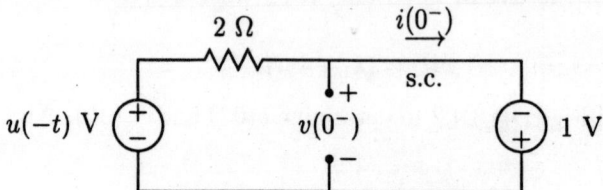

Fig. DP 3.5.1: $t = 0^-$ circuit.

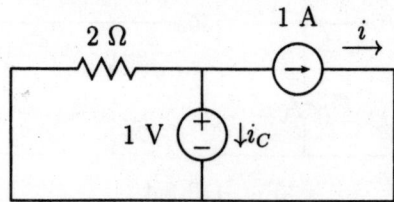

Fig. DP 3.5.2: $t = 0^+$ circuit.

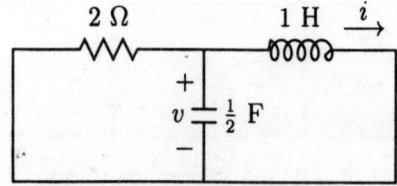

Fig. DP 3.5.3: $t > 0$ circuit.

Hints : For $t > 0$, the circuit is parallel with $\alpha = 1/2$, $\omega_n = \sqrt{2}$ and $\omega_d = 1.32$. Hence

$$v(t) = (k_1 \cos 1.32t + k_2 \sin 1.32t)\, e^{-t/2}$$

Find $v(0^+)$ and $\frac{dv}{dt}(0^+)$ with the help of Fig. DP 3.5.1 and Fig. DP 3.5.2 to determine k_1 and k_2.

DP 3.6 Find $i(t)$ for $t > 0$ for the circuit of Fig. DP 3.6.

Hints : For $t > 0$, the circuit is parallel RLC with $\alpha = 2$, $\omega_n = \sqrt{6}$ and $\omega_d = \sqrt{2}$. Hence

$$v(t) = e^{-2t}(A \cos \sqrt{2}\, t + B \cos \sqrt{2}\, t)$$

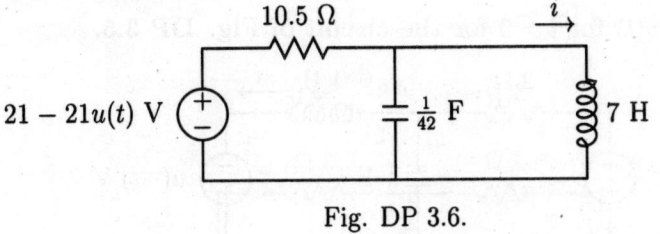

Fig. DP 3.6.

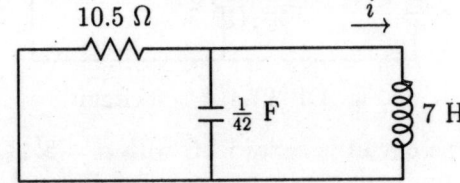

Fig. DP 3.6.1: $t > 0$ circuit.

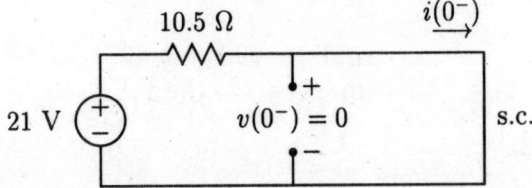

Fig. DP 3.6.2: $t = 0^-$ circuit.

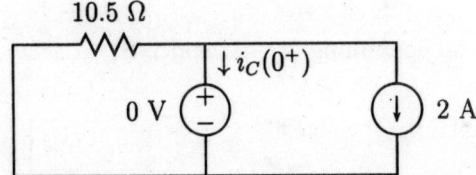

Fig. DP 3.6.3: $t = 0^+$ circuit.

Make use of Fig. DP 3.6.2 and Fig. DP 3.6.3 to determine $v(0^+)$ and $\frac{dv}{dt}(0^+)$ and so find constants A and B.

DP 3.7 Determine $i_2(t)$ for $t > 0$ for the circuit of Fig. DP 3.7.

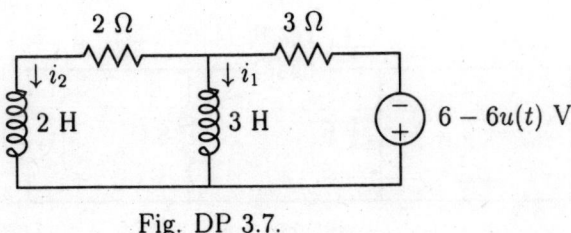

Fig. DP 3.7.

Hints : By KVL around the mesh on the right,

$$3(i_1 + i_2) + 3\frac{di_1}{dt} = 0$$

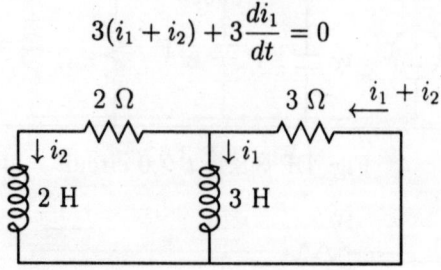

Fig. DP 3.7.1: $t > 0$ circuit.

Hints : For $t > 0$, the circuit is series RLC with $\alpha = 5$, $\omega_n = 4$, and $s_{1,2} = -2, -8$. Hence

$$i(t) = k_1 e^{-2t} + k_2 e^{-8t}$$

From Fig. DP 3.8.2,

$$v_C(0^-) = 20 \text{ V} = v_C(0^+)$$
and $$i(0^-) = 5 \text{ A} = i(0^+)$$

By KVL in Fig. DP **3.8.3**,

$$v_L(0^+) + 5 \times 5 - 20 = 0 \qquad \Rightarrow \ v_L(0^+) = -5 \text{ V}$$

Hence $$\frac{di}{dt}(0^+) = \frac{v_L(0^+)}{L} = -10 \text{ A/s}$$

By KVL around the loop containing 3Ω, 2Ω and 2 H,

$$3(i_1 + i_2) + 2i_2 + 2\frac{di_2}{dt} = 0$$

$$3i_1 = -5i_2 - 2\frac{di_2}{dt} \qquad \Rightarrow \ i_1 = -\frac{5}{3}i_2 - \left(\frac{2}{3}\right)\frac{di_2}{dt}$$

Putting this value of i_1 into first equation, we get

$$\frac{d^2 i_2}{dt^2} + 3.5\frac{di_2}{dt} + i_2 = 0$$

DP 3.8 Find $i(t)$ for $t > 0$ for the circuit of Fig. DP **3.8**.

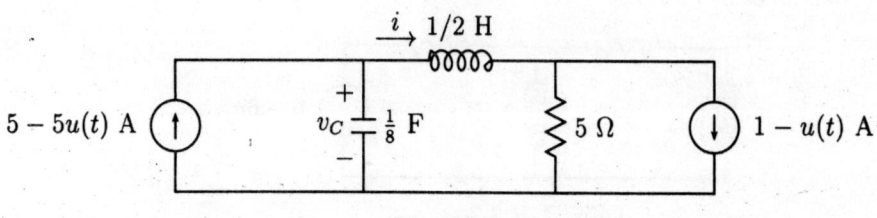

Fig. DP **3.8**.

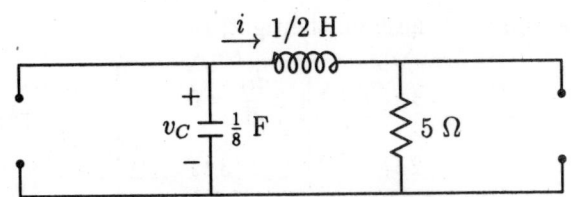

Fig. DP 3.8.1: $t > 0$ circuit.

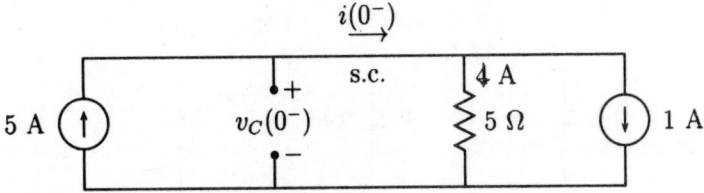

Fig. DP 3.8.2: $t = 0^-$ circuit.

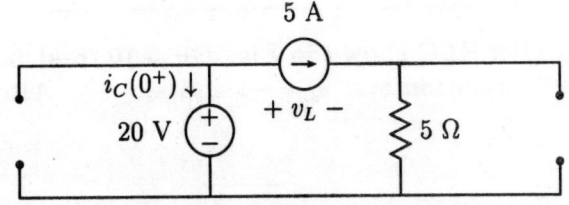

Fig. DP 3.8.3: $t = 0^+$ circuit.

DP 3.9 Determine $v_2(t)$ for $t > 0$ for the circuit of Fig. DP 3.9.

Hints : By KCL at the node v_2 of Fig. DP 3.9.1,

$$-\frac{v_2}{3} = 2\frac{dv_2}{dt} + \frac{v_2 - v_1}{2}$$

In p – domain

$$-\frac{v_2}{3} = 2pv_2 + \frac{v_2 - v_1}{2}$$

By KCL at the node v_1 of Fig. DP 3.9.1,

$$\frac{v_2 - v_1}{2} = 1\frac{dv_1}{dt}$$

In p – domain

$$\frac{v_2 - v_1}{2} = pv_1 \qquad \Rightarrow v_1 = \frac{v_2}{1 + 2p}$$

Putting this value of v_1 into first equation and then simplifying, we get

$$(12p^2 + 11p + 1)v_2 = 0$$

And differential equation

$$12\frac{d^2v_2}{dt} + 11\frac{dv_2}{dt} + v_2 = 0$$

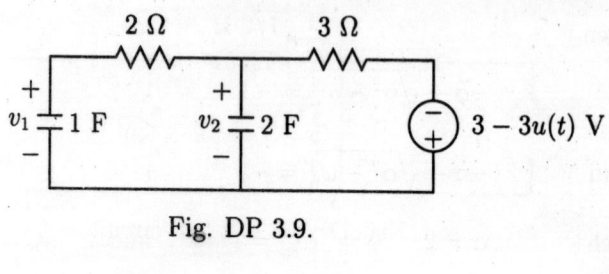

Fig. DP 3.9.

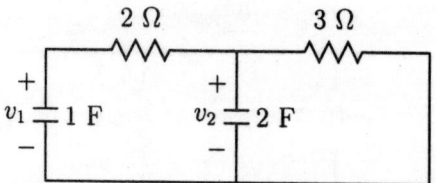

Fig. DP 3.9.1: $t > 0$ circuit.

DP 3.10 For the parallel RLC shown in Fig. DP 3.10, find R, L, and C if the roots of characteristic equation are: $s_1 = -1$ and $s_2 = -6$. Also $\omega_n C = 0.058$.

Hints :

$$-\alpha + \sqrt{\alpha^2 - \omega_n^2} = -1$$

$$-\alpha - \sqrt{\alpha^2 - \omega_n^2} = -6$$

Adding	$-2\alpha = -7$	$\Rightarrow \alpha = 3.5$
Subtracting	$2\sqrt{\alpha^2 - \omega_n^2} = 5$	$\Rightarrow \omega_n = \sqrt{6}$
Given	$\omega_n C = 0.058$	$\Rightarrow C = 23.68$ mF

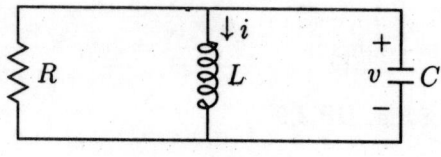

Fig. DP 3.10.

$$i_R(t) = v(t)/R = 3e^{-2t} - 8e^{-5t}$$

$$i_C(t) = -i_R(t) - i(t)$$

DP 3.11 Find the element values of a parallel RLC circuit so that $s_1 = -1$, $s_2 = -3$ and the initial current through resistor (in A) is numerically three times the initial voltage across the capacitor (in V).

Hints : Given

$$-\alpha + \sqrt{\alpha^2 - \omega_n^2} = -1$$

and

$$-\alpha - \sqrt{\alpha^2 - \omega_n^2} = -3$$

from which $\quad \alpha = 2 \quad \Rightarrow RC = 1/4 \quad$ and $\quad \omega_n = 3 \quad \Rightarrow LC = 1/3$

The voltage expression is

$$v(t) = k_1 e^{-t} + k_2 e^{-3t}$$

Hence $\qquad i(t) = v(t)/R = \dfrac{1}{R}(k_1 e^{-t} + k_2 e^{-3t})$

Here $\qquad v(0) = k_1 + k_2$

and $\qquad i(0) = (k_1 + k_2)/R = v(0)/R \qquad \Rightarrow 3x = x/R \qquad \Rightarrow R = 1/3 \ \Omega$

DP 3.12 For the circuit of Fig. DP 3.12, the inductor current is given by $i = 3e^{-2t} - 8e^{-5t}$ A if $L = 2$ H, find (a) $v(t)$ (b) $i_R(t)$ (c) $i_C(t)$

Hints :

$$v(t) = L\frac{di}{dt} = -12e^{-t} + 80e^{-3t}$$

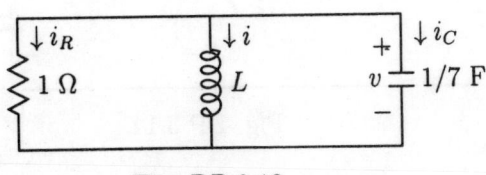

Fig. DP 3.12.

DP 3.13 An RLC series circuit has $L = 125$ mH, $R = 125 \ \Omega$ and $C = 40\mu$F. A sinusoidal voltage $v(t) = 125 \sin(1000t + \alpha)$ is applied to the circuit at an instant when $\alpha = 90^0$. Find the current response $i(t)$, assuming that the capacitor has no initial charge.

Hints: By KVL

$$125i + 125 \times 10^{-3}\frac{di}{dt} + \frac{10^6}{40} \int_0^t i\,dt = 125 \sin(1000t + 90^0)$$

Differentiating and manipulating

$$\frac{d^2i}{dt^2} + 10^3\frac{di}{dt} + 2 \times 10^5 i = 10^6\cos(1000t + 90^0)$$

Transient solution

$$i_{tr} = k_1 e^{-723.61t} + k_2 e^{-276.39t}$$

Steady-state solution

$$i_{ss} = A\cos(1000t) + B\sin(1000t)$$

The initial conditions are:

$$i(0^+) = 0$$
$$\frac{di}{dt}(0^+) = \frac{125\sin(90^0)}{125 \times 10^{-3}} = 10^3$$

DP 3.14 The switch closes in the circuit of Fig. 3.14 at $t = 0$. Assuming a relaxed circuit at the time for switching, determine the current i for $t > 0$. Also find the voltage v_L across the inductance for $t > 0$.

IES-98(EC)

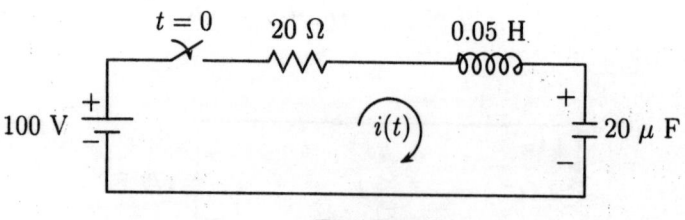

Fig. DP 3.14.

Hints: By KVL

$$20i + 0.05\frac{di}{dt} + \frac{10^6}{20}\int_0^t i\,dt = 100$$

or $\qquad \frac{d^2i}{dt^2} + 400\frac{di}{dt} + 10^6 i = 0$

Transient solution

$$i(t) = e^{-200t}\left[A\cos(979.8t) + B\sin(979.8t)\right]$$

DP 3.15 In the circuit of Fig. DP 3.15, the switch is closed at $t = 0$. Find the expressions for $v_C(t)$ and $i_L(t)$ for $t > 0$. Given that $i_L(0) = 1$ A and $v_C(0) = 3$ V.

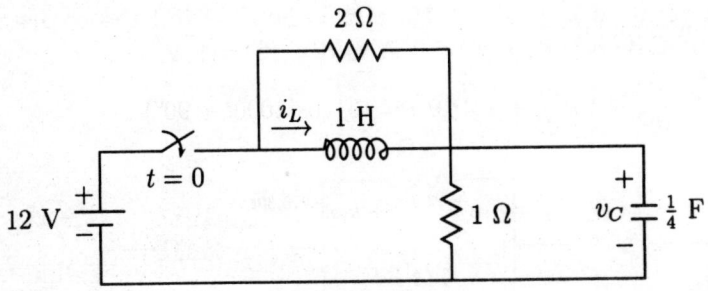

Fig. DP 3.15.

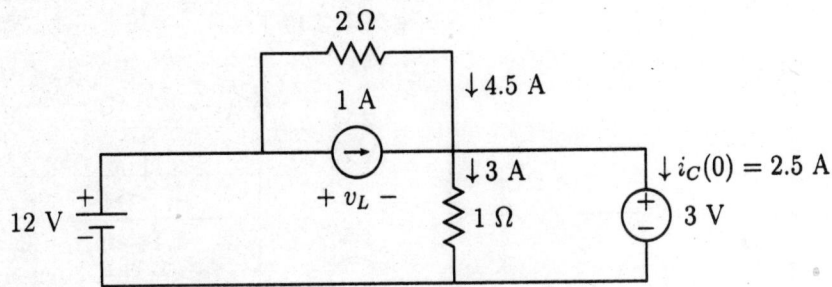

Fig. DP 3.15.1 Circuit for initial conditions.

Hints: By KCL at the upper right node of Fig. DP 3.15,

$$\frac{12 - v_C}{2} + i(0) + \frac{1}{1}\int_0^t (12 - v_C)dt = \frac{v_C}{1} + \frac{1}{4}\frac{dv_C}{dt}$$

Differentiating and rearranging, we get

$$\frac{d^2 v_C}{dt^2} + 6\frac{dv_C}{dt} + 4v_C = 48$$

The initial conditions may be determined from the Fig. DP 3.15.1 as:

$$i_C(0) = 2.5 \text{ A}$$
$$v_L(0) = 9 \text{ V}$$
$$\frac{dv_C}{dt}(0) = \frac{i_C(0)}{C} = 10 \text{ V/sec.}$$
$$\frac{di_L}{dt}(0) = \frac{v_L(0)}{L} = 9 \text{ A/sec.}$$

The inductor current may be determined from the equation

$$i_L(t) = i(0) + \frac{1}{1}\int_0^t (12 - v_C)dt$$

DP 3.16 In the circuit of Fig. DP 3.16, $v_s(t) = 5u(t)$. **Find the expression for** v_C **and** i_L **for** $t > 0$. **Given that** $i_L(0) = 1$ **A and** $v_C(0) = 1$ **V.**

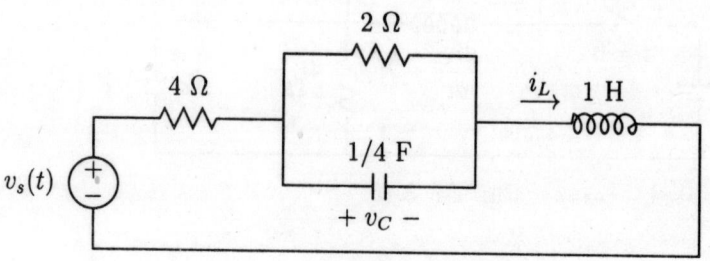

Fig. DP 3.16.

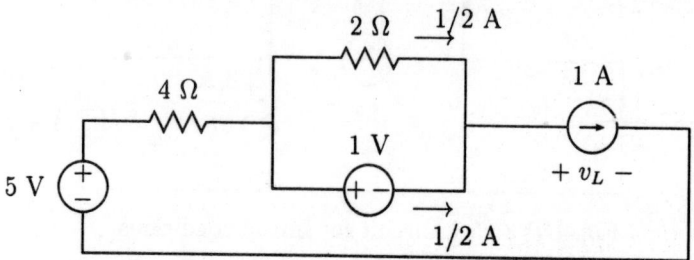

Fig. DP 3.16.1 Circuit for initial conditions.

Hints: The inductor current from p-domain circuit is

$$i_L = \frac{5}{4 + (2\|4/p) + p} = \frac{5p + 10}{p^2 + 6p + 12}$$

Thus, the corresponding differential equation is

$$\frac{d^2 i_L}{dt^2} + 6\frac{di_L}{dt} + 12 i_L = 10$$

The initial conditions are determined from the Fig. DP 3.16.1 as:

$$v_L(0) = 0$$

$$i_L(0) = 1/2 \text{ A}$$

$$\frac{di_L}{dt} = \frac{v_L(0)}{L} = 0$$

$$\frac{dv_C}{dt}(0) = \frac{i_C(0)}{C} = \frac{1/2}{1/4} = 2 \text{ V/sec}.$$

The differential equation in v_C may be given as:

$$v_C = (2\|4/p)\, i_L = \frac{4}{p+2} i_L$$

Thus

$$\frac{dv_C}{dt} + 2v_C = 4i_L$$

****

The Laplace Transform

SUMMARY OF THIS CHAPTER

4.1 DEFINITION

The Laplace transform is one of the mathematical tools used for the solution of linear ordinary differential equations. In comparison with the classical method of solving linear differential equations, the Laplace transform method has following advantages:

1. The complementary function and particular integral are obtained in one operation.
2. The Laplace transform converts the differential equation into an algebraic equation in s.
3. Initial conditions are automatically taken care of in the beginning steps. The final solution is obtained by taking the inverse Laplace transform.

Given a function of time $f(t)$, we define its Laplace Transform - designated either $\mathcal{L}[f(t)]$ or $F(s)$- to be

$$\mathcal{L}[f(t)] = F(s) = \int_0^\infty f(t)e^{-st}dt$$

where the variable s is referred to as the **Laplace operator**, which is complex variable; that is $s = \sigma + j\omega$.

The inverse Laplace transform formula is given by

$$\mathcal{L}^{-1}[F(s)] = f(t) = \frac{1}{2\pi j} \int_{c-j\infty}^{c+j\infty} F(s)e^{st}ds$$

where c is a real constant that is greater than the real part of all the singularities of $F(s)$. This equation represents a line integral that is to be evaluated in the s-plane. However, for most engineering purposes, the inverse Laplace transform operation can be carried out simply by referring to the Laplace transform table given below.

Table 4.1: **Some Important Laplace Transform Pairs**

S. No.	f(t)	F(s)
1.	Unit step function $u(t) = 1$ for $t \geq 0$	$\dfrac{1}{s}$
2.	$u(t - T)$	$\dfrac{e^{-sT}}{s}$
3.	Unit impulse function $\delta(t)$	1
4.	$\delta(t - T)$	e^{-Ts}
5.	$\dfrac{d\delta(t)}{dt}$	s
6.	$\dfrac{d\delta(t - T)}{dt}$	se^{-Ts}
7.	Unit ramp function $f(t) = t$	$\dfrac{1}{s^2}$
8.	$e^{-\alpha t}$	$\dfrac{1}{s + \alpha}$
9.	$te^{-\alpha t}$	$\dfrac{1}{(s + \alpha)^2}$
10.	t^n	$\dfrac{n!}{s^{n+1}}$
11.	$\sin \omega t$	$\dfrac{\omega}{s^2 + \omega^2}$
12.	$\cos \omega t$	$\dfrac{s}{s^2 + \omega^2}$
13.	$\sin(\omega t + \theta)$	$\dfrac{s \sin \theta + \omega \cos \theta}{s^2 + \omega^2}$
14.	$\cos(\omega t + \theta)$	$\dfrac{s \cos \theta - \omega \sin \theta}{s^2 + \omega^2}$
15.	$e^{-\alpha t} \sin \omega t$	$\dfrac{\omega}{(s + \alpha)^2 + \omega^2}$
16.	$e^{-\alpha t} \cos \omega t$	$\dfrac{s + \alpha}{(s + \alpha)^2 + \omega^2}$
17.	$\sinh \alpha t$	$\dfrac{\alpha}{s^2 - \alpha^2}$
18.	$\cosh \alpha t$	$\dfrac{s}{s^2 - \alpha^2}$

Table 4.2: **Properties of Laplace Transform**

S. No.	f(t)	Properties	F(s)
1.	$f(t)$	Definition	$\int\limits_{0}^{\infty} f(t)e^{-st}dt$
2.	$f_1(t) + f_2(t)$	Linearity	$F_1(s) + F_2(s)$
3.	$kf(t)$	Linearity	$kF(s)$
4.	$\dfrac{df(t)}{dt}$	Differentiation	$sF(s) - f(0)$
5.	$\dfrac{d^n f(t)}{dt^n}$	-do-	$s^n F(s) - s^{n-1}f(0) - \cdots$ $-\frac{d^{n-1}f(0)}{dt^{n-1}}$
6.	$\dfrac{d^2 f(t)}{dt^2}$	-do-	$s^2 F(s) - sf(0) - f'(0)$
7.	$\int_0^t f(t)dt$	Integration	$\dfrac{F(s)}{s}$
8.	$tf(t)$	Complex differentiation	$-\dfrac{dF(s)}{ds}$
9.	$e^{-\alpha t}f(t)$	Complex translation	$F(s + \alpha)$
10.	$f(t-a)u(t-a)$	Real translation	$e^{-sa}F(s)$
11.	$f(t)$	Periodic function	$\dfrac{F_1(s)}{1 - e^{-sT}}$
12.	$\int_0^t x_1(\tau)x_2(t-\tau)d\tau$	Convolution	$X_1(s)X_2(s)$

Initial Value Theorem

$$f(0) = \lim_{s\to\infty}\{sF(s)\}$$

Final Value Theorem

$$f(\infty) = \lim_{s\to 0} sF(s)$$

4.2   s-DOMAIN CIRCUITS

Figure 4.1 explains the method for converting a time-domain circuit into a s-domain circuit.

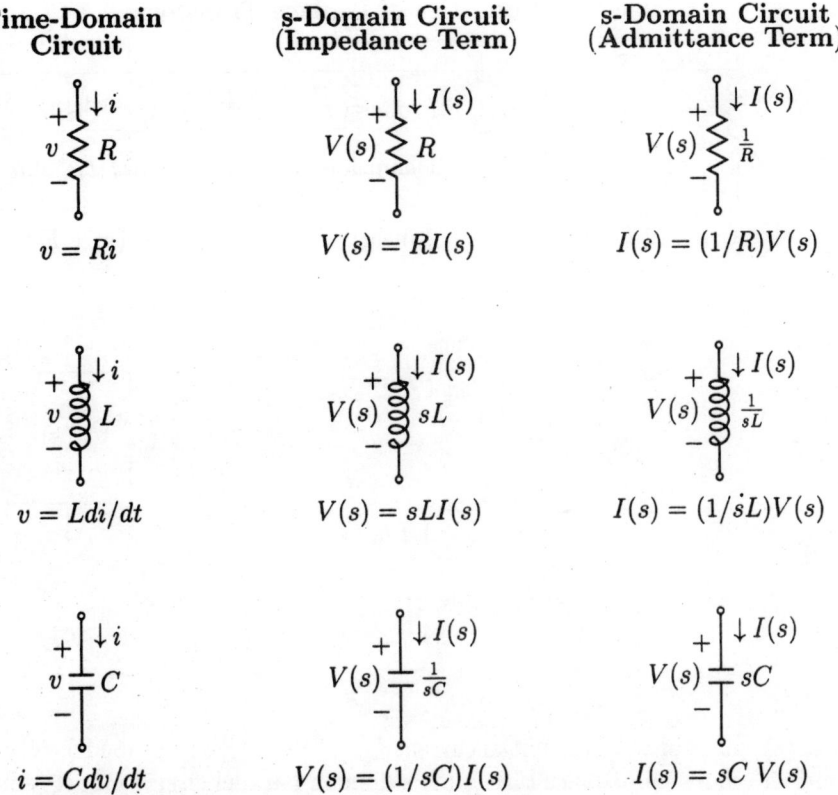

Time-Domain Circuit	s-Domain Circuit (Impedance Term)	s-Domain Circuit (Admittance Term)

$v = Ri$ | $V(s) = RI(s)$ | $I(s) = (1/R)V(s)$

$v = Ldi/dt$ | $V(s) = sLI(s)$ | $I(s) = (1/sL)V(s)$

$i = Cdv/dt$ | $V(s) = (1/sC)I(s)$ | $I(s) = sC\,V(s)$

Figure 4.1: Time-domain circuits with their s-domain circuits containing zero initial conditions for the inductor and the capacitor.

The Inductor with an Initial Condition

Figure 4.2 shows the elements needed to construct the s-domain circuit model for the inductor having an initial current from the given time-domain circuit.

The Capacitor with an Initial Condition

Figure 4.3 shows the elements needed to construct the s-domain circuit model for the capacitor having an initial voltage or charge.

4.3 PARTIAL-FRACTION EXPANSION

In a majority of problems, the evaluation of the inverse Laplace transform does not rely on the use of the inversion integral formula. Rather, the inverse Laplace transform operation involving rational functions can be carried out using Laplace transform table and partial-fraction expansion. The solution to a differential equation by Laplace transform in term

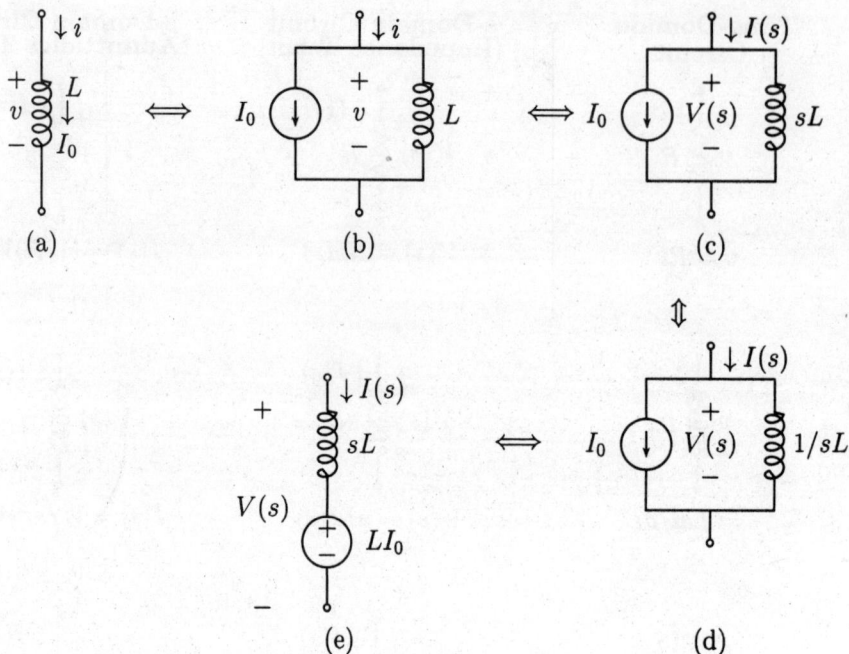

Figure 4.2: (a) Inductor with the initial current I_0; (b) Parallel circuit model; (c) s-domain parallel circuit model (impedance term); (d) s-domain parallel circuit model (admittance term); (e) Series circuit model in s-domain.

of s can be written in general form as:

$$F(s) = \frac{N(s)}{D(s)} = \frac{b_0 s^m + b_1 s^{m-1} + \cdots + b_m}{a_0 s^m + a_1 s^{n-1} + \cdots + a_n}$$

where $b_0, b_1, \cdots, b_m, a_0, a_1, \cdots, a_n$ are the real constants. The methods of partial fraction expansion will now be given for the cases simple, multiple-order roots, complex conjugate roots of $D(s)$.

Case 1 DISTINCT ROOTS: When the denominator of function $F(s)$ has all distinct roots, the corresponding partial-fraction terms are of the form

$$F(s) = \frac{\text{Numerator polynomial}}{(s + p_1)(s + p_2) \cdots (s + p_n)}$$

$$= \frac{k_1}{s + p_1} + \frac{k_2}{s + p_2} + \cdots + \frac{k_n}{s + p_n}$$

where

$$k_i = (s + p_i) F(s)|_{s=-p_i}$$

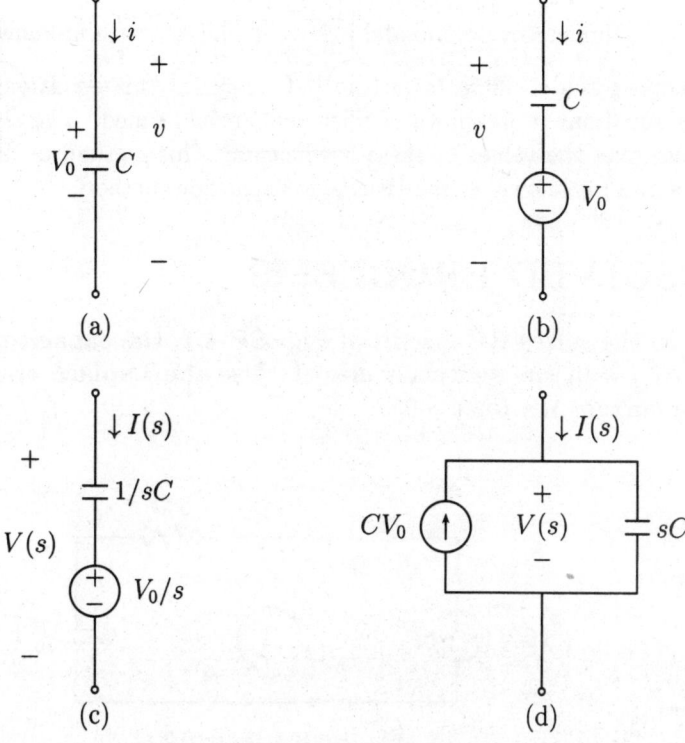

Figure 4.3: (a) Capacitor with the initial voltage V_0; (b) Series circuit model; (c) s-domain series circuit model; (d) s-domain parallel circuit model

Case 2 REPEATED ROOTS: When the denominator of function $F(s)$ has repeated roots, the corresponding partial-fraction terms are of the form

$$F(s) = \frac{\text{Numerator polynomial}}{(s+p_1)^r(s+p_2)\cdots(s+p_n)}$$

$$= \frac{k_1}{s+p_1} + \frac{k_2}{(s+p_1)2} + \cdots + \frac{k_r}{(s+p_1)^r}$$

+terms for other different roots.

where

$$k_i = \frac{1}{(r-i)!}\left[\frac{d^{r-i}}{dt^{r-i}}[(s+p_1)^r F(s)]\right]\Bigg|_{s=-p_1}$$

Case 3 COMPLEX-CONJUGATE ROOTS: When the denominator of function $F(s)$ has complex-conjugate roots, $-\alpha \pm j\omega$, the corresponding partial-fraction terms are of the form

$$F(s) = \frac{\text{Numerator polynomial}}{[(s+\alpha)^2+\omega^2](s+p_3)\cdots(s+p_n)}$$

$$= \frac{k_1 s + k_2}{(s+\alpha)^2+\omega^2} + \frac{k_3}{s+p_3} + \cdots + \frac{k_n}{(s+p_n}$$

After simplification, we get

Numerator polynomial = Polynomial in s with unknown coefficients

Equating the coefficients on the both sides of this equation, a set of simultaneous equations in unknown coefficients can be formed. The solution of these equations gives the values of these coefficients. This method of finding the coefficients (residues) is known as the Heaviside expansion method.

4.4 SOLVED PROBLEMS

SP 4.1 In the series RC circuit of Fig. SP 4.1, the capacitor has initial charge 0.2 C. At $t = 0$, the switch is closed. Use the Laplace transform method to find the current $i(t)$ for $t > 0$.

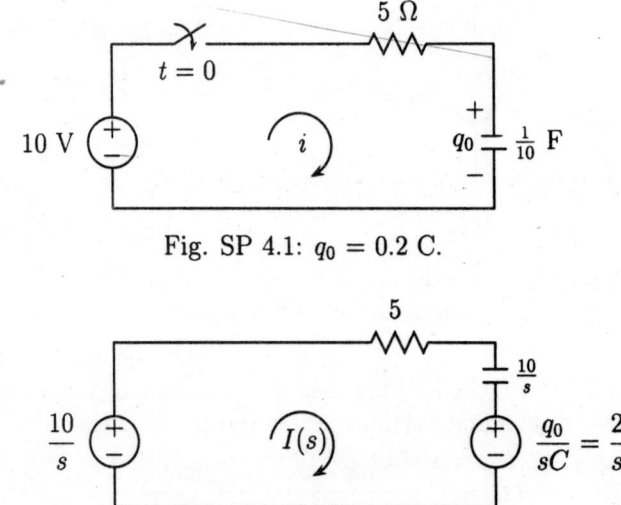

Fig. SP 4.1: $q_0 = 0.2$ C.

Fig. 4.1.1: s-domain circuit.

SOLUTION:

By KVL in the circuit of Fig. 4.1.1

$$(5 + 10/s)I(s) + 2/s = 10/s$$

or $$I(s) = \frac{10/s - 2/s}{5 + 10/s} = \frac{8}{5s + 10} = \frac{8/5}{s + 2}$$

Therefore $$i(t) = \mathcal{L}^{-1}[I(s)] = \frac{8}{5}e^{-2t}u(t) \text{ A}$$

SP 4.2 In the network shown in the Fig. SP 4.2, capacitor is charged to 10 V, and the switch is closed at $t = 0$. Solve for the current $i(t)$ using Laplace transformation.

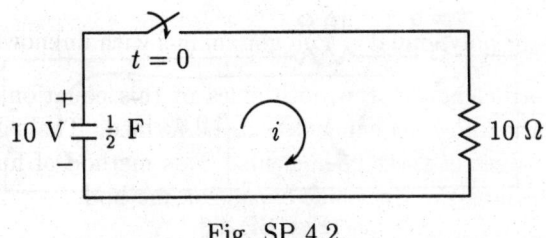

Fig. SP 4.2.

Fig. 4.2.1: s-domain circuit.

SOLUTION:

By KVL in the circuit of Fig. 4.2.1,

$$(2/s + 10)I(s) = 10/s$$

or $\qquad I(s) = \dfrac{10/s}{2/s + 10} = \dfrac{1}{s + 0.2}$

Therefore $\qquad i(t) = \mathcal{L}^{-1}[I(s)] = e^{-0.2t}u(t)$

SP 4.3 Find the loop current i_1 and i_2 for the circuit of Fig. SP 4.3 when the switch is closed at $t = 0$.

SOLUTON:

From Fig. 4.3.1, KVL equations are

$$10I_1(s) + \frac{5}{s}I_1(s) + 10I_2(s) = \frac{50}{s} \qquad \text{or, } (10 + 5/s)I_1(s) + 10I_2(s) = \frac{50}{s} \qquad [4.3.1]$$

and $\quad 10I_1(s) + 10I_2(s) + 40I_2(s) = \dfrac{50}{s} \qquad \text{or, } 10I_1(s) + 50I_2(s) = \dfrac{50}{s} \qquad$ [4.3.2]

Using Cramer's rule to eqns [4.3.1] and [4.3.2], we have

$$I_1(s) = \dfrac{\begin{vmatrix} 50/s & 10 \\ 50/s & 50 \end{vmatrix}}{\begin{vmatrix} 10 + 5/s & 10 \\ 10 & 50 \end{vmatrix}} = \dfrac{2000/s}{(400s + 250)/s} = \dfrac{5}{s + 0.625}$$

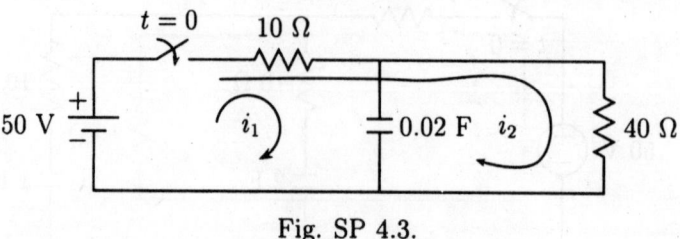

Fig. SP 4.3.

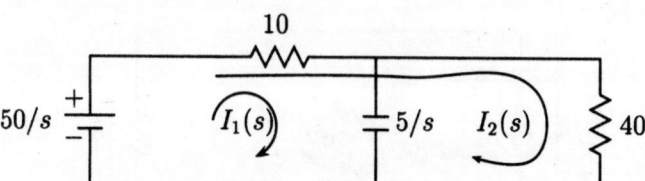

Fig. 4.3.1: s-domain circuit.

Therefore

$$i_1(t) = \mathcal{L}^{-1}\left[\frac{5}{s + 0.625}\right] = 5e^{-0.625t}$$

Similarly

$$I_2(s) = \frac{\begin{vmatrix} 10 + 5/s & 50/s \\ 10 & 50/s \end{vmatrix}}{\begin{vmatrix} 10 + 5/s & 10 \\ 10 & 50 \end{vmatrix}} = \frac{0.625}{s(s + 0.625)} = \frac{1}{s} - \frac{1}{s + 0.625}$$

and

$$i_2(t) = \mathcal{L}^{-1}\left[\frac{1}{s} - \frac{1}{s + 0.625}\right] = (1 - e^{-0.625t})u(t) \text{ A}$$

SP 4.4 Find $i(t)$ in the circuit of Fig. SP 4.4, if the initial conditions are all zero and the switch is closed $t = 0$.

SOLUTION:

It is observed from the circuit of Fig. 4.4.1 that input impedance is

$$Z(s) = 20 + \frac{(10 + 1/2s)(10 + 1/s)}{10 + 1/2s + 10 + 1/s} = 20 + \frac{200s^2 + 30s + 1}{s(40s + 3)} = \frac{1000s^2 + 90s + 1}{s(40s + 3)}$$

So $$I(s) = \frac{V(s)}{Z(s)} = \frac{50/s}{\frac{1000s^2 + 90s + 1}{s(40s + 3)}} = \frac{2(s + 3/40)}{(s^2 + 0.09s + 0.001)} = \frac{2(s + 3/40)}{(s + 0.013)(s + 0.077)}$$

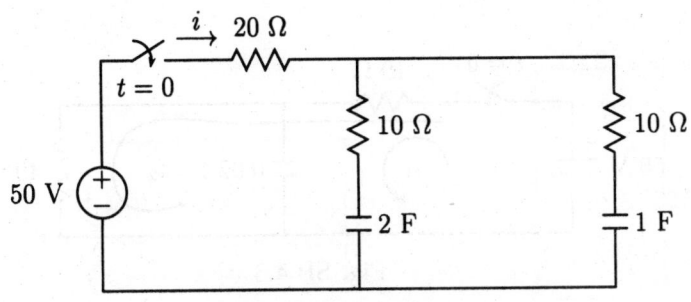

Fig. SP 4.4.

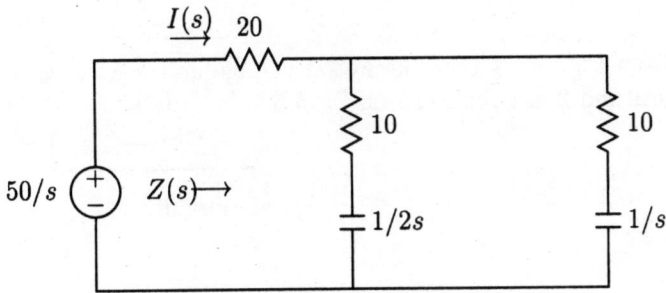

Fig. 4.4.1 : s-domain circuit.

However
$$\frac{2(s+3/40)}{(s+0.013)(s+0.077)} = \frac{A}{s+0.013} + \frac{B}{s+0.077}$$

where
$$A = \frac{2(s+3/40)}{s+0.077} \bigg|_{s=-0.013} = 1.9375$$

and
$$B = \frac{2(s+3/40)}{s+0.013} \bigg|_{s=-0.077} = 0.0625$$

Hence
$$I(s) = \frac{1.938}{s+0.013} + \frac{0.063}{s+0.077}$$

Therefore,
$$i(t) = \mathcal{L}^{-1}[I(s)] = \left(1.938e^{-0.013t} + 0.063e^{-0.077t}\right) u(t) \text{ A}$$

SP 4.5 With switch in a position a, the network shown in Fig. SP 4.5 attains equilibrium. At $t = 0$, the switch is moved to position b. Find the voltage across R_2 as a function of time.

SOLUTION:

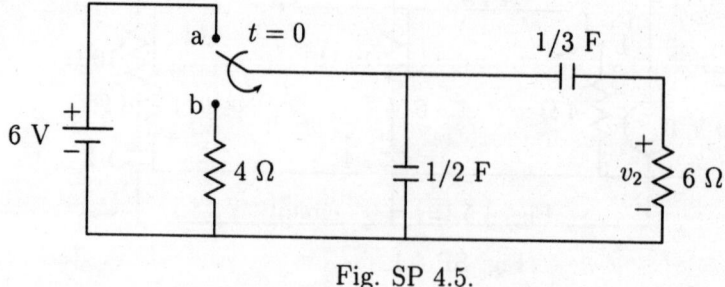

Fig. SP 4.5.

It observed from Fig. 4.5.1 that the initial voltage across each capacitor is 6 V. By KCL at the node 1 and 2 in circuit of Fig. SP 4.5.2

Node 1 :
$$\frac{V_1(s)}{4} + \frac{V_1(s)}{2/s} + \frac{V_1(s) - V_2(s)}{3/s} = 3 + 2$$

Thus
$$(10s + 3)V_1(s) - 4sV_2(s) = 60 \qquad\qquad [4.5.1]$$

Node 2 :
$$\frac{V_1(s) - V_2(s)}{3/s} = 2 + \frac{V_2(s)}{6}$$

Thus
$$2sV_1(s) - (2s + 1)V_2(s) = 12 \qquad\qquad [4.5.2]$$

The $V_2(s)$ can be determined by Cramer's rule as

$$V_2(s) = \frac{\begin{vmatrix} 10s + 3 & 60 \\ 2s & 12 \end{vmatrix}}{\begin{vmatrix} 10s + 3 & -4s \\ 2s & -(2s + 1) \end{vmatrix}} = \frac{120s + 36 - 120s}{-(10s + 3)(2s + 1) + 8s^2}$$

or
$$V_2(s) = \frac{-3}{(s + 0.226)(s + 1.107)} = \frac{A}{s + 0.226} + \frac{B}{s + 1.107}$$

where
$$A = \text{-3.41}$$

and
$$B = 3.41. \text{ Thus}$$

Hence
$$V_2(s) = \frac{-3.41}{s + 0.226} + \frac{3.41}{s + 1.107}$$

and
$$v_2(t) = \mathcal{L}^{-1}[V_2(s)] = 3.41(e^{-1.107t} - e^{-0.226t})u(t) \text{ V}$$

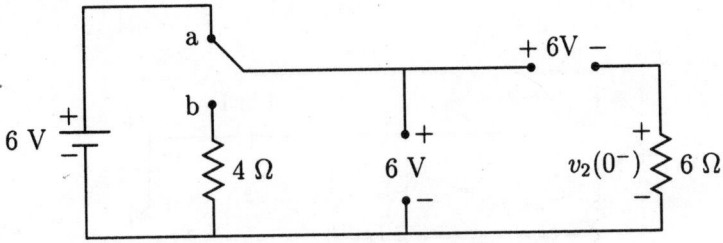

Fig. 4.5.1: $t = 0^-$ circuit.

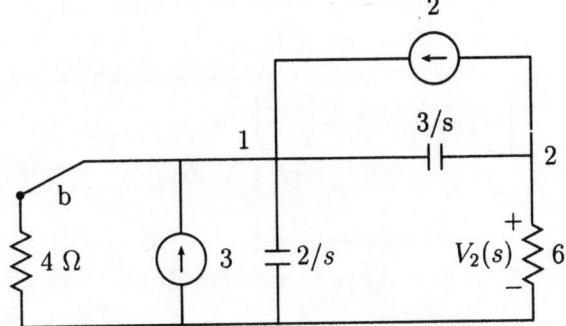

Fig. 4.5.2: s-domain circuit for $t > 0$.

SP 4.6 Find $v(t)$ for the circuit of Fig. SP 4.6 subject to the initial $v(0) = 2$ V.

SOLUTION:

By KCL at the node $V_1(s)$

$$I(s) + 2I(s) = \frac{V_1(s) - V(s)}{5}$$

or

$$3I(s) = \frac{V_1(s) - V(s)}{5}$$

or

$$3\left[\frac{-4/s - V_1(s)}{3}\right] = \frac{V_1(s) - V_2(t)}{5}$$

Thus

$$6V_1(s) - V(s) = -20/s \qquad [4.6.1]$$

By KCL at the node $V(s)$

$$\frac{V_1(s) - V(s)}{5} = \frac{V(s) - 2/s}{1/s}$$

or

$$V_1(s) - (1 + 5s)V(s) = -10 \qquad [4.6.2]$$

From eqn [4.6.1] and eqn [4.6.2] , we can determine V(s) as

$$V(s) = \frac{2(s - 1/3)}{s(s + 1/6)} = \frac{-4}{s} + \frac{6}{s + 1/6}$$

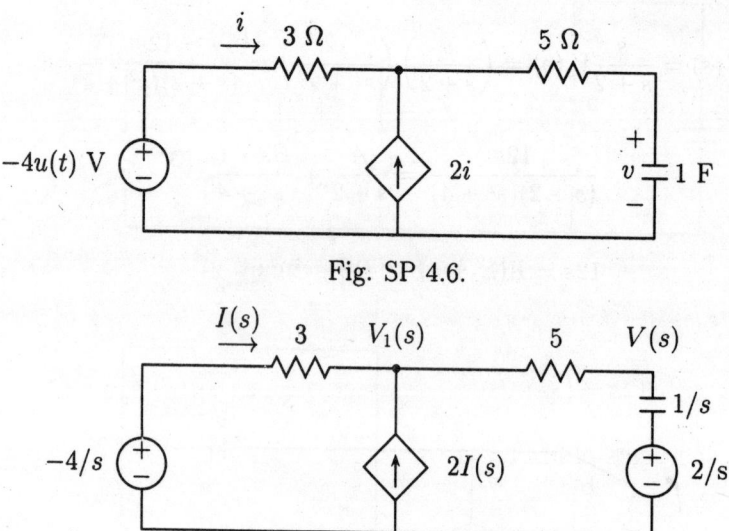

Fig. SP 4.6.

Fig. 4.6.1: s-domain circuit for $t > 0$.

Hence

$$v(t) = -4u(t) + 6e^{-t/6}u(t) = (-4 + 6e^{-t/6})u(t) \text{ V}$$

SP 4.7 The circuit in Fig. SP 4.7 has a sinusoidal voltage excitation which begins at $t = 0$. Find $i(t)$.

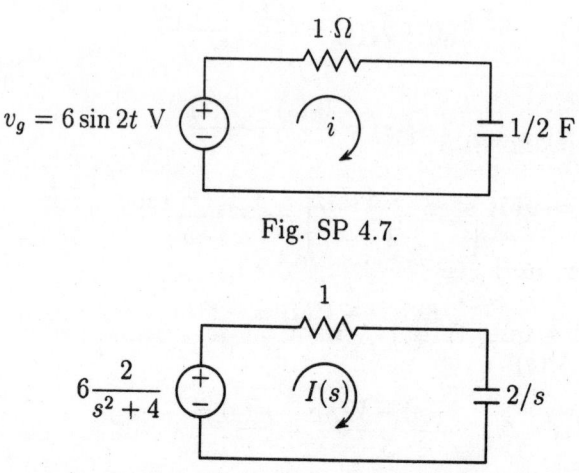

Fig. SP 4.7.

Fig. 4.7.1: s-domain circuit for $t > 0$.

SOLUTION:

By KVL in Fig. 4.7.1,

$$1I(s) + (2/s)I(s) = V_g(s)$$

Thus

$$I(s) = \frac{s}{s+2}V_g(s) = \left(\frac{s}{s+2}\right)\left(\frac{12}{s^2+2^2}\right) = \frac{12s}{(s+2)(s^2+4)} \qquad [4.7.1]$$

However

$$\frac{12s}{(s+2)(s^2+4)} = \frac{A}{s+2} + \frac{Bs+C}{s^2+4}$$

or

$$12s = A(s^2+4) + (Bs+C)(s+2)$$

On rearranging, we get

$$12s = (A+B)s^2 + (2B+C)s + (4A+2C) \qquad [4.7.2]$$

Comparing the coefficients of like terms in eqn [4.7.2], we get

$$A+B = 0 \qquad [4.7.3]$$

$$2B+C = 12 \qquad [4.7.4]$$

$$4A+2C = 0 \qquad [4.7.5]$$

Solving the above set of equations, we get $A = -3$, $B = 3$, and $C = 6$.

Hence
$$I(s) = -\frac{3}{s+2} + \frac{3s}{s^2+2^2} + \frac{3(2)}{s^2+2^2}$$

and the resulting current is

$$i(t) = -3e^{-2t}u(t) + 3\cos 2t\, u(t) + 3\sin 2t\, u(t) \text{ A}$$

SP 4.8 For the circuit shown in Fig. SP 4.8, find

(a) $v_C(t)$ if $v_g = u(t) =$ unit step function (or, unit step response for $v_C(t)$)

(b) $i(t)$ if $v_g = u(t)$(or, unit step response for $i(t)$

(c) $v_C(t)$ if $v_g = \delta(t)=$ unit impulse function (or, unit impulse response for $v_C(t)$)

(d) i(t) if $v_g = u(t-1)$

(e) $v_C(t)$ if $v_g = d\delta(t)/dt$

(f) $v_C(t)$ if $v_g = d\delta(t-1)/dt$

(g) $v_C(t)$ if $v_g = tu(t)$

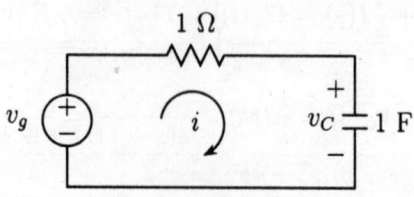

Fig. SP 4.8.

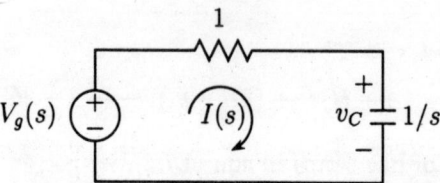

Fig. 4.8.1: s-domain circuit for $t > 0$.

SOLUTION:

(a) By KVL in Fig. 4.8.1

$$1I(s) + \frac{1}{s}I(s) = \mathcal{L}[u(t)] = \frac{1}{s} \quad \Rightarrow \quad I(s) = \frac{1}{s+1} \qquad [4.8.1]$$

Hence

$$V_C(s) = \frac{1}{s}I(s) = \frac{1}{s(s+1)} = \frac{A}{s} + \frac{B}{s+1} \qquad [4.8.2]$$

$$\text{where} \qquad A = \left.\frac{1}{s+1}\right|_{s=0} = 1$$

$$\text{and} \qquad B = \left.\frac{1}{s}\right|_{s=-1} = -1$$

$$\text{Hence} \qquad V_C(s) = \frac{1}{s} - \frac{1}{s+1}$$

and the resulting capacitor voltage is

$$v_C(t) = u(t) - e^{-t}u(t) = (1 - e^{-t})u(t) \text{ V}$$

(b) We know from eqn [4.8.1] that

$$I(s) = \frac{1}{s+1}$$

and the resulting current is

$$i(t) = \mathcal{L}^{-1}[i(t)] = e^{-t}\, u(t) \text{ A}$$

(c) By KVL in Fig. 4.8.1, when $v_g(t) = \delta(t)$

$$1I(s) + \frac{1}{s}I(s) = \mathcal{L}[\delta(t)] = 1 \qquad \Rightarrow \quad I(s) = \frac{s}{s+1}$$

Hence

$$V_C(s) = \frac{1}{s}I(s) = \frac{s}{s(s+1)} = \frac{1}{s+1}$$

Therefore

$$v_C(t) = \mathcal{L}^{-1}[V_C(s)] = e^{-t}u(t) \text{ V}$$

(d) By KVL in Fig. 4.8.1, when $v_g(t) = u(t-1)$

$$1I(s) + \frac{1}{s}I(s) = \mathcal{L}[u(t-1)] = \frac{e^{-s}}{s} \qquad \Rightarrow \quad I(s) = \frac{e^{-s}}{s+1}$$

and

$$i(t) = \mathcal{L}^{-1}[I(s)] = e^{-(t-1)}u(t-1) \text{ A}$$

(e) By KVL in Fig. 4.8.1, when $v_g(t) = \frac{\delta(t)}{dt}$

$$1I(s) + \frac{1}{s}I(s) = \mathcal{L}\left[\frac{d\delta}{dt}\right] = s \qquad \Rightarrow \quad I(s) = \frac{s^2}{s+1} \qquad\qquad [4.8.3]$$

Hence

$$V_C(s) = \frac{1}{s}I(s) = \frac{s}{s+1} = \frac{s+1-1}{s+1} = 1 - \frac{1}{s+1}$$

Therefore

$$v_C(t) = \mathcal{L}^{-1}[V_C(s)] = \delta(t) - e^{-t}u(t) \text{ V}$$

(f) By KVL in Fig. 4.8.1, when $v_g(t) = \frac{d\delta}{dt}(t-1)$

$$1I(s) + \frac{1}{s}I(s) = \mathcal{L}\left[\frac{d\delta(t-1)}{dt}\right] = se^{-s} \qquad \Rightarrow \quad I(s) = \frac{s^2 e^{-s}}{s+1} \qquad [4.8.4]$$

Hence

$$V_C(s) = \frac{1}{s}I(s) = \frac{1}{s}\left(\frac{s^2 e^{-s}}{s+1}\right) = \frac{se^{-s}}{s+1} = \left(1 - \frac{1}{s+1}\right)e^{-s}$$

Therefore

$$v_C(t) = \mathcal{L}^{-1}[V_C(s)] = \delta(t-1) - e^{-(t-1)}u(t-1) \text{ V}$$

(g) By KVL in Fig. 4.8.1, when $v_g(t) = tu(t)$

$$1I(s) + \frac{1}{s}I(s) = \mathcal{L}[tu(t)] = \frac{1}{s^2} \qquad \Rightarrow \quad I(s) = \frac{1}{s(s+1)}$$

Hence

$$V_C(s) = \frac{1}{s}I(s) = \frac{1}{s^2(s+1)} = \frac{A}{s^2} + \frac{B}{s} + \frac{C}{s+1}$$

where
$$A = \frac{1}{s+1}\bigg|_{s=0} = 1$$

$$B = \frac{d}{dt}\left(\frac{1}{s+1}\right)\bigg|_{s=0} = -\frac{1}{(s+1)^2}\bigg|_{s=0} = -1$$

$$C = \frac{1}{s^2}\bigg|_{s=-1} = 1$$

Hence

$$V_C(s) = \frac{1}{s^2(s+1)} = \frac{1}{s^2} - \frac{1}{s} + \frac{1}{s+1}$$

Therefore

$$v_C(t) = \mathcal{L}^{-1}[V_C(s)] = (t - 1 + e^{-t})u(t) \text{ V}$$

SP 4.9 In the RL circuit shown in Fig. SP 4.9, the switch is moved from position a to position b at $t = 0$, a steady state having been established at position a. Solve for $i(t)$, using Laplace Transform method.

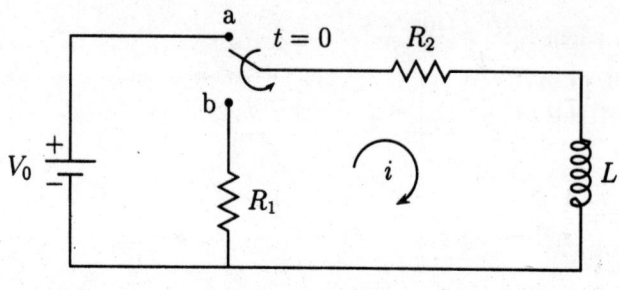

Fig. SP 4.9.

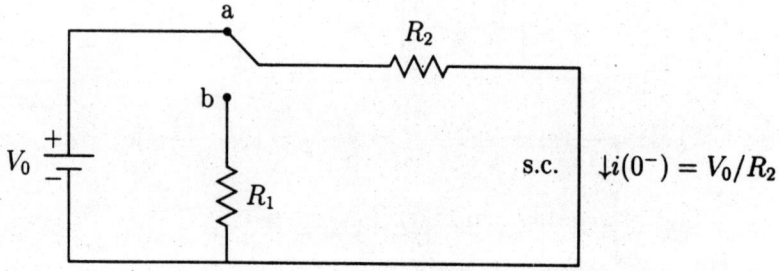

Fig. 4.9.1: $t = 0^-$ circuit.

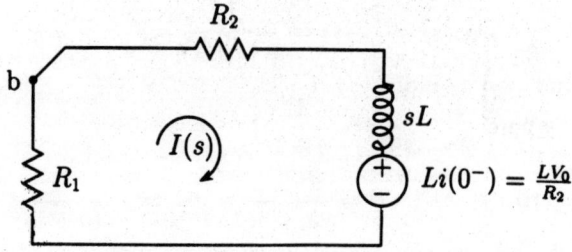

Fig. 4.9.2 : s-domain circuit for $t > 0$.

SOLUTION:

From Fig.4.9.1

$$i(0^-) = V_0/R_2 = i(0^+)$$

Applying KVL in Fig. 4.9.2

$$(R_1 + R_2 + sL)I(s) - Li(0^+) = 0$$

Thus

$$I(s) = \frac{Li(0^-)}{R_1 + R_2 + sL} = \frac{L(V_0/R_2)}{L(s + \frac{R_1+R_2}{L})} = \frac{V_0/R_2}{s + \frac{R_1+R_2}{L}}$$

and

$$i(t) = \frac{V_0}{R_2} e^{-(R_1+R_2)t/L} u(t)$$

SP 4.10 In the RL circuit shown in Fig. SP 4.10, the switch is in position 1 for long enough to establish steady state condition. At $t = 0$ it is switched to position 2. Find the resulting current $i(t)$.

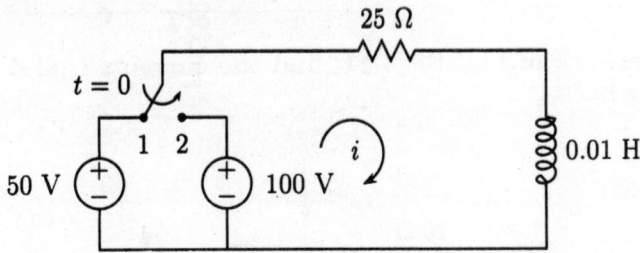

Fig. SP 4.10.

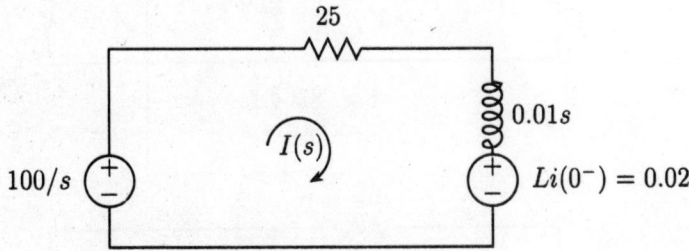

Fig. 4.10.1: s-domain circuit for $t > 0$.

SOLUTION:

The initial condition as indicated in the Fig. SP 4.10 is given by

$$i(0^-) = 50/25 = 2 \text{ A} = i(0^+)$$

By KVL in Fig. 4.10.1,

$$25I(s) + 0.01sI(s) - 0.02 - 100/s = 0$$

Hence $\qquad I(s) = \dfrac{0.02}{0.01s + 25} + \dfrac{100/s}{0.01s + 25} = \dfrac{2}{s + 2500} + \dfrac{10^4}{s(s + 2500)}$

However $\qquad\qquad \dfrac{10^4}{s(s + 2500)} = \dfrac{A}{s} + \dfrac{B}{s + 2500}$

where

$$A = \dfrac{10^4}{s + 2500}\bigg|_{s=0} = 4$$

$$B = \dfrac{10^4}{s}\bigg|_{s=-2500} = -4$$

Thus

$$I(s) = \dfrac{2}{s + 2500} + \dfrac{4}{s} - \dfrac{4}{s + 2500} = -\dfrac{2}{s + 2500} + \dfrac{4}{s}$$

and

$$i(t) = \mathcal{L}^{-1}[I(s)] = (4 - 2e^{-2500t})u(t) \text{ V}$$

SP 4.11 In the network of Fig. SP 4.11, find the currents i_1 and i_2 when the switch is closed at $t = 0$.

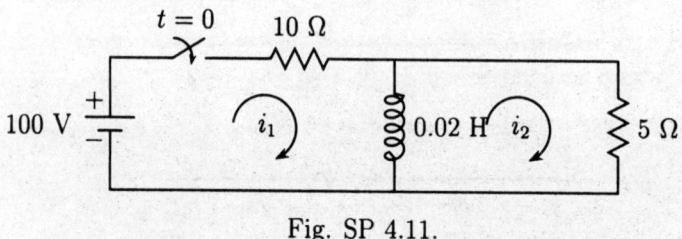

Fig. SP 4.11.

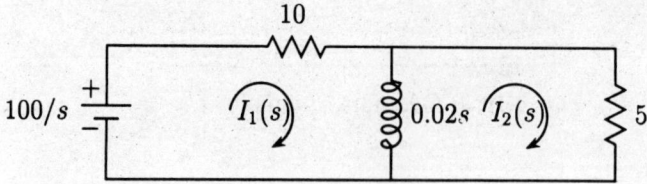

Fig. 4.11.1: s-domain circuit.

SOLUTION:

Writing mesh equations for the circuit in Fig. 4.11.1
Mesh 1:

$$(10 + 0.02s)I_1(s) - 0.02sI_2(s) = 100/s \qquad\qquad [4.11.1]$$

Mesh 2:

$$-0.02sI_1(s) + (5 + 0.02s)I_2(s) = 0 \qquad [4.11.2]$$

Applying Cramer's rule to eqns [4.11.1] and [4.11.2] as

$$I_1(s) = \frac{\begin{vmatrix} 100/s & -0.02s \\ 0 & 5 + 0.02s \end{vmatrix}}{\begin{vmatrix} 10 + 0.02s & -0.02s \\ -0.02s & 5 + 0.02s \end{vmatrix}}$$

$$= \frac{100(5 + 0.02s)/s}{(10 + 0.02s)(5 + 0.02s) - (0.02s)^2} = \frac{6.67(s + 250)}{s(s + 166.7)} = \frac{10}{s} - \frac{3.33}{s + 166.7}$$

Therefore

$$i_1(t) = \mathcal{L}^{-1}[I_1(s)] = (10 - 3.33e^{-166.7t})u(t) \text{ A}$$

$$I_2(s) = \frac{\begin{vmatrix} 10 + 0.02s & 100/s \\ -0.02s & 0 \end{vmatrix}}{\begin{vmatrix} 10 + 0.02s & -0.02s \\ -0.02s & 5 + 0.02s \end{vmatrix}} = \frac{6.67}{s + 166.7}$$

Therefore

$$i_2(t) = \mathcal{L}^{-1}[I_2(s)] = 6.67e^{-166.7t} u(t) \text{ A}$$

SP 4.12 Find $i(t)$ using Laplace Transform method for the circuit shown in Fig. SP 4.12 when switch is closed at $t = 0$.

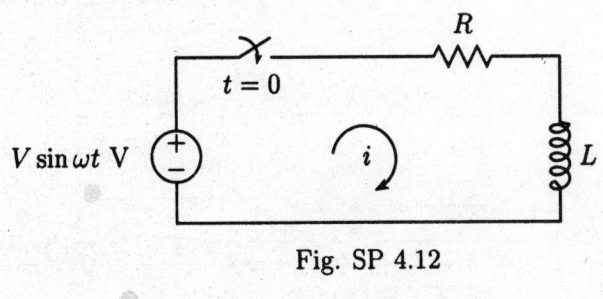

Fig. SP 4.12

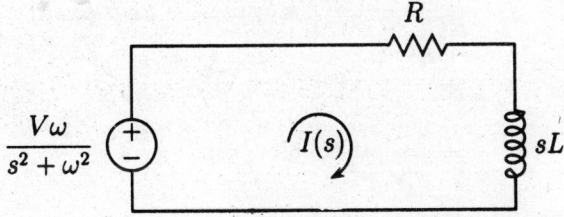

Fig. 4.12.1: s-domain circuit.

SOLUTION:

By KVL in the circuit of Fig. 4.12.1

$$RI(s) + sLI(s) = \frac{V\omega}{s^2 + \omega^2}$$

Therefore

$$I(s) = \frac{V\omega}{(s^2 + \omega^2)(sL + R)} = \frac{V\omega}{L(s^2 + \omega^2)(s + R/L)}$$

However

$$\frac{1}{(s^2 + \omega^2)(s + R/L)} = \frac{A}{s + R/L} + \frac{Bs + C}{s^2 + \omega^2}$$

or

$$1 = A(s^2 + \omega^2) + (Bs + C)(s + R/L)$$

Hence

$$1 = (A + B)s^2 + (BR/L + C)s + RC/L + A\omega^2$$

Comparing the coefficients of like terms

$$A + B = 0$$
$$BR/L + C = 0$$
$$RC/L + A\omega^2 = 1$$

From the above sets of equation, we can get

$$A = \frac{L^2}{R^2 + \omega^2 L^2}; \quad B = -\frac{L^2}{R^2 + \omega^2 L^2} \quad \text{and} \quad C = \frac{RL}{R^2 + \omega^2 L^2}$$

Hence

$$I(s) = \frac{V\omega}{L}\left[\left(\frac{L^2}{R^2 + \omega^2 L^2}\right)\frac{1}{s + R/L} - \left(\frac{L^2}{R^2 + \omega^2 L^2}\right)\frac{s}{s^2 + \omega^2} + \left(\frac{RL}{R^2 + \omega^2 L^2}\right)\frac{1}{s^2 + \omega^2}\right]$$

$$= \left(\frac{\omega LV}{R^2 + \omega^2 L^2}\right)\frac{1}{s + R/L} - \left(\frac{\omega LV}{R^2 + \omega^2 L^2}\right)\frac{s}{s^2 + \omega^2} + \left(\frac{RV}{R^2 + \omega^2 L^2}\right)\frac{\omega}{s^2 + \omega^2}$$

Therefore

$$
\begin{aligned}
i(t) &= \frac{\omega LV}{R^2 + \omega^2 L^2}e^{-\frac{R}{L}t} - \frac{\omega LV}{R^2 + \omega^2 L^2}\cos\omega t + \frac{RV}{R^2 + \omega^2 L^2}\sin\omega t \\
&= \frac{\omega LV}{R^2 + \omega^2 L^2}e^{-\frac{R}{L}t} - \frac{V}{R^2 + \omega^2 L^2}(\omega L\cos\omega t - R\sin\omega t) \\
&= \frac{\omega LV}{R^2 + \omega^2 L^2}e^{-\frac{R}{L}t} - \frac{V\sqrt{R^2 + \omega^2 L^2}}{R^2 + \omega^2 L^2}\left[\frac{\omega L}{\sqrt{R^2 + \omega^2 L^2}}\cos\omega t - \frac{R}{\sqrt{R^2 + \omega^2 L^2}}\sin\omega t\right] \\
&= \frac{\omega LV}{R^2 + \omega^2 L^2}e^{-\frac{R}{L}t} - \frac{V\sqrt{R^2 + \omega^2 L^2}}{R^2 + \omega^2 L^2}[\sin\theta\cos\omega t - \cos\theta\sin\omega t] \\
&= \frac{\omega LV}{R^2 + \omega^2 L^2}e^{-\frac{R}{L}t} + \frac{V}{\sqrt{R^2 + \omega^2 L^2}}[\sin(\omega t - \theta)] \\
&= \frac{\omega LV}{R^2 + \omega^2 L^2}e^{-\frac{R}{L}t} + \frac{V}{\sqrt{R^2 + \omega^2 L^2}}\sin(\omega t - \tan^{-1}\frac{\omega L}{R})
\end{aligned}
$$

where $\tan\theta = \frac{\omega L}{R}$ from the impedance triangle.

SP 4.13 In the network shown in Fig. SP 4.13, the switch is closed at $t = 0$. Find i_1 and i_2

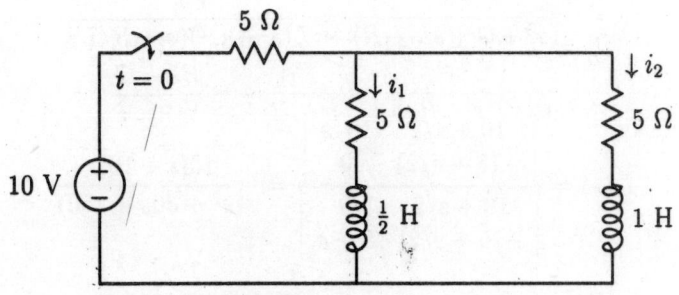

Fig. SP 4.13.

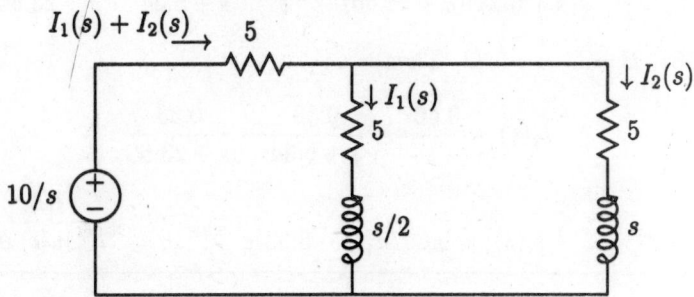

Fig. 4.13.1 : s-domain circuit $t > 0$.

SOLUTION:

By KVL in mesh 1 and mesh 2 for the circuit in Fig. 4.13.1
Mesh 1:

$$5[I_1(s) + I_2(s)] + 5I_1(s) + (s/2)I_1(s) = 10/s$$

or

$$(10 + s/2)I_1(s) + 5I_2(s) = 10/s \qquad [4.13.1]$$

Mesh 2:

$$-(5 + s/2)I_1(s) + +(5 + s)I_2(s) = 0 \qquad [4.13.2]$$

Applying Cramer's rule to determine $I_1(s)$ and $I_2(s)$ as

$$I_1(s) = \frac{\begin{vmatrix} 10/s & 5 \\ 0 & s+5 \end{vmatrix}}{\begin{vmatrix} 10 + s/2 & 5 \\ -(5 + s/2) & 5 + s \end{vmatrix}} = \frac{20(s + 5)}{s(s^2 + 30s + 150)}$$

or $\qquad I_1(s) = \dfrac{20(s + 5)}{s(s + 6.34)(s + 23.66)} = \dfrac{A}{s} + \dfrac{B}{s + 6.34} + \dfrac{C}{s + 23.66}$

where $A = 0.667$, $B = 0.244$, and $C = -0.910$. Then

$$I_1(s) = \frac{0.667}{s} + \frac{0.244}{s + 6.34} - \frac{0.910}{s + 23.66}$$

and

$$i_1(t) = (0.667 + 0.244e^{-6.34t} - 0.910e^{-23.66t})u(t) \text{ A}$$

Similarly

$$I_2(s) = \frac{\begin{vmatrix} 10 + s/2 & 10/s \\ -(5 + s/2) & 0 \end{vmatrix}}{\begin{vmatrix} 10 + s/2 & 5 \\ -(5 + s/2) & 5 + s \end{vmatrix}} = \frac{10(s + 10)}{s(s^2 + 30s + 150)}$$

or

$$I_2(s) = \frac{10(s + 10)}{s(s + 6.34)(s + 23.66)} = \frac{A_1}{s} + \frac{B_1}{s + 6.34} + \frac{C_1}{s + 23.66}$$

where $A = 0.667$, $B = C = -0.33$. Then

$$I_2(s) = \frac{0.667}{s} - \frac{0.33}{s + 6.34} - \frac{0.33}{s + 23.66}$$

Therefore

$$i_2(t) = \mathcal{L}^{-1}[I_2(s)] = 0.667u(t) - 0.33(e^{-6.34t} + e^{-23.66t})u(t) \text{ A}$$

SP 4.14 In the network shown in Fig. SP 4.14, capacitor is initially charged to 10 V. The switch is closed at $t = 0$. Solve for $i(t)$.

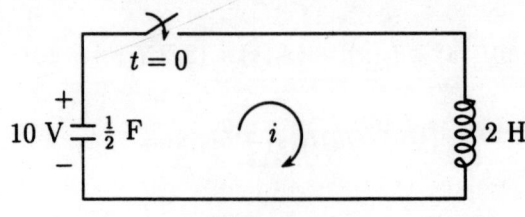

Fig. SP 4.14.

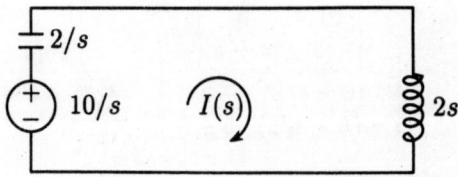

Fig. 4.14.1: s-domain circuit.

SOLUTION:

By KVL in the circuit of Fig. 4.14.1

$$-\frac{10}{s} + (2/s + 2s)I(s) = 0$$

or

$$I(s) = \frac{10/s}{2/s + 2s} = \frac{5}{s^2 + 1}$$

Therefore

$$i(t) = \mathcal{L}^{-1}[I(s)] = 5\sin t\, u(t)$$

SP 4.15 The switch in the circuit of Fig. SP 4.15 was at position 1 for a long time. And at $t = 0$, the switch is moved to position 2. Find $i(t)$ for $t > 0$.

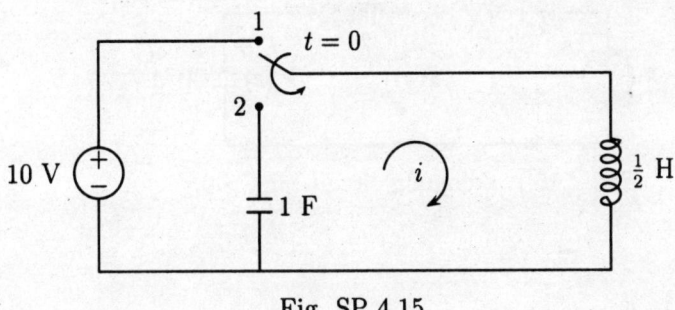

Fig. SP 4.15.

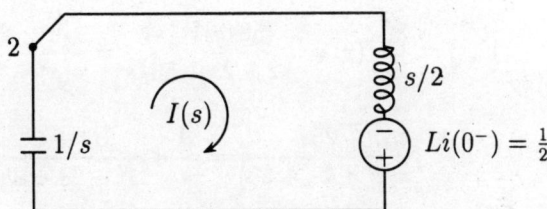

Fig. 4.15.1: s-domain circuit for $t > 0$.

SOLUTION:

The inductor current at $t = 0^-$ is

$$i(0^-) = 10/10 = 1\,\text{A} = i(0^+)$$

KVL in the circuit of Fig. 4.15.1 gives

$$\bar{I}(s) = \frac{1/2}{s/2 + 1/s} = \frac{s}{s^2 + 2}$$

Hence

$$i(t) = \mathcal{L}^{-1}[I(s)] = \cos\sqrt{2}t\, u(t)$$

SP 4.16 The series RLC circuit shown in Fig. SP 4.16 has zero initial conditions, find unit step response for $v(t)$.

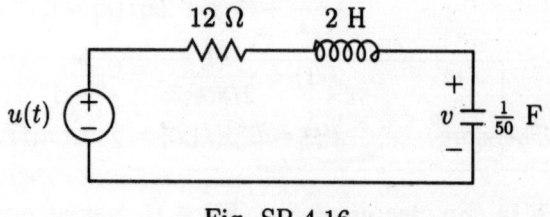

Fig. SP 4.16.

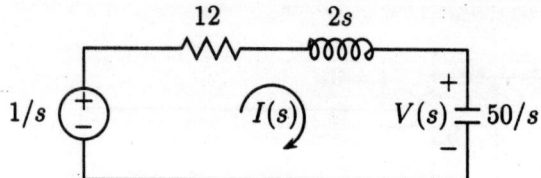

Fig. 4.16.1: s-domain circuit for $t > 0$.

SOLUTION:

By KVL in Fig. 4.16.1

$$12I(s) + 2sI(s) + \frac{50}{s}I(s) = V_g(s) \qquad\qquad \Rightarrow I(s) = \frac{V_g(s)}{12 + 2s + 50/s} \qquad [4.16.1]$$

Hence

$$V(s) = \frac{50}{s}I(s) = \frac{(50/s)V_g(s)}{12 + 2s + 50/s} = \frac{25}{s(s^2 + 6s + 25)} \qquad [4.16.2]$$

However

$$\frac{25}{s(s^2 + 6s + 25)} = \frac{A}{s} + \frac{Bs + C}{s^2 + 6s + 25}$$

or

$$25 = A(s^2 + 6s + 25) + Bs^2 + Cs$$

Rearranging

$$25 = (A + B)s^2 + (6A + C)s + 25A \qquad [4.16.3]$$

Comparing coefficients of like terms of eqn [4.16.3], we have $25A = 25$ $\quad 6A + C = 0$ and $A + B = 0$. Solving the above set of equations, we get $A = 1$, $B = -1$, and $C = -6$. Hence

$$V(s) = \frac{25}{s(s^2 + 6s + 25)} = \frac{1}{s} - \frac{s + 6}{s^2 + 6s + 25} = \frac{1}{s} - \frac{s + 3}{(s + 3)^2 + 4^2} - \left(\frac{3}{4}\right)\frac{4}{(s + 3)^2 + 4^2}$$

and

$$v(t) = \mathcal{L}^{-1}[V(s)] = u(t) - e^{-3t}\cos 4t\, u(t) - \frac{3}{4}e^{-3t}\sin 4t\, u(t)\ A$$

SP 4.17 In Fig. SP 4.17, $R = L = C = 1$, $v_C(0^-) = 1$ V, $i_L(0^-) = 1$ A and $i_1(t) = \delta(t)$; **using Laplace transform, find** $v_2(t)$ **for** $t > 0$.

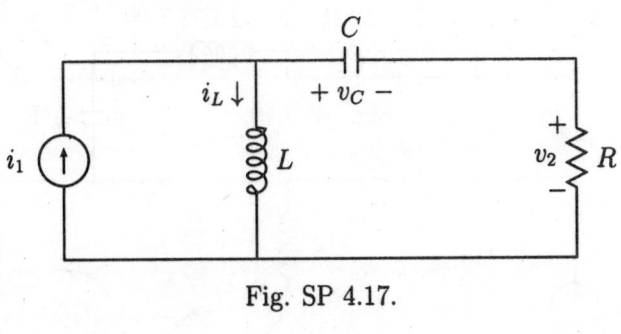

Fig. SP 4.17.

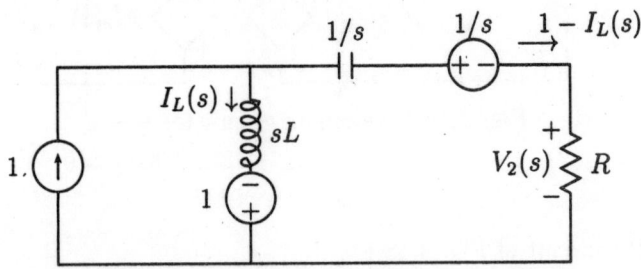

Fig. 4.17.1: s-domain circuit for $t > 0$.

SOLUTION:

Taking KVL in the right mesh of the circuit of Fig. 4.17.1 as

$$1 - sI_L(s) + \frac{1}{s}[1 - I_L(s)] + \frac{1}{s} + 1[1 - I_L(s)] = 0$$

Thus
$$I_L(s) = \frac{2s + 2}{s^2 + s + 1} = \frac{2(s + 1)}{s^2 + 2s(\frac{1}{2}) + \frac{1}{2^2} + 1 - \frac{1}{2^2}} = \frac{2(s + 1/2 + 1/2)}{(s + 1/2)^2 + (\frac{\sqrt{3}}{2})^2}$$

$$= \frac{2(s + 1/2)}{(s + 1/2)^2 + (\frac{\sqrt{3}}{2})^2} + \left(\frac{2}{\sqrt{3}}\right)\frac{\frac{\sqrt{3}}{2}}{(s + 1/2)^2 + (\frac{\sqrt{3}}{2})^2}$$

Hence
$$i_L(t) = e^{-t/2}\left(2\cos\frac{\sqrt{3}}{2}t + \frac{2}{\sqrt{3}}\sin\frac{\sqrt{3}}{2}t\right)u(t)$$

Therefore, $v_2(t) = 1 \times [1 - i_L(t)] = 1 - 2e^{-t/2}\left(\cos\frac{\sqrt{3}}{2}t + \frac{1}{\sqrt{3}}\sin\frac{\sqrt{3}}{2}t\right)u(t)$ A

SP 4.18 Find $i_1(t)$ **and** $i_2(t)$ **if switch in Fig. SP 4.18 is closed at** $t = 0$ **with the circuit previously unenergised.**

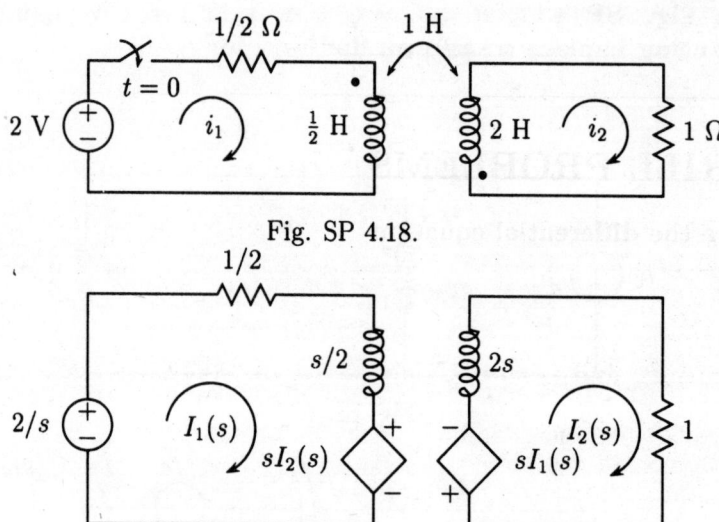

Fig. SP 4.18.

Fig. 4.18.1; s-domain circuit for $t > 0$.

SOLUTION:

By KVL in the circuit of Fig. 4.18.1

Loop 1:

$$(1/2 + s/2)I_1(s) + sI_2(s) = 2/s \qquad [4.18.1]$$

Loop 2:

$$sI_1(s) + (2s + 1)I_2(s) = 0 \qquad [4.18.2]$$

Using Cramer's rule to eqn [4.18.1] and eqn [4.18.2] as:

$$I_1(s) = \frac{\begin{vmatrix} 2/s & s \\ 0 & 2s+1 \end{vmatrix}}{\begin{vmatrix} 1/2+s/2 & s \\ s & 2s+1 \end{vmatrix}} = \frac{2(2s+1)/s}{(1/2+s/2)(2s+1) - s^2}$$

or

$$I(s) = \left(\frac{8}{3}\right) \frac{(s+1/2)}{s(s+1/3)} = \frac{8}{3} \left[\frac{A}{s} + \frac{B}{s+1/3}\right]$$

where

$$A = \left.\frac{s+1/2}{s+1/3}\right|_{s=0} = 3/2 \qquad B = \left.\frac{s+1/2}{s}\right|_{s=-1/3} = -1/2$$

Thus

$$I_1(s) = \frac{4}{3}\left[\frac{3}{s} - \frac{1}{s+1/3}\right] \qquad \text{and} \qquad i_1(t) = \frac{4}{3}\left[3 - e^{-t/3}\right]u(t)$$

Next

$$I_2(s) = \frac{\begin{vmatrix} 1/2+s/2 & 2/s \\ s & 0 \end{vmatrix}}{\begin{vmatrix} 1/2+s/2 & s \\ s & 2s+1 \end{vmatrix}} = -\frac{4}{3}\frac{1}{s+1/3}$$

Therefore

$$i_2(t) = -\frac{4}{3}e^{-\frac{t}{3}}u(t)$$

4.5   DRILL PROBLEMS

DP 4.1 Solve the differential equation

$$\frac{d^2y}{dt^2} + 3\frac{dy}{dt} + 2y = \frac{d^2w}{dt^2} + 2\frac{dw}{dt} + w$$

where y = output, and w = unit impulse input

Hints: Take Laplace Transform on both sides of the equation as

$$s^2Y(s) - sy(0) - y'(0) + 3[sY(s) - y(0)] + 2Y(s) = s^2\mathcal{L}[\delta(t)] + 2s\mathcal{L}[\delta(t)] + \mathcal{L}[\delta(t)]$$

Under zero initial conditions

$$[s^2 + 3s + 2]Y(s) = s^2 + 2s + 1$$

DP 4.2 Find $i_L(t)$ for $t > 0$ in the circuit of Fig. DP 4.2.

Hints : Take KCL at the node V_L of Fig. DP 4.2.2.

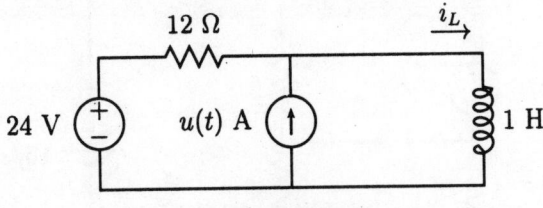

Fig. DP 4.2.

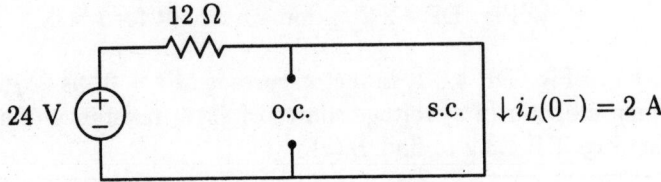

Fig. DP 4.2.1: $t = 0^-$ circuit.

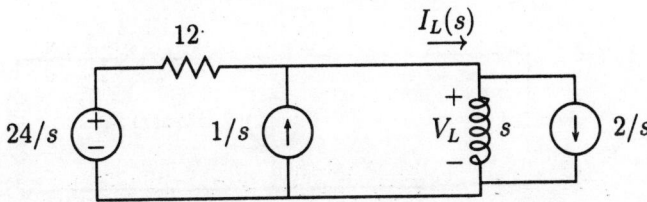

Fig. DP 4.2.2: s-domain circuit.

DP 4.3 Find $i_L(t)$ for $t > 0$ for the circuit of Fig. DP 4.3, if switch is opened at $t = 0$.

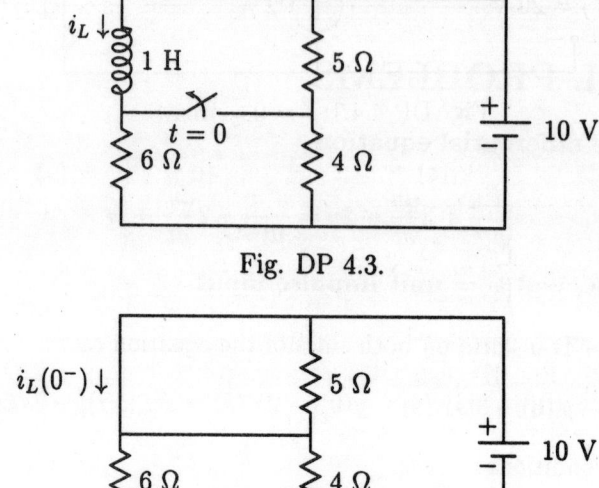

Fig. DP 4.3.

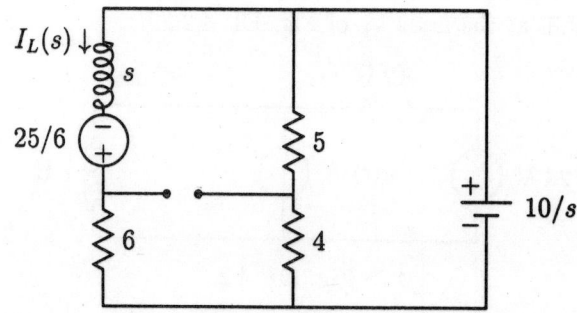

Fig. DP 4.3.1: $t = 0^-$ circuit.

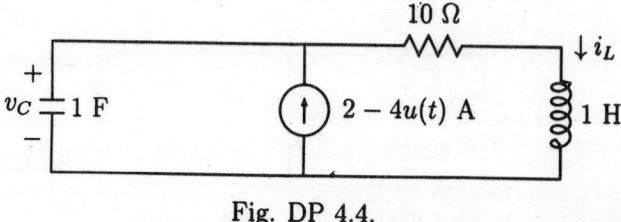

Fig. DP 4.3.2: s-domain circuit for $t > 0$.

Hints : Form Fig. DP 4.3.1, inductor current at $t = 0^-$ is 25/6 A. Use KVL in the loop containing inductor of s, voltage source of 25/6, resistance of 6, and voltage source of $10/s$ in the Fig. DP 4.3.2 to find $I_L(s)$.

DP 4.4 Determine $i_L(t)$ for $t > 0$ for the circuit of Fig. DP 4.4.

Fig. DP 4.4.

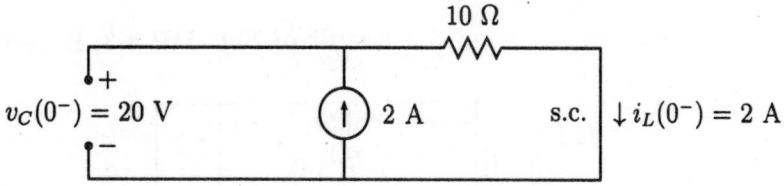

Fig. DP 4.4.1: $t = 0^-$ circuit.

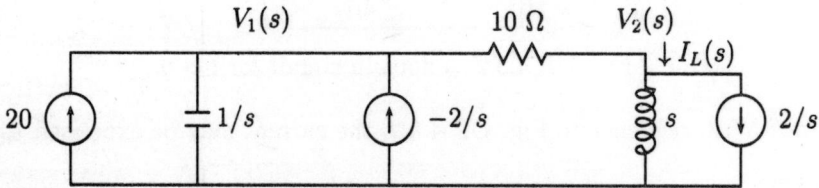

Fig. DP 4.4.2: s-domain circuit for $t > 0$.

Hints :

$$2[1 - 2u(t)] \quad = \quad -2 \text{ A} \quad \text{for } t \geq 0$$
$$= \quad 2 \text{ A} \quad \text{for } t < 0$$

The circuit at t = 0^- gives $i_L(0^-) = 2$ A and $v_C(0^-) = 20$ V. Use nodal method to determine $V_2(s)$ and then $I_L(s) = V_2(s)/s + 2/s$ from Fig. DP 4.4.2.

DP 4.5 Determine $i(t)$ for $t > 0$ for the circuit of Fig. DP 4.5.

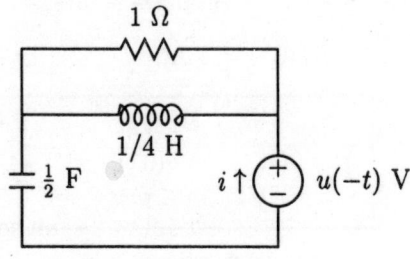

Fig. DP 4.5.

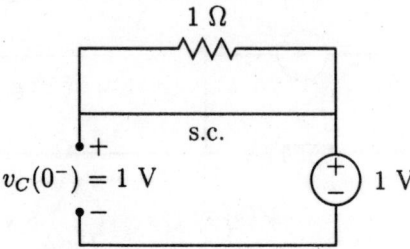

Fig. DP 4.5.1: $t = 0^-$ circuit.

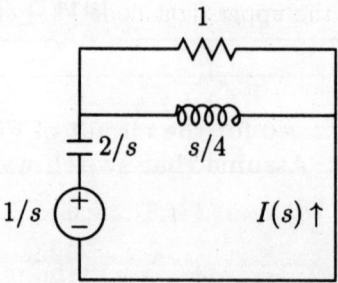

Fig. DP 4.5.2: s-domain circuit for $t > 0$.

Hints : With reference to Fig. DP 4.5.2, the current may be expressed as :

$$-I(s) = \frac{1/s}{[1||(s/4)] + 2/s}$$

DP 4.6 Determine $v(t)$ for $t > 0$ for the circuit of Fig. DP 4.6.

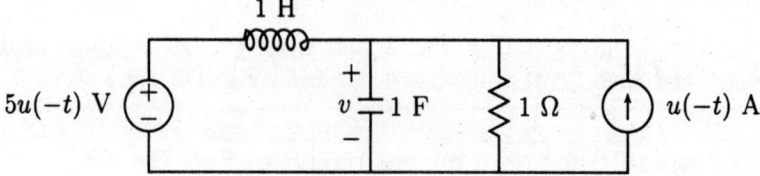

Fig. DP 4.6.

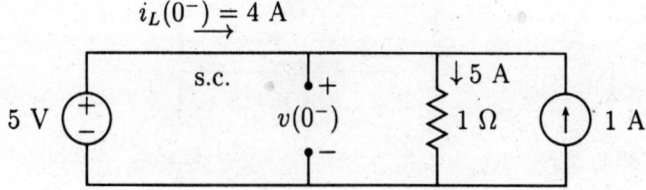

Fig. DP 4.6.1: $t = 0^-$ circuit.

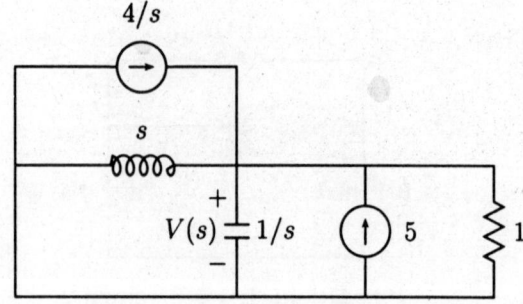

Fig. DP 4.6.2: s-domain circuit for $t > 0$.

Hints : Write the KCL at the upper right node $V(s)$ of Fig. DP 4.6.2 to determine $V(s)$.

DP 4.7 Determine $v_C(t)$ for $t > 0$ for the circuit of Fig. DP 4.7. The switch is moved from a to b at $t = 0$. Assume that switch was at position a for a long time.

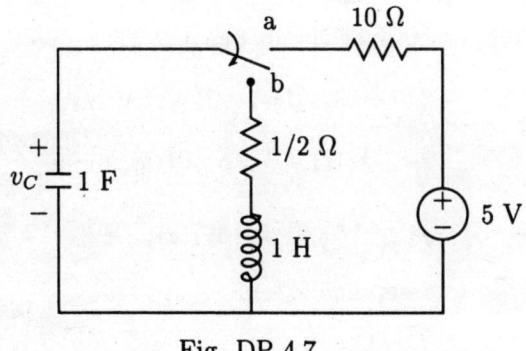

Fig. DP 4.7.

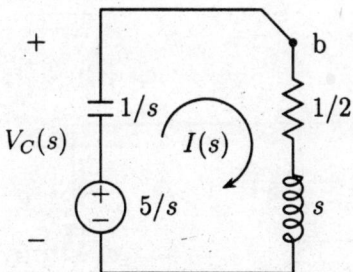

Fig. DP 4.7.1: s-domain circuit for $t > 0$.

Hints : With reference to the circuit in Fig. DP 4.7.1, determine $I(s)$ and then find $V_C(s)$ as:

$$V_C(s) = 5/s - (1/s)I(s)$$

DP 4.8 Determine $v_0(t)$ for $t > 0$ in the circuit of Fig. DP 4.8.

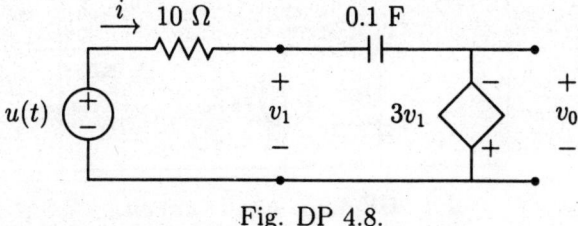

Fig. DP 4.8.

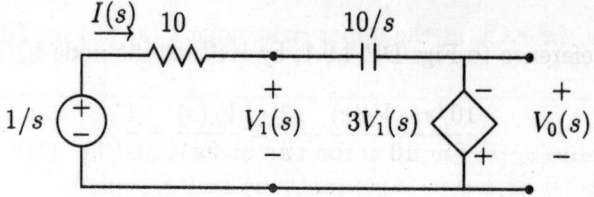

Fig. DP 4.8.1: s-domain circuit for $t > 0$.

Hints : KVL in the transformed circuit Fig. DP 4.8.1 gives

$$(10 + 10/s)I(s) - 3V_1(s) = 1/s$$

$$V_1(s) = 1/s - 10I(s)$$

and $$V_0(s) = -3V_1(s)$$

Eliminate $I(s)$ and $V_1(s)$ to determine $V_0(s)$.

DP 4.9 Determine $v_L(t)$ for $t > 0$ for the circuit of Fig. DP 4.9. If the switch is opened at $t = 0$.

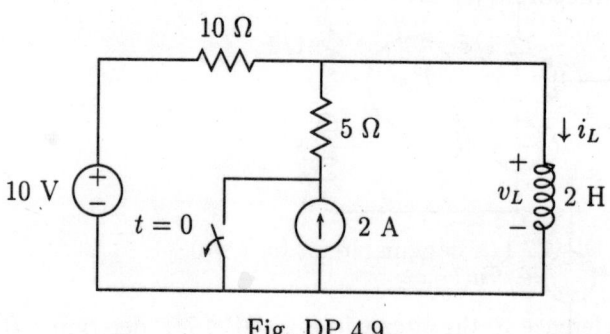

Fig. DP 4.9.

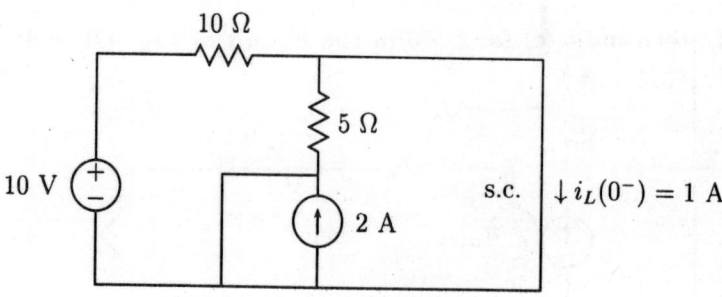

Fig. DP 4.9.1: $t = 0^-$ circuit.

Hints : With reference to Fig. DP 4.9.1, by KCL at the node $V_L(s)$ gives

$$\frac{10/s - V_L(s)}{10} + \frac{2}{s} = \frac{V_L(s)}{2s} + \frac{1}{s}$$

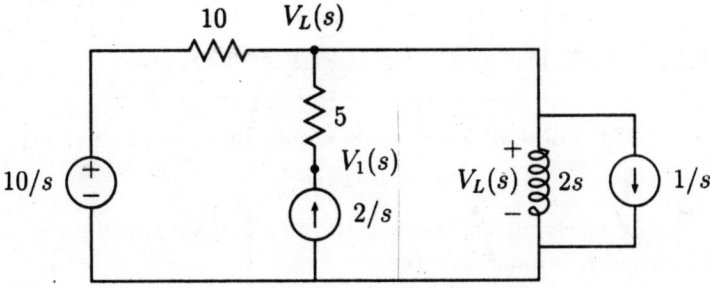

Fig. DP 4.9.2: s-domain circuit for $t > 0$.

DP 4.10 In the given network of Fig. DP 4.10, steady-state is reached with the switch at position a. At $t = 0$ switch is moved to position b. Find the current through inductor, $i_L(t)$ for $t > 0$.

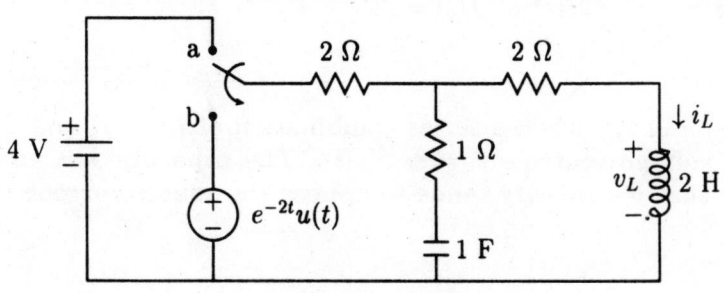

Fig. DP 4.10.

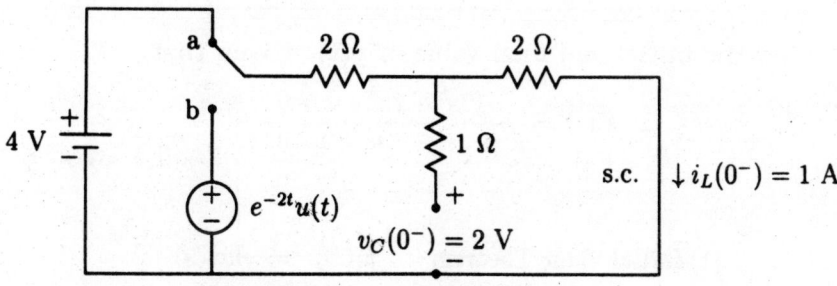

Fig. DP 4.10.1: $t = 0^-$ circuit.

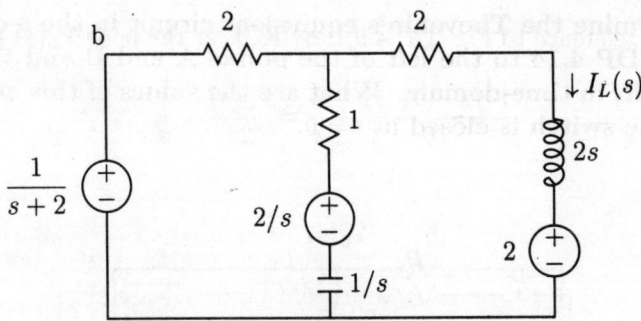

Fig. DP 4.10.2: s-domain circuit for $t > 0$.

Hints : Apply mesh-current method to the s-domain transformed circuit of Fig. DP 4.10.2 to determine $I_L(s)$ and so $i_L(t)$.

DP 4.11 Find an expression for the current $i(t)$ that will flow out of a unit impulse voltage source which is connected across the series combination of a 2-Ω resistor and a 1/4-F capacitor. The voltage across the capacitor is initially zero.

Hint:

$$2I(s) + \frac{4}{s}I(s) = \mathcal{L}[\delta(t)] = 1$$

DP 4.12 Find $I(s)$ and $i(t)$ when a series combination of $R = 4 \, \Omega$ and $C = 1/5$ F is driven by a voltage source $v(t) = 8e^{-2t}u(t)$. The capacitor has an initial voltage of 2 V so that its polarity tends to oppose the source current.

Hint:

$$4I(s) + \frac{5}{s}I(s) + \frac{2}{s} = \frac{8}{s+2}$$

DP 4.13 Find the initial and final value of y(t). Given that

$$Y(s) = \frac{17s^3 + 7s^2 + s + 6}{s^5 + 3s^4 + 5s^3 + 4s^2 + 2s}$$

Hints:

(1) Initial Value Theorem $y(0^+) \;=\; \lim_{s \to \infty} sY(s)$

(2) Final Value Theorem $y(\infty) \;=\; \lim_{s \to 0} sY(s)$

DP 4.14 Determine the Thevenin's equivalent circuit in the s-domain for the circuit of Fig. DP 4.14 to the left of the points A and B and then determine the current in R_3 in time-domain. What are the values of this current for $t = 0$ and $t = \infty$? The switch is closed at $t = 0$.

IES-96(EC)

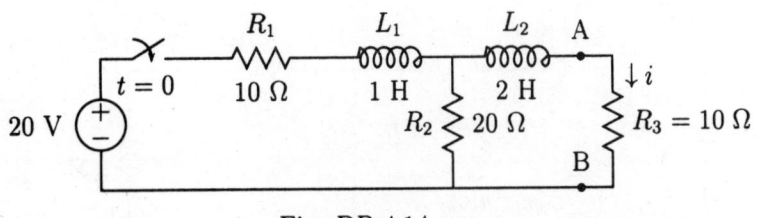

Fig. DP 4.14.

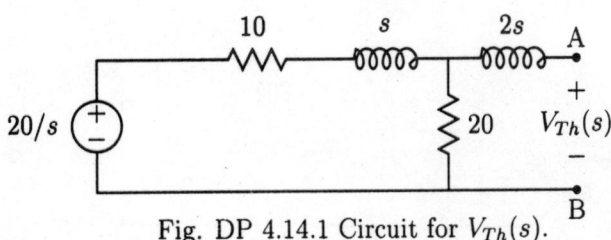

Fig. DP 4.14.1 Circuit for $V_{Th}(s)$.

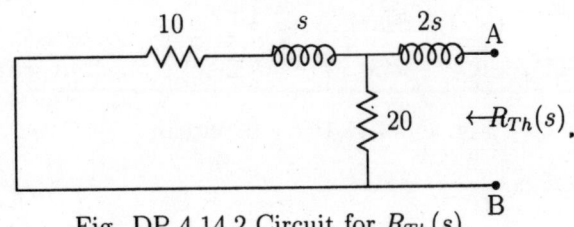

Fig. DP 4.14.2 Circuit for $R_{Th}(s)$.

Hints: From Fig. DP 4.14.1,

$$V_{Th}(s) = \frac{20}{10 + 20 + s}$$

From Fig. DP 4.14.2,

$$R_{Th}(s) = [(10 + s)\|20] + 2s$$

Thus, Thevenin's equivalent circuit with the load R_3 will give

$$I(s) = \frac{V_{Th}(s)}{R_3 + R_{Th}(s)}$$

DP 4.15 In the network shown in Fig. DP 4.15, switch K is in position 1 until steady-state is reached. Switch K is moved to position 2 at $t = 0$. With switch in position 2, determine the transform of current through the inductor.

IES-94(EE)

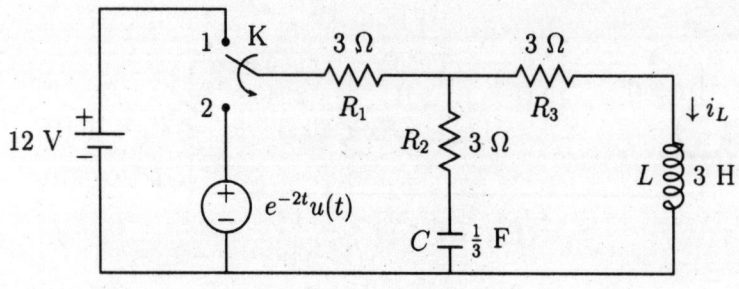

Fig. DP 4.15.

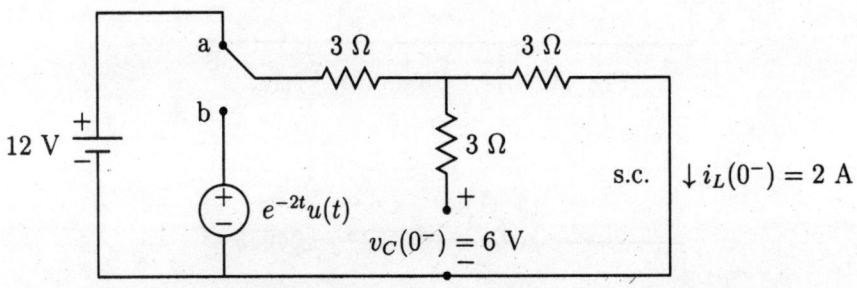

Fig. DP 4.15.1: $t = 0^-$ circuit.

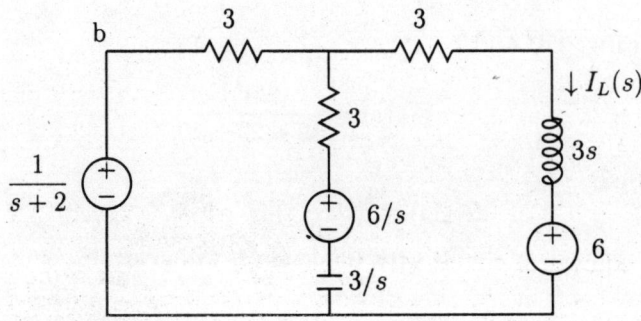

Fig. DP 4.15.2: s-domain circuit for $t > 0$.

Hints: The initial conditions from Fig. DP 4.15.1 are

$$i_L(0^-) = \frac{12}{3+3} = 2 \text{ A}$$
$$v_C(0^-) = 3i_L(0^-) = 6 \text{ V}$$

From Fig. DP 4.15.2, determine $I_2(s$ by mesh analysis.

****

5

The Laplace Transform
of Signal Waveforms

SUMMARY OF THIS CHAPTER

5.1   THE UNIT STEP FUNCTION

The unit step function is defined as:

$$u(t) \; = \; 1 \qquad \text{for } t \geq 0$$
$$= \; 0 \qquad \text{for } t < 0$$

A function $Ku(t)$ is called step function of strength K. The Laplace transform of the unit step function is

$$\mathcal{L}[u(t)] = \frac{1}{s}$$

Figure 5.1 shows the graphical representations of unit step functions.

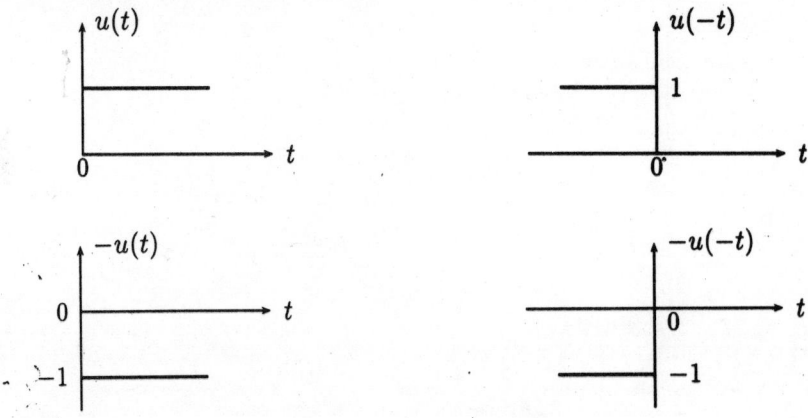

Figure 5.1: Unit step functions.

5.2 THE SHIFTED UNIT STEP FUNCTION

The shifted unit step function is defined as:

$$u(t - \tau) = 1 \quad \text{for } t \geq \tau$$
$$= 0 \quad \text{for } t < \tau$$

The Laplace transform of this shifted unit step function is

$$\mathcal{L}[u(t - \tau)] = \frac{e^{-s\tau}}{s}$$

Figure 5.2 shows the different shifted unit step functions.

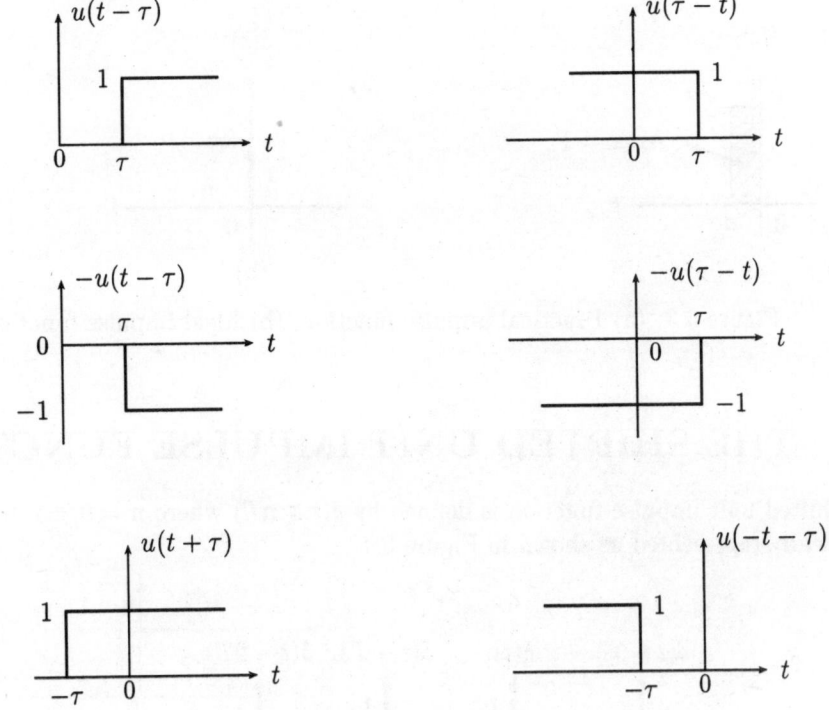

Figure 5.2: Shifted unit step functions.

5.3 THE UNIT IMPULSE FUNCTION

The unit impulse function is defined as a signal which has zero value everywhere except at $t = 0$, where its magnitude is infinite. It is generally called the δ-function and has following properties:

$$\delta(t) = 0 \qquad \text{for } t \neq 0$$

$$\int_{-t_0}^{t_0} \delta(t)dt = 1$$

where t_0 tends to zero.

Since an ideal impulse can not be obtained in practice, it is usually approximated by a pulse of small width but unit area as shown in Figure 5.3(a). Mathematically, an impulse function is derivative of a step function, i.e.,

$$\delta(t) = \lim_{\Delta \to 0} \frac{1}{\Delta}[u(t) - u(t - \Delta)] = \dot{u}(t)$$

The Laplace transform of a unit-impulse is

$$\mathcal{L}[\delta(t)] = 1$$

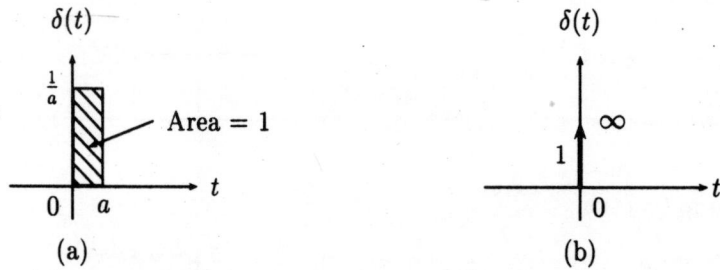

Figure 5.3: (a) Practical impulse function; (b) Ideal impulse function.

5.4   THE SHIFTED UNIT IMPULSE FUNCTION

The shifted unit impulse function is defined by $\delta(t \mp nT)$ where $n = 0, \pm 1, \pm 2 \cdots$. It is graphically represented as shown in Figure 5.4

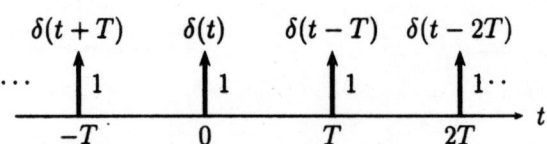

Figure 5.4: Shifted unit impulse functions.

5.5   THE RAMP FUNCTION

The ramp function which starts at a value of zero and increases linearly with time. Mathematically,

$$f(t) = \begin{cases} Kt & \text{for } t \geq 0 \\ 0 & \text{for } t < 0 \end{cases}$$

The Laplace transform of this function is

$$F(s) = K/s^2$$

The graphical representation of the ramp function is given in Figure 5.5(a).

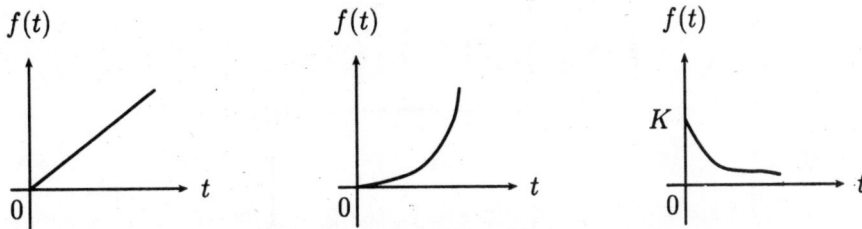

Figure 5.5: (a) A ramp function; (b) A parabolic function; (c) An exponential function.

5.6   THE PARABOLIC FUNCTION

The mathematical representation of the parabolic signal is

$$f(t) = \begin{cases} Kt^2/2 & \text{for } t \geq 0 \\ 0 & \text{for } t < 0 \end{cases}$$

This function in s-domain is

$$F(s) = K/s^3$$

The graphical representation of the parabolic function is given in Figure 5.5(b).

5.7   THE EXPONENTIAL FUNCTION

The exponential function, as shown in Figure 5.5(c), is defined as:

$$f(t) = \begin{cases} Ke^{at} & \text{for } t \geq 0 \\ 0 & \text{for } t < 0 \end{cases}$$

5.8   THE PULSE FUNCTION

A function comprising two unit step functions of the form

$$F_\tau(T) = u(t - \tau) - u(t - \tau - T)$$

represents a rectangular pulse of unit height which starts at $t = \tau$ and lasts for a time T, as shown in Figure 5.6. This is also known as the gate function.

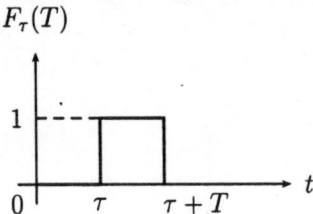

Figure 5.6: Pulse function.

5.9   LAPLACE TRANSFORM OF PERIODIC FUNCTIONS

The Laplace transform of a periodic function with periodic T is equal to $1/(1 - e^{-Ts})$ times the Laplace transform of the first cycle, i.e.

$$F(s) = \frac{1}{1 - e^{-Ts}} F_1(s)$$

where $F_1(s)$ is Laplace transform of first cycle and T is time period of the function.

5.10   SOLVED PROBLEMS

SP 5.1 Find $v(t)$ and $V(s)$ for the following waveforms shown in Fig. SP 5.1(a to g).

SOLUTION:

(a) The equation of the waveform of Fig. SP 5.1(a) is obtained graphically by the Fig. 5.1.1a as:

$$v(t) = u(t) - u(t - 1)$$

and $$V(s) = \frac{1}{s} - \frac{e^{-s}}{s} = \frac{1}{s}(1 - e^{-s})$$

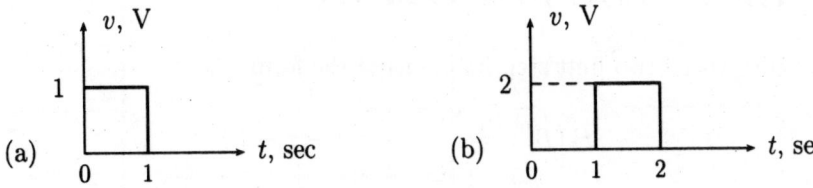

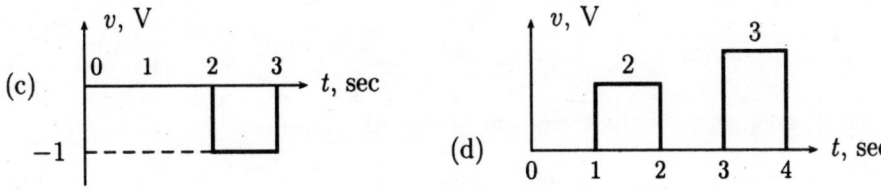

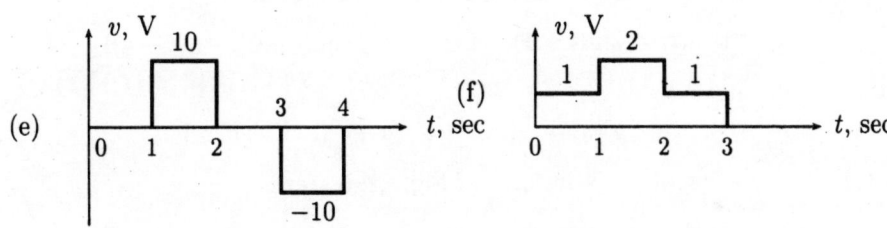

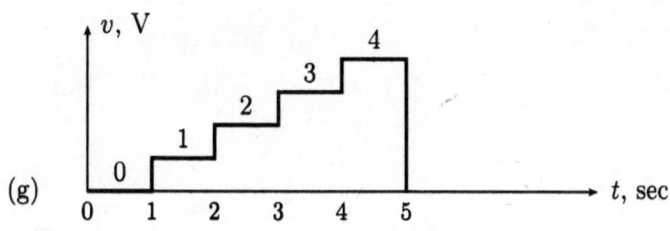

Fig. SP 5.1.

(b) The equation of the waveform of Fig. SP 5.1(b) is

$$v(t) = 2u(t - 1) - 2u(t - 2)$$

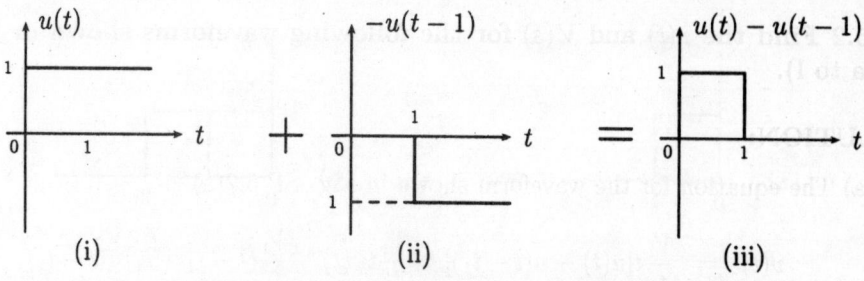

Fig. 5.1.1a: Graphical representation of the Fig. SP 5.1(a).

$$\text{and} \quad V(s) = \frac{2e^{-s}}{s} - \frac{2e^{-2s}}{s} = \frac{2}{s}(e^{-s} - e^{-2s})$$

(c) The equation of the waveform of Fig. SP 5.1(c) is

$$v(t) = -u(t - 2) + u(t - 3)$$

$$\text{and} \quad V(s) = -\frac{e^{-2s}}{s} + \frac{e^{-3s}}{s} = \frac{1}{s}(e^{-3s} - e^{-2s})$$

(d) The equation of the waveform of Fig. SP 5.1(d) is

$$v(t) = 2[u(t - 1) - u(t - 2)] + 3[u(t - 3) - u(t - 4)]$$

and

$$V(s) = \frac{2}{s}(e^{-s} - e^{-2s}) + \frac{3}{s}(e^{-3s} - e^{-4s})$$

(e) The equation of the waveform of Fig. SP 5.1(e) is

$$v(t) = 10[u(t - 1) - u(t - 2)] - 10[u(t - 3) - u(t - 4)]$$

and

$$V(s) = \frac{10}{s}(e^{-s} - e^{-2s}) - \frac{10}{s}(e^{-3s} - e^{-4s}) = \frac{10}{s}(e^{-s} - e^{-2s} - e^{-3s} + e^{-4s})$$

(f) The equation of the waveform of Fig. SP 5.1(f) is

$$\begin{aligned} v(t) &= [u(t) - u(t - 1)] + 2[u(t - 1) - u(t - 2)] + 1[u(t - 2) - u(t - 3)] \\ &= u(t) + u(t - 1) - u(t - 2) - u(t - 3) \end{aligned}$$

and

$$V(s) = \frac{1}{s}(1 + e^{-s} - e^{-2s} - e^{-3s})$$

(g) The equation of the waveform of Fig. SP 5.1(g) is

$$\begin{aligned} v(t) &= 0 + [u(t - 1) - u(t - 2)] + 2[u(t - 2) - u(t - 3)] + 3[u(t - 3) - u(t - 4)] \\ &\quad + 4[u(t - 4) - u(t - 5)] \\ &= u(t - 1) + u(t - 2) + u(t - 3) + u(t - 4) - 4u(t - 5) \end{aligned}$$

and

$$V(s) = \frac{1}{s}(e^{-s} + e^{-2s} + e^{-3s} + e^{-4s} - 4e^{-5s})$$

SP 5.2 Find the $v(t)$ and $V(s)$ for the following waveforms shown in Fig. SP 5.2(a to l).

SOLUTION:

(a) The equation for the waveform shown in Fig. SP 5.2(a) is

$$v(t) = \frac{V_1}{t_1}t[u(t) - u(t - t_1)] = \frac{V_1}{t_1}tu(t) - \frac{V_1}{t_1}(t - t_1 + t_1)u(t - t_1)$$

$$= \frac{V_1}{t_1}tu(t) - \frac{V_1}{t_1}(t - t_1)u(t - t_1) - V_1u(t - t_1)$$

and

$$V(s) = \frac{V_1}{t_1}\frac{1}{s^2} - \frac{V_1}{t_1}\frac{e^{-t_1 s}}{s^2} - \frac{V_1 e^{-t_1 s}}{s} = \frac{V_1}{t_1}\left[\frac{1}{s^2} - (\frac{1}{s^2} + \frac{t_1}{s})e^{-t_1 s}\right]$$

(b) The equation for the waveform shown in Fig. SP 5.2(b) is

$$v(t) = \frac{1}{t_1}t[u(t) - u(t - t_1)] + u(t - t_1)$$

$$= \frac{1}{t_1}tu(t) - \frac{1}{t_1}(t - t_1 + t_1)u(t - t_1) + u(t - t_1)$$

$$= \frac{1}{t_1}tu(t) - \frac{1}{t_1}(t - t_1)u(t - t_1) - u(t - t_1) + u(t - t_1)$$

$$= \frac{1}{t_1}[tu(t) - (t - t_1)u(t - t_1)]$$

and

$$V(s) = \frac{1}{t_1}\left[\frac{1}{s^2} - \frac{e^{-t_1 s}}{s^2}\right]$$

(c) The equation for the waveform shown in Fig. SP 5.2(c) is

$$v(t) = \frac{V_1}{t_1}t[u(t) - u(t - t_1)] + V_1[u(t - t_1) - u(t - 2t_1)]$$

$$= \frac{V_1}{t_1}tu(t) - \frac{V_1}{t_1}(t - t_1 + t_1)u(t - t_1) + V_1[u(t - t_1) - u(t - 2t_1)]$$

$$= \frac{V_1}{t_1}tu(t) - \frac{V_1}{t_1}(t - t_1)u(t - t_1) - V_1u(t - 2t_1)$$

and
$$V(s) = \frac{V_1}{t_1}\frac{1}{s^2} - \frac{V_1}{t_1}\frac{e^{-t_1 s}}{s^2} - \frac{V_1}{s}e^{-2t_1 s} = \frac{V_1}{t_1}\left[\frac{1}{s^2} - \frac{e^{-t_1 s}}{s^2} - t_1\frac{e^{-2t_1 s}}{s}\right]$$

(d) The equation for the waveform shown in Fig. SP 5.2(d) is

$$v(t) = (t - 1)[u(t - 1) - u(t - 2)]$$

$$= [(t - 1)u(t - 1) - (t - 2 + 1)u(t - 2)]$$

$$= [(t - 1)u(t - 1) - (t - 2)u(t - 2) - u(t - 2)]$$

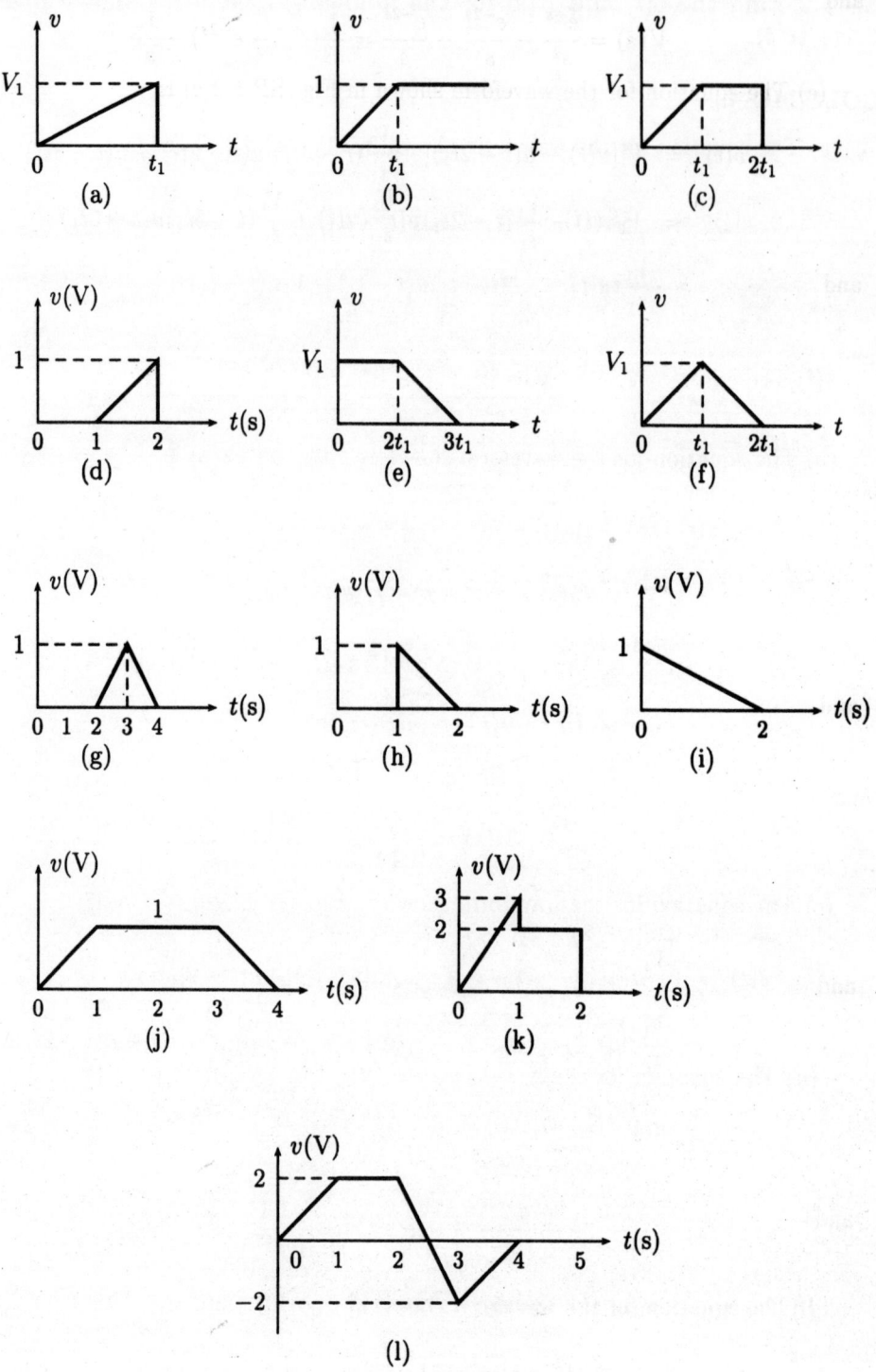

Fig.SP 5.2.

and

$$V(s) = \frac{e^{-s}}{s^2} - \frac{e^{-2s}}{s^2} - \frac{e^{-2s}}{s} = \frac{1}{s^2}(e^{-s} - e^{-2s}) - \frac{1}{s}e^{-2s}$$

(e) The equation for the waveform shown in Fig. SP 5.2(e) is

$$
\begin{aligned}
v(t) &= V_1[u(t) - u(t - 2t_1)] - \frac{V_1}{t_1}(t - 3t_1)[u(t - 2t_1) - u(t - 3t_1)] \\
&= V_1 u(t) - \frac{V_1}{t_1}(t - 2t_1)u(t - 2t_1) + \frac{V_1}{t_1}(t - 3t_1)u(t - 3t_1)
\end{aligned}
$$

and

$$V(s) = \frac{V_1}{s} - \frac{V_1}{t_1}\frac{e^{-2t_1 s}}{s^2} + \frac{V_1}{t_1}\frac{e^{-3t_1 s}}{s^2}$$

(f) The equation for the waveform shown in Fig. SP 5.2(f) is

$$
\begin{aligned}
v(t) &= \frac{V_1}{t_1}t[u(t) - u(t - t_1)] - \frac{V_1}{t_1}(t - 2t_1)[u(t - t_1) - u(t - 2t_1)] \\
&= \frac{V_1}{t_1}tu(t) - \frac{V_1}{t_1}(t - t_1)u(t - t_1) - V_1 u(t - t_1) - \frac{V_1}{t_1}(t - t_1 - t_1)u(t - t_1) \\
&\quad + \frac{V_1}{t_1}(t - 2t_1)u(t - 2t_1) \\
&= \frac{V_1}{t_1}[tu(t) - 2(t - t_1)u(t - t_1) + (t - 2t_1)u(t - 2t_1)]
\end{aligned}
$$

and

$$V(s) = \frac{V_1}{t_1}\left[\frac{1}{s^2} - \frac{2e^{-t_1 s}}{s^2} + \frac{e^{-2t_1 s}}{s^2}\right] = \frac{V_1}{t_1 s^2}(1 - 2e^{-t_1 s} + e^{-2t_1 s})$$

(g) The equation for the waveform shown in Fig. SP 5.2(g) is

$$
\begin{aligned}
v(t) &= (t - 2)[u(t - 2) - u(t - 3)] - (t - 4)[u(t - 3) - u(t - 4)] \\
&= (t - 2)u(t - 2) - (t - 3 + 1)u(t - 3) - (t - 3 - 1)u(t - 3) + (t - 4)u(t - 4) \\
&= (t - 2)u(t - 2) - 2(t - 3)u(t - 3) + (t - 4)u(t - 4)
\end{aligned}
$$

and

$$V(s) = \frac{e^{-2s}}{s^2} - \frac{2e^{-3s}}{s^2} + \frac{e^{-4s}}{s^2} = \frac{1}{s^2}(e^{-2s} - 2e^{-3s} + e^{-4s})$$

(h) The equation for the waveform shown in Fig. SP 5.2(h) is

$$
\begin{aligned}
v(t) &= -(t - 2)[u(t - 1) - u(t - 2)] \\
&= -(t - 1)u(t - 1) + u(t - 1) + (t - 2)u(t - 2)
\end{aligned}
$$

and

$$V(s) = -\frac{e^{-s}}{s^2} + \frac{e^{-s}}{s} + \frac{e^{-2s}}{s^2}$$

(i) The equation of the waveform shown in Fig. SP 5.2(i) is

$$
\begin{aligned}
v(t) &= -\frac{1}{2}(t - 2)[u(t) - u(t - 2)] \\
&= -\frac{1}{2}tu(t) + u(t) + \frac{1}{2}(t - 2)u(t - 2)
\end{aligned}
$$

and

$$V(s) = -\frac{1}{2s^2} + \frac{1}{s} + \frac{e^{-2s}}{2s^2}$$

(j) The equation for the waveform shown in Fig. SP 5.2(j) is

$$
\begin{aligned}
v(t) &= t[u(t) - u(t-1)] + [u(t-1) - u(t-3)] - (t-4)[u(t-3) - u(t-4)] \\
&= tu(t) - (t-1+1)u(t-1) + u(t-1) - u(t-3) \\
&\quad -(t-3-1)u(t-3) + (t-4)u(t-4) \\
&= tu(t) - (t-1)u(t-1) - (t-3)u(t-3) + (t-4)u(t-4)
\end{aligned}
$$

and

$$V(s) = \frac{1}{s^2} - \frac{e^{-s}}{s^2} - \frac{e^{-3s}}{s^2} + \frac{e^{-4s}}{s^2} = \frac{1}{s^2}(1 - e^{-s} - e^{-3s} + e^{-4s})$$

(k) The equation for the waveform shown in Fig. SP 5.2(k) is

$$
\begin{aligned}
v(t) &= 3t[u(t) - u(t-1)] + 2[u(t-1) - u(t-2)] \\
&= 3tu(t) - 3(t-1+1)u(t-1) + 2u(t-1) - 2u(t-2) \\
&= 3tu(t) - 3(t-1)u(t-1) - u(t-1) - 2u(t-2)
\end{aligned}
$$

and

$$V(s) = \frac{3}{s^2} - \frac{3e^{-s}}{s^2} - \frac{e^{-s}}{s} - \frac{2e^{-2s}}{s} = \frac{3}{s^2}(1 - e^{-s}) - \frac{1}{s}(e^{-s} + 2e^{-2s})$$

(l) The equation for the waveform shown in Fig. SP 5.2(l) is

$$
\begin{aligned}
v(t) &= 2t[u(t) - u(t-1)] + 2[u(t-1) - u(t-2)] \\
&\quad -4(t-5/2)[u(t-2) - u(t-3)]' + 2(t-4)[u(t-3) - u(t-4)] \\
&= 2tu(t) - 2(t-1+1)u(t-1) + 2u(t-1) - 2u(t-2) \\
&\quad -4(t-2-1/2)u(t-2) + 4(t-3+1/2)u(t-3) \\
&\quad +2(t-3-1)u(t-3) - 2(t-4)u(t-4) \\
&= 2tu(t) - 2(t-1)u(t-1) - 4(t-2)u(t-2) \\
&\quad +6(t-3)u(t-3) - 2(t-4)u(t-4)
\end{aligned}
$$

and

$$
\begin{aligned}
V(s) &= \frac{2}{s^2} - \frac{2e^{-s}}{s^2} - \frac{4e^{-2s}}{s^2} + + \frac{6e^{-3s}}{s^2} - \frac{2e^{-4s}}{s^2} \\
&= \frac{2}{s^2}(1 - e^{-s} - 2e^{-2s} + 3e^{-3s} - e^{-4s})
\end{aligned}
$$

SP 5.3 Find $v(t)$ and $V(s)$ for the following waveforms shown in Fig. SP 5.3(a to g).

SOLUTION:

(a) The equation for the waveform shown in Fig. SP 5.3(a) may be obtained from the Fig. 5.3.1a as:

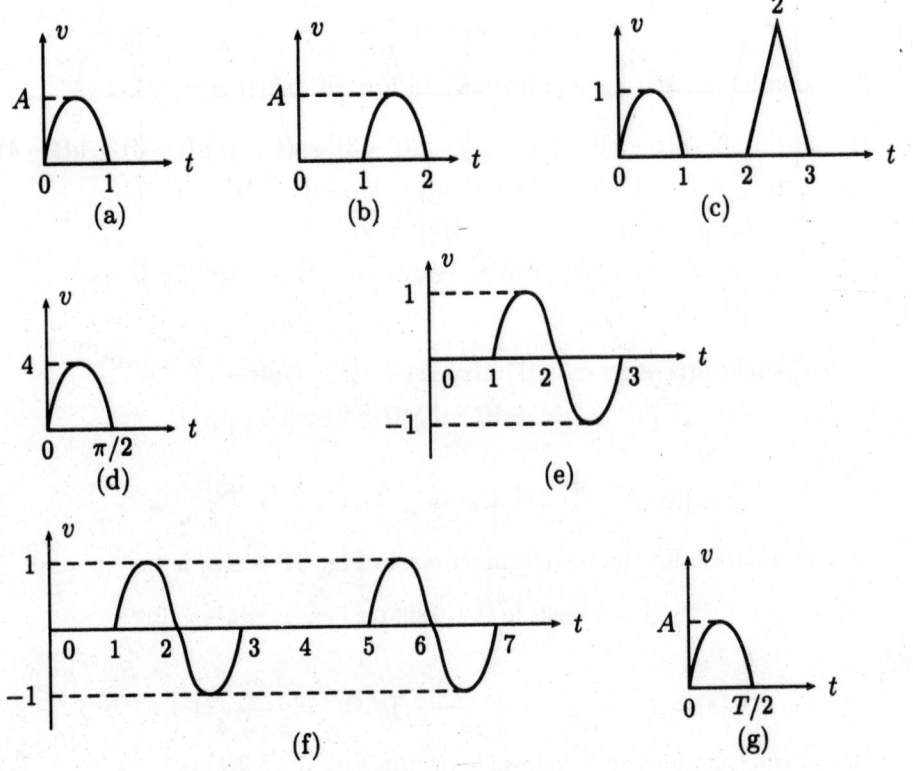

Fig. SP 5.3.

$$A \sin \pi t \, u(t) + A \sin \pi(t-1) \, u(t-1)$$

and

$$V(s) = A \frac{\pi}{s^2 + \pi^2} + A \frac{\pi}{s^2 + \pi^2} e^{-s} = \frac{A\pi}{s^2 + \pi^2}(1 + e^{-s})$$

(b) The equation for the waveform shown in Fig. SP 5.3(b) is

$$A \sin \pi(t-1) \, u(t-1) + A \sin \pi(t-2) \, u(t-2)$$

and

$$V(s) = A \frac{\pi}{s^2 + \pi^2} e^{-s} + A \frac{\pi}{s^2 + \pi^2} e^{-2s} = \frac{A\pi}{s^2 + \pi^2}(e^{-s} + e^{-2s})$$

(c) The equation for the waveform shown in Fig. SP 5.3(c) is

$$
\begin{aligned}
v(t) &= \sin \pi t \, u(t) + \sin \pi(t-1) \, u(t-1) + 4(t-2)[u(t-2) - u(t-2.5)] \\
&\quad -4(t-3)[u(t-2.5) - u(t-3)] \\
&= \sin \pi t \, u(t) + \sin \pi(t-1) \, u(t-1) + 4(t-2)u(t-2) \\
&\quad -4(t-2.5+0.5)u(t-2.5) - 4(t-2.5-0.5)u(t-2.5) + 4(t-3)u(t-3)
\end{aligned}
$$

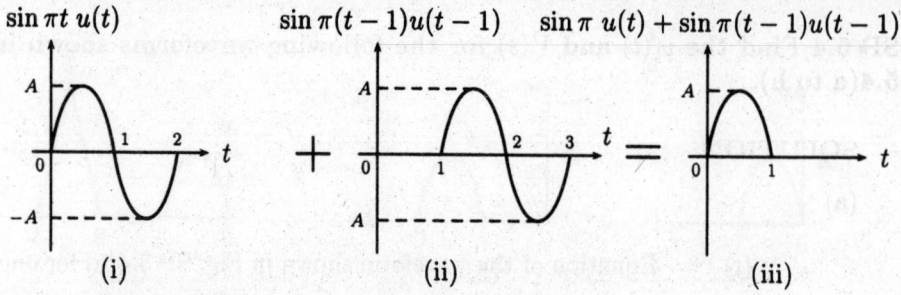

Fig. 5.3.1a: Graphical representation of the Fig. SP 5.3(a).

$$= \quad \sin \pi t \, u(t) + \sin \pi (t-1) \, u(t-1) + 4(t-2)u(t-2)$$
$$-8(t-2.5)u(t-2.5) + 4(t-3)u(t-3)$$

And

$$V(s) = \frac{\pi}{s^2 + \pi^2}(1 + e^{-s}) + \frac{4}{s^2}(e^{-2s} - 2e^{-2.5s} + e^{-3s})$$

(d) The equation for the waveform shown in Fig. SP 5.3(d) is

$$v(t) = 4\sin 2t \, u(t) + 4\sin 2(t - \pi/2) \, u(t - \pi/2)$$

and

$$V(s) = 4(\frac{2}{s^2 + 2^2}) + 4(\frac{2}{s^2 + 2^2})e^{-\pi s/2} = \frac{8}{s^2 + 4}(1 + e^{-\pi s/2})$$

(e) The equation for the waveform shown in Fig. SP 5.3(e) is

$$. \, v(t) = \sin \pi (t-1) \, u(t-1) - \sin \pi (t-3) \, u(t-3)$$

and

$$V(s) = \frac{\pi}{s^2 + \pi^2}e^{-s} - \frac{\pi}{s^2 + \pi^2}e^{-3s} = \frac{\pi}{s^2 + \pi^2}(e^{-s} - e^{-3s})$$

(f) The equation for the waveform shown in Fig. SP 5.3(f) is

$$v(t) \quad = \quad [\sin \pi (t-1) \, u(t-1) - \sin \pi (t-3) \, u(t-3)]$$
$$+[\sin \pi (t-5) \, u(t-5) - \sin \pi (t-7) \, u(t-7)]$$

and

$$V(s) \quad = \quad (\frac{\pi}{s^2 + \pi^2}e^{-s} - \frac{\pi}{s^2 + \pi^2}e^{-3s}) + (\frac{\pi}{s^2 + \pi^2}e^{-5s} - \frac{\pi}{s^2 + \pi^2}e^{-7s})$$
$$= \quad \frac{\pi}{s^2 + \pi^2}(e^{-s} - e^{-3s} + e^{-5s} - e^{-7s})$$

(g) The equation for the waveform shown in Fig. SP 5.3(g) is

$$v(t) = A\sin \frac{2\pi}{T}t \, u(t) + A\sin \frac{2\pi}{T}(t - T/2) \, u(t - T/2)$$

and

$$V(s) = \frac{A(2\pi/T)}{s^2 + (2\pi/T)^2} + \frac{A(2\pi/T)}{s^2 + (2\pi/T)^2}e^{-Ts/2} = \frac{A(2\pi/T)}{s^2 + (2\pi/T)^2}(1 + e^{-Ts/2})$$

SP 5.4 Find the $v_1(t)$ and $V(s)$ for the following waveforms shown in Fig. SP 5.4(a to h).

SOLUTION:

(a)

$$v_1(t) = \text{Equation of the waveform shown in Fig. SP 5.4(a) for one time period}$$
$$= [u(t) - u(t - c)] + [-u(t - c) + u(t - 2c)]$$
$$= u(t) - 2u(t - c) + u(t - 2c)$$

and

$$V_1(s) = \frac{1}{s} - \frac{2e^{-cs}}{s} + \frac{e^{-2cs}}{s} = \frac{1}{s}(1 - 2e^{-cs} + e^{-2cs}) = \frac{(1 - e^{-cs})^2}{s}$$

Therefore, $\quad V(s) = \dfrac{V_1(s)}{1 - e^{-Ts}} = \dfrac{1}{1 - e^{-2cs}} \cdot \dfrac{(1 - e^{-cs})^2}{s} = \dfrac{1 - e^{-cs}}{s(1 + e^{-cs})} = \dfrac{1}{s} \tanh\left(\dfrac{cs}{2}\right)$

(b)

$$v_1(t) = \text{Equation of the waveform shown in Fig. SP 5.4(a) for one time period}$$
$$= 4[u(t) - u(t - 1)] + [-4u(t - 1) + 4u(t - 3)] + 4[u(t - 3) - u(t - 4)]$$
$$= 4u(t) - 8u(t - 1) + 8u(t - 3) - 4u(t - 4)$$

and

$$V_1(s) = \frac{4}{s} - \frac{8e^{-s}}{s} + \frac{8e^{-3s}}{s} - \frac{4e^{-4s}}{s} = \frac{4}{s}(1 - 2e^{-s} + 2e^{-3s} - e^{-4s})$$

Therefore $\quad V(s) = \dfrac{V_1(s)}{1 - e^{-Ts}} = \dfrac{1}{1 - e^{-4s}} \cdot \dfrac{4}{s}(1 - 2e^{-s} + 2e^{-3s} - e^{-4s})$

$$= \frac{4(1 - 2e^{-s} + 2e^{-3s} - e^{-4s})}{s(1 - e^{-4s})}$$

(c)

$$v_1(t) = \text{Equation of the waveform shown in Fig. SP 5.4(a) for one time period}$$
$$= \frac{1}{c}t[u(t) - u(t - c)] = \frac{1}{c}tu(t) - \frac{1}{c}(t - c + c)u(t - c)$$
$$= \frac{1}{c}tu(t) - \frac{1}{c}(t - c)u(t - c) - u(t - c)$$

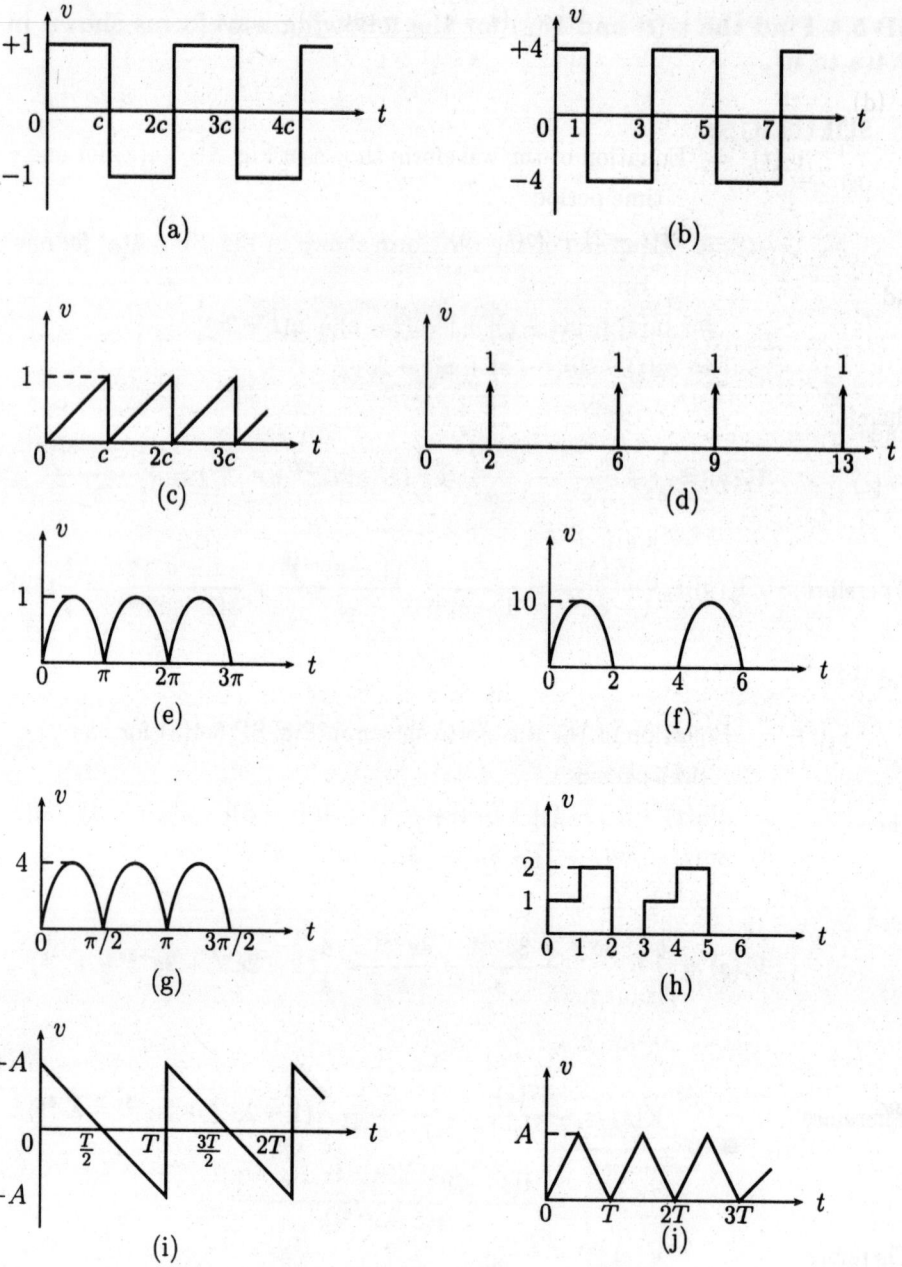

Fig. SP 5.4.

and

$$V_1(s) = \frac{1}{cs^2} - \frac{e^{-cs}}{cs^2} - \frac{e^{-cs}}{s} = \frac{1}{cs^2}(1 - e^{-cs} - cse^{-cs})$$

Therefore

$$V(s) = \frac{V_1(s)}{1 - e^{-Ts}} = \frac{1}{1 - e^{-cs}} \cdot \frac{1}{cs^2}(1 - e^{-cs} - cse^{-cs})$$

$$= \frac{(1 - e^{-cs} - cse^{-cs})}{cs^2(1 - e^{-cs})}$$

(d)

$$
\begin{aligned}
v_1(t) &= \text{Equation of the waveform shown in Fig. SP 5.4(a) for one} \\
&\quad \text{time period} \\
&= 2\delta(t - 2) + 2\delta(t - 6)
\end{aligned}
$$

and

$$V_1(s) = 2e^{-2s} + 2e^{-6s} = 2(e^{-2s} + e^{-6s})$$

Therefore, $$V(s) = \frac{V_1(s)}{1 - e^{-Ts}} = \frac{2(e^{-2s} + e^{-6s})}{1 - e^{-7s}} \quad \text{where T} = 7.$$

(e)

$$
\begin{aligned}
v_1(t) &= \text{Equation of the waveform shown in Fig. SP 5.4(a) for one} \\
&\quad \text{time period} \\
&= \sin t\, u(t) + \sin(t - \pi)\, u(t - \pi)
\end{aligned}
$$

and

$$V_1(s) = \frac{1}{s^2 + 1^2} + \frac{e^{-\pi s}}{s^2 + 1^2} = \frac{1 + e^{-\pi s}}{s^2 + 1}$$

Therefore $$V(s) = \frac{V_1(s)}{1 - e^{-Ts}} = \frac{1}{1 - e^{-\pi s}} \left[\frac{1 + e^{-\pi s}}{s^2 + 1} \right] = \frac{1}{s^2 + 1} \coth\left(\frac{\pi s}{2} \right)$$

(f)

$$
\begin{aligned}
v_1(t) &= \text{Equation of the waveform shown in Fig. SP 5.4(a) for one} \\
&\quad \text{time period} \\
&= 10 \sin \frac{\pi}{2} t\, u(t) + 10 \sin \frac{\pi}{2}(t - 2)\, u(t - 2)
\end{aligned}
$$

and

$$V_1(s) = 10\frac{\pi/2}{s^2 + (\pi/2)^2} + 10\frac{\pi/2}{s^2 + (\pi/2)^2}e^{-2s} = 10\frac{\pi/2}{s^2 + (\pi/2)^2}(1 + e^{-2s})$$

Therefore $$V(s) = \frac{V_1(s)}{1 - e^{-Ts}} = \frac{1}{1 - e^{-4s}} \left[10\frac{\pi/2}{s^2 + (\pi/2)^2}(1 + e^{-2s}) \right]$$

$$= \frac{10(\pi/2)}{(1 - e^{-2s})(s^2 + (\pi/2)^2)}$$

(g)

$$
\begin{aligned}
v_1(t) &= \text{Equation of the waveform shown in Fig. SP 5.4(a) for one} \\
&\quad \text{time period} \\
&= 4 \sin 2t\, u(t) + 4 \sin 2(t - \pi/2)\, u(t - \pi/2)
\end{aligned}
$$

and

$$V_1(s) = 4\frac{2}{s^2 + 2^2} + 4\frac{2}{s^2 + 2^2}e^{-\pi s/2} = \frac{8(1 + e^{-\pi s/2})}{s^2 + 2^2}$$

Therefore $$V(s) = \frac{V_1(s)}{1 - e^{-Ts}} = \frac{1}{1 - e^{-\pi s/2}}\left[\frac{8(1 + e^{-\pi s/2})}{s^2 + 2^2}\right] = \frac{8}{s^2 + 2^2}\coth\left(\frac{\pi s}{4}\right)$$

(h)

$v_1(t)$ = Equation of the waveform shown in Fig. SP 5.4(a) for one
time period

$= u(t) - u(t-1) + 2[u(t-1) - u(t-2)] = u(t) + u(t-1) - 2u(t-2)$

and

$$V_1(s) = \frac{1}{s} + \frac{e^{-s}}{s} - \frac{2e^{-2s}}{s} = \frac{1}{s}(1 + e^{-s} - 2e^{-2s})$$

Therefore $$V(s) = \frac{V_1(s)}{1 - e^{-Ts}} = \frac{1}{1 - e^{-3s}}\left[\frac{1}{s}(1 + e^{-s} - 2e^{-2s})\right]$$

(i)

$v_1(t)$ = Equation of the waveform shown in Fig. SP 5.4(a) for one
time period

$$= -\frac{2A}{T}(t - T/2)[u(t) - u(t-T)]$$

$$= -\frac{2A}{T}(t - T/2)u(t) + \frac{2A}{T}(t - T + T/2)u(t-T)$$

$$= -\frac{2A}{T}(t - T/2)u(t) - [\frac{2A}{T}(t-T) + A]u(t-T)$$

and

$$V_1(s) = -\frac{2A}{T}\left[\frac{1}{s^2} - \frac{T}{2s}\right] - \left[\frac{2A}{Ts^2} + \frac{A}{s}\right]e^{-Ts}$$

$$= \frac{2A}{Ts}\left[\frac{T}{2}(1 + e^{-Ts}) - \frac{1}{s}(1 - e^{-Ts})\right]$$

Therefore

$$V(s) = \frac{V_1(s)}{1 - e^{-Ts}} = \frac{1}{1 - e^{-Ts}}\cdot\frac{2A}{Ts}\left[\frac{T}{2}(1 + e^{-Ts}) - \frac{1}{s}(1 - e^{-Ts})\right]$$

$$= \frac{2A}{Ts}\left[\frac{T}{2}(\frac{1 + e^{-Ts}}{1 - e^{-Ts}}) - \frac{1}{s}\right] = \frac{2A}{Ts}\left[\frac{T}{2}\coth\left(\frac{Ts}{2}\right) - \frac{1}{s}\right]$$

(j)

$v_1(t)$ = Equation of the waveform shown in Fig. SP 5.4(a) for one
time period

$$= \frac{2A}{T}t[u(t) - u(t - T/2)] - \frac{2A}{T}(t - T)[u(t - T/2) - u(t - T)]$$

$$= \frac{2A}{T}tu(t) - \frac{2A}{T}(t - T/2 + T/2)u(t - T/2) - \frac{2A}{T}(t - T/2 - T/2)u(t - T/2)$$

$$+ \frac{2A}{T}(t - T)u(t - T)$$

$$= \frac{2A}{T}tu(t) - \frac{4A}{T}(t - T/2)u(t - T/2) + \frac{2A}{T}(t - T)u(t - T)$$

and

$$V_1(s) = \frac{2A}{Ts^2} - \frac{4A}{Ts^2}e^{-Ts/2} + \frac{2A}{Ts^2}e^{-Ts}$$

$$= \frac{2A}{Ts^2}\left[1 - 2e^{-Ts/2} + e^{-Ts}\right] = \frac{2A}{Ts^2}(1 - e^{-Ts/2})^2$$

Therefore

$$V(s) = \frac{V_1(s)}{1 - e^{-Ts}} = \frac{2A}{Ts^2} \cdot \frac{(1 - e^{-Ts/2})^2}{(1 - e^{-Ts})}$$

$$= \frac{2A}{Ts^2}\left[\frac{1 - e^{-Ts/2}}{1 + e^{-Ts/2}}\right] = \frac{2A}{Ts^2}\tanh\left(\frac{Ts}{4}\right)$$

SP 5.5 A pulse of voltage of 5 V magnitude and 1 sec. duration is applied to the RL network shown in Fig. SP 5.5. Find and plot $i(t)$.

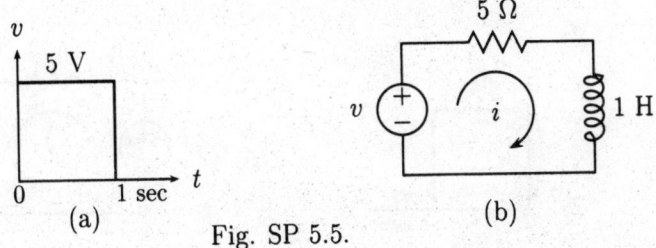

Fig. SP 5.5.

SOLUTION:

The equation for the input waveform shown in Fig. SP 5.5a is

$$v(t) = 5[u(t) - u(t - 1)] \quad \text{and its} \quad V(s) = \frac{5}{s}(1 - e^{-s})$$

By KVL in the circuit of Fig. SP 5.5.1

$$(5 + s)I(s) = V(s) = \frac{5}{s}(1 - e^{-s})$$

Hence

$$I(s) = \frac{5(1 - e^{-s})}{s(s + 5)}$$

However

$$\frac{5}{s(s + 5)} = \frac{A}{s} + \frac{B}{s + 5}$$

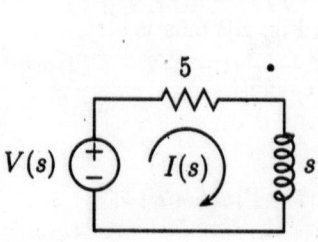

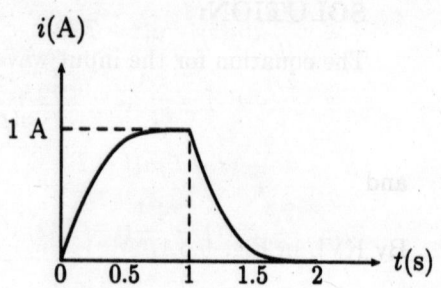

Fig. 5.5.1: s-domain circuit.

Fig. 5.5.2: Plot of $i(t)$.

where $A = 1$ and $B = -1$
Thus

$$I(s) = \left(\frac{1}{s} - \frac{1}{s+5}\right)(1 - e^{-s})$$

$$= \left(\frac{1}{s} - \frac{1}{s+5}\right) - \left(\frac{1}{s} - \frac{1}{s+5}\right)e^{-s}$$

Therefore $i(t) = \mathcal{L}^{-1}[I(s)] = (1 - e^{-5t})u(t) - (1 - e^{-5(t-1)})u(t-1)$

SP 5.6 A voltage pulse of 5 V magnitude and 1 sec. duration is applied to the network shown in Fig. SP 5.6. Find and plot i(t).

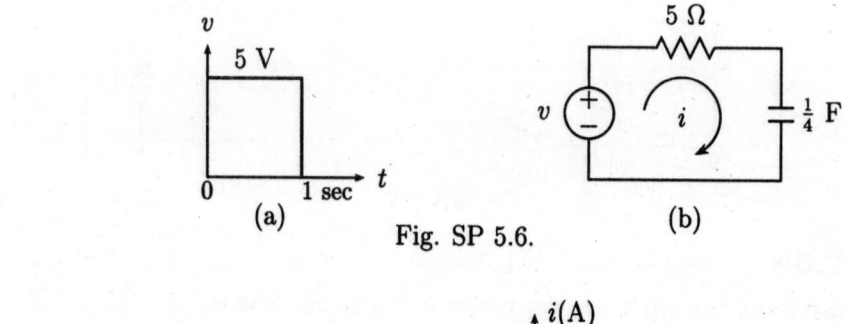

Fig. SP 5.6.

(a)   (b)

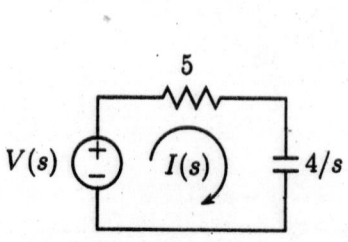

Fig. 5.6.1: s-domain circuit.

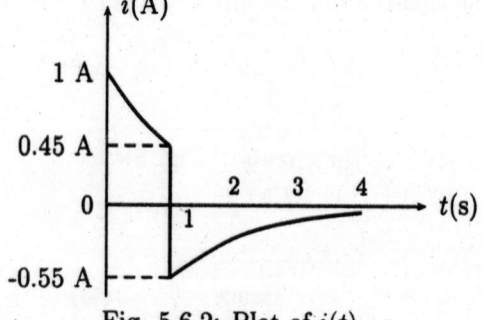

Fig. 5.6.2: Plot of $i(t)$.

SOLUTION:

The equation for the input waveform shown in Fig. SP 5.6a is

$$v(t) = 5[u(t) - u(t-1)]$$

and
$$V(s) = \frac{5}{s}(1 - e^{-s})$$

By KVL in Fig. 5.6.1

$$(5 + 4/s)I(s) = V(s) = \frac{5}{s}(1 - e^{-s})$$

Hence
$$I(s) = \frac{5(1 - e^{-s})/s}{5 + 4/s} = \frac{1}{s + 4/5} - \frac{1}{s + 4/5}e^{-s}$$

Therefore
$$i(t) = \mathcal{L}^{-1}[I(s)] = e^{-4t/5}u(t) - e^{-4(t-1)/5}u(t-1)$$

SP 5.7 A series RL circuit is excited by a truncated ramp signal waveform as shown in Fig. SP 5.7. Find i(t).

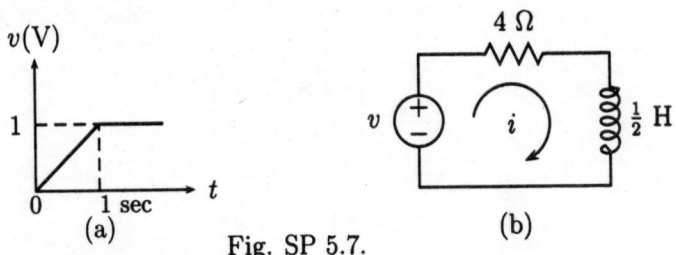

Fig. SP 5.7.

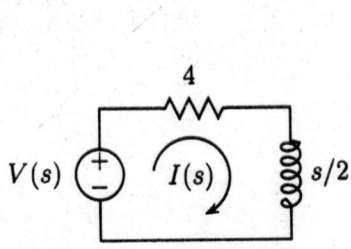

Fig. 5.7.1: s-domain circuit.

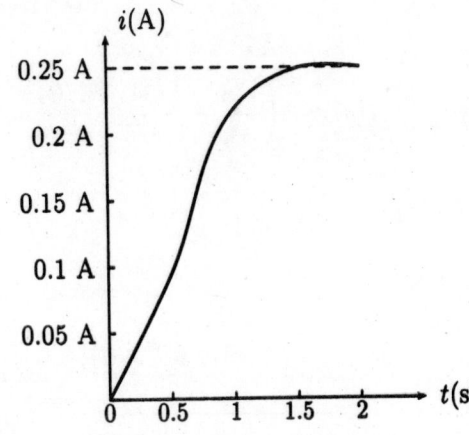

Fig. 5.7.2: Plot of $i(t)$.

SOLUTION:

The equation of the input waveform shown in Fig. SP 5.7a is

$$v(t) = t[u(t) - u(t-1)] + u(t-1)$$
$$= tu(t) - (t-1)u(t-1)$$

and
$$V(s) = \frac{1}{s^2}(1 - e^{-s})$$

By KVL in Fig. SP 5.7.1

$$(4 + s/2)I(s) = V(s) = \frac{1}{s^2}(1 - e^{-s})$$

or
$$I(s) = 2\left[\frac{1}{s^2(s+8)} - \frac{1}{s^2(s+8)}e^{-s}\right]$$

However
$$\frac{1}{s^2(s+8)} = \frac{A}{s^2} + \frac{B}{s} + \frac{C}{s+8}$$

where
$$A = \left[\frac{1}{s+8}\right]_{s=0} = 1/8$$

$$B = \left[\frac{d}{ds}\left(\frac{1}{s+8}\right)\right]_{s=0} = -\left[\frac{1}{(s+8)^2}\right]_{s=0} = -1/64$$

and
$$C = \left[\frac{1}{s^2}\right]_{s=-8} = 1/64$$

Hence
$$I(s) = 2\left(\frac{1/8}{s^2} - \frac{1/64}{s} + \frac{1/64}{s+8}\right) - 2\left(\frac{1/8}{s^2} - \frac{1/64}{s} + \frac{1/64}{s+8}\right)e^{-s}$$

$$= \frac{1}{32}\left[\left(\frac{8}{s^2} - \frac{1}{s} + \frac{1}{s+8}\right) - \left(\frac{8}{s^2} - \frac{1}{s} + \frac{1}{s+8}\right)e^{-s}\right]$$

Therefore

$$i(t) = \frac{1}{32}\left[(8t - 1 + e^{-8t})u(t)\right] - \frac{1}{32}\left[\{8(t-1) - 1 + e^{-8(t-1)}\}u(t-1)\right]$$

SP 5.8 A series RL circuit is excited by a waveform shown in Fig. SP 5.8a. Find and plot for $i(t)$.

SOLUTION:
The equation of the input waveform shown in Fig. SP 5.8a is

$$v(t) = -2(t-1)[u(t) - u(t-1)] = -2tu(t) + 2u(t) + 2(t-1)u(t-1)$$

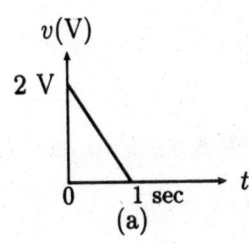

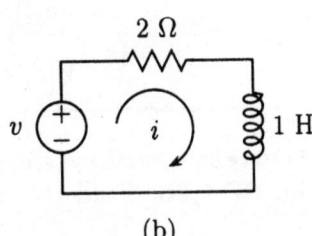

Fig. SP 5.8.

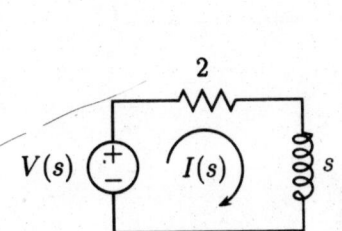

Fig. 5.8.1: s-domain circuit.

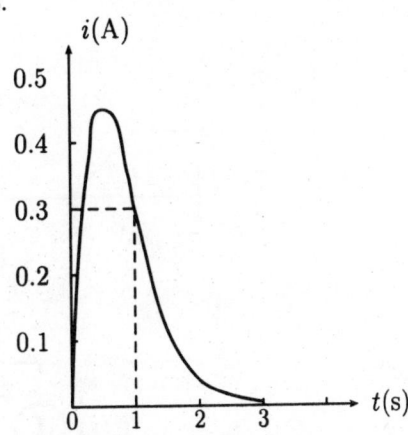

Fig. 5.8.2: Plot of $i(t)$.

and
$$V(s) = \frac{-2}{s^2} + \frac{2}{s} + \frac{2}{s^2}e^{-s}$$

By KVL in the circuit of Fig. 5.8.1

$$(2+s)I(s) = V(s) = \frac{-2}{s^2} + \frac{2}{s} + \frac{2}{s^2}e^{-s}$$

Hence
$$I(s) = \frac{-2}{s^2(s+2)} + \frac{2}{s(s+2)} + \frac{2}{s^2(s+2)}e^{-s}$$

However
$$\frac{1}{s(s+2)} = \frac{A_1}{s} + \frac{B_1}{s+2}$$

where
$$A_1 = 1/2 \qquad \text{and} \qquad B_1 = -1/2$$

and
$$\frac{1}{s^2(s+2)} = \frac{A_2}{s^2} + \frac{B_2}{s} + \frac{C_2}{s+2}$$

where $\qquad A_2 = 1/2, B_2 = -1/4 \qquad$ and $\qquad C_1 = 1/4$

$$I(s) = 2\left[\left(\frac{1/2}{s} - \frac{1/2}{s+2}\right) - \left(\frac{1/2}{s^2} - \frac{1/4}{s} + \frac{1/4}{s+2}\right) + \left(\frac{1/2}{s^2} - \frac{1/4}{s} + \frac{1/4}{s+2}\right)e^{-s}\right]$$

Therefore

$$i(t) = \left(1 - e^{-2t}\right)u(t) - \left(t - \frac{1}{2} + \frac{1}{2}e^{-2t}\right)u(t) + \left[(t-1) - \frac{1}{2} + \frac{1}{2}e^{-2(t-1)}\right]u(t-1)$$

SP 5.9 The current waveform shown in Fig. SP 5.9a is applied to a RC circuit shown in Fig. SP 5.9b. Find the $v(t)$.

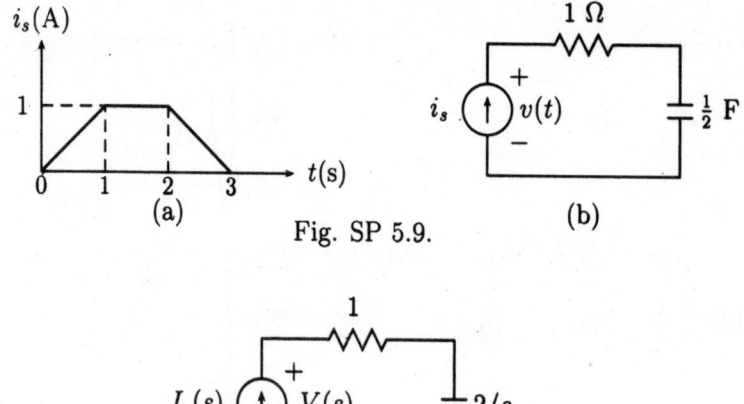

Fig. SP 5.9.

Fig. 5.9.1: s-domain circuit.

SOLUTION:

The equation of current waveform is

$$
\begin{aligned}
i_s(t) &= t[u(t) - u(t-1)] + u(t-1) - u(t-2) - (t-3)[u(t-2) - u(t-3)] \\
&= tu(t) - (t-1)u(t-1) - (t-2)u(t-2) + (t-3)u(t-3)
\end{aligned}
$$

and $\qquad\qquad I_s(s) = \dfrac{1}{s^2}\left[1 - e^{-s} - e^{-2s} + e^{-3s}\right]$

By KVL in the circuit of Fig. 5.9.2

$$
\begin{aligned}
V(s) &= (1 + 2/s)I_s(s) \\
&= \frac{s+2}{s}\cdot\frac{1}{s^2}\left(1 - e^{-s} - e^{-2s} + e^{-3s}\right) \\
&= \left(\frac{1}{s^2} + \frac{2}{s^3}\right) - \left(\frac{1}{s^2} + \frac{2}{s^3}\right)e^{-s} - \left(\frac{1}{s^2} + \frac{2}{s^3}\right)e^{-2s} + \left(\frac{1}{s^2} + \frac{2}{s^3}\right)e^{-3s}
\end{aligned}
$$

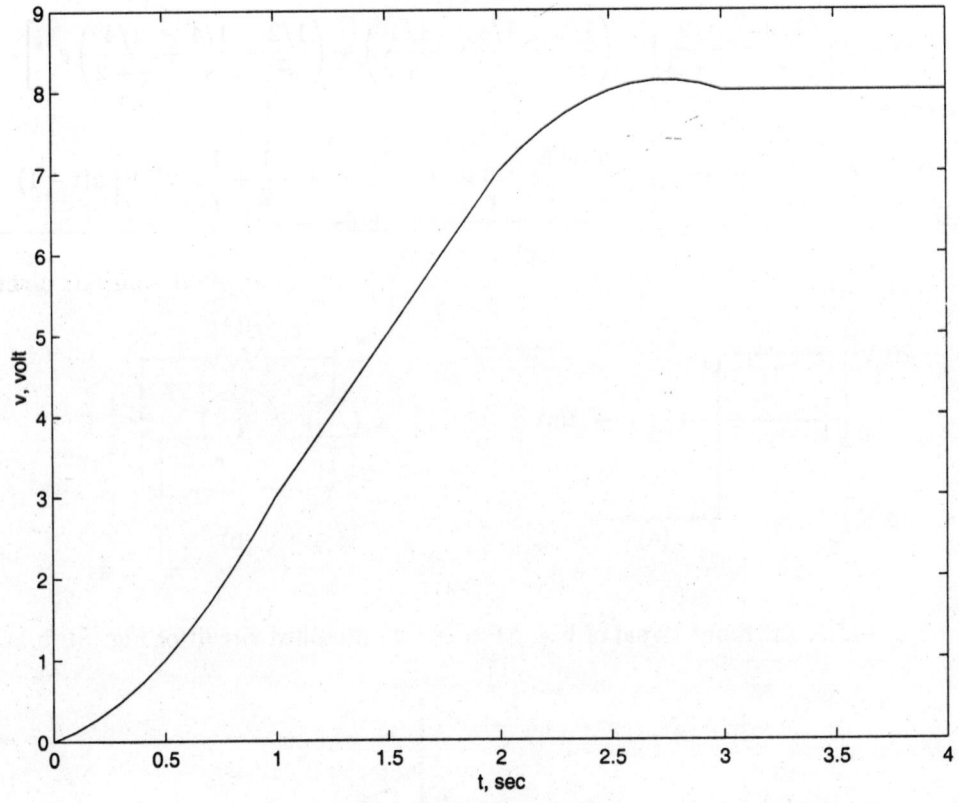

Fig. 5.9.2: Plot of $v(t)$.

Therefore

$$v(t) = (t + 2t^2)u(t) - \left[(t-1) + 2(t-1)^2\right]u(t-1) - \left[(t-2) + 2(t-2)^2\right]u(t-2)$$
$$+ \left[(t-3) + 2(t-3)^2\right]u(t-3)$$

SP 5.10 Find and plot $i(t)$ for the circuit shown in Fig. SP 5.10. If switch is closed at $t = 0$, and then moved to 2 after one time constant at $t = \tau = 2$ sec.

SOLUTION:

The equation of the input waveform shown in Fig. 5.10.1 is

$$v(t) = 20[u(t) - u(t-2)] - 30u(t-2) = 20u(t) - 50u(t-2)$$

and

$$V(s) = \frac{10}{s}(2 - 5e^{-2s})$$

By KVL in Fig. 5.10.2

$$(10 + 5/s)I(s) = V(s) = \frac{10}{s}(2 - 5e^{-2s})$$

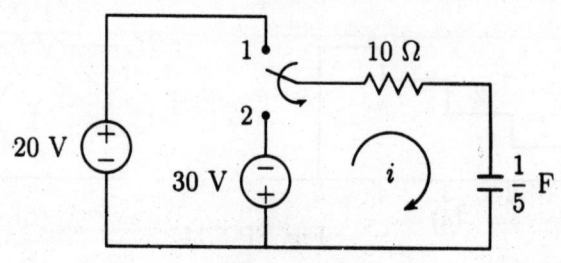

Fig. SP 5.10.

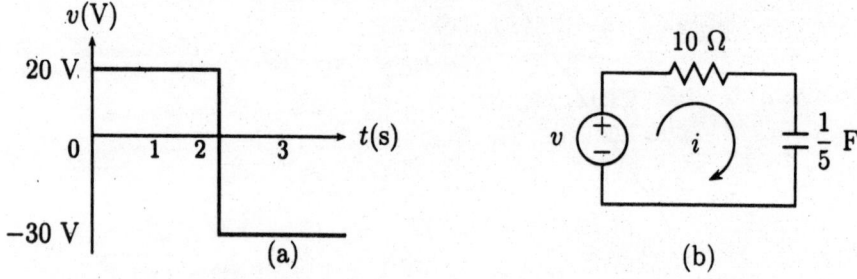

Fig.5.11.1: (a) Input signal of Fig. SP 5.10; (b) Modified circuit of Fig. SP 5.11.

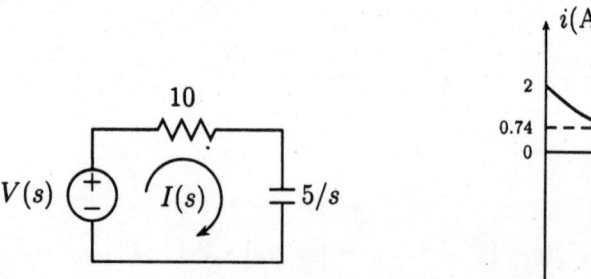

Fig. 5.10.2: s-domain circuit.

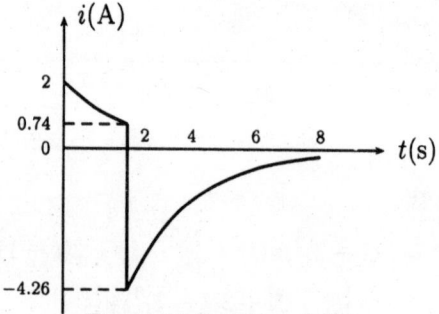

Fig. 5.10.2: Plot of $i(t)$.

or
$$I(s) = \frac{10(2 - 5e^{-2s})}{s(10 + 5/s)} = \frac{2}{s + 1/2} - \frac{5}{s + 1/2}e^{-2s}$$

Therefore
$$i(t) = 2e^{-t/2}u(t) - 5e^{-(t-2)/2}u(t - 2)$$

SP 5.11 A series RL circuit is excited by the staircase waveform as shown in Fig. SP 5.11. Find and plot $i(t)$.

SOLUTION:

The equation for the input waveform shown in Fig. SP 5.11 is

$$v(t) = u(t-1) - u(t-2) + 2[u(t-2) - u(t-3)] + 3[u(t-3) - u(t-4)]$$
$$+ 4[u(t-4) - u(t-5)]$$
$$= u(t-1) + u(t-2) + u(t-3) + u(t-4) - 4u(t-5)$$

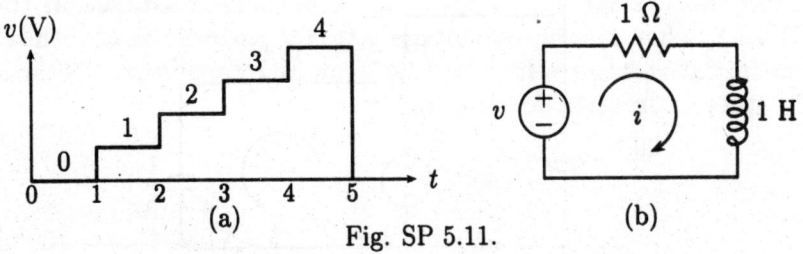

Fig. SP 5.11.

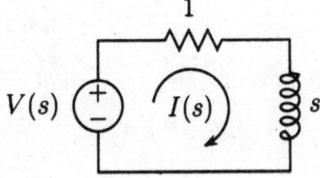

Fig. 5.11.1: s-domain circuit.

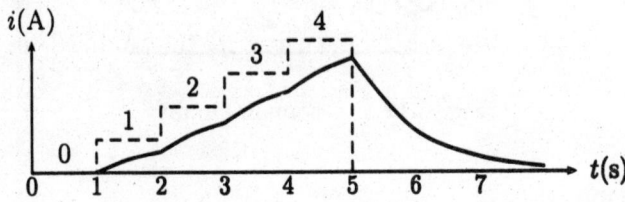

Fig. SP 5.11.2: Plot of $i(t)$.

and
$$V(s) = \frac{1}{s}(e^{-s} + e^{-2s} + e^{-3s} + e^{-4s} - 4e^{-5s})$$

By KVL in Fig. 5.11.1

$$(1+s)I(s) = V(s) = \frac{1}{s}(e^{-s} + e^{-2s} + e^{-3s} + e^{-4s} - 4e^{-5s})$$

Therefore

$$
\begin{aligned}
I(s) &= \frac{1}{s(s+1)}(e^{-s} + e^{-2s} + e^{-3s} + e^{-4s} - 4e^{-5s}) \\
&= \left[\frac{1}{s} - \frac{1}{s+1}\right](e^{-s} + e^{-2s} + e^{-3s} + e^{-4s} - 4e^{-5s})
\end{aligned}
$$

and

$$
\begin{aligned}
i(t) &= [1 - e^{-(t-1)}]u(t-1) + [1 - e^{-(t-2)}]u(t-2) + [1 - e^{-(t-3)}]u(t-3) \\
&\quad + [1 - e^{-(t-4)}]u(t-4) - 4[1 - e^{-(t-5)}]u(t-5)
\end{aligned}
$$

SP 5.12 Find the voltage $v_R(t)$ appearing across the resistance in the circuit of Fig. SP 5.12, when the input voltage $v(t)$ is a periodic sawtooth wave, and there is an initial voltage $v_C(0^+) = 1/2$ V on the capacitor. Determine also the steady state expression for for v_R.

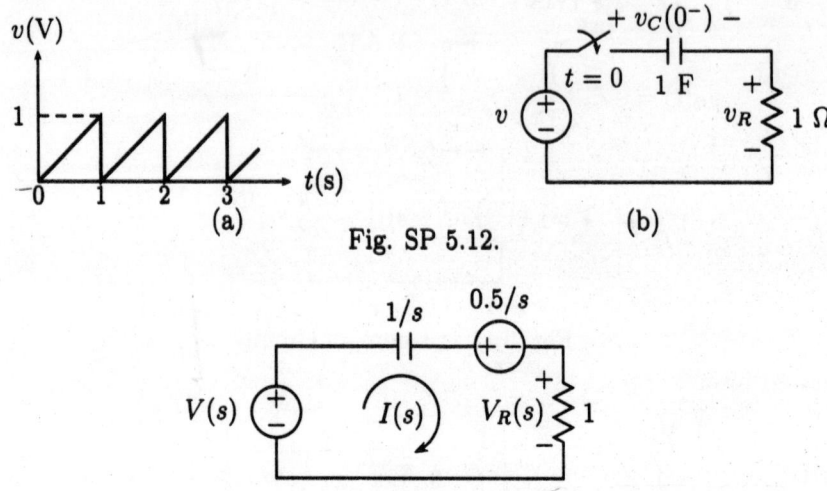

Fig. SP 5.12.

Fig. 5.12.1: s-domain circuit.

SOLUTION:

The equation of the input waveform shown in Fig. SP 5.12 as

$$
\begin{aligned}
v_1(t) &= \text{equation of the waveform for a period} \\
&= t[u(t) - u(t-1)] \\
&= tu(t) - (t-1)u(t-1) - u(t-1)
\end{aligned}
$$

and

$$
V_1(s) = \frac{1}{s^2} - \frac{e^{-s}}{s^2} - \frac{e^{-s}}{s} = \frac{1}{s^2}\left[1 - (1+s)e^{-s}\right]
$$

Therefore

$$
\begin{aligned}
V(s) &= \frac{V_1(s)}{1 - e^{-s}} = \frac{[1 - (1+s)e^{-s}]}{s^2(1 - e^{-s})} = \frac{(1+s)(1 - e^{-s}) - s}{s^2(1 - e^{-s})} \\
&= \left[\frac{1+s}{s^2} - \frac{1}{s(1 - e^{-s})}\right]
\end{aligned}
$$

By KVL in Fig. 5.12.1

$$
(1 + 1/s)I(s) + 1/2s = V(s)
$$

Hence

$$
I(s) = \frac{s}{s+1}\left[V(s) - 1/2s\right]
$$

and

$$V_R(s) = 1I(s) = \frac{s}{s+1}\left[\frac{1+s}{s^2} - \frac{1}{s(1-e^{-s})} - \frac{1}{2s}\right]$$

$$= \frac{1}{2}\left[\frac{s+2}{s(s+1)} - \frac{2}{(s+1)(1-e^{-s})}\right]$$

Thus

$$v_R(t) = \frac{1}{2}\mathcal{L}^{-1}\left[\frac{s+2}{s(s+1)}\right] - \mathcal{L}^{-1}\left[\frac{1}{(s+1)(1-e^{-s})}\right]$$

However

$$\frac{1}{2}\mathcal{L}^{-1}\left[\frac{s+2}{s(s+1)}\right] = \frac{1}{2}\mathcal{L}^{-1}\left[\frac{2}{s} - \frac{1}{s+1}\right]$$

$$= \frac{1}{2}\left[2 - e^{-t}\right] \quad \text{for } t \geq 0$$

And the second term is

$$\mathcal{L}^{-1}\left[\frac{1}{(s+1)(1-e^{-s})}\right] = \mathcal{L}^{-1}\frac{1}{s+1}\left[1 + e^{-s} + e^{-2s} + ...\right]$$

$$= e^{-t}u(t) + e^{-(t-1)}u(t-1) + e^{-(t-2)}u(t-2) + ...$$

It is important to note that the right side of above equation is not an infinite series because in the interval $0 < t < 1$, only first term exists and other terms vanish, in the interval $1 < t < 2$, only the first two terms exist, and so on. Hence, in the interval of the nth tooth, $(n-1) < t < n$

$$\mathcal{L}^{-1}\left[\frac{1}{(s+1)(1-e^{-s})}\right] = e^{-t}\left[1 + e + e^2 + - - - - - + e^{(n-1)}\right]$$

$$= e^{-t}\left[\frac{e^n - 1}{e - 1}\right]$$

Therefore

$$v_R(t) = \left[1 - \frac{e^{-t}}{2}\right] - \frac{e^n - 1}{e - 1}e^{-t} \quad \text{for } (n-1) < t < n$$

To determine $(v_R)_{ss}$, we rearrange above equation as follows:

$$v_R(t) = 1 - \frac{e^{-t}}{2} - \frac{e^{-(t-n)}}{e - 1} + \frac{e^{-t}}{e - 1}$$

$$= \left[-\frac{1}{2} + \frac{1}{e - 1}\right]e^{-t} + \left[1 - \frac{e^{-(t-n)}}{e - 1}\right]$$

$$\text{for } (n-1) < t < n$$

Now it is clear that

$$(v_R)_{tr} = \left[-\frac{1}{2} + \frac{1}{e-1}\right]e^{-t} \quad t \geq 0$$

And

$$(v_R)_{ss} = \left[1 - \frac{e^{-(t-n)}}{e-1}\right] \quad (n-1) < t < n,$$

By setting n = 1, we obtain

$$(v_R)_{ss} = 1 - \frac{e^{-(t-1)}}{e-1} \quad 0 < t < 1$$

which gives the same curve in the first period as it does in the nth period

---

5.11   DRILL PROBLEMS

DP 5.1 Sketch the waveforms for the following functions in the interval
$-1 \leq t \leq 1$.

a. $tu(t)$

b. $tu(-t)$

c. $(1-t)u(t)$

d. $(1-t) + u(t)$

e. $(t-1)u(t)$

f. $(t-1)u(t+1)$

g. $tu(t+1)u(1-t)$

Hints : For (g) part, the sketch is given in Fig. DP 5.1(g).

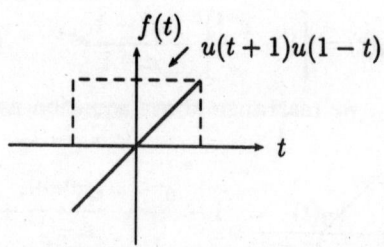

Fig. DP 5.1(g).

DP 5.2 Write time-domain equations for the following waveforms shown in Fig. DP 5.2.

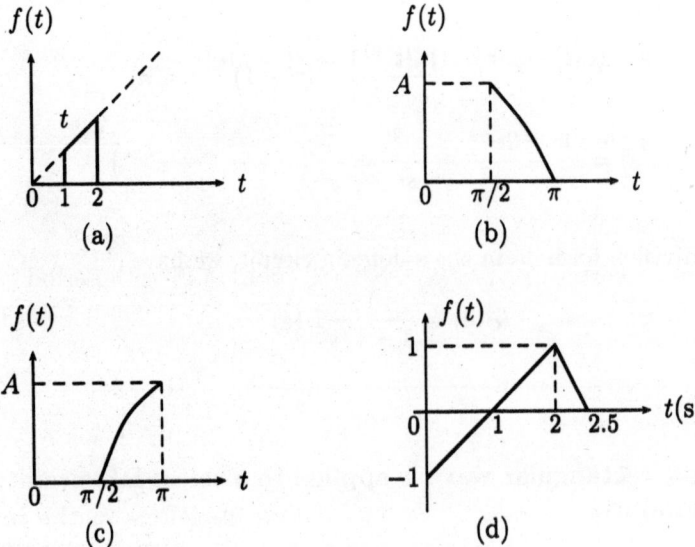

Fig. DP 5.2.

Hints :

(a) $t\,[u(t-1) - u(t-2)]$

(b) $A\sin t\ [u(t-\pi/2) - u(t-\pi)]$

(c) $-A\cos t\ [u(t-\pi/2) - u(t-\pi)]$

DP 5.3 A triangular current wave shown in Fig. DP 5.3a is applied to the RC- circuit shown in Fig. DP 5.3b. Find $i_C(t)$.

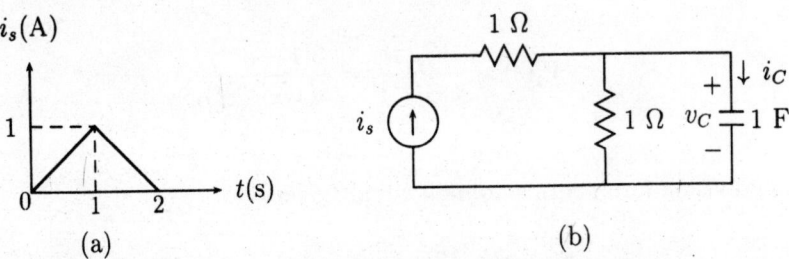

Fig. DP 5.3.

Hints : The equation of given current waveform is

$$i_s(t) \;=\; t\left[u(t) - u(t-1)\right] - (t-2)\left[u(t-1) - u(t-2)\right]$$

$$=\; tu(t) - 2(t-1)u(t-1) + (t-2)u(t-2)$$

$$I_s(s) = \frac{1}{s^2} - \frac{2e^{-s}}{s^2} + \frac{e^{-2s}}{s^2} = \frac{1}{s^2}\left(1 - 2e^{-s} + e^{-2s}\right)$$

Applying current division formula in the s-domain circuit, we have

$$I_C(s) = \frac{1}{1 + 1/s} I_s(s)$$

DP 5.4 A periodic rectangular wave is applied to a series RL circuit as shown in fig. DP 5.4. Find v_R.

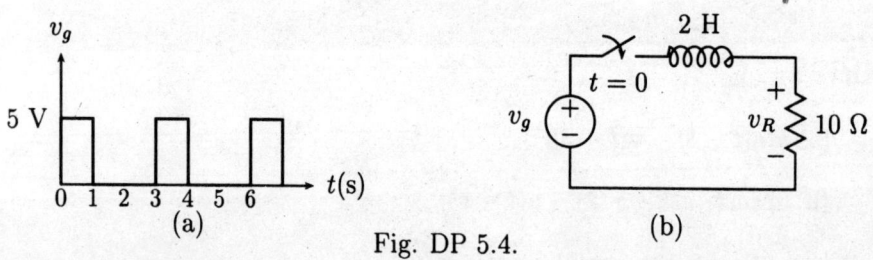

Fig. DP 5.4.

Hints : The equation of one pulse is

$$v_{1g}(t) = 5\left[u(t) - u(t-1)\right] \qquad \text{and} \qquad V_{1g}(s) = \frac{5}{s}(1 - e^{-s})$$

Hence
$$V_g(s) = \frac{V_{1g}(s)}{1 - e^{-3s}} = \frac{5(1 - e^{-s})}{s(1 - e^{-3s})}$$

The voltage division formula in s-domain circuit gives

$$V_R(s) = \frac{10}{2s + 10} V_g(s) = \frac{25}{s(s+5)}\left(\frac{1 - e^{-s}}{1 - e^{-3s}}\right)$$

Hence

$$v_R = \mathcal{L}^{-1}\left[\frac{25(1-e^{-3s})^{-1}}{s(s+5)}\right] - \mathcal{L}^{-1}\left[\frac{25e^{-s}(1-e^{-3s})^{-1}}{s(s+5)}\right]$$

The first term on right side of $v_R(t)$ is

First term $= \mathcal{L}^{-1}\left[\dfrac{25(1-e^{-3s})^{-1}}{s(s+5)}\right]$

$$= 5\mathcal{L}^{-1}\left(\frac{1}{s} - \frac{1}{s+5}\right)\left(1 + e^{-3s} + e^{-6s} + \ldots\right)$$

$$= 5\left[u(t) + u(t-3) + u(t-6) + \ldots\right] - 5\left[e^{-5t}u(t) + e^{-5(t-3)}u(t-3) + \ldots\right]$$

$$= 5\left[u(t) + u(t-3) + u(t-6) + \ldots\right] - 5e^{-5t}\left[1 + e^{15} + e^{30} + \ldots\right]$$

$$= 5\left[u(t) + u(t-3) + u(t-6) + \ldots\right] - 5e^{-5t}\left[\frac{e^{15n}-1}{e^{15}-1}\right] \text{ for n } - \text{ th period}$$

The second term on right side of $v_R(t)$ is

2nd term $= \mathcal{L}^{-1}\left[\dfrac{25e^{-s}(1-e^{-3s})^{-1}}{s(s+5)}\right]$

$$= 5\mathcal{L}^{-1}\left(\frac{1}{s} - \frac{1}{s+5}\right)e^{-s}\left(1 + e^{-3s} + e^{-6s} + \ldots\right)$$

$$= 5\mathcal{L}^{-1}\left(\frac{1}{s} - \frac{1}{s+5}\right)\left(e^{-s} + e^{-4s} + e^{-7s} + \ldots\right)$$

$$= 5\left[u(t-1) + u(t-4) + \ldots\right] - 5\left[e^{-5(t-1)}u(t-1) + e^{-5(t-4)}u(t-4) + \ldots\right]$$

$$= 5\left[u(t-1) + u(t-4) + \ldots\right] - 5e^{-5t}\left[e^5 + e^{20} + e^{35} + \ldots\right]$$

$$= 5\left[u(t-1) + u(t-4) + \ldots\right] - 5e^{-5t}\left[\frac{e^5(e^{15n}-1)}{e^{15}-1}\right] \text{ for n } - \text{ th period}$$

On combining

$$v_R(t) = 5e^{-5t}\frac{1-e^5}{e^{15}-1} + v_g(t) + \frac{5}{e^{15}-1}\left[e^{-5(t-3n+1)} - e^{-5(t-3n)}\right]$$

for n-th period

DP 5.5 A periodic exponential voltage wave which is $2e^{-5t}$ Volts in any one period is applied to a series RC- circuit as shown in Fig. DP 5.5. Find $v_C(t)$ for $0.2 < t < 0.4$ sec.

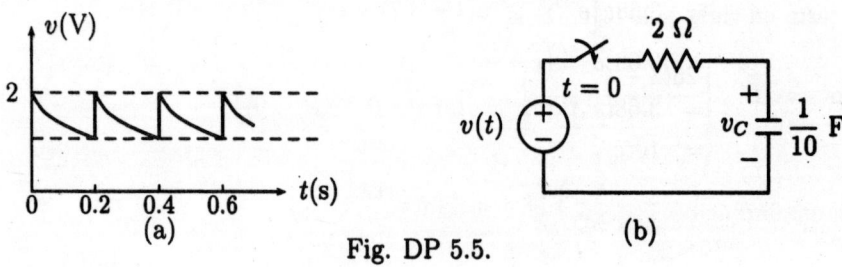

Fig. DP 5.5.

Hints : The equation for the first period of the waveform is

$$v_1(t) = 2e^{-5t}\left[u(t) - u(t-0.2)\right] = 2e^{-5t}u(t) - 2e^{-5(t-0.2)}e^{-1}u(t-0.2)$$

and

$$V_1(s) = \frac{2}{s+5}\left(1 - 0.368e^{-0.2s}\right)$$

Therefore

$$V(s) = \frac{V_1(s)}{1 - e^{-0.2s}} = \frac{2}{s+5}\left(\frac{1 - 0.368e^{-0.2s}}{1 - e^{-0.2s}}\right)$$

The voltage division formula in s-domain circuit gives

$$V_C(s) = \frac{10/s}{2 + 10/s}V(s) = \frac{10}{(s+5)^2}\left[\frac{1 - 0.368e^{-0.2s}}{1 - e^{-0.2s}}\right]$$

and

$$v_R(t) = \mathcal{L}^{-1}\left[\frac{10}{(s+5)^2}(1 - e^{-0.2s})^{-1}\right] - \mathcal{L}^{-1}\left[\frac{3.68}{(s+5)^2}e^{-0.2s}(1 - e^{-0.2s})^{-1}\right]$$

The first term on the right side of equation $v_R(t)$ is

$$\text{First term} = \mathcal{L}^{-1}\left[\frac{10}{(s+5)^2}(1 - e^{-0.2s})^{-1}\right]$$

$$= \frac{10}{(s+5)^2}(1 + e^{-0.2s} + e^{-0.4s} + \dots\dots)$$

$$= 10t\left[e^{-5t}u(t) + e^{-5(t-0.2)}u(t-0.2) + e^{-5(t-0.4)}u(t-0.4) + \dots\right]$$

$$= 10te^{-5t}\left(1 + e + e^2 + \dots\dots\right)$$

$$= 10te^{-5t}\left(1 + e\right) \qquad \text{for} \qquad 0.2 < t < 0.4$$

$$= 37.2te^{-5t}$$

$$\text{Second term} = \mathcal{L}^{-1}\left[\frac{3.68}{(s+5)^2}e^{-0.2s}(1-e^{-0.2s})^{-1}\right]$$

$$= \mathcal{L}^{-1}\left[\frac{3.68}{(s+5)^2}\left(e^{-0.2}+e^{-0.4}+e^{-0.6}+......\right)\right]$$

$$= 3.68t\left[e^{-5(t-0.2)}u(t-0.2)+e^{-5(t-0.4)}u(t-0.4)+......\right]$$

$$= 3.68te^{-5t}\left(e+e^2+e^3+.......\right)$$

$$= 3.68te^{-5t}(e)\qquad\text{for}\qquad 0.2 < t < 0.4$$

$$= 10te^{-5t}$$

Therfore on combining

$$v_C(t) = 27.2te^{-5t}$$

DP 5.6 A periodic rectangular current wave is applied to the parallel RC-circuit as shown in Fig. DP 5.6. find steady-state value of $v_C(t)$.

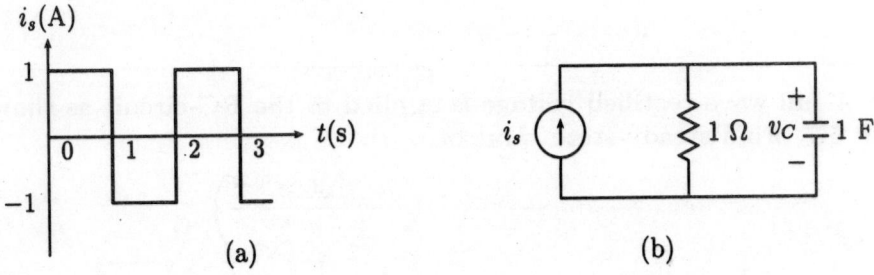

(a)   (b)

Fig. DP 5.6.

Hints :
Due to the symmetry of given current waveform, one may conclude that under steady-state conditions the response in the second half-cycle will be same as the response in the first half-cycle except for a change in sign and shift of 1 sec. in t. Hence it is important to test only the steady-state expression for $v_C(t)$ in the first half-cycle , $0 < t < 1$ sec.

The differential equation for $v_{C,ss}$ is

$$\frac{dv_{C,ss}}{dt} + \frac{v_{Css}}{1} = 1$$

Laplace transform gives

$$sV_{C,ss}(s) - v_{C,ss}(0) + V_{C,ss}(s) = 1/s$$

or

$$V_{C,ss}(s) = \frac{1}{s(s+1)} + \frac{v_{C,ss}(0)}{s+1}$$

Hence
$$v_{C,ss}(t) = (1 - e^{-t}) + v_{C,ss}(0)e^{-t}$$

Use $v_{C,ss}(1) = -v_{C,ss}(0)$ in the last equation to determine $v_{C,ss}(0)$ as

$$v_{C,ss}(1) = -v_{C,ss}(0) = (1 - e^{-1}) + v_{C,ss}(0)e^{-1}$$

from which $v_{C,ss}(0) = -0.462$. Thus $v_{C,ss}$ may be obtained by putting the value of $v_{C,ss}(0)$ in the expression of $v_{C,ss}(t)$ for $0 < t < 1$ sec.

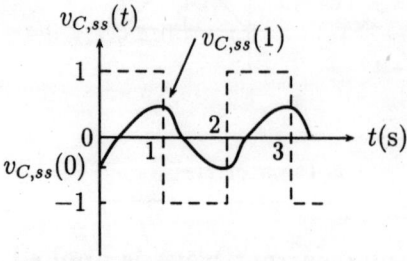

Fig. DP 5.6.1: Steady-state response.

DP 5.7 A full wave rectified voltage is applied to the RC-circuit as shown in Fig. DP 5.7. Find steady-state value of $v_C(t)$.

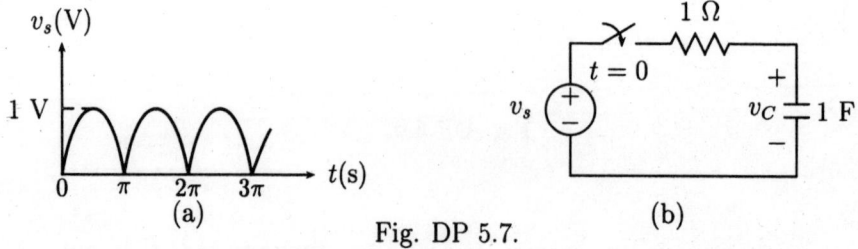

Fig. DP 5.7.

Hints : Follow the method outlined in problem DP 5.6 with a slight modification in initial condition that is $v_{C,ss}(\pi) = v_{C,ss}(0)$.

DP 5.8 A rectangular voltage pulse of magnitude V and duration T is applied to a series combination of resistance R and capacitance C. Find the maximum voltage developed across the capacitor.

Hints : The expression for voltage across the capacitor will be

$$v_C(t) = V(1 - e^{-t/RC})$$

The maximum voltage can be developed at t = T, i.e.,

$$v_{C,max.} = v_C(T) = V(1 - e^{-T/RC})$$

DP 5.9 Explain the ramp function. Find the equation for triangular wave in terms of ramp functions for Fig. DP 5.9.

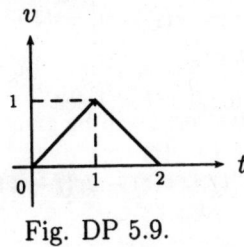

Fig. DP 5.9.

SOLUTION:

The ramp function is defined as:

$$Kr(t) = \begin{cases} Kt & \text{for } t \geq 0 \\ 0 & \text{for } t < 0 \end{cases}$$

where K is the slope of ramp function, $r(t)$.

For $K = 1$, the function is called unit ramp function. The shifted ramp function is defined as:

$$Kr(t - \tau) = \begin{cases} 0 & \text{for } -\infty \leq t < 0 \\ K(t - \tau) & \text{for } \tau \leq t < \infty \end{cases}$$

We may differentiate the ramp function to obtain the step function as:

$$\frac{d}{dt}[Kr(t)] = Ku(t)$$

and $$\frac{d}{dt}[Kr(t - \tau)] = Ku(t - \tau)$$

The Fig DP 5.9.1 shows the graphical representations of ramp and shifted functions.

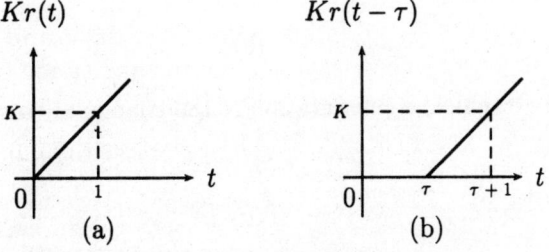

Fig. DP 5.9.1 (a) Ramp function; (b) Shifted ramp function.

The equation for the triangular wave form may be obtained by the graphical representation in a number of steps as shown in Fig. DP 5.9.2. Here the slope of ramp function is 1. Firstly, the $r(t)$ and $-r(t-1)$ is added to give $r(t)-r(t-1)$ and its wave form is shown in Fig. 5.9.2(c). Secondly, $r(t)-r(t-1$ and $-r(t-1)$ is added to give $r(t)-2r(t-1)$ and its wave form is shown in Fig. 5.9.2(f). Finally, $r(t)-2r(t-1)$ and $r(t-2)$ is added to give $r(t)-2r(t-1)+r(t-2)$ and its wave form is shown in Fig 5.9.2(i). Thus, the equation of the triangular wave form will be given from Fig. 5.9.2(i) as:

$$v(t) = r(t) - 2r(t-1) + r(t-2)$$

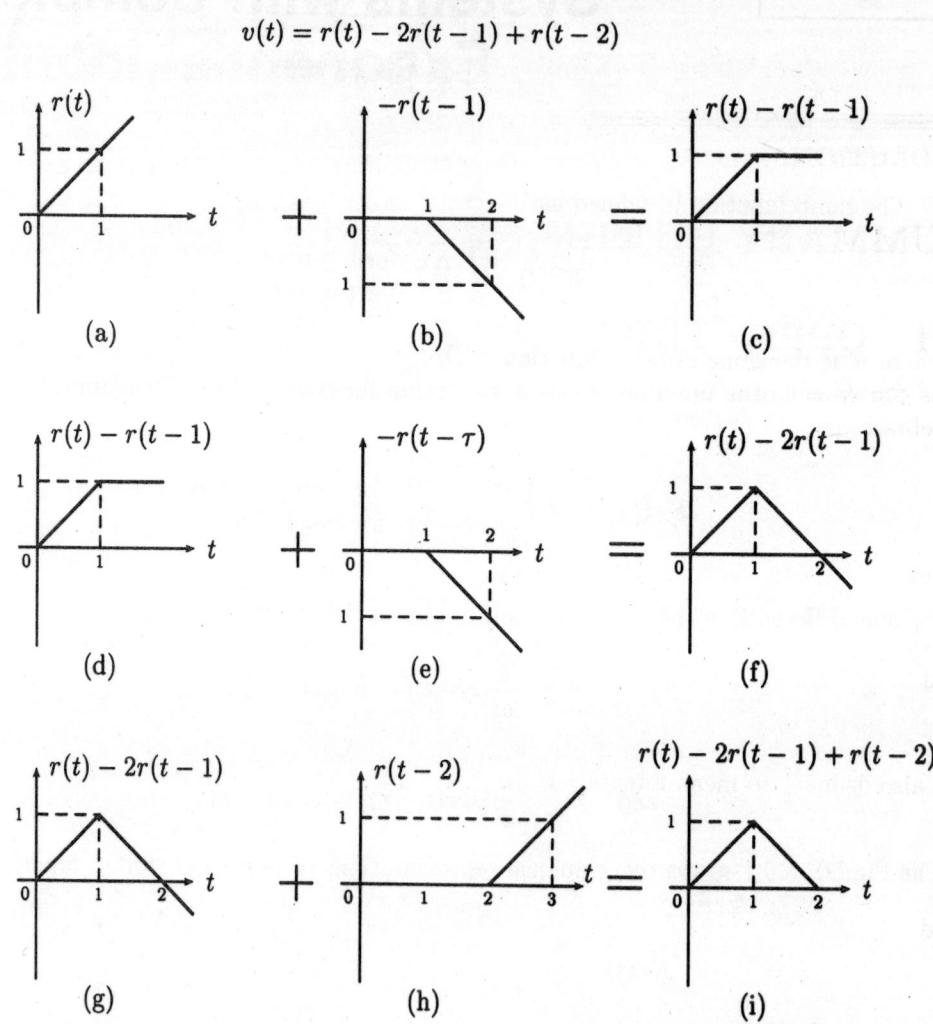

Fig. DP 5.9.2: Graphical representation of the triangular wave of Fig. DP 5.9.

CHAPTER
6

Systems with Complex
Exponential Inputs

SUMMARY OF THIS CHAPTER

6.1 OPERATOR p

It is convenient to manipulate integrodifferential equations in the same way that we do algebraic ones. In order to do this we define the operator p as:

$$p = \frac{d}{dt}$$

Thus

$$pf = \frac{df}{dt}$$

and

$$p^n f = \frac{d^n f}{dt^n}$$

we also define $\frac{1}{p}$ to mean integration such that

$$\frac{1}{p}pf = p\frac{f}{p} = f$$

and

$$\frac{1}{p}f = \int_0^t f(t)dt$$

The terminal relations of the resistor, inductor, and capacitor can be written as

Resistor	$v_R(t) = Ri_R(t)$
Inductor	$v_L(t) = L\dfrac{di_L}{dt} = Lpi_L(t)$
Capacitor	$v_C(t) = \dfrac{1}{C}\displaystyle\int_0^t i_C(\tau)d\tau = \dfrac{1}{Cp}i_C(t)$

Note that the inductor may be considered to have a "resistance" of Lp and capacitance a "resistance" of $\frac{1}{Cp}$. These will be referred to as impedances and denoted as Z:

$$Z_R = R$$
$$Z_L = Lp$$
$$Z_C = \frac{1}{Cp}$$

So that $v(t) = Zi(t)$

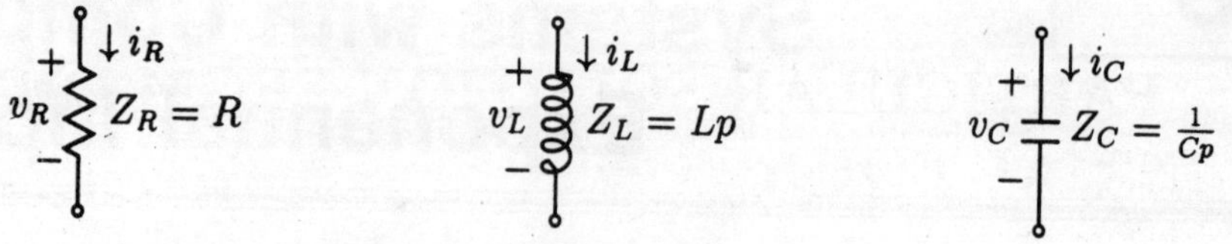

Figure 6.1: Impedances in terms of operator p

Hence, the terminal equations in term of the operator impedance for all elements are similar to Ohm's law for the resistor:

$$v_R(t) = Z_R i_R(t)$$
$$v_L(t) = Z_L i_L(t)$$
$$v_C(t) = Z_C i_C(t)$$

Alternately, we consider the dynamic elements as having an admittance similar to the conductance of a resistor. This admittance is simply the reciprocal of the element impedance and is denoted by Y:

$$Y_R = \frac{1}{Z_R} = \frac{1}{R} = G$$
$$Y_L = \frac{1}{Z_L} = \frac{1}{Lp}$$
$$Y_C = \frac{1}{Z_C} = Cp$$

Thus, for the each element, we can write $i(t) = Yv(t)$.

Steps to Derive Differential Equation from the Circuit

Step 1: Convert the given circuit into p-domain.

Step 2: Find variable of interest in terms of p to the existing ideal source(s). (Assuming that we have a resistive circuit with the elements replaced by their impedances). For example

$$y(t) = \frac{N(p)}{D(p)}x(t)$$

where $y(t)$ is the output function, $x(t)$ is the input function and $N(p)/D(p)$ is the transfer function.

Step 3: Separate the input and output variables after cross multiplication. For **example**:
$$D(p)\, y(t) = N(p)\, x(t)$$

Step 4: Write the terms in increasing order of p.

Step 5: Replace p by $\frac{d}{dt}$ which yields required differential equation.

6.2 PARTICULAR SOLUTION FROM DIFFERENTIAL EQUATIONS

The following steps may be used to determine the particular solution:

Step 1: Write the differential equation in the form of: $y(t) = \frac{N(p)}{D(p)} x(t)$.

Step 2: Replace $x(t)$ by its exponential form with the help of the Table 6.1.

Step 3: Replace p by the coefficient of t in exponential term.

Step 4: Simplify the coefficient of exponential term.

Step 5: Remove Re or, Im from the terms, if any. Which yields particular solution.

Table 6.1: **Functions and their exponential forms**

S. No.	Input Function $x(t)$	Exponential Form e^{pt}
1.	Constant A	Ae^{0t} where $p = 0$
2.	$Ae^{-\alpha t}$	$Ae^{-\alpha t}$ where $p = -\alpha$
3.	$A\cos\omega t$	$\mathrm{Re}[Ae^{j\omega t}]$ where $p = j\omega$
4.	$A\sin\omega t$	$\mathrm{Im}[Ae^{j\omega t}]$ where $p = j\omega$
5.	$A\cos(\omega t \pm \phi)$	$\mathrm{Re}[A\angle \pm \phi e^{j\omega t}]$ where $p = j\omega$
6.	$A\sin(\omega t \pm \phi)$	$\mathrm{Im}[A\angle \pm \phi e^{j\omega t}]$ where $p = j\omega$
7.	$Ae^{\pm\alpha t}\cos(\omega t)$	$\mathrm{Re}[Ae^{(j\omega\pm\alpha)t}]$ where $p = j\omega \pm \alpha$
8.	$Ae^{\pm\alpha t}\sin(\omega t)$	$\mathrm{Im}[Ae^{(j\omega\pm\alpha)t}]$ where $p = j\omega \pm \alpha$
9.	$Ae^{\pm\alpha t}\cos(\omega t \pm \phi)$	$\mathrm{Re}[A\angle \pm \phi e^{(j\omega\pm\alpha)t}]$ where $p = j\omega \pm \alpha$
10.	$Ae^{\pm\alpha t}\sin(\omega t \pm \phi)$	$\mathrm{Im}[A\angle \pm \phi e^{(j\omega\pm\alpha)t}]$ where $p = j\omega \pm \alpha$

6.3 SOLVED PROBLEMS

SP 6.1 Find the particular (steady state) response for the current $i(t)$ in the circuit shown in Fig. SP 6.1, if

(a) $v(t) = 10$ V,

(b) $v(t) = 60e^{-4t}$ V,

(c) $v(t) = 10\sin(2t + 45^\circ)$ V,

(d) $v(t) = 10e^{-5t}\sin(3t + 60^\circ)$ V,

(e) $v(t) = 10 + 60e^{-4t} + 10e^{-5t}\sin(3t + 60^\circ)$ V.

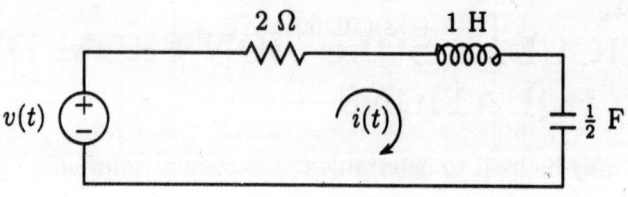

Fig. SP 6.1.

SOLUTION: By KVL in Fig. 6.1

$$2i + \frac{di}{dt} + 2\int_0^t i\,dt = v(t)$$

Writing above equation in p-domain as

$$2i + pi + \frac{2}{p}i = v(t)$$

Thus
$$i(t) = \frac{1}{2 + p + 2/p}v(t) = \frac{p}{p^2 + 2p + 2}v(t)$$

(a) Given $v(t) = 10V = 10e^{0t}$ thus, the particular response is given by

$$i_P(t) = \frac{0}{0^2 + 2(0) + 2}10e^0 = 0 \qquad \text{(setting } p = 0\text{)}$$

(b) Here $v(t) = 60e^{-4t}$ thus, the particular response is

$$i_P(t) = \frac{-4}{(-4)^2 + 2(-4) + 2}(60e^{-4t}) = -24e^{-4t} \qquad \text{(setting } p = -4\text{)}$$

(c) Here $v(t) = 10\sin(2t + 45^\circ) = \text{Im}[10\angle 45^\circ e^{j2t}]$ thus, the particular response is

$$
\begin{aligned}
i_P(t) &= \text{Im}\left[\frac{j2}{(j2)^2 + 2(j2) + 2}(10\angle 45^\circ)e^{j2t}\right] \qquad \text{(setting } p = j2\text{)} \\
&= \text{Im}\left[\frac{j2}{-2 + j4}(10\angle 45^\circ)e^{j2t}\right] \\
&= \text{Im}\left[\frac{(2\angle 90^\circ)(10\angle 45^\circ)}{4.472\angle 116.6^\circ}e^{j2t}\right] \\
&= \text{Im}[4.472\angle 18.4^\circ e^{j2t}] \\
&= \text{Im}[4.472e^{j(2t + 18.4^\circ)}] \\
&= 4.472\sin(2t + 18.4^\circ)
\end{aligned}
$$

(d) Here $v(t) = 10e^{-5t}\sin(3t + 60^0) = \text{Im}[10e^{-5t}e^{j(3t+60^0)}] = \text{Im}[10\angle60^0 e^{(-5+j3)t}]$

Thus, the particular response is

$$i_P(t) = \text{Im}\left[\frac{(-5+j3)(10\angle60^0)}{(-5+j3)^2 + 2(-5+j3) + 2}e^{(-5+j3)t}\right]$$

$$(\text{Setting } p = -5 + j3)$$

$$= \text{Im}\left[\frac{(-5+j3)(10\angle60^0)}{8 - j24}e^{(-5+j3)t}\right]$$

$$= \text{Im}\left[\frac{(5.83\angle149.04^0)(10\angle60^0)}{25.298\angle - 71.57^0}e^{(-5+j3)t}\right]$$

$$= \text{Im}[2.3\angle - 79.4^0 e^{(-5+j3)t}] = 2.3e^{-5t}\sin(3t - 79.4^0)$$

(e) Using the superposition theorem together with the result of parts (a), (b) and (d), we find the particular response to be

$$i(t) = 0 - 24e^{-4t} + 2.3e^{-5t}\sin(3t - 79.4^0) = -24e^{-4t} + 2.3e^{-5t}\sin(3t - 79.4^0) \text{ A}$$

SP 6.2 An ideal voltage source $v_s = 10\cos(10t + 20^0)u(t)$ drives the series combination of $R = 3\ \Omega$ and $C = 1/40$ F. Find :

(a) the steady state sinusoidal response $i_{ss}(t)$,

(b) the natural response $i_n(t)$,

(c) the force (or, particular) response $i_P(t)$,

(d) the complete response $i(t)$ for all t, under zero initial conditions.

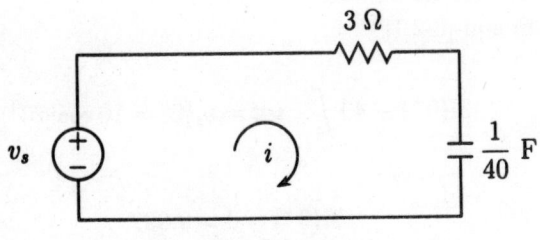

Fig. SP 6.2.

SOLUTION:

By KVL in Fig. SP 6.2,

$$3i + 40\int_0^t i\,dt = v_s(t) \qquad [6.2.1]$$

Writing above equation in p-domain as

$$3i + \frac{40}{p}i = v_s(t)$$

$$\text{thus} \qquad i(t) = \frac{1}{3 + 40/p}v_s(t) = \frac{p}{40 + 3p}v_s(t) \qquad\qquad [6.2.2]$$

(a) Given $v_s(t) = 10\cos(10t + 20^0) = \text{Re}\,[10\angle 20^0 e^{j10t}]$

    Thus, the steady state response is given by

$$
\begin{aligned}
i_{ss}(t) &= \text{Re}\left[\frac{j10}{3(j10) + 40}10\angle 20^0 e^{j10t}\right] \qquad \text{(setting } p = j10\text{)}\\[2mm]
&= \text{Re}\left[2\angle 73.13^0 e^{j10t}\right]
\end{aligned}
$$

Thus

$$i_{ss}(t) = 2\cos(10t + 73.13^0) \qquad\qquad [6.2.3]$$

(b) The characteristic equation is

$$(3p + 40)i = 0 \qquad \text{Root is } p = -40/3$$

Thus, the natural response is

$$i_n(t) = ke^{-40t/3} \qquad\qquad [6.2.4]$$

(c) The force response is same as the steady state response.

    (d) The complete response is

$$i(t) = ke^{-40t/3} + 2\cos(10t + 73.13^0) \qquad\qquad [6.2.5]$$

Now $i(0^+)$ are determined as follows:

    Setting $t = 0^+$ in eqn [6.2.1]

$$3i(0^+) + 40\int_0^{0^+} i\,dt = v_s(0) = 10\cos(20^0)$$

or $$3i(0^+) + 0 = 9.397$$

Hence

$$i(0^+) = 3.132\text{ A}$$

Using eqn [6.2.5], we have

$$i(0^+) = 3.132 = k + 2\cos 73.13^\circ \quad \text{or} \quad k = 3.132 - 0.580 = 2.55$$

Therefore

$$i(t) = 2.55e^{-40t/3} + 2\cos(10t + 73.13^\circ)\text{ A} \qquad\qquad Ans$$

SP 6.3 In the network shown in Fig. SP 6.3, a steady state is reached with the switch **K** open. At $t = 0$, the switch is closed. Determine current $i(t)$ for $t > 0$.

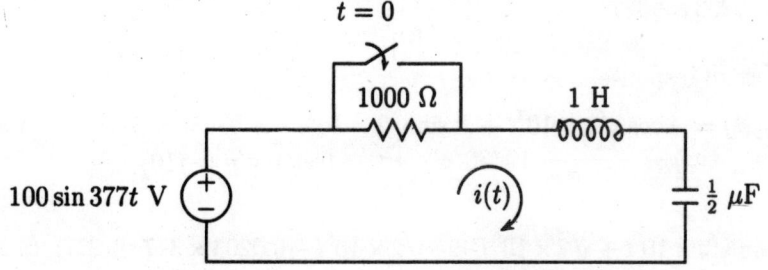

Fig. SP 6.3.

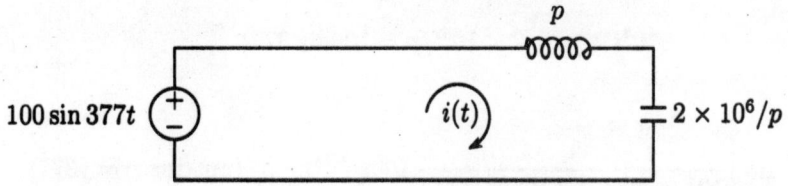

Fig. 6.3.1: p-domain circuit for $t > 0$.

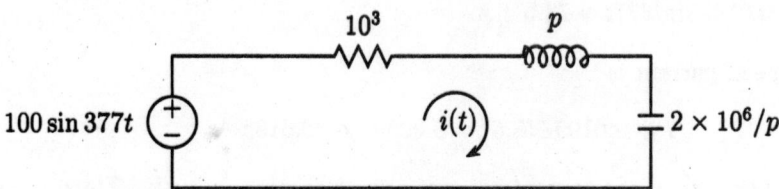

Fig. 6.3.2: p-domain circuit for $t < 0$.

SOLUTION:

By KVL in Fig. 6.3.1

$$pi + \frac{2 \times 10^6}{p}i = 100 \sin 377t$$

The characteristic equation is

$$pi + \frac{2 \times 10^6}{p}i = 0$$

Roots are :

$$p_1, p_2 = \pm j\sqrt{2} \times 10^3$$

Thus the natural response is

$$i_n(t) = A \cos \sqrt{2} \times 10^3 t + B \sin \sqrt{2} \times 10^3 t \qquad [6.3.2]$$

The steady state response is

$$i_{ss} = \text{Im}\left[\frac{1}{j377 + 2 \times 10^6/j377}100e^{j377t}\right] \qquad (\text{setting } p = j377)$$

$$= \operatorname{Im}\left[0.0203\angle+90^0 e^{j377t}\right]$$

$$= 0.0203 \sin(377t + 90^0)$$

$$= 0.0203 \cos 377t$$

The complete response is

$$i(t) = i_n(t) + i_{ss}(t) = A \cos\sqrt{2} \times 10^3 t + B \sin\sqrt{2} \times 10^3 t + 0.0203 \cos 377t \qquad [6.3.3]$$

and

$$\frac{di}{dt} = -\sqrt{2} \times 10^3 A \sin\sqrt{2} \times 10^3 t + \sqrt{2} \times 10^3 B \cos\sqrt{2} \times 10^3 t - 0.0203 \times 377 \sin 377t \quad [6.3.4]$$

The initial conditions are determined as From Fig. 6.3.2, we have

$$10^3 i + pi + 2 \times 10^6 i/p = 100 \sin 377t$$

Hence

$$i_{ss} = \operatorname{Im}\left[\frac{1}{10^3 + j377 + 2 \times 10^6/j377} \cdot 100 e^{j377t}\right] \qquad \text{(putting } p = j377\text{)}$$

$$= \operatorname{Im}[0.0199\angle 78.5^0 e^{j377t}]$$

$$= 0.0199 \sin(377t + 78.5^0) \text{ A}$$

The Phasor value of current is

$$I = 0.0199\angle 78.5^0 = 0.00397 + j0.0195 \text{ A}$$

The reactive component of the current can not change instantaneously. Thus

$$i(0^-) = i(0^+) = 0.0195 \text{ A}$$

The steady state voltage across the capacitor is

$$v_C(t) = \frac{2 \times 10^6}{p} i_{ss}(t) = \frac{2 \times 10^6}{p}\{0.0199 \sin(377t + 78.5^0)\}$$

$$= \frac{-2 \times 10^6 \times 0.0199}{377} \cos(377t + 78.5^0)$$

$$= -105 \cos(377t + 78.5^0) \text{ V}$$

The Phasor value of v_C is

$$V_C = 105.57\angle(78.5^0 - 90^0) = 105.57\angle(-11.5) = 103.5 - j21 \text{ V}$$

Again the reactive component of capacitor voltage can not change abruptly. Hence,

$$v_C(0^+) = v_C(0^-) = -21 \text{ V}$$

By KVL in Fig. 6.3.3

$$v_L(0^+) - 21 = 0 \qquad \Rightarrow v_L(0^+) = 21$$

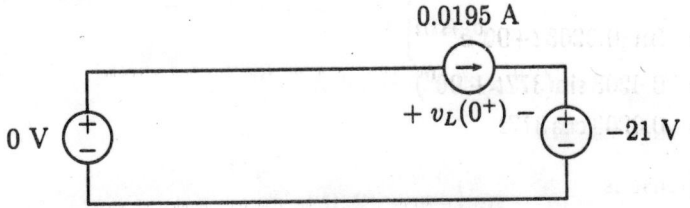

Fig. 6.3.3: t=0$^+$ circuit.

Hence
$$\frac{di}{dt}(0^+) = \frac{v_L(0^+)}{L} = \frac{21}{1} = 21 \text{ A/sec}$$

Applying initial condition $i(0^+)$ to eqn [6.3.3] as

$$i(0^+) = A + 0.0203 = 0.0195$$

Hence
$$A = -0.8 \times 10^{-3}$$

Next

$$\frac{di}{dt}(0^+) = \sqrt{2} \times 10^3 B = 21$$

Hence
$$B = 14.85 \times 10^{-3}$$

Therefore

$$i(t) = -0.8 \cos \sqrt{2} \times 10^3 t + 14.85 \sin \sqrt{2} \times 10^3 t + 20.3 \cos(377t) \text{ mA}$$

SP 6.4 For the circuit of Fig. SP 6.4, given that $i_s(t) = 1 + 2 \sin 2t$. **Determine the steady state voltage** $v(t)$.

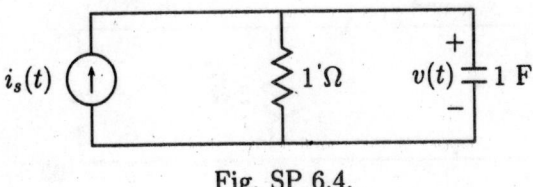

Fig. SP 6.4.

SOLUTION:

By KCL in Fig. SP 6.4

$$\frac{v}{1} + 1\frac{dv}{dt} = i_s = 1 + 2\sin 2t \qquad\qquad [6.4.1]$$

Writing the above equation in p-domain as

$$pv + v = 1 + 2\sin 2t \qquad [6.4.2]$$

The steady state response is

$$
\begin{aligned}
v_{ss}(t) &= \frac{1}{0+1} 1 e^0 + \text{Im}\left[\frac{1}{j2+1}(2\angle 0^0)e^{j2t}\right] \\
&\quad \text{(setting } p = 0 \text{ in the first term and } p = j2 \text{ in second term)} \\
&= 1 + \text{Im}[0.894\angle - 63.43^0 e^{j2t}] \\
&= 1 + 0.894\sin(2t - 63.43^0)
\end{aligned}
$$

SP 6.5 For the circuit shown in Fig. SP 6.5, determine
   (a) the natural response $i_n(t)$,
   (b) the particular response $i_P(t)$,
   (c) the complete response $i(t)$ for $t \geq 0$.

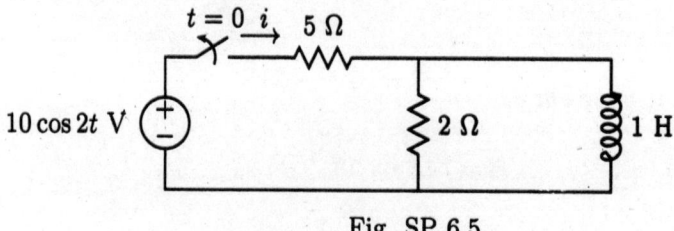

Fig. SP 6.5.

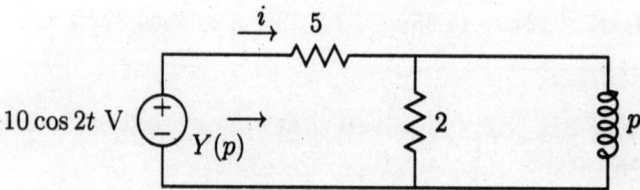

Fig. 6.5.1: p-domain circuit for $t > 0$.

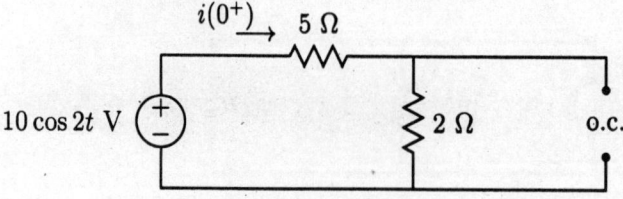

Fig. 6.5.2: $t = 0^+$ circuit.

SOLUTION:

With reference to Fig. 6.5.1, the admittance $Y(p)$ seen by $v_s(t)$ is

$$Y(p) = \frac{1}{Z(p)}$$

where

$$Z(p) = 5 + (2||p) = \frac{10 + 7p}{2 + p}$$

Hence

$$Y(p) = \frac{2 + p}{10 + 7p}$$

Thus

$$i(t) = Y(p)v_s(t) = \frac{2 + p}{10 + 7p} v_s(t) \qquad [6.5.1]$$

(a) The characteristic equation is

$$(7p + 10)i(t) = 0 \qquad \text{and root is} \quad p = \text{-}10/7$$

Thus, the natural solution is

$$i_n(t) = ke^{-10t/7} \qquad [6.5.2]$$

(b) Given that $v_s(t) = 10 \cos 2t = \text{Re}[10e^{j2t}]$

Thus the particular response corresponding to eqn [6.5.1] is

$$\begin{aligned}
i_P(t) &= \text{Re}\left[\frac{2 + j2}{10 + 7(j2)} 10e^{j2t}\right] \qquad \text{setting } t = j2 \\
&= \text{Re}\left[1.64\angle\text{-}9.46^0 e^{j2t}\right]
\end{aligned}$$

Thus

$$i_P(t) = 1.64 \cos(2t - 9.46^0) \text{ A} \qquad [6.5.3]$$

(c) The complete response is

$$i(t) = i_n(t) + i_P(t) = ke^{-10t/7} + 1.64 \cos(2t - 9.46^0) \qquad [6.5.4]$$

It is observe from Fig. 6.5.2 that

$$i(0^+) = \frac{10}{5 + 2} = 10/7 \text{ A}$$

Thus, from eqn [6.5.4]

$$i(0^+) = 10/7 = k + 1.64 \cos(-9.46^0) \qquad \Rightarrow \quad k = -0.189$$

and

$$i(t) = -0.189e^{-10t/7} + 1.64 \cos(2t - 9.46^0) \text{ A}$$

SP 6.6 Derive a differential equation relating $i(t)$ to the voltage source $v_s(t)$ in the circuit shown in Fig. SP 6.6.

SOLUTION:

This problem may be solved using a Thevenin's equivalent at the terminals a–b. The Thevenin impedance Z_{TH} is computed from Fig. 6.6.1 as

$$Z_{TH} = (2/p)||(1 + 2p) = \frac{4p + 2}{2p^2 + p + 2}$$

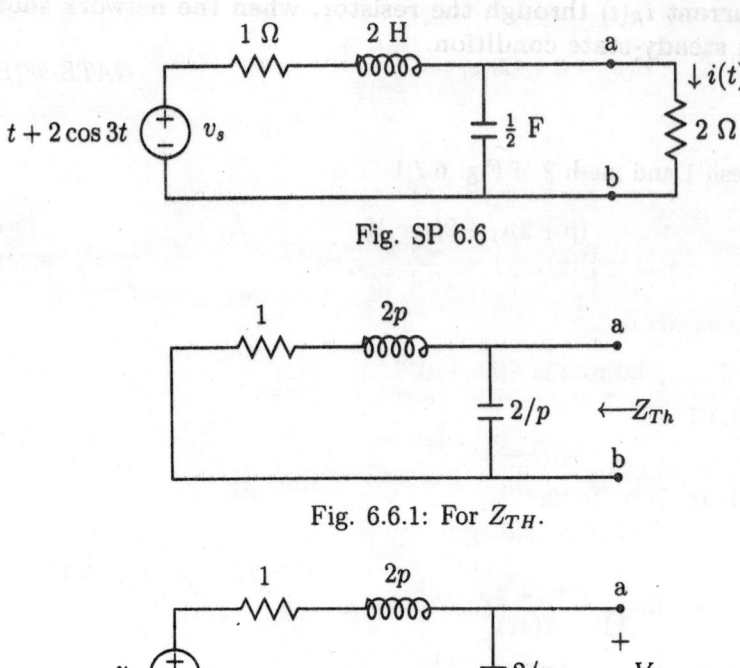

Fig. SP 6.6

Fig. 6.6.1: For Z_{TH}.

Fig. 6.6.2: For V_{Th}

The Thevenin open circuit voltage is determined from Fig. 6.6.2 as

$$V_{Th} = \frac{2/p}{1 + 2p + 2/p} v_s = \frac{2}{2p^2 + p + 2} v_s$$

From Fig. 6.6.3, we obtain

$$i(t) = \frac{V_{Th}}{Z_{TH} + 2} = \frac{[2/(2p^2 + p + 2)]v_s}{(4p + 2)/(2p^2 + p + 2) + 2} = \frac{1}{2p^2 + 3p + 3} v_s$$

or, $(2p^2 + 3p + 3)i(t) = v_s$ and the differential equation becomes

$$2\frac{d^2i}{dt^2} + 3\frac{di}{dt} + 3i = t + 2\cos 3t$$

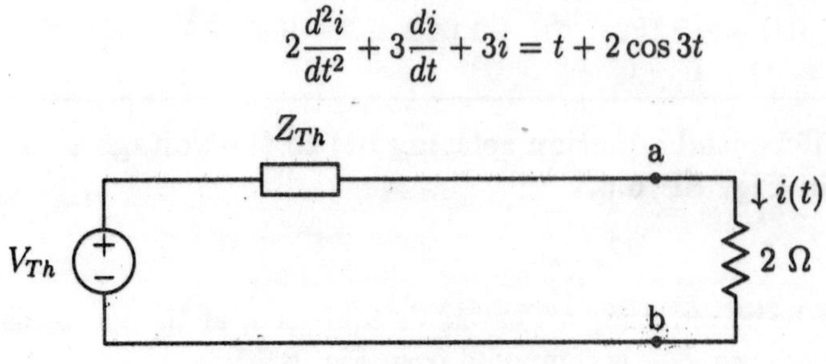

Fig. 6.6.3: For $i(t)$.

SP 6.7 Find the current $i_R(t)$ through the resistor, when the network shown in Fig. SP 6.7 is in steady-state condition.

GATE-98(EE)

SOLUTION:
By KVL in the mesh 1 and mesh 2 of Fig. 6.7.1

$$(p+2)i_1 + 2i_2 = 10 \qquad [6.7.1]$$

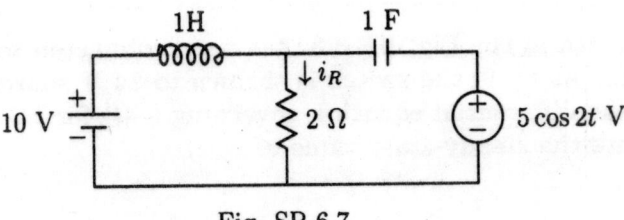

Fig. SP 6.7.

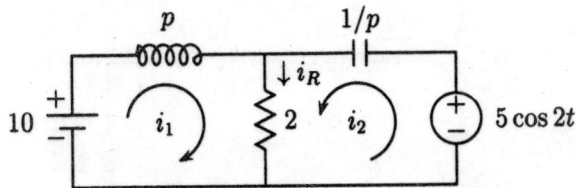

Fig. 6.7.1: p-domain circuit for $t > 0$.

and

$$2i_1 + (1/p+2)i_2 = 5\cos 2t \qquad [6.7.2]$$

Applying Cramer's rule to determine i_1 and i_2 as:

$$i_1 = \frac{\begin{vmatrix} 10 & 25\cos 2t & 2+1/p \\ p+2 & 22 & 1/p+2 \end{vmatrix}}{} = \frac{20 + 10/p - 10\cos 2t}{1 + 2p + 2/p}$$

and

$$i_2 = \frac{\begin{vmatrix} p+2 & 102 & 5\cos 2t \\ p+2 & 22 & 1/p+2 \end{vmatrix}}{} = \frac{5p\cos 2t + 10\cos 2t - 20}{1 + 2p + 2/p}$$

$$i_R(t) = i_1 + i_2 = \frac{10/p + 5p\cos 2t}{1 + 2p + 2/p} = \frac{10 + 5p^2\cos 2t}{2p^2 + p + 2}$$

Hence

$$i_R(t) = \frac{10}{2p^2 + p + 2}e^{0t} + \frac{5p^2}{2p^2 + p + 2}Re[e^{j2t}]$$

Therefore the steady-state solution for $i_R(t)$ is

$$i_{R,ss} = \frac{10}{2(0)^2 + 0 + 2}e^0 + Re\left[\frac{5(j2)^2}{2(j2)^2 + j2 + 2}e^{j2t}\right]$$

(setting $p = 0$ in the first term and $p = j2$ in the second term.)

$$= 5 + \mathrm{Re}\left[\frac{-20}{-6 + j2}e^{j2t}\right]$$

$$= 5 + \mathrm{Re}\left[\frac{20\angle 180^0}{6.325\angle 161.57^0}e^{j2t}\right]$$

$$= 5 + \mathrm{Re}\left[3.16\angle 18.43^0 e^{j2t}\right]$$

$$= 5 + 3.16\cos(2t + 18.43^0)\ \mathrm{A}$$

SP 6.8 The switch in the Fig. SP 6.8, has been connected to the 12 V source for a long time. At $t = 0$, the switch is thrown to 24 V source.
  (a) Write the differential equation governing $v_C(t)$ for $t > 0$.
  (b) Compute the steady-state value of $v_C(t)$.

GATE-98(EE)

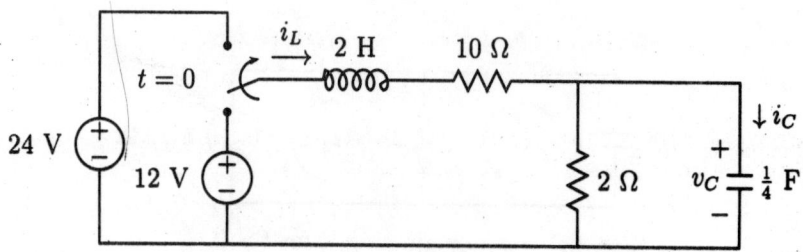

Fig. SP 6.8.

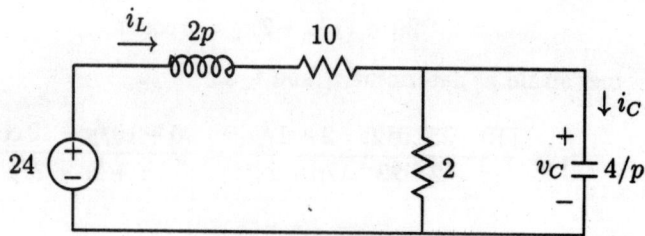

Fig. 6.8.1: p-domain circuit for $t > 0$.

SOLUTION:
  (a) With reference to Fig. 6.8.1, the voltage division formula gives

$$v_C(t) = \frac{2\|(4/p)}{(2p + 10) + [2\|(4/p)]}(24) = \frac{2}{p^2 + 7p + 12}(24)$$

or

$$(p^2 + 7p + 12)v_C(t) = 48$$

And differential equation becomes

$$\frac{d^2 v_C}{dt^2} + 7\frac{dv_C}{dt} + 12v_C = 48$$

(b) The steady-state solution for $v_C(t)$ is

$$v_{C,ss} = \frac{48}{(0)^2 + 7(0) + 12}e^{0t} = 4 \text{ V}$$

SP 6.9 Determine the steady-state current $i(t)$ in the circuit of Fig. SP 6.9.

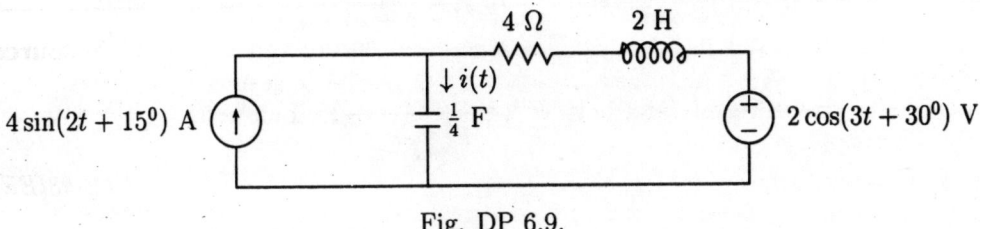

Fig. DP 6.9.

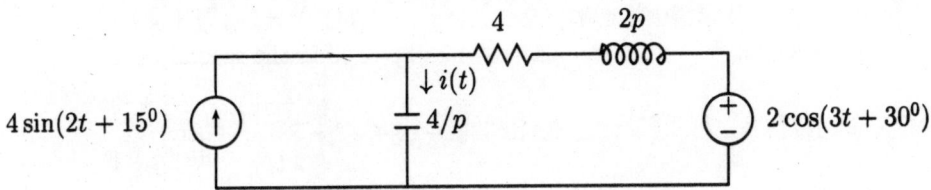

Fig. 6.9.1: p-domain circuit for $t > 0$.

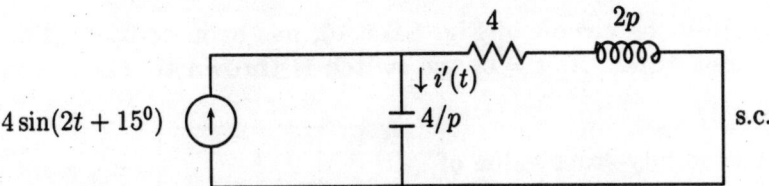

Fig. 6.9.2: Circuit excited by current source alone.

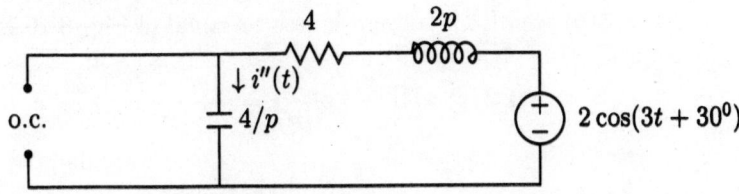

Fig. 6.9.3: Circuit excited by voltage source alone.

SOLUTION:

With reference to Fig. 6.9.2, current division formula gives

$$i'(t) = \frac{4 + 2p}{4 + 2p + 4/p}\left[4\sin(2t + 15^0)\right]$$

and with reference to Fig. 6.9.3, KVL gives

$$i''(t) = \frac{2}{4 + 2p + 4/p} \left[\cos(3t + 30^0) \right]$$

By superposition

$$i(t) = i'(t) + i''(t)$$

$$= \frac{4 + 2p}{4 + 2p + 4/p} \left[4 \sin(2t + 15^0) \right] + \frac{2}{4 + 2p + 4/p} \left[\cos(3t + 30^0) \right]$$

$$= \frac{8p + 4p^2}{p^2 + 2p + 2} \text{Im}[1\angle 15^0 e^{j2t}] + \frac{p}{p^2 + 2p + 2} \text{Re}[1\angle 30^0 e^{j3t}]$$

The steady-state solution for $i(t)$ is

$$i_{ss}(t) = \text{Im} \left[\frac{8(j2) + 4(j2)^2}{(j2)^2 + 2(j2) + 2} \left(1\angle 15^0 e^{j2t} \right) \right] + \text{Re} \left[\frac{j3}{(j3)^2 + 2(j3) + 2} \left(1\angle 30^0 e^{j3t} \right) \right]$$

(setting $p = j2$ in the first term and $p = j3$ in the second term)

$$= \text{Im} \left[5.06\angle 33.43^0 e^{j2t} \right] + \text{Re} \left[0.325\angle -19.4^0 e^{j3t} \right]$$

$$= 5.06 \sin(2t + 33.43^0) + 0.325 \cos(3t - 19.4^0)$$

SP 6.10 The switch in the circuit of Fig. SP 6.10, has been connected to the 5 A source for a long time. At $t = 0$, the switch is thrown to $2 \sin 2t$ voltage source.

(a) compute the steady-state value of $i_C(t)$
(b) Find the complete solution of $i_C(t)$.

SOLUTION:
Impedance $Z(p)$ seen from voltage source terminal of Fig. 6.10.1 is

$$Z(p) = 10 + \left(p \| \frac{2}{p} \right) = \frac{10p^2 + 2p + 20}{p^2 + 2}$$

and
$$i(t) = \frac{2 \sin 2t}{Z(p)} = \frac{(p^2 + 2)}{5p^2 + p + 10} \sin 2t$$

Therefore the current division formula gives

$$i_C(t) = \frac{p}{p + 2/p} i(t) = \frac{p^2}{5p^2 + p + 10} \sin 2t \qquad [6.10.1]$$

And

$$(5p^2 + p + 10)i_C(t) = p^2 \sin 2t$$

The differential equation is

$$5\frac{d^2 i_C}{dt^2} + \frac{di_C}{dt} + 10i_C = \frac{d^2}{dt^2}(\sin 2t) = -4\sin 2t \qquad [6.10.2]$$

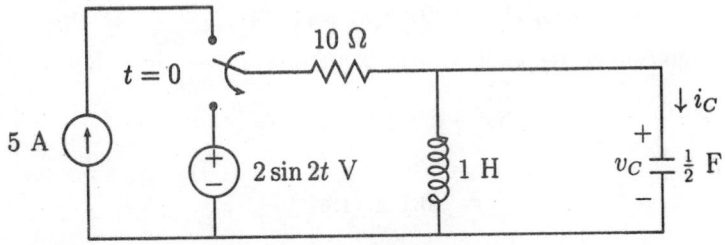

Fig. SP 6.10.

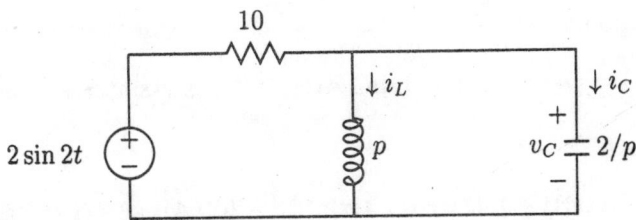

Fig. 6.10.1: *p*-domain circuit for $t > 0$.

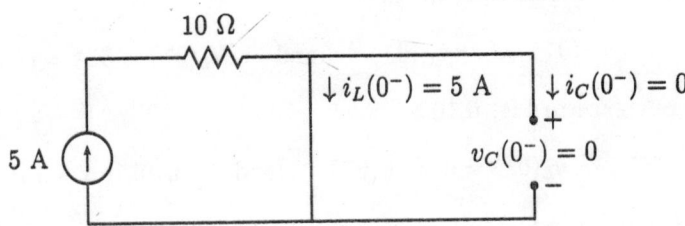

Fig. 6.10.2: $t = 0^-$ circuit.

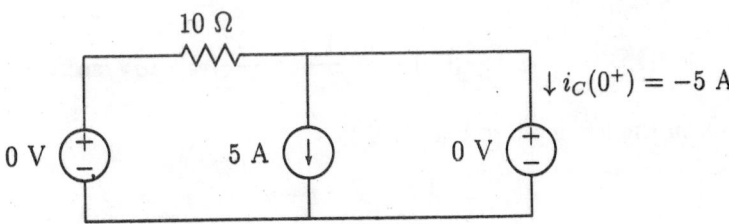

Fig. 6.10.3: $t = 0^+$ circuit.

The steady-state value of $i_C(t)$ using eqn [6.10.1] is

$$i_{C,ss}(t) = \text{Im}\left[\frac{(j2)^2}{5(j2)^2 + j2 + 10}e^{j2t}\right]$$

$$= \text{Im}\left[0.39\angle 11.3^0 e^{j2t}\right]$$

Hence

$$i_{C,ss}(t) = 0.39\sin(2t + 11.3^0) \qquad\qquad [6.10.3]$$

The characteristic equation corresponds to eqn [6.10.2] is

$$5s^2 + s + 10 = 0$$

Roots are :

$$s_{1,2} = -0.1 \pm j1.41$$

Hence the transient solution is

$$(i_C)_{tr} = (A\cos 1.41t + B\sin 1.41t)e^{-0.1t} \qquad\qquad [6.10.4]$$

The complete solution is

$$i_C(t) = (A\cos 1.41t + B\sin 1.41t)e^{-0.1t} + 0.39\sin(2t + 11.3^0) \qquad [6.10.5]$$

It's derivative is

$$\frac{di_C}{dt} = (-1.41A\sin 1.41t + 1.41B\cos 1.41t)e^{-0.1t} - 0.1(A\cos 1.41t + B\sin 1.41t)e^{-t/10}$$

$$+0.78\cos(2t + 11.3^0) \qquad\qquad [6.10.6]$$

With reference to Fig. 6.10.2, we have

$$v_C(0^-) = 0 = v_C(0^+) \qquad \text{and} \qquad i_L(0^-) = 5\ A = i_L(0^+)$$

Now with reference to Fig. 6.10.3

$$v_L(0^+) = 0 = v_C(0^+) \qquad \text{and} \qquad i_C(0^+) = -5\ A$$

Hence for the inductor:

$$\frac{di_L}{dt}(0^+) = \frac{v_L(0^+)}{L} = 0$$

And for the capacitor:

$$\frac{dv_C}{dt}(0^+) = \frac{i_C(0^+)}{C} = \frac{-5}{1/2} = -10\text{V/sec.}$$

By KVL in the left mesh of Fig. 6.10.1

$$10(i_C + i_L) + v_C = 2\sin 2t$$

or $\qquad\qquad\qquad\qquad 10\dfrac{di_C}{dt} + 10\dfrac{di_L}{dt} + \dfrac{dv_C}{dt} = 4\cos 2t$

or
$$10\frac{di_C}{dt}(0^+) + 10\frac{di_L}{dt}(0^+) + \frac{dv_C}{dt}(0^+) = 4$$

$$10\frac{di_C}{dt}(0^+) + 10 \times 0 - 10 = 4 \qquad \Rightarrow \qquad \frac{di_C}{dt}(0^+) = 1.4$$

From eqn [6.10.5]

$$i_C(0^+) = -5 = A + 0.39\sin 11.3^0 \qquad \Rightarrow \quad A = -5.08$$

From eqn [6.10.6]

$$\frac{di_C}{dt}(0^+) = 1.4 = 1.41B - 0.1A + 0.76 \qquad \Rightarrow \quad B = 0.094$$

and finally

$$i_C(t) = (-5.08\cos 1.41t + 0.094\sin 1.41t)e^{-0.1t} + 0.39\sin(2t + 11.3^0)$$

SP 6.11 For the circuit shown in Fig. SP 6.11, find the steady state solution for $i(t)$

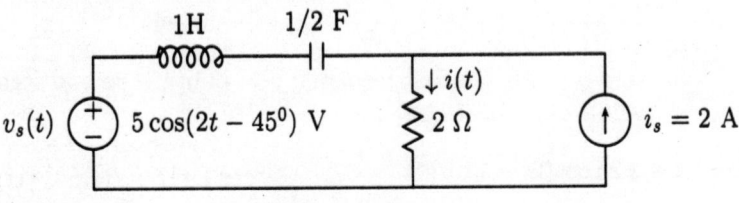

Fig. SP 6.11.

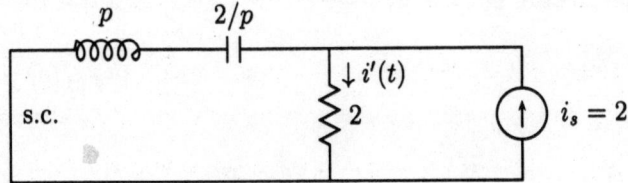

Fig. 6.11.1: p-domain circuit with $i_s(t)$ alone.

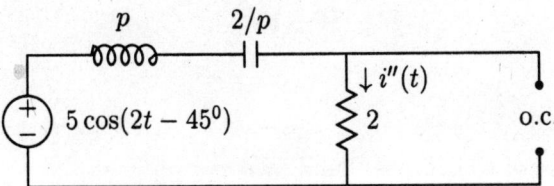

Fig. 6.11.2: p-domain circuit with $v_s(t)$ alone.

SOLUTION:

The $i(t)$ can be found by superposition as:

$$i(t) = i'(t) + i''(t)$$

where $i'(t)$ = current in 2-Ω resistor due to dc current source alone,
  $i''(t)$ = current in 2-Ω resistor due to sinusoidal voltage alone.

With reference to Fig. 6-7.1, current division formula gives

$$i'(t) = \frac{p + 2/p}{2 + p + 2/p} i_s(t) = \frac{p^2 + 2}{p^2 + 2p + 2} 2e^0$$

From Fig. 6-7.3, KVL gives

$$i''(t) = \frac{v_s(t)}{p + 2/p + 2} = \frac{p}{p^2 + 2p + 2} 5\cos(2t - 45^0)$$

By superposition

$$
\begin{aligned}
i(t) &= i'(t) + i''(t) \\
&= \frac{p^2 + 2}{p^2 + 2p + 2} 2e^0 + \frac{p}{p^2 + 2p + 2} 5\cos(2t - 45^0)
\end{aligned}
$$

Therefore, steady-state solution is determined as:

$$
\begin{aligned}
i_{ss}(t) &= \frac{0^2 + 2}{0^2 + 2(0) + 2} 2e^0 + \text{Re}\left[\frac{j2}{(j2)^2 + 2(j2) + 2} 5\angle{-45^0} e^{j2t} \right] \\
&\qquad \text{(setting } p = 0 \text{ in the first term and } p = j2 \text{ in the second term)} \\
&= 2 + \text{Re}\left[2.24\angle{-71.57^0} e^{j2t} \right] \\
&= 2 + 2.24\cos(2t - 71.57^0)
\end{aligned}
$$

SP 6.12 For the circuit shown in Fig. SP 6.12, determine $i(t)$ for $t > 0$.

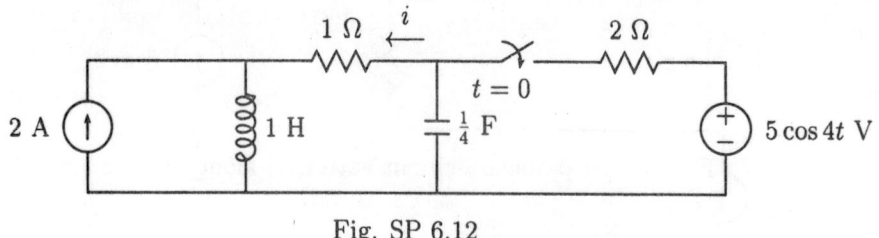

Fig. SP 6.12

SOLUTION:

The steady state value of $i(t)$ can be found by super position:

$$i(t) = i'(t) + i''(t)$$

where $i'(t)$ is current specified in 1 Ω resistor due to 2 A dc current source alone and $i''(t)$ is current specified in 1 Ω resistor due to $5\cos 4t$ sinusoidal voltage source alone.

From Fig. 6.12.2, $i'(t)$ may be found as :

$$i_1(t) = \frac{5\cos 4t}{(1+p)||(4/p)+2} = \frac{5(p^2+p+4)}{2p^2+6p+12}\cos 4t$$

Hence, by current division formula

$$i'(t) = \frac{4/p}{1+p+4/p}i_1(t)$$

$$= \frac{4}{p^2+p+4}\left(\frac{5(p^2+p+4)}{2p^2+6p+12}\cos 4t\right)$$

$$= \frac{10}{p^2+3p+6}\cos 4t$$

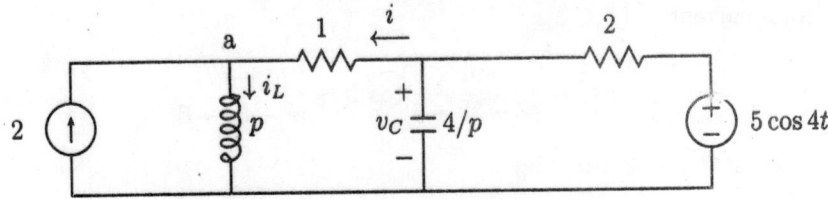

Fig. 6.12.1: p-domain circuit for $t > 0$

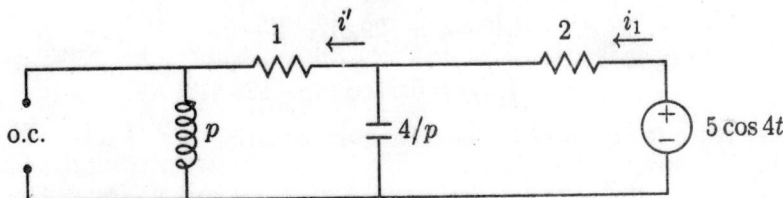

Fig. 6.12.2: Circuit excited by voltage source alone.

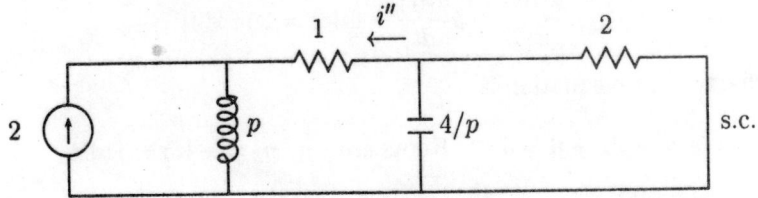

Fig. 6.12.3: Circuit excited by current source alone.

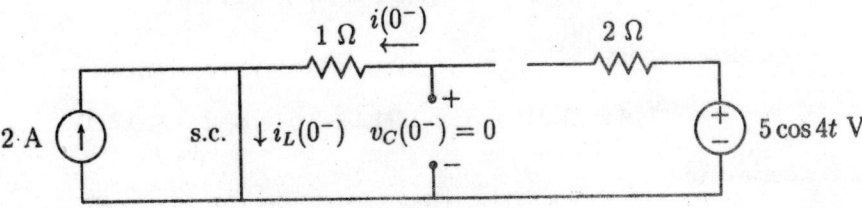

Fig. 6.12.4: t = 0^- circuit.

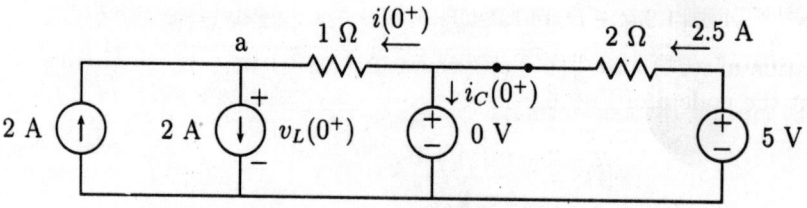

Fig. 6.12.5: $t = 0^+$ circuit.

From Fig. 6.12.3, again by current division formula

$$i''(t) = -\frac{p}{p + [1 + (4/p)\|2]}(2) = -\frac{p(p+2)}{p^2 + 3p + 6}2e^0$$

Thus, total current:

$$i(t) = \frac{10}{p^2 + 3p + 6}\cos 4t - \frac{p(p+2)}{p^2 + 3p + 6}2e^0 \qquad [6.12.1]$$

Therefore, the steady-state value is

$$
\begin{aligned}
i_{ss}(t) &= \text{Re}\left[\frac{10}{(j4)^2 + 3(j4) + 6}e^{j4t}\right] - \frac{0(0+1)}{0^2 + 3(0) + 6}2e^0 \\
&= \text{Re}[0.64\angle - 129.81^0 e^{j4t}] - 0
\end{aligned}
$$

$$i_{ss}(t) = 0.64\cos(4t - 129.81^0) \text{ A} \qquad [6.12.2]$$

With reference to eqn [6.12.1], the differential equation is

$$(p^2 + 3p + 6)i(t) = 10\cos 4t - p(p+2)2e^0 = 10\cos 4t$$

or

$$\frac{d^2 i(t)}{dt^2} + 3\frac{di(t)}{dt} + 6i(t) = 10\cos 4t$$

Thus, the characteristic equation is

$$p^2 + 3p + 6 = 0 \qquad \text{Roots are} \quad p_1, p_2 = -1.5 \pm j1.94$$

Thus, the natural response is

$$i_n(t) = e^{-1.5t}(A\cos 1.94t + B\sin 1.94t) \qquad [6.12.3]$$

and the complete response is

$$i(t) = e^{-1.5t}(A\cos 1.94t + B\sin 1.94t) + 0.64\cos(4t - 129.81^0) \qquad [6.12.4]$$

And it's derivative is

$$\frac{di}{dt} = e^{-1.5t}(-1.94A\sin 1.94t + 1.94B\cos 1.94t)$$

$$+e^{-1.5t}(A\cos 1.94t + B\sin 1.94t)(-1.5) - 4 \times 0.64\sin(4t - 129.81^0) \qquad [6.12.5]$$

Determination of $i(0^+)$ and $\frac{di}{dt}(0^+)$ as follows:
KCL at the node **a** of Fig. 6.12.5 gives

$$2 - 2 + i(0^+) = 0 \qquad \Rightarrow i(0^+) = 0$$

By KVL in the middle mesh of Fig. 6.12.5

$$-v_L(0^+) - 1i(0^+) + 0 \qquad \Rightarrow v_L(0^+) = 0$$

Hence for inductor:

$$\frac{di_L}{dt}(0^+) = \frac{v_L(0^+)}{L} = 0$$

Taking KCL at the node **a** of Fig. 6.12.1

$$i(t) = -2 + i_L(t) \qquad \text{for } t \geq 0$$

$$\frac{di}{dt} = \frac{di_L}{dt} + \frac{d}{dt}(-2) \qquad \text{for } t \geq 0$$

Hence
$$\frac{di}{dt}(0^+) = \frac{di_L}{dt}(0^+) + 0 = 0$$

From eqn [6.12.4]

$$i(0^+) = 0 = A + 0.64\cos(-129.81^0) \qquad \Rightarrow A = 0.41$$

From eqn [6.12.5]

$$\frac{di}{dt}(0^+) = 0 = 1.94B - 1.5A - 4 \times 0.64\sin(-129.81^0) \qquad \Rightarrow B = -0.7$$

Thus
$$i(t) = e^{-1.5t}(0.41\cos 1.94t - 0.7\sin 1.94t) + 0.64\cos(4t - 129.81^0) \text{ A}$$

SP 6.13 A voltage source $100\sin(5t + \alpha)$ V is applied to the RC circuit of Fig. SP 6.13 at $t = 0$. The capacitor is initially charged to a voltage $v_C(0) = -2$ **V**. Determine
 (a) the voltage $v_C(t)$,
 (b) the phase angle α for which there will be no transient upon closing the switch.

SOLUTION:
With reference to Fig. 6.13.1, the voltage division formula gives

$$v_C(t) = \frac{2/p}{10 + 2/p}.100\sin(5t + \alpha) = \frac{2}{10p + 2}[100\sin(5t + \alpha)] \qquad [6.13.1]$$

and

$$(10p + 2)v_C = 200\sin(5t + \alpha)$$

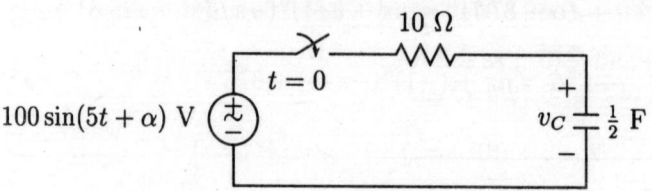

Fig. SP 6.13.

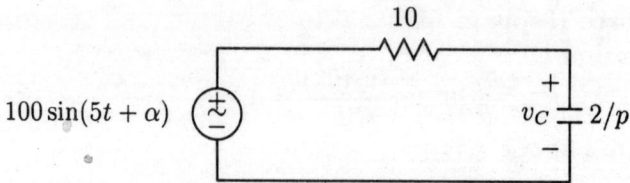

Fig. 6.13.1: p-domain circuit for $t > 0$.

The differential equation is

$$10\frac{dv_C}{dt} + 2v_C = 200\sin(5t + \alpha) \qquad\qquad [6.13.2]$$

The natural solution to eqn [6.13.2] is

$$(v_C)_n = Ke^{-0.2t} \qquad\qquad [6.13.3]$$

The steady-state solution to eqn [6.13.1] is

$$
\begin{aligned}
v_{C,ss} &= \operatorname{Im}\left[\frac{2}{10(j5) + 2}.100\angle\alpha e^{j5t}\right] \\
&= \operatorname{Im}\left[4\angle(\alpha - 87.71^0)e^{j5t}\right]
\end{aligned}
$$

Thus

$$v_{C,ss} = 4\sin(5t + \alpha - 87.71^0) \qquad\qquad [6.13.4]$$

Complete solution

$$v_C(t) = Ke^{-0.2t} + 4\sin(5t + \alpha - 87.71^0) \qquad\qquad [6.13.5]$$

Applying initial condition

$$v_C(0) = -2 = K + 4\sin(\alpha - 87.71^0) \qquad \Rightarrow K = -2 - 4\sin(\alpha - 87.71)$$

And finally

$$v_C(t) = -\left[2 + 4\sin(\alpha - 87.71^0)\right]e^{-0.2t} + 4\sin(5t + \alpha - 87.71^0)$$

For no transient

$$2 + 4\sin(\alpha - 87.71^0) = 0$$

or

$$\sin(\alpha - 87.71^0) = -1/2$$

or
$$\alpha - 87.71^0 = n\pi + (-1)^n(-\pi/6)$$

or
$$\alpha = n\pi + (-1)^n(-\pi/6) + 87.71^0$$

where $n = 0, 1, 2, \ldots\ldots$

6.4 DRILL PROBLEMS

DP 6.1 Find the force response for $t > 0$ to the input $5e^{-4t}u(t)$ from a system whose transfer function is

$$T(p) = \frac{2(p+1)}{p+2}.$$

Hints :

Output
$$y(t) = T(p)x(t)$$

$$y(t) = \frac{2(p+1)}{p+2}(5e^{-4t})$$

The force response is

$$y_{ss}(t) = \frac{2(-4+1)}{-4+2}5e^{-4t} = 15e^{-4t}$$

DP 6.2 Given the transfer function

$$T(p) = \frac{5p(p+1)}{p^2 + p + 1}$$

Find:
(a) **undamped natural frequency,**
(b) **damping ratio,**
(c) **natural response,**
(d) **steady-state response to input $u(t)$.**

Hints : Denominator polynomial of transfer function when equated to zero is called characteristic equation. Comparing the characteristic equation, $p^2 + p + 1 = 0$ with the general second order characteristic equation, $p^2 + 2\zeta\omega_n p + \omega_n^2 = 0$, we find that undamped natural frequency, $\omega_n = \sqrt{1} = 1$ rad/sec. and $2\zeta\omega_n = 1$ from which damping ratio, $\zeta = 1/2\omega_n = 1/2$.

Roots of the characteristic equation are:

$$p_{1,2} = -1/2 \pm j\sqrt{3}/2$$

Natural response

$$y_n(t) = [A\cos(\sqrt{3}/2)t + B\sin(\sqrt{3}/2)t]e^{-t/2}$$

The input $u(t)$ may be treated as $1e^{0t}$.

DP 6.3 Find $i_L(t)$ in the circuit of Fig. DP 6.3.

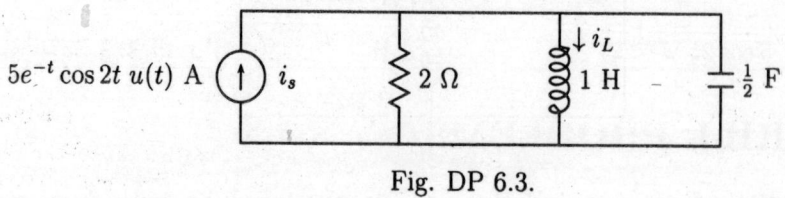

Fig. DP 6.3.

Hints : Use current division formula to find $i_L(t)$ as:

$$i_L(t) = \frac{1/p}{1/2 + 1/p + p/2} i_s$$

DP 6.4 Find the differential equation in $i_L(t)$ when switch is opened at $t = 0$ in the circuit of Fig. DP 6.4.

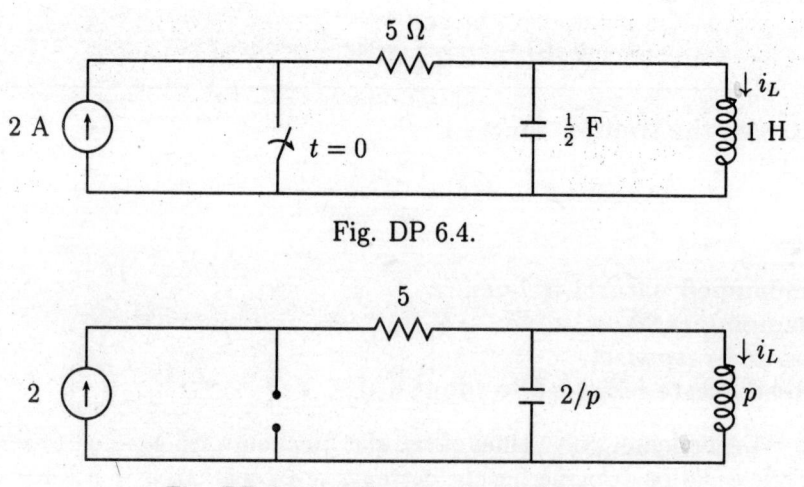

Fig. DP 6.4.

Fig. DP 6.4.1: p-domain circuit for $t > 0$.

Hints : Use current division formula to find $i_L(t)$ as:

$$i_L(t) = \frac{2/p}{2/p + p}(2e^{0t})$$

DP 6.5 Find the steady-state value of $i(t)$ for circuit shown in Fig. DP 6.5.

Hint : See SP 6.7.

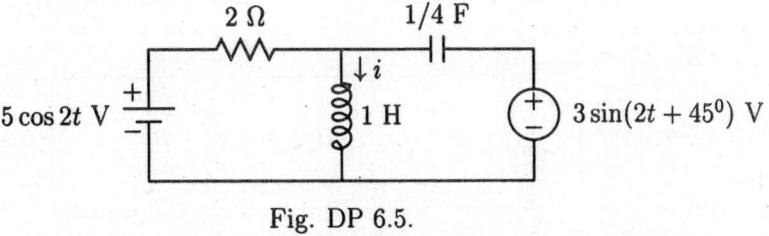

Fig. DP 6.5.

DP 6.6 Find the differential equation for the circuit of Fig. DP 6.6 in
  (a) $i(t)$ **for** $t > 0$,
  (b) $v_C(t)$ **for** $t > 0$,
  (c) $v_L(t)$ **for** $t > 0$,
  (d) $v_R(t)$ **for** $t > 0$.

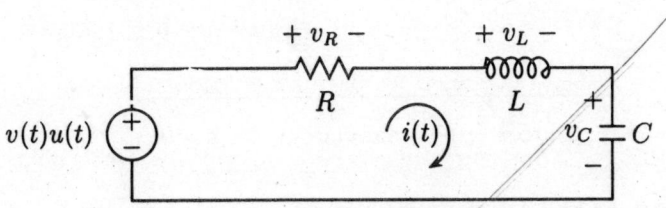

Fig. DP 6.6.

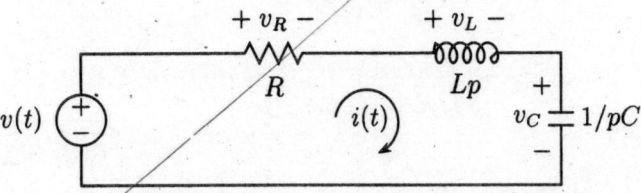

Fig. DP 6.6.1: *p*-domain circuit for $t > 0$.

Hints : From the circuit of Fig. DP 6.6.1,

$$i(t) = \frac{1}{R + Lp + 1/Cp} v(t)$$

$$v_C(t) = \frac{1}{Cp} i(t) = \frac{1}{Cp}\left(\frac{1}{R + Lp + 1/Cp} v(t)\right)$$

$$v_L(t) = Lpi(t) = Lp\left(\frac{1}{R + Lp + 1/Cp} v(t)\right)$$

and $$v_R(t) = Ri(t) = R\left(\frac{1}{R + Lp + 1/Cp} v(t)\right)$$

DP 6.7 Find the differential equation for the circuit of Fig. DP 6.7 in
   (a) $v(t)$ for $t > 0$,
   (b) $i_C(t)$ for $t > 0$,
   (c) $i_L(t)$ for $t > 0$,
   (d) $i_R(t)$ for $t > 0$.

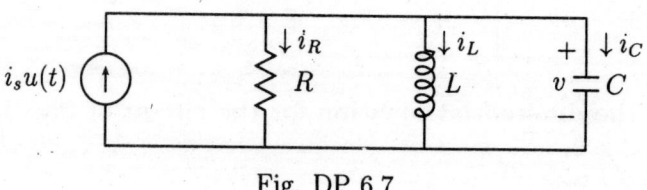

Fig. DP 6.7.

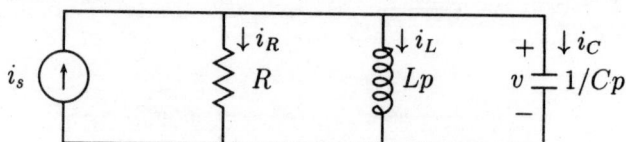

Fig. DP 6.7.1: p-domain circuit for $t > 0$.

Hints : With reference to Fig. DP 6.7.1, KCL gives

$$\frac{v(t)}{R} + \frac{v(t)}{Lp} + \frac{v(t)}{1/Cp} = i_s(t)$$

$$v(t) = \frac{1}{1/R + 1/Lp + Cp} i_s(t)$$

Hence $\qquad i_C(t) = \dfrac{v}{1/Cp} \qquad i_L(t) = \dfrac{v}{Lp} \qquad$ and $\qquad i_R(t) = \dfrac{v}{R}$

DP 6.8 Find the steady-state response for $v_{ab}(t)$ for $t > 0$ for the circuit of Fig. DP 6.8.

Hints : Impedance seen from the voltage source terminal of Fig. DP 6.8.1 is

$$Z(p) = 3 + (2 + p)\|(5 + 1/p)$$

Hence $\qquad i(t) = \dfrac{1}{3 + (2 + p)\|(5 + 1/p)}[10e^{-4t}\sin(2t + 30^0)]$

Now i_1 and i_2 may be determined by current division method as

$$i_1(t) = \frac{5 + 1/p}{(2 + p) + (5 + 1/p)}i(t)$$

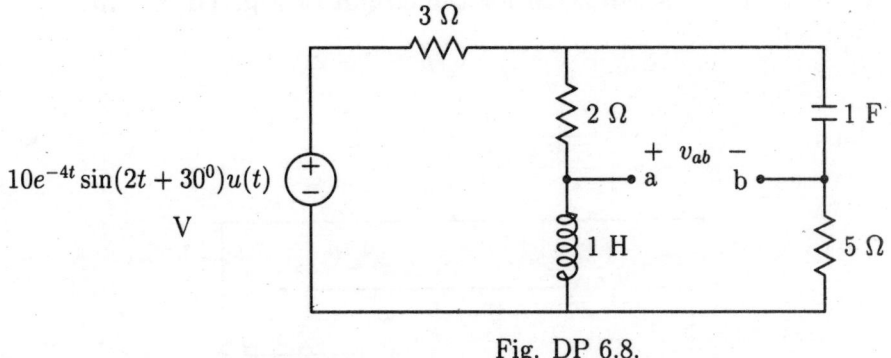

Fig. DP 6.8.

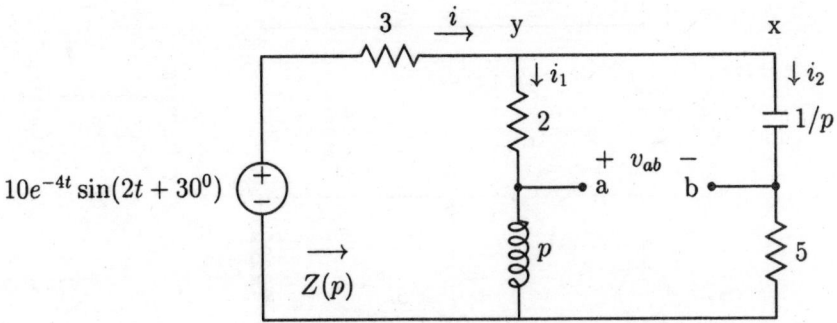

Fig. DP 6.8.1: p-domain circuit for $t > 0$.

and
$$i_2(t) = \frac{2+p}{(2+p) + (5 + 1/p)} i(t)$$

Using KVL in the loop x-y-b-a-x as

$$(1/p)i_2 - v_{ab}(t) - 2i_1 = 0$$

or

$$v_{ab}(t) = (1/p)i_2 - 2i_1$$

DP 6.9 Determine the steady-state value of $i_0(t)$ for the circuit of Fig. DP 6.9.

Hints : Take KCL at **a** of Fig. DP 6.9.1 as:

$$i = 3i + i_0 \qquad \Rightarrow i = -i_0/2$$

KVL in the Fig. DP 6.9.1 around the major loop containing $2\cos 2t$, $1\,\Omega$, $p/5$, and $1/p$ gives

$$i + (p/5)i_0 + i_0/p = 2\cos 2t$$

or
$$-i_0/2 + (p/5)i_0 + i_0/p = 2\cos 2t$$

or
$$i_0 = \frac{10p}{2p^2 - 5p + 10}2\cos 2t$$

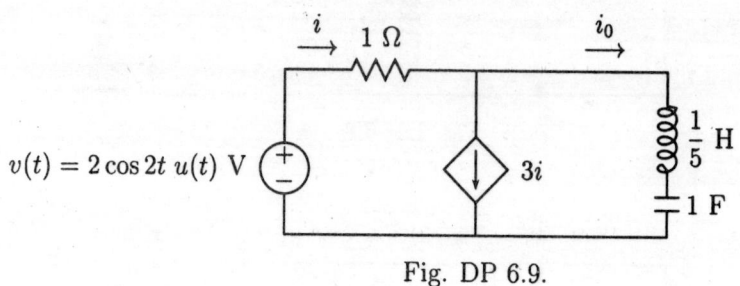

Fig. DP 6.9.

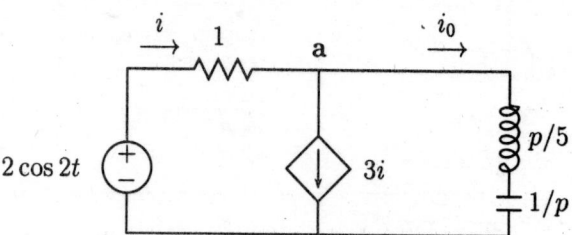

Fig. DP 6.9.1: p-domain circuit for $t > 0$.

DP 6.10 Find the steady-state value of $v_0(t)$ for the circuit of Fig. DP 6.10.

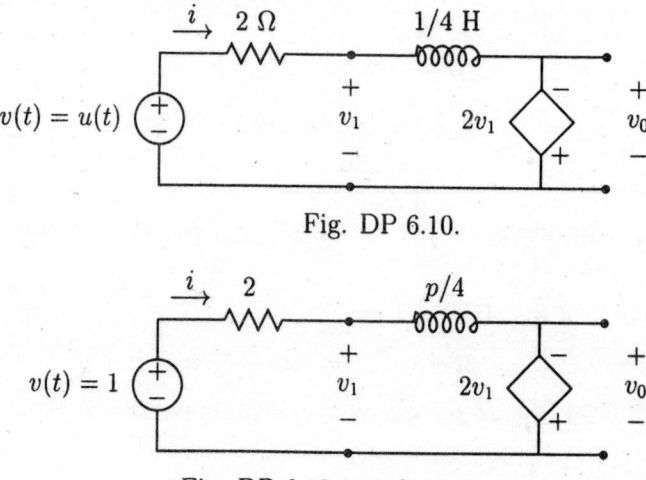

Fig. DP 6.10.

Fig. DP 6.10.1: p-domain circuit.

Hints : With reference to Fig. DP 6.10.1, following equations may be found as:

$$1 = (2 + p/4)i - 2v_1$$

$$i = \frac{1 - v_1}{2}$$

and $$v_0 = -2v_1 \qquad \Rightarrow \quad v_1 = -v_0/2$$

Eliminate i and v_1 from the first equation with the help of second and third equations to get v_0 as a function of p.

****

Steady-State Sinusoidal Analysis

SUMMARY OF THIS CHAPTER

For a linear time-invariant circuit whose input is a sinusoidal, the steady-state response will be sinusoidal with modified amplitude and phase angle without change in frequency.

7.1   PHASOR

The sinusoidal signal $v(t) = V_m \cos(\omega t + \phi)$ has an angular frequency ω, amplitude V_m, and phase angle ϕ. Similarly its complex sinusoidal signal

$$
\begin{aligned}
v(t) &= Re\left[V_m e^{j(\omega t + \phi)}\right] \\
&= Re\left[\left(V_m e^{j\phi}\right) e^{j\omega t}\right]
\end{aligned}
$$

has an angular frequency ω, amplitude V_m, and phase angle ϕ. We denote the complex number $V = V_m e^{j\phi}$ by $V = V_m \angle \phi$ which is known as phasor of sinusoid $v(t)$. The $V e^{j\omega t}$ is a rotating phasor rotating in the counterclockwise.

Note: usually cosine signal is transformed into phasor form or vice-versa.

Representation of Phasor in Three Ways

$$
\begin{array}{ll}
\text{Polar form} & V = V_m \angle \phi \\
\text{Rectangular form} & V = V_m \cos \phi + j V_m \sin \phi \\
\text{Exponential form} & V = V_m e^{j\phi}
\end{array}
$$

7.2   PHASOR MANIPULATIONS

Addition of Complex Numbers

We add complex numbers by adding the real and imaginary parts separately.

For example:

$$(a_1 + jb_1) + (a_2 + jb_2) = (a_1 + a_2) + j(b_1 + b_2)$$

Subtraction of Complex Numbers

Subtraction is carried out exactly the same way as addition after changing the sign of the real and imaginary parts of the subtrahend. For example:

$$(a_1 + jb_1) - (a_2 + jb_2) = (a_1 - a_2) + j(b_1 - b_2)$$

Note: addition and subtraction are accomplished in rectangular forms only.

Multiplication of Complex Numbers

Multiplication of complex numbers is defined as:

$$(V_1 \angle \theta_1)(V_2 \angle \theta_2) = V_1 V_2 \angle(\theta_1 + \theta_2)$$

Division of Complex Numbers

Division of complex numbers is defined as:

$$\frac{V_1 \angle \theta_1}{V_2 \angle \theta_2} = \frac{V_1}{V_2} \angle(\theta_1 - \theta_2)$$

7.3 SINUSOIDAL CONVERSIONS

$$A \sin(\omega t + \phi) = A \cos(\omega t + \phi - 90^0)$$
$$A \cos(\omega t + \theta) = A \sin(\omega t + \theta + 90^0)$$
$$-A \sin(\omega t + \alpha) = A \cos(\omega t + \alpha + 90^0)$$
$$-A \cos(\omega t + \beta) = A \cos(\omega t + \beta \pm 180^0)$$

7.4 IMPEDANCE AND ADMITTANCE

Impedance is defined as:

$$Z = \frac{\text{phasor voltage}}{\text{phasor current}} = \frac{V}{I}$$

The impedance values of resistance R, inductance L, and capacitance C are:

$$Z_R = R \qquad Z_L = j\omega L = jX_L \qquad Z_C = \frac{1}{j\omega C} = -jX_C$$

where $X_L = \omega L = 2\pi f L$ and $X_C = 1/\omega C = 1/2\pi f C$. The X_L is called the inductive reactance and the X_C is called the capacitive reactance. The f is the frequency in Hz.
   For the series RLC circuit, the total impedance is

$$Z = R + j\omega L + 1/j\omega C = R + j(\omega L - 1/\omega C) = R + j(X_L - X_C) = R + jX$$

where $X = X_L - X_C$ and the X is the total reactance of the sinusoidal series RLC-circuit. Admittance is defined as:

$$Y = \frac{\text{phasor current}}{\text{phasor voltage}} = \frac{I}{V} = \frac{1}{Z}$$

The admittance values of resistance R, inductance L, and capacitance C are:

$$Y_R = \frac{1}{R} = G \qquad Y_L = \frac{1}{j\omega L} = -jB_L \qquad Y_C = j\omega C = jB_C$$

where $B_L = 1/\omega L = 1/2\pi fL$ and $B_C = \omega C = 2\pi fC$. The B_L is the inductive susceptance and the B_C is the capacitive susceptance.

For the parallel RLC circuit, the total admittance is

$$Y = G + 1/j\omega L + j\omega C = G + j(\omega C - 1/\omega L) = G + j(B_C - B_L) = G + jB$$

where $B = B_C - B_L$ and the B is the total susceptance of the sinusoidal parallel RLC circuit.

7.5   COMBINATION OF IMPEDANCES

Impedances in series	$Z_{eq} = Z_1 + Z_2 + ...$
Impedances in parallel	$\dfrac{1}{Z_{eq}} = \dfrac{1}{Z_1} + \dfrac{1}{Z_2} + ...$
Two parallel impedances	$Z_{eq} = \dfrac{Z_1 Z_2}{Z_1 + Z_2}$

7.6   COMBINATION OF ADMITTANCES

Admittances in series	$Y_{eq} = Y_1 + Y_2 + ...$
Admittances in parallel	$\dfrac{1}{Y_{eq}} = \dfrac{1}{Y_1} + \dfrac{1}{Y_2} + ...$
Two series admittances	$Y_{eq} = \dfrac{Y_1 Y_2}{Y_1 + Y_2}$

7.7   VOLTAGE AND CURRENT DIVISION

With reference to Figure 7.1(a),

$$V_1 = \frac{Z_1}{Z_1 + Z_2} V$$

$$V_2 = \frac{Z_2}{Z_1 + Z_2} V$$

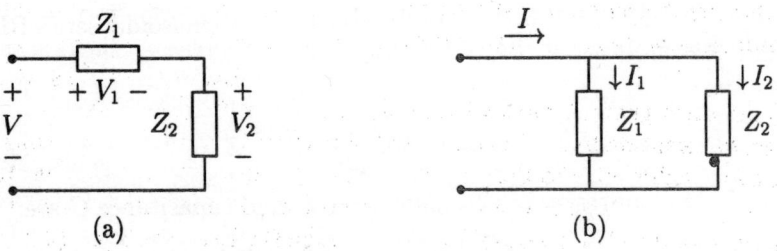

Figure 7.1: (a) Voltage division; (b) Current division.

With reference to Figure 7.1(b),

$$I_1 = \frac{Z_2}{Z_1 + Z_2} I$$

$$I_2 = \frac{Z_1}{Z_1 + Z_2} I$$

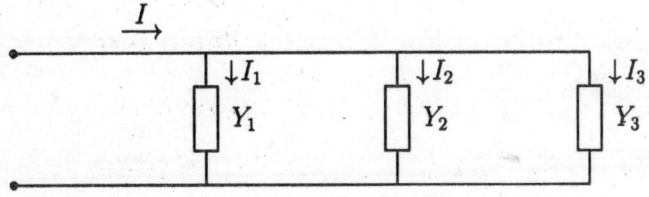

Figure 7.2: Admittances in parallel.

With reference to Figure 7.2,

$$I_1 = \frac{Y_1}{Y_1 + Y_2 + Y_3} I$$

$$I_2 = \frac{Y_2}{Y_1 + Y_2 + Y_3} I$$

$$I_3 = \frac{Y_3}{Y_1 + Y_2 + Y_3} I$$

7.8 PHASOR DIAGRAM

Graphically, a phasor may be represented in a complex plane by a vector whose length is proportional to the magnitude of the phasor, measured from the origin. The phase angle of the vector is equal to the phase angle of the phasor, measured with respect to the horizontal positive axis. Positive angles are measured in the counterclockwise direction and negative angles are measured in the clockwise direction with respect to the reference horizontal line. Such a graphical representation of different phasor quantities of specific circuit is called a **phasor diagram**. For example, voltage phasor is $V = 230\angle 30^0$ and current phasor is $I = 5\angle -45^0$ for a specific circuit. Its phasor diagram is shown in Figure 7.3.

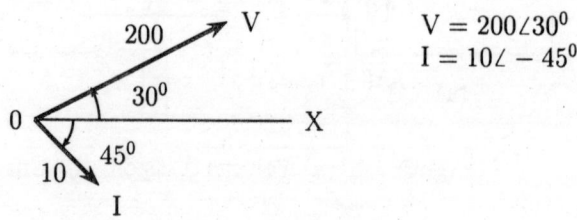

Figure 7.3: Showing a phasor diagram.

7.9   NODAL ANALYSIS

In nodal analysis, we determine node voltages with respect to a reference node in a circuit.

7.9.1   Nodal Analysis For Circuits With No Voltage Sources

The nodal analysis steps are:

1. Choose any one node as a reference node (zero-potential node) from the circuit.

2. Label the remaining nodes as shown in Figure 7.4.

3. Use KCL at each node except reference node and express the current through each resistor in terms of node voltages.

4. Solve the resulting set of simultaneous equations by any method.

Consider a circuit containing only current sources as shown in Figure 7.4. Applying KCL at the nonreference nodes, we have

$$\text{Node 1}: \qquad \frac{V_1 - V_2}{Z_1} + \frac{V_1 - V_3}{Z_4} = I_1 \qquad\qquad (7.1)$$

$$\text{Node 2}: \qquad \frac{V_2 - V_1}{Z_1} + \frac{V_2}{Z_2} + \frac{V_2 - V_3}{Z_3} = 0 \qquad\qquad (7.2)$$

$$\text{Node 3}: \qquad \frac{V_3 - V_2}{Z_3} + \frac{V_3 - V_1}{Z_4} = I_2 \qquad\qquad (7.3)$$

Rearranging the terms of these equations gives the node-voltage equations in standard form:

$$\text{Node 1}: \qquad \left(\frac{1}{Z_1} + \frac{1}{Z_4}\right) V_1 - \frac{1}{Z_1} V_2 - \frac{1}{Z_4} V_3 = I_1 \qquad\qquad (7.4)$$

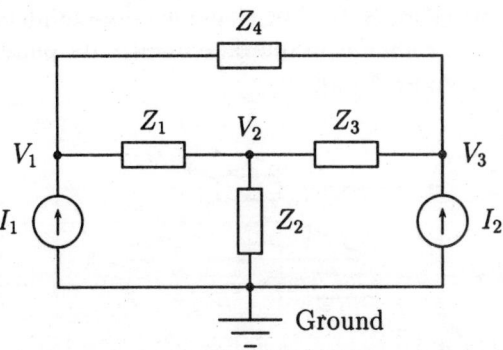

Figure 7.4: A typical circuit containing no voltage sources.

Node 2 :
$$-\frac{1}{Z_1}V_1 + \left(\frac{1}{Z_1} + \frac{1}{Z_2} + \frac{1}{Z_3}\right)V_2 - \frac{1}{Z_3}V_3 = 0 \qquad (7.5)$$

Node 3 :
$$-\frac{1}{Z_4}V_1 - \frac{1}{Z_3}V_2 + \left(\frac{1}{Z_3} + \frac{1}{Z_4}\right)V_3 = I_2 \qquad (7.6)$$

These equations may be solved for V_1, V_2, and V_3 by Cramer's rule. If these node voltages are known, any circuit variable can be calculated.

7.9.2 Nodal Analysis For Circuits With Voltage Sources

Let us consider a circuit containing a voltage source connected between a nonreference node and reference node as shown in Figure 7.5.

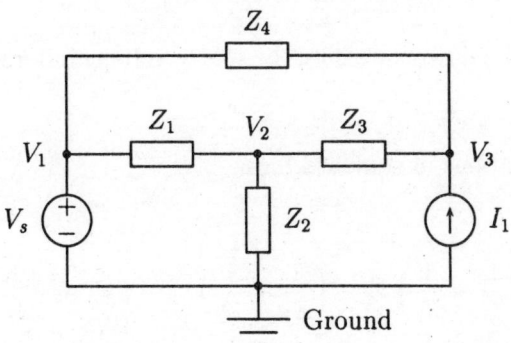

Figure 7.5: A typical circuit containing a voltage source connected between nodes V_1 and reference node.

With reference to Figure 7.5, we find that the first node voltage, V_1 is equal to the source voltage, V_s. Write the nodal equations only at the remaining nodes, V_2 and V_3 as:

Node 1 :
$$V_1 = V_s \text{ (known)} \qquad (7.7)$$

Node 2 :
$$-\frac{1}{Z_1}V_s + \left(\frac{1}{Z_1} + \frac{1}{Z_2} + \frac{1}{Z_3}\right)V_2 - \frac{1}{Z_3}V_3 = 0 \qquad (7.8)$$

Node 3 :
$$-\frac{1}{Z_4}V_s - \frac{1}{Z_3}V_2 + \left(\frac{1}{Z_3} + \frac{1}{Z_4}\right)V_3 = I_2 \qquad (7.9)$$

Thus, the remaining node voltages, V_2 and V_3 can be determined easily.

Let us now consider a circuit containing a voltage source connected between two nonreference nodes as shown in Figure 7.6.

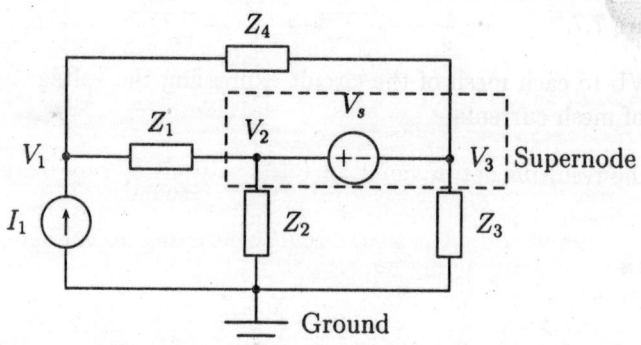

Figure 7.6: A typical circuit containing a voltage source connected between nonreference nodes V_2 and V_3.

Here, nodes V_2 and V_3 are merged to form a larger node. This node is called a **supernode**.

First equation is obtained by KCL at the node V_1:

$$\frac{V_1 - V_2}{Z_1} + \frac{V_1 - V_3}{Z_4} = I_1 \tag{7.10}$$

Second equation may be obtained by KCL at the supernode:

$$\frac{V_2 - V_1}{Z_1} + \frac{V_2}{Z_2} + \frac{V_3}{Z_3} + \frac{V_3 - V_1}{Z_4} = 0 \tag{7.11}$$

Third equation is obtained by expressing the source voltage in terms of node voltages, V_2 and V_3 as:

$$V_2 - V_3 = V_s \tag{7.12}$$

Rearranging these equations in standard form:

$$\left(\frac{1}{Z_1} + \frac{1}{Z_4}\right) V_1 - \frac{1}{Z_1} V_2 - \frac{1}{Z_4} V_3 = I_1 \tag{7.13}$$

$$-\left(\frac{1}{Z_1} + \frac{1}{Z_4}\right) V_1 + \left(\frac{1}{Z_1} + \frac{1}{Z_2}\right) V_2 + \left(\frac{1}{Z_3} + \frac{1}{Z_4}\right) V_3 = 0 \tag{7.14}$$

$$V_2 - V_3 = V_s \tag{7.15}$$

Thus, these simultaneous equations can be solved by Cramer's rule.

Note: In case of transformable voltage source(s) (a voltage source in series with a resistance), each voltage source is transformed into a current source and usual method of nodal analysis can be followed.

7.10   MESH ANALYSIS

In mesh analysis, we determine a circulating current in each mesh of a circuit. The meshes are closed paths which do not enclosed other circuit elements. The loops may contain other loops or meshes inside them.

7.10.1   Mesh Analysis For Circuits With No Current Sources

We summarize the method in the following steps:

1. Assign clockwise mesh current in each mesh of the circuit (planner circuit) as shown in Figure 7.7.

2. Use KVL to each mesh of the circuit expressing the voltage across each resistor in terms of mesh currents.

3. Solve the resulting set of simultaneous equations by any method for the mesh currents.

Consider a circuit containing no current sources as shown in Figure 7.7.
By KVL in Figure 7.7, we obtain

$$\text{Mesh 1}: \qquad Z_1(I_1 - I_3) + Z_2(I_1 - I_2) = V_1 \qquad\qquad (7.16)$$
$$\text{Mesh 2}: \qquad Z_2(I_2 - I_1) + Z_3(I_2 - I_3) = -V_2 \qquad\qquad (7.17)$$
$$\text{Mesh 3}: \qquad Z_4 I_3 + Z_3(I_3 - I_2) + Z_1(I_3 - I_1) = 0 \qquad\qquad (7.18)$$

Rearranging the terms of these equations gives the mesh-current equations in standard form:

$$\text{Mesh 1}: \qquad (Z_1 + Z_2)I_1 - Z_2 I_2 - Z_1 I_3 = V_1 \qquad\qquad (7.19)$$
$$\text{Mesh 2}: \qquad -Z_2 I_1 + (Z_2 + Z_3)I_2 - Z_3 I_3 = -V_2 \qquad\qquad (7.20)$$
$$\text{Mesh 3}: \qquad -Z_1 I_1 - Z_3 I_2 + (Z_1 + Z_3 + Z_4)I_3 = 0 \qquad\qquad (7.21)$$

These equations may be solved for I_1, I_2, and I_3 by Cramer's rule. If these mesh currents are known, any circuit variable can be calculated.

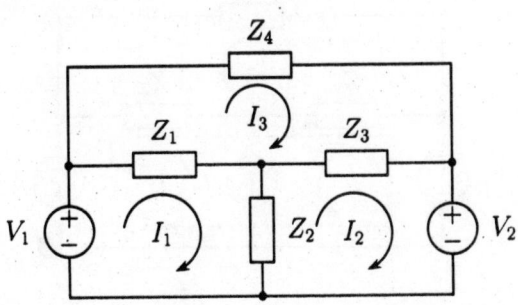

Figure 7.7: A typical circuit containing no current sources.

7.10.2   Mesh Analysis For Circuits With Current Sources

Let us consider a circuit containing a current source which is not common to any mesh as shown in Figure 7.8. For the first mesh, mesh current I_1 is equal to the source current I_s.

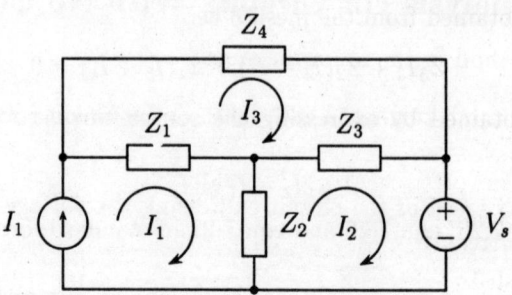

Figure 7.8: A typical circuit containing a current source which is not common to any mesh.

For the remaining meshes, write the usual mesh current equations. Thus from Figure 7.8, we can obtain,

Mesh 1 : $\qquad I_1 = I_s$ (known) $\qquad\qquad$ (7.22)

Mesh 2 : $\qquad -Z_2 I_s + (Z_2 + Z_3) I_2 - Z_3 I_3 = -V_s \qquad$ (7.23)

Mesh 3 : $\qquad -Z_1 I_s - Z_3 I_2 + (Z_1 + Z_3 + Z_4) I_3 = 0 \qquad$ (7.24)

Thus, we see that the presence of current sources reduces our labour in solving these equations.

Let us now consider a circuit containing a current source which is common to mesh 1 and mesh 2 as shown in Figure 7.9.

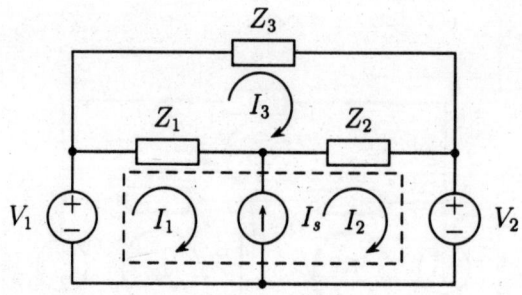

Figure 7.9: A typical circuit containing a current source which is common to mesh 1 and mesh 2.

Here, mesh1 and mesh2 are merged to get a larger mesh. This mesh is called a **supermesh**.

First equation is obtained from the supermesh by KVL around the dotted loop as shown in Figure 7.9 as:

$$Z_1(I_1 - I_3) + Z_2(I_2 - I_3) + V_2 - V_1 = 0 \qquad (7.25)$$

Second equation is obtained from the mesh 3 as:

$$Z_3 I_3 + Z_2(I_3 - I_2) + Z_1(I_3 - I_1) = 0 \tag{7.26}$$

Third equation is obtained by expressing the source current I_s in terms of the mesh-currents i_1 and i_2 as:

$$I_2 - I_1 = I_s \tag{7.27}$$

Rearranging the eqn 7.25, eqn 7.26 and eqn 7.27 in standard form as

Mesh 1 : $\qquad Z_1 I_1 + Z_2 I_2 - (Z_1 + Z_2)I_3 = V_1 - V_2 \tag{7.28}$

Mesh 2 : $\qquad -Z_1 I_1 - Z_2 I_2 + (Z_1 + Z_2 + Z_3)I_3 = 0 \tag{7.29}$

Mesh 3 : $\qquad -I_1 + I_2 = I_s \tag{7.30}$

These last set of three equations can solved by Cramer's rule.

Note: In case of transformable current source(s) (a current source in parallel with a resistance), each current source is transformed into a voltage source and usual method of mesh analysis can be followed.

7.11   THEVENIN'S AND NORTON'S THEOREMS

The basic concepts of Thevenin's and Norton's theorems explained for resistive circuits in chapter 1 are also applicable to the sinusoidal circuits if such networks are expressed in terms of phasors and impedances (or admittances). Figure 7.10 shows the Thevenin's and Norton's equivalent circuits. The equivalent circuit parameters are defined as:

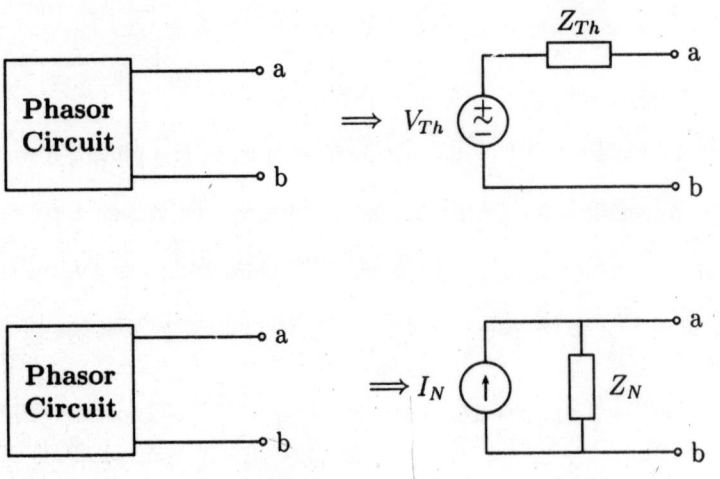

Figure 7.10: Thevenin's and Norton's equivalent circuit.

$V_{Th} = V_{oc}$ Open circuit phasor voltage across the terminals a and b.

$V_N = V_{sc}$ Short circuit phasor current through the terminals a and b.

$Z_{Th} = Z_{ab}$ Impedance across the terminals a and b with all independent sources killed and dependent sources unchanged.

7.12   SUPERPOSITION THEOREM

This theorem states that any branch voltage V or current I in a linear circuit containing many independent sources is composed of algebraic phasor sum of the voltage V or current I in that branch due to each of the sources acting separately in circuit, while all the other source in the circuit are killed.

The independent voltage sources are killed by short circuits and the independent current sources are killed by open circuits. If there are N independent sources, we have to perform N computations. Each independent source is working only in one computation. Note that the dependent sources are acting in every computation. This method will be more clear in the solved problems section.

7.13   SOLVED PROBLEMS

SP 7.1 Convert each of the following time functions into phasor form:

   (a) $2\cos 4t$
   (b) $3\cos(2t + 45^0)$
   (c) $5\sin 10t$
   (d) $-5\sin 10t$
   (e) $-10\sin(10t - 45^0)$
   (f) $4\cos(4t - 30^0) + 2\sin(4t - 100^0)$

SOLUTION:

(a) $2\cos 4t = \text{Re}[2\angle 0^0 e^{j4t}]$ thus, the phasor form is $2\angle 0^0$

(b) $3\cos(2t + 45^0) = \text{Re}[3e^{j(2t+45^0)}] = \text{Re}[3\angle 45^0 e^{j2t}]$ thus, the phasor form is $3\angle 45^0$

(c) $5\sin 10t = 5\cos(10t - 90^0) = \text{Re}[5\angle -90^0 e^{j10t}]$ thus, the phasor form is $5\angle -90^0$

(d) $-5\sin 10t = 5\cos(10t + 90^0) = \text{Re}[5\angle 90^0 e^{j10t}]$ thus, the phasor form is $5\angle 90^0$

(e) $-10\sin(10t - 45^0) = 10\cos(10t + 45^0) = \text{Re}[10\angle 45^0 e^{j10t}]$ thus, the phasor form is $10\angle 45^0$

(f)
$$
\begin{aligned}
4\cos(4t - 30^0) + 2\sin(4t - 100^0) &= 4\cos(4t - 30^0) + 2\cos(4t - 190^0) \\
&= \text{Re}[4\angle -30^0 e^{j4t} + 2\angle -190^0 e^{j4t}] \\
&= \text{Re}[(3.46 - j2 - 1.97 + j0.347)e^{j4t}] \\
&= \text{Re}[(1.49 - j1.653)e^{j4t}] \\
&= \text{Re}[2.23\angle -48^0 e^{j4t}]
\end{aligned}
$$

Thus, the phasor form is $2.23\angle -48^0$

SP 7.2 Determine the steady-state values of $i(t)$, $v_R(t)$, $v_L(t)$ and $v_C(t)$ for the circuit in Fig. SP 7.2.

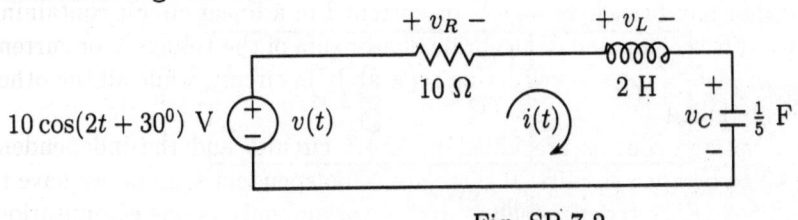

Fig. SP 7-2.

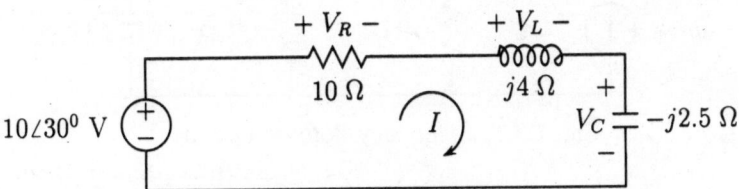

Fig. 7.2.1: Frequency-domain circuit.

SOLUTION: The Fig. 7.2.1 shows the frequency-domain circuit wherein $\omega = 2$ rad/sec, $Z_R = 10\ \Omega$, $Z_L = j\omega L = j4\ \Omega$, and $Z_C = 1/j\omega C = -j2.5\ \Omega$. The phasor value of $v(t)$ is $10\angle30^0$.

From Fig. 7.2.1,

$$I = \frac{10\angle30^0}{10 + j4 - j2.5} = \frac{10\angle30^0}{10 + j1.5} = \frac{10\angle30^0}{10.1\angle8.53^0} = 0.99\angle21.47^0$$

then

$$V_R = 10 \times 0.99\angle21.47^0 = 9.9\angle21.47^0$$
$$V_L = j4 \times 0.99\angle21.47^0 = 3.96\angle111.47^0$$
$$V_C = -j2.5 \times 0.99\angle21.47^0 = 2.475\angle-68.53^0$$

In time-domain

$$i(t) = 0.99\cos(2t + 21.47^0)\ \text{A}$$
$$v_R(t) = 9.9\cos(2t + 21.47^0)\ \text{V}$$
$$v_L(t) = 3.96\cos(2t + 111.47^0)\ \text{V}$$
$$v_C(t) = 2.475\cos(2t - 68.53^0)\ \text{V}$$

SP 7.3 Find the steady-state values of $i_R(t)$, $i_L(t)$, $i_C(t)$ and $v(t)$ for the circuit of Fig. 7.3.

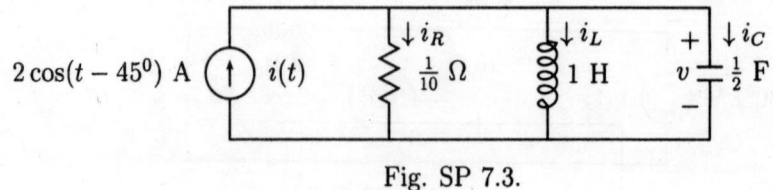

Fig. SP 7.3.

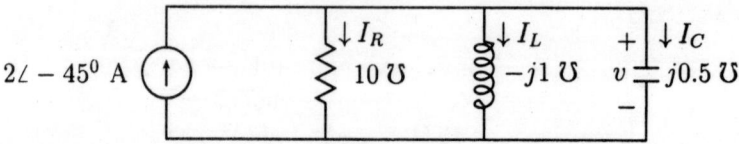

Fig. 7.3.1: Frequency-domain circuit.

SOLUTION: The Fig. 7.3.1 shows the frequency-domain circuit wherein the circuit parameters are: $\omega = 1$ rad/sec, $Y_R = 10$ ℧, $Y_L = 1/j\omega L = -j1$ ℧, and $Y_C = j\omega C = j0.5$ ℧. The phasor value of current is $2\angle - 45^0$.

From Fig. 7.3.1, using current division method

$$I_R = \frac{10}{10 - j1 + j0.5}(2\angle - 45^0) = \frac{10}{10 - j0.5}(2\angle - 45^0) = \frac{10 \times 2\angle - 45^0}{10\angle - 2.86^0} = 2\angle - 42.14^0 \text{ A}$$

$$I_L = \frac{-j1}{10 - j1 + j0.5}(2\angle - 45^0) = \frac{1\angle - 90^0}{10\angle - 2.86^0}(2\angle - 45^0) = 0.2\angle - 132.14^0 \text{ A}$$

$$I_C = \frac{j0.5}{10 - j1 + j0.5}(2\angle - 45^0) = \frac{0.5\angle 90^0}{10\angle - 2.86^0}(2\angle - 45^0) = 0.1\angle 47.86^0 \text{ A}$$

and $$V = \frac{I_C}{j0.5} = \frac{0.1\angle 47.86^0}{0.5\angle 90^0} = 0.2\angle - 42.14^0 \text{ V}$$

In time-domain

$$i_R(t) = 2\cos(t - 42.14^0) \text{ A}$$

$$i_L(t) = 0.2\cos(t - 132.14^0) \text{ A}$$

$$i_C(t) = 0.1\cos(t + 47.86^0) \text{ A}$$

$$v(t) = 0.2\cos(t - 42.14^0) \text{ V}$$

SP 7.4 Determine the steady-state values of i_s, i_a, i_b, and v_{ab} for the circuit of Fig. SP 7.4.

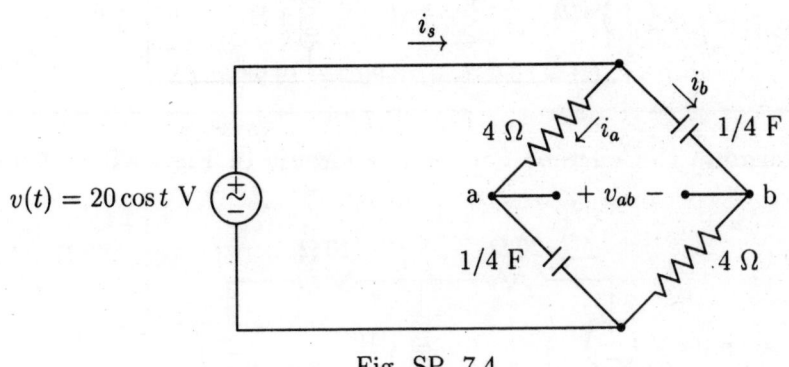

Fig. SP 7.4.

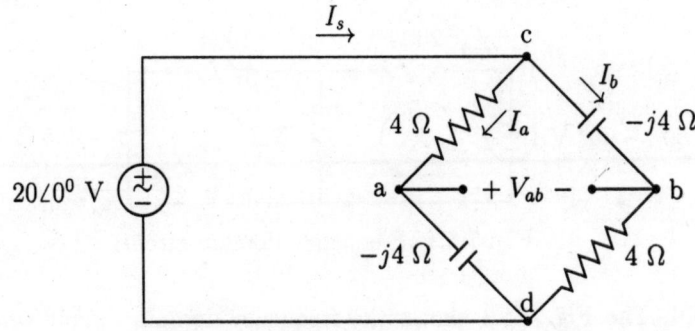

Fig. 7.4.1: Frequency-domain circuit.

SOLUTION: Fig. 7.4.1 shows the frequency-domain circuit with $\omega = 1$ rad/sec. The circuit can be treated having two parallel paths connected across the voltage source $20\angle 0^0$ V.

From Fig. 7.4.1,

$$I_a = I_b = \frac{20\angle 0^0}{4 - j4} = \frac{20\angle 0^0}{5.66\angle - 45^0} = 3.53\angle 45^0 \text{ A}$$

$$I_s = I_a + I_b = (2)(3.53\angle 45^0) = 7.06\angle 45^0 \text{ A}$$

By KVL around the loop c-a-b-c,

$$V_{ab} = -4I_a - j4I_b = (-4 - j4)I_a = (5.66\angle - 135^0)(3.53\angle 45^0) = 19.98\angle - 90^0 \text{ V}$$

In time-domain

$$i_a(t) = i_b(t) = 3.53 \cos(t + 45^0) \text{ A}$$

$$i_s(t) = 7.06 \cos(t + 45^0) \text{ A}$$

$$v_{ab}(t) = 19.98 \cos(t - 90^0) = 19.98 \sin t \text{ V}$$

SP 7.5 Determine the current $i(t)$ for the circuit in Fig. SP 7.5 by mesh analysis.

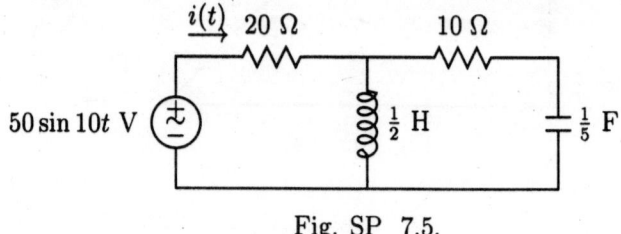

Fig. SP 7.5.

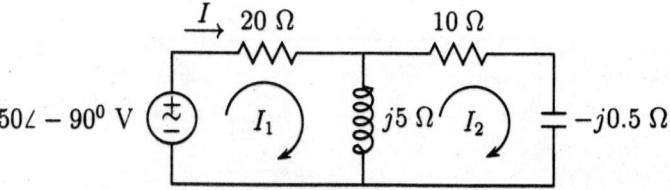

Fig. 7.5.1: Frequency-domain circuit.

SOLUTION: The Fig. 7.5.1 shows the frequency-domain circuit wherein circuit parameters are shown with elements. The voltage source $50 \sin 10t$ is transformed into $50 \cos(10t - 90^0)$ to obtain the phasor value as $50\angle - 90^0$. By mesh analysis, we need to determine I_1 only to get I as $I_1 = I$.

Mesh-current equations for the circuit of Fig. 7.5.1 are:

$$M_1: \quad (20 + j5)I_1 - j5I_2 = 50\angle - 90^0 = -j50$$
$$M_2: \quad -j5I_1 + (10 + j4.5)I_2 = 0$$

By Cramer's rule

$$I_1 = \frac{\begin{vmatrix} -j50 & -j5 \\ 0 & 10 + j4.5 \end{vmatrix}}{\begin{vmatrix} 20 + j5 & -j5 \\ -j5 & 10 + j4.5 \end{vmatrix}} = \frac{225 - j500}{202.5 + j140} = \frac{548.29\angle -65.77^0}{246.18\angle 34.66^0} = 2.23\angle -100.43^0$$

In time domain

$$i(t) = i_1(t) = 2.23 \cos(10t - 100.43^0) \text{ A}$$

SP 7.6 Solve for mesh currents $i_1(t)$ **and** $i_2(t)$ **in the circuit of Fig. SP 7.6.**

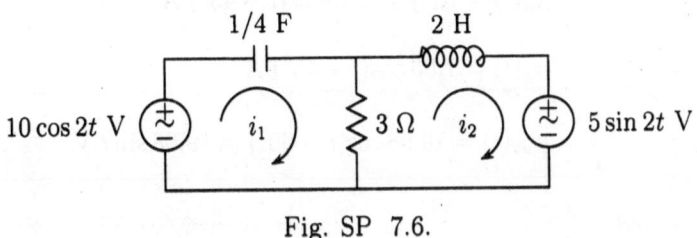

Fig. SP 7.6.

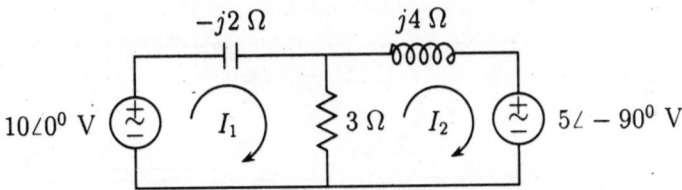

Fig. 7.6.1: Frequency domain circuit.

SOLUTION: The Fig. 7.6.1 shows the frequency-domain circuit wherein circuit parameters are shown with elements. The frequency for both the voltage source is 2 rad/sec. The phasor value for the left voltage source is directly written as $10\angle0^0$ whereas the phasor value for the right voltage source is written as $5\angle -90^0$ from the conversion $5\cos(2t - 90^0) = 5\sin 2t$.

Mesh equations for circuit of Fig. 7.6.1 are:

$$M_1: \quad (3 - j2)I_1 - 3I_2 = 10\angle0^0 = 10$$
$$M_2: \quad -3I_1 + (3 + j4)I_2 = -5\angle -90^0 = j5$$

By Cramer's rule

$$I_1 = \frac{\begin{vmatrix} 10 & -3 \\ j5 & 3+j4 \end{vmatrix}}{\begin{vmatrix} 3-j2 & -3 \\ -3 & 3+j4 \end{vmatrix}} = \frac{30 + j55}{8 + j6} = \frac{62.65\angle61.39^0}{10\angle36.9^0} = 6.265\angle24.5^0$$

$$I_2 = \frac{\begin{vmatrix} 3-j2 & 10 \\ -3 & j5 \end{vmatrix}}{\begin{vmatrix} 3-j2 & -3 \\ -3 & 3+j4 \end{vmatrix}} = \frac{40 + j15}{8 + j6} = \frac{42.7\angle20.6^0}{10\angle36.9^0} = 4.27\angle -16.3^0$$

Hence

$$i_1(t) = 6.27\cos(2t + 24.5^0)\ \text{A} \quad \text{and} \quad i_2(t) = 4.27\cos(2t - 16.3^0)\ \text{A}$$

SP 7.7 Determine the steady-state value of i_0 for circuit in Fig. SP 7.7 using mesh equations.

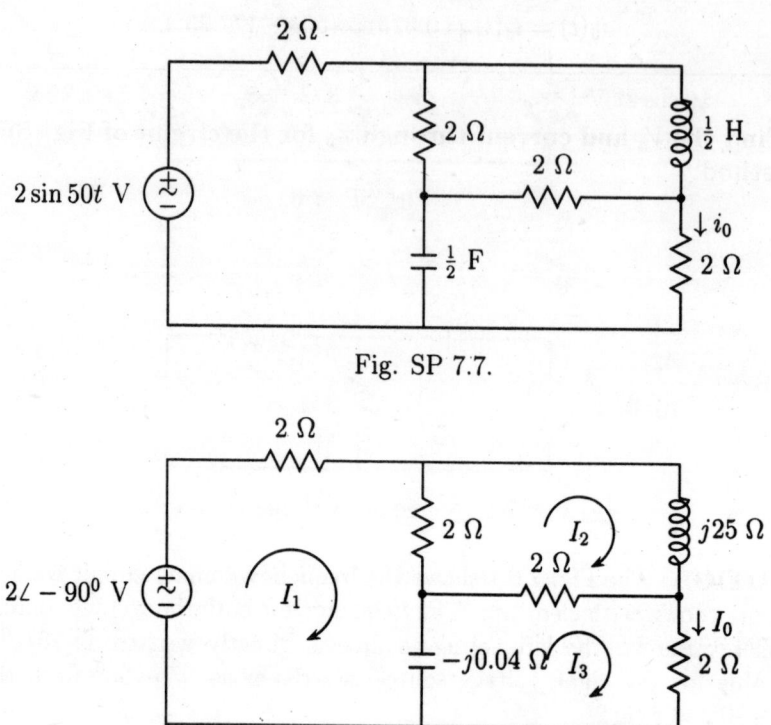

Fig. SP 7.7.

Fig. 7.7.1: Frequency domain circuit.

SOLUTION: The Fig. 7.7.1 shows the frequency-domain circuit of Fig. SP 7.7. Here, we have to determine I_3 only to obtain I_0.

Mesh equations for the circuit of Fig. 7.7.1 are:

$$M_1 : \quad (4 - j0.04)I_1 - 2I_2 + j0.04I_3 = 2\angle - 90^0 = -j2$$
$$M_2 : \quad -2I_1 + (4 + j25)I_2 - 2I_3 = 0$$
$$M_3 : \quad j0.04I_1 - 2I_2 + (4 - j0.04)I_3 = 0$$

By Cramer's rule

$$I_3 = \frac{\begin{vmatrix} 4 - j0.04 & -2 & -j2 \\ -2 & 4 + j25 & 0 \\ j0.04 & -2 & 0 \end{vmatrix}}{\begin{vmatrix} 4 - j0.04 & -2 & j0.04 \\ -2 & 4 + j25 & -2 \\ j0.04 & -2 & 4 + j0.04 \end{vmatrix}}$$

$$= \frac{-0.32 - j10}{32.0128 + j400.4} = \frac{10\angle - 91.8^0}{401.68\angle 85.43^0} = 0.025\angle - 177.23^0$$

Therefore

$$i_0(t) = i_3(t) = 0.025 \cos(50t - 177.23^0) \text{ A}$$

SP 7.8 Find the V_a and current through Z_3 for the circuit of Fig. SP 7.8 using nodal method.

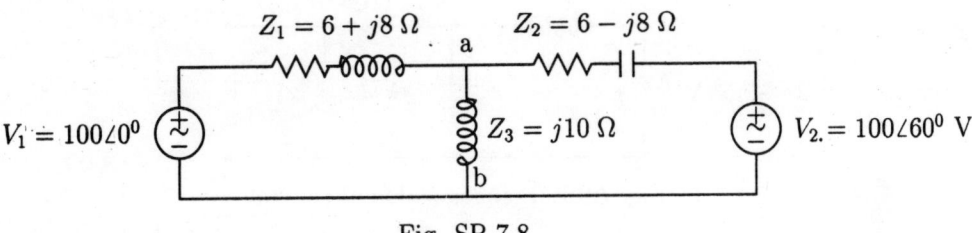

Fig. SP 7.8.

SOLUTION:

Required node equation is

$$(Y_1 + Y_2 + Y_3)V_a - Y_1V_1 - Y_2V_2 = 0$$

where

$$Z_1 = 6 + j8 = 10\angle 53.13^0 \qquad Y_1 = 0.1\angle - 53.13^0 = 0.06 - j0.08$$

$$Z_2 = 6 - j8 = 10\angle - 53.13^0 \qquad Y_2 = 0.1\angle 53.13^0 = 0.06 + j0.08$$

$$Z_3 = j10 = 10\angle 90^0 \qquad Y_3 = 0.1\angle - 90^0 = -j0.1$$

then

$$(0.12 - j0.1)V_a = (0.1\angle - 53.13^0)(100\angle 0^0) + (0.1\angle 53.1^0)(100\angle 60^0)$$
$$= 6 - j8 - 3.93 + j9.2 = 2.07 + j1.2$$

Hence

$$V_a = \frac{2.07 + j1.2}{0.12 - j0.1} = \frac{2.39\angle 30.1^0}{0.156\angle - 39.8^0} = 15.32\angle 69.9^0 \text{ V}$$

and current through Z_3 is

$$I_{ab} = Y_3V_a = (0.1\angle - 90^0)(15.32\angle 69.9^0) = 1.5\angle - 20.1^0 \text{ A}$$

SP 7.9 find the steady-state value of v_0 for circuit of Fig. SP 7.9 by nodal method.

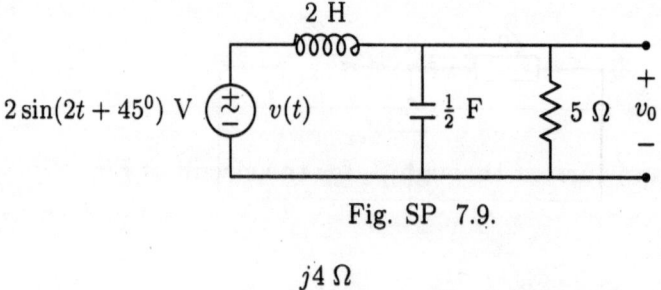

Fig. SP 7.9.

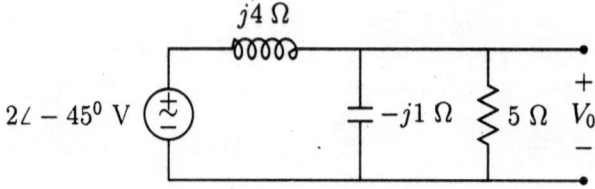

Fig. 7.9.1: Frequency domain circuit.

SOLUTION: The Fig. 7.9.1 shows the frequency-domain circuit wherein $\omega = 2$ rad/sec. The phasor value of $v(t)$ is $2\angle -45^0$ obtained form the conversion $2\sin(2t+45^0) = 2\cos(2t + 45^0 - 90^0) = 2\cos(2t - 45^0)$.

Required nodal equation is

$$\left(\frac{1}{j4} + \frac{1}{-j1} + \frac{1}{5}\right) V_0 - \left(\frac{1}{j4}\right)(2\angle - 45^0) = 0$$

or $\qquad (0.2 + j0.75)V_0 = -(j0.25)(2\angle - 45^0) = 0.5\angle - 135^0$

Hence $\qquad V_0 = \dfrac{0.5\angle - 135^0}{0.78\angle 75^0} = 0.64\angle - 210^0$

In time domain

$$v_0(t) = 0.64\cos(2t - 210^0) \text{ V}$$

SP 7.10 Find current through Z_3 from a to b for the circuit shown in Fig. SP 7.8 using Thevenin's equivalent circuit.

SOLUTION: To determine Thevenin impedance, Z_{Th} seen through the Z_3, remove the Z_3 and kill both the voltage sources by short circuits as shown in Fig. 7.10.1. The equivalent impedance, Z_{eq} is equal to Z_{Th}. The Fig. 7.10.2 determines the open circuit voltage, V_{oc} which is equal to V_{Th}. The Fig. 7.10.3 is the Thevenin's equivalent circuit from which current through Z_3 can be easily obtained.

From Fig. 7.10.1, Z_{eq} is

$$Z_{eq} = Z_1||Z_2 = \frac{Z_1 Z_2}{Z_1 + Z_2} = \frac{(6 + j8)(6 - j8)}{6 + j8 + 6 - j8} = 8.33$$

Thus
$$Z_{Th} = Z_{eq} = 8.33 \ \Omega$$

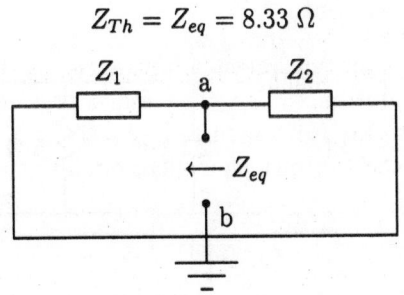

Fig. 7.10.1: Finding Z_{eq}.

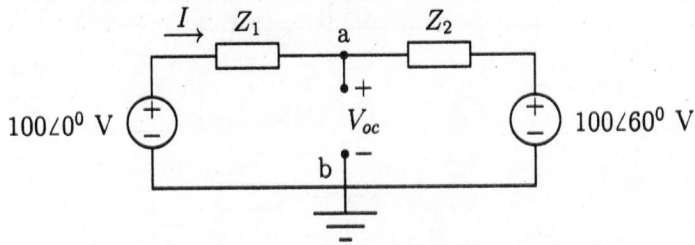

Fig. 7.10.2: Finding V_{oc}.

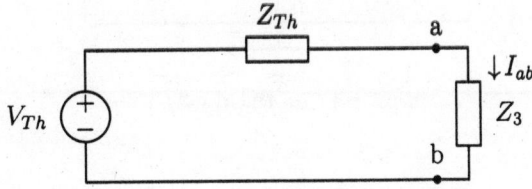

Fig. 7.10.3: Thevenin's equivalent circuit.

From Fig. 7.10.2, current I will be

$$I = \frac{100\angle 0^0 - 100\angle 60^0}{6 + j8 + 6 - j8} = \frac{100 - (50 + j86.6)}{12} = \frac{100\angle - 60^0}{12} = 8.33\angle - 60^0$$

From Fig. 7.10.2, by KVL around the loop containing $100\angle 0^0$ V, Z_1, and V_{oc} ;

$$V_{oc} = 100\angle 0^0 - Z_1 I = 100 - (6 + j8)(8.33\angle - 60^0) = 17.3 + j9.96 = 19.96\angle 29.93^0$$

Hence
$$V_{Th} = V_{oc} = 19.96\angle 29.93^0 \text{ V}$$

From Fig. 7.10.3,

$$I_{ab} = \frac{V_{Th}}{Z_{Th} + Z_3} = \frac{19.96\angle 29.93^0}{8.33 + j10} = \frac{19.96\angle 29.93^0}{13\angle 50.19^0} = 1.5\angle - 20.26^0 \text{ A}$$

SP 7.11 Find current through Z_3 from a to b for the circuit shown in Fig. SP 7.8 using Norton equivalent.

SOLUTION: We determine equivalent admittance, Y_{eq} from Fig. 7.11.1. This Y_{eq} is equal to the Thevenin admittance, Y_{Th}. The Fig. 7.11.2 determines the short circuit current, I_{sc} which is equal to the Norton current, I_N. Norton current, I_N and Thevenin admittance, Y_{Th} constitute the Norton's equivalent circuit as shown in Fig. 7.11.3.

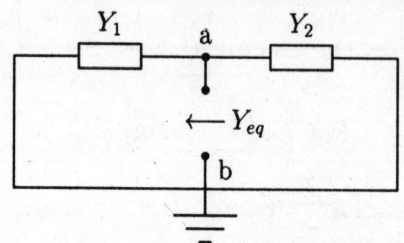

Fig. 7.11.1: Finding Y_{eq}.

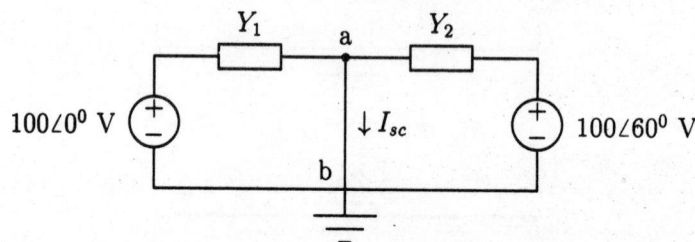

Fig. 7.11.2: Finding I_{sc}.

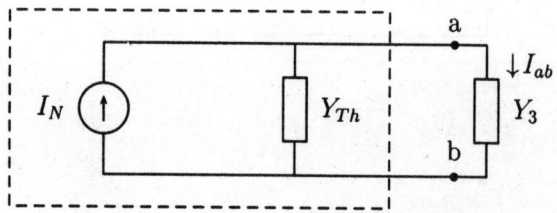

Fig. 7.11.3: Norton's equivalent circuit.

Here

$$Z_1 = 6 + j8 = 10\angle 53.13^0 \ \Omega \qquad Y_1 = 0.1\angle -53.13^0 = 0.06 - j0.08 \ \mho$$
$$Z_2 = 6 - j8 = 10\angle -53.13^0 \ \Omega \quad Y_2 = 0.1\angle 53.13^0 = 0.06 + j0.08 \ \mho$$
$$Z_3 = j10 = 10\angle 90^0 \ \Omega \qquad Y_3 = 0.1\angle -90^0 = -j0.1 \ \mho$$

From Fig. 7.11.1,

$$Y_{eq} = Y_1 + Y_2 = 0.06 - j0.08 + 0.06 + j0.08 = 0.12$$

Thus $$Y_{Th} = Y_{eq} = 0.12 \ \mho$$

From Fig. 7.11.2,

$$I_1 = 100\angle 0^0(Y_1) = (100\angle 0^0)(0.1\angle -53.13^0) = 10\angle -53.13^0 = 6 - j8$$
$$I_2 = 100\angle 60^0(Y_2) = (100\angle 60^0)(0.1\angle 53.13^0) = 10\angle 113.13^0 = -3.93 + j9.2$$
$$I_{sc} = I_1 + I_2 = 6 - j8 - 3.93 + j9.2 = 2.07 + j1.2 = 2.39\angle 30.1^0$$

and
$$I_N = I_{sc} = 2.39\angle 30.1^0 \text{ A}$$

From Fig. 7.11.3, using current division method

$$I_{ab} = \frac{Y_3}{Y_{Th} + Y_3} I_N = \frac{0.1\angle - 90^0}{0.156\angle - 39.8^0}(2.39\angle 30.1^0) = 1.53\angle - 20.1^0 \text{ A}$$

SP 7.12 Determine $i(t)$ in the circuit of **Fig. SP 7.12** by using superposition in the frequency domain.

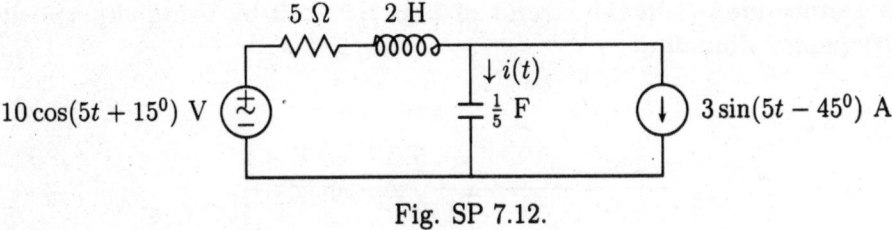

Fig. SP 7.12.

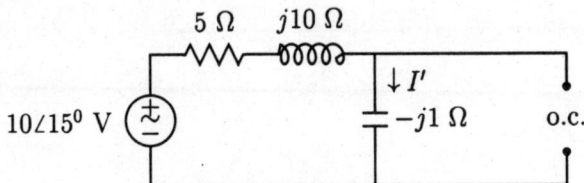

Fig. 7.12.1: Frequency domain circuit with left source alone.

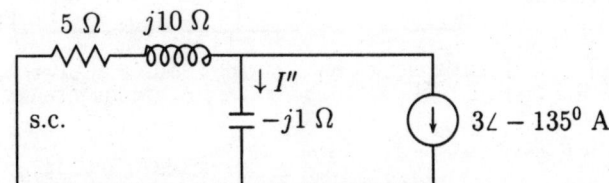

Fig. 7.12.2: Frequency domain circuit with right source alone.

SOLUTION: The Fig. 7.12.1 is the frequency-domain circuit with only the voltage source $10\angle 15^0$ is acting and the current source is killed by open circuit. The Fig. 7.12.2 is the frequency-domain circuit with only current source is acting and the voltage source is killed by short circuit.

From Fig. 7.12.1,

$$I' = \frac{10\angle 15^0}{5 + j10 - j1} = \frac{10\angle 15^0}{5 + j9} = \frac{10\angle 15^0}{10.3\angle 60.95^0} = 0.97\angle - 45.95^0$$

From Fig. 7.12.2, using current division method

$$I'' = -\frac{5+j10}{5+j10-j1}(3\angle-135^0) = -\frac{11.18\angle63.44^0}{10.3\angle60.95^0}(3\angle-135^0) = -3.26\angle-132.51^0$$

By superposition $I = I' + I''$

$$= 0.97\angle-45.95^0 - 3.26\angle-132.51^0 = 0.67-j0.7-(-2.2-j2.4) = 2.87+j1.7 = 3.34\angle30.64^0$$

In time domain

$$i(t) = 3.34\cos(5t + 30.64^0)\ \text{A}$$

SP 7.13 Determine $i(t)$ in the circuit of Fig. SP 7.13 by using superposition in the frequency domain.

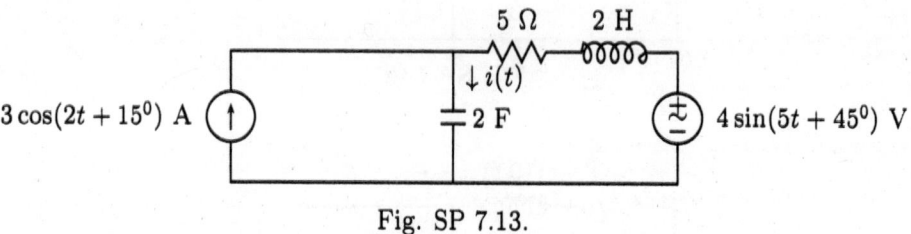

Fig. SP 7.13.

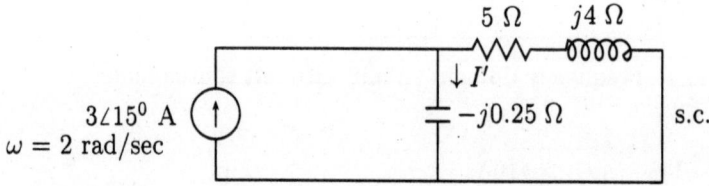

Fig. 7.13.1: Phasor circuit with current source at $\omega = 2$ rad./sec. alone.

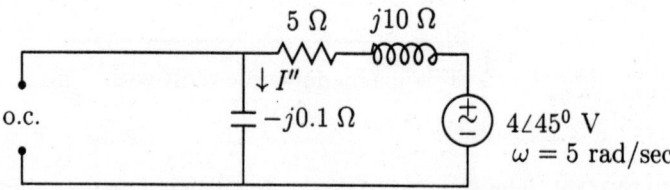

Fig. 7.13.2: Phasor circuit with voltage source at $\omega = 5$ rad./sec. alone.

SOLUTION: The superposition theorem is the only tool available for networks that contain sources with different frequencies; mesh and node analysis techniques can not be applied because they involve phasor and impedance representations, which are suitable only at a single frequency. Because the superposition theorem gives analysis of the network with each source acting alone at its specific operating frequency, then for each source, ac network analysis involving phasor and impedance representation can be performed.

The Fig. 7.13.1 shows the frequency-domain circuit with current source acting at $\omega = 2$ rad/sec alone while the voltage source is short circuited. And the Fig. 7.13.2 shows the frequency-domain circuit with voltage source acting at $w = 5$ rad/sec while current source is open circuited.

From Fig. 7.13.1, using current division method

$$I' = \frac{5+j4}{5+j4-j0.25}(3\angle15^0) = \frac{5+j4}{5+j3.75}(3\angle15^0) = \frac{6.4\angle38.66^0}{6.25\angle36.87^0}(3\angle15^0) = 3.1\angle16.79^0$$

In time domain

$$i'(t) = 3.1\cos(2t + 16.79^0) \text{ A}$$

From Fig. 7.13.2,

$$I'' = \frac{4\angle45^0}{5+j10-j0.1} = \frac{4\angle45^0}{5+j9.9} = \frac{4\angle45^0}{11.1\angle63.2^0} = 0.36\angle-18.2^0$$

In time domain

$$i''(t) = 0.36\sin(5t - 18.2^0) \text{ A}$$

Thus, by superposition

$$i(t) = i'(t) + i''(t) = 3.1\cos(2t + 16.79^0) + 0.36\sin(5t - 18.2^0) \text{ A}$$

Note : It is not necessary to transform sinusoidal sources into cosinusoidal sources when they have different frequencies

---

SP 7.14 Determine $i(t)$ in the circuit of Fig. SP 7.14.

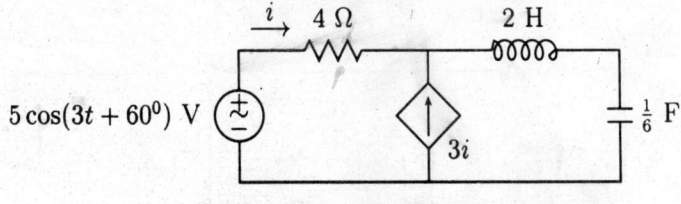

Fig. SP 7.14.

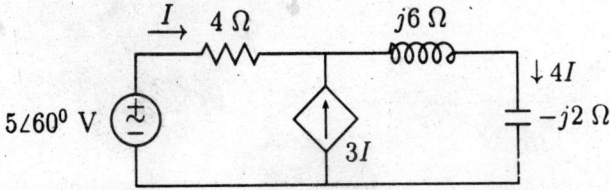

Fig. 7.14.1: Frequency-domain circuit.

SOLUTION: By KCL at the middle node, it is noted from Fig. 7.14.1 that the current in the inductor and capacitor branch is $4I$.

By KVL around the loop containing voltage source of $5\angle 60^0$ V, ihductive reactance of $j6$ Ω, and capacitive reactance of $-j2$ Ω in Fig. 7.14.1,

$$4I + (j6 - j2)(4I) = 5\angle 60^0$$

or

$$(4 + j16)I = 5\angle 60^0$$

Hence

$$I = \frac{5\angle 60^0}{4 + j16} = \frac{5\angle 60^0}{16.49\angle 75.96^0} = 0.3\angle - 15.96^0$$

In time domain

$$i(t) = 0.3\cos(3t - 15.96^0) \text{ A}$$

SP 7.15 Find the V_{th} and Z_{th} for the circuit shown in Fig. SP 7.15.

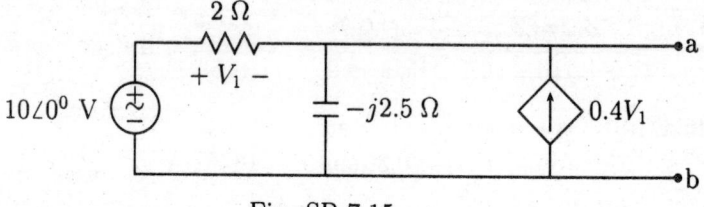

Fig. SP 7.15.

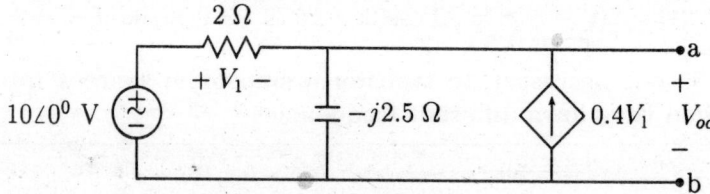

Fig. 7.15.1: Finding V_{oc}.

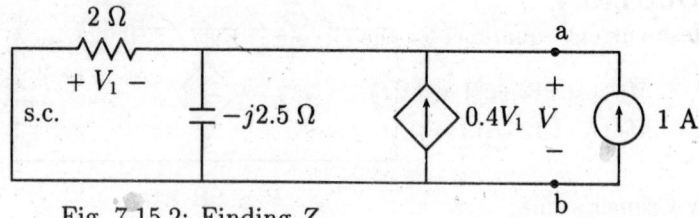

Fig. 7.15.2: Finding Z_{eq}.

SOLUTION:

From Fig. 7.15.1, by KCL at node V_{oc}

$$\frac{10\angle 0^0 - V_{oc}}{2} = \frac{V_{oc}}{-j2.5} - 0.4V_1$$

or

$$\frac{10\angle 0^0 - V_{oc}}{2} = \frac{V_{oc}}{-j2.5} - 0.4(10\angle 0^0 - V_{oc})$$

or $(0.9 + j0.4)V_{oc} = 9$

Hence $V_{th} = V_{oc} = \dfrac{9}{0.9 + j0.4} = \dfrac{9}{0.98\angle 23.96^0} = 9.18\angle -23.96^0$

From Fig. 7.15.2, by KCL at node **V**

$$\frac{V}{2} + \frac{V}{-j2.5} = 0.4V_1 + 1 = 0.4(-V) + 1$$

or $(0.9 + j0.4)V = 1$ $\cdots\cdots$

Hence $V = \dfrac{1}{0.9 + j0.4} = \dfrac{1}{0.98\angle 23.96^0} = 1.02\angle -23.96^0 = 0.93 - j0.41$

and $Z_{th} = Z_{eq} = \dfrac{V}{1} = 0.93 - j0.41 \ \Omega$

SP 7.16 Determine the impedance seen by the source $V_s = 24\angle 0^0$ in the network shown in Fig. SP 7.16.

GATE-98(EE)

Fig. SP 7.16.

SOLUTION:
Mesh-current equations for the circuit of Fig. SP 7.16 are :

M_1 : $(2 - j1 + j3)I_1 + j1 I_2 = 24\angle 0^0$ $\Rightarrow (2 + j2)I_1 + j1 I_2 = 24\angle 0^0$

M_2 : $(2 + j3 - j2)I_2 + j1 I_1 = 0$ $\Rightarrow j1 I_1 + (2 + j1)I_2 = 0$

By Cramer's rule

$$I_1 = \frac{\begin{vmatrix} 24\angle 0^0 & j1 \\ 0 & 2 + j1 \end{vmatrix}}{\begin{vmatrix} 2 + j2 & j1 \\ j1 & 2 + j1 \end{vmatrix}} = \frac{48 + j24}{3 + j6} = \frac{53.67\angle 26.57^0}{6.71\angle 63.43^0} = 8.0\angle -36.86^0$$

Input impedance

$$Z_{in} = \frac{V_s}{I_1} = \frac{24\angle 0^0}{8.0\angle -36.86^0} = 3.0\angle 36.86^0 = 2.4 + j1.8 \ \Omega$$

SP 7.17 Find $i(t)$ in the circuit of Fig. SP 7.17.

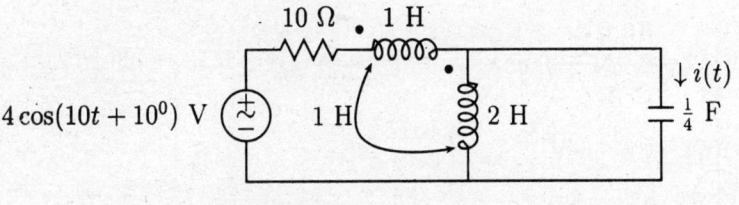

Fig. SP 7.17.

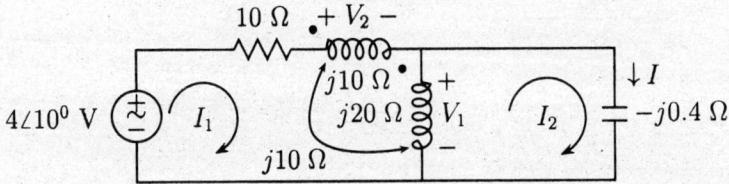

Fig. 7.17.1: Phasor circuit.

SOLUTION:

The voltage across the two inductors are :

$$V_1 = j20(I_1 - I_2) + j10I_1 = j30I_1 - j20I_2$$
$$V_2 = j10I_1 + j10(I_1 - I_2) = j20I_1 - j10I_2$$

Thus the mesh-current equations become

$$M_1: \qquad 10I_1 + V_2 + V_1 = 4\angle 10^0$$
$$M_2: \qquad -V_1 + (-j0.4)I_2 = 0$$

Setting for V_1 and V_2 and rearranging gives

$$M_1: \qquad (10 + j50)I_1 - j30I_2 = 4\angle 10^0$$
$$M_2: \qquad j30I_1 - j19.6I_2 = 0$$

Solving for I_2 by Cramer's rule

$$I_2 = \cfrac{\begin{vmatrix} 10 + j50 & 4\angle 10^0 \\ j30 & 0 \end{vmatrix}}{\begin{vmatrix} 10 + j50 & -j30 \\ j30 & -j19.6 \end{vmatrix}} = \frac{20.84 - j118.18}{80 - j196} = \frac{120\angle -80^0}{211.7\angle -67.8^0} = 0.57\angle -12.2^0$$

Thus

$$I = I_2 = 0.57\angle -12.2^0$$

In time domain

$$i(t) = 0.57\cos(10t - 12.2^0) \text{ A}$$

SP 7.18 Determine the voltage across capacitor $v_C(t)$ in the circuit of Fig. SP 7.18.

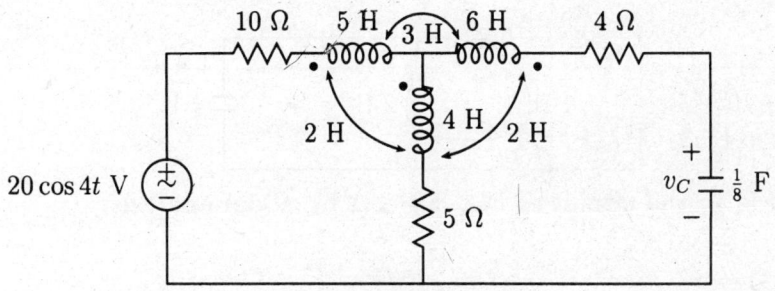

Fig. SP 7.18.

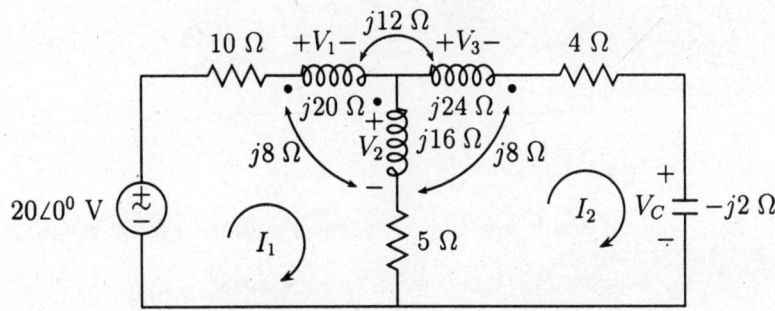

Fig. 7.18.1 : Phasor circuit.

SOLUTION:

The phasor voltage across the three inductors are :

$$V_1 = j20I_1 + j8(I_1 - I_2) - j12I_2 = j28I_1 - j20I_2$$
$$V_2 = j16(I_1 - I_2) + j8I_1 - j8I_2 = j24I_1 - j24I_2$$
$$V_3 = j24I_2 - j12I_1 - j8(I_1 - I_2) = -j20I_1 + j32I_2$$

Thus the mesh-current equations become

$$M_1 : \qquad 10I_1 + V_1 + V_2 + 5(I_1 - I_2) = 20\angle 0^0$$
$$M_2 : \qquad V_3 + 4I_2 - j2I_2 + 5(I_2 - I_1) - V_2 = 0$$

Setting for V_1, V_2 and V_3 and rearranging terms gives

$$M_1 : \qquad (15 + j52)I_1 - (5 + j44)I_2 = 20\angle 0^0$$
$$M_2 : \qquad -(5 + j44)I_1 + (9 + j54)I_2 = 0$$

Solving for I_2 by Cramer's rule

$$I_2 = \frac{\begin{vmatrix} 5 + j52 & 20\angle 0^0 \\ -(5 + j44) & 0 \end{vmatrix}}{\begin{vmatrix} 5 + j52 & -(5 + j44) \\ -(5 + j44) & 9 + j54 \end{vmatrix}} = \frac{100 + j880}{-852 + j298} = \frac{885.66\angle 83.52^0}{902.61\angle 160.72^0} = 0.98\angle -77.2^0$$

Thus $\qquad V_C = -j2I_2 = -j2(0.98\angle - 77.2^0) = 1.96\angle - 167.2^0$

In time domain

$$v_C(t) = 1.96\cos(4t - 167.2^0) \text{ V}$$

7.14 DRILL PROBLEMS

DP 7.1 Find V_0 in the circuit of Fig. DP 7.1 by nodal analysis.

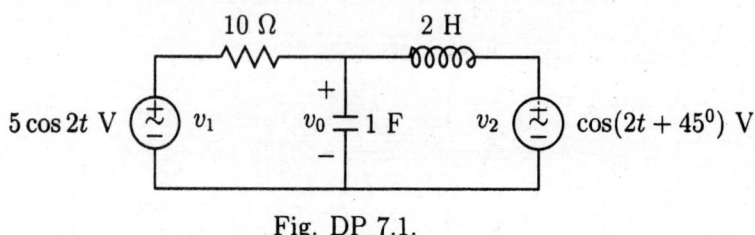

Fig. DP 7.1.

Hints : Only one required nodal equation in frequency domain to be solved is

$$\frac{5\angle 0^0 - V_0}{10} = \frac{V_0}{-j0.5} + \frac{V_0 - 1\angle 45^0}{j4}$$

DP 7.2 Use nodal analysis to determine V_1 and V_2 for the circuit in Fig. DP 7.2.

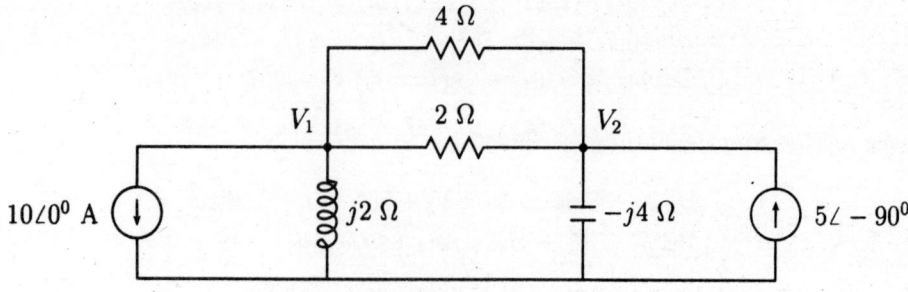

Fig. DP 7.2.

Hints : Nodal equations are :

$$\left(\frac{1}{j2} + \frac{1}{2} + \frac{1}{4}\right) V_1 - \left(\frac{1}{2} + \frac{1}{4}\right) V_2 = -10\angle 0^0$$

$$-\left(\frac{1}{2} + \frac{1}{4}\right) V_1 + \left(\frac{1}{-j4} + \frac{1}{2} + \frac{1}{4}\right) V_2 = 5\angle - 90^0$$

DP 7.3 Find the Norton equivalent of the network shown in Fig. DP 7.3.

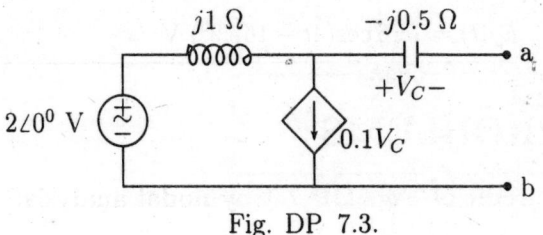

Fig. DP 7.3.

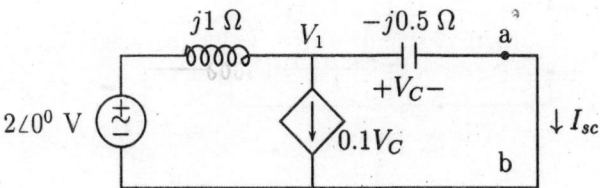

Fig. DP 7.3.1 : Finding I_{sc}.

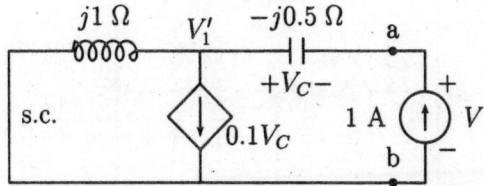

Fig. DP 7.3.2 : Finding Z_{eq}.

Hints : With reference to Fig. DP 7.3.1, find I_{sc} from the following equations :

$$\frac{2\angle 0^0 - V}{j1} = 0.1V_C + \frac{V - 0}{j0.5}$$

$$V = V_C$$

and
$$I_{sc} = \frac{V}{-j0.5}$$

With reference to Fig. DP 7.3.2, determine Z_{eq} from the following equations :

$$\frac{V'}{j1} + 0.1V_C = 1$$

$$V' - V = V_C$$

$$\frac{V' - V}{-j0.5} = -1$$

and
$$Z_{eq} = \frac{V}{1}$$

DP 7.4 Find the Thevenin's equivalent circuit from the network shown in Fig. DP 7.4.

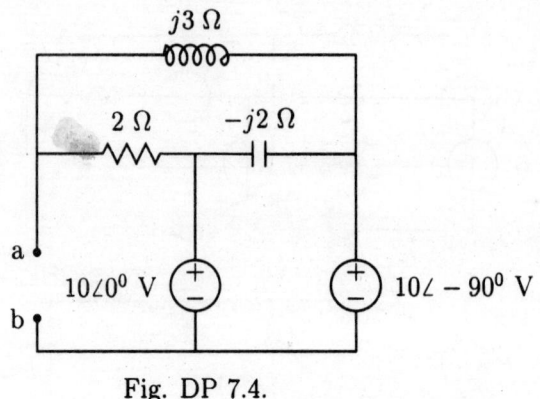

Fig. DP 7.4.

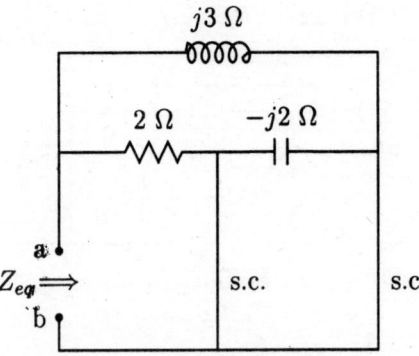

Fig. DP 7.4.1 : Finding Z_{eq}.

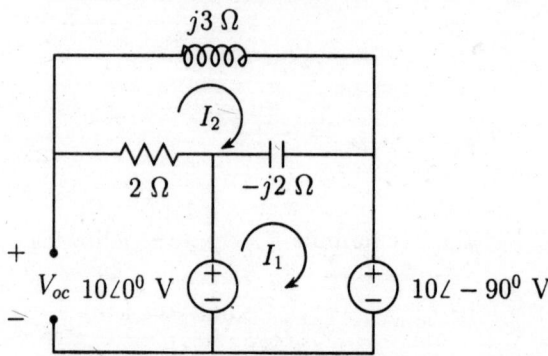

Fig. DP 7.4.2 : Finding V_{oc}.

Hints : From Fig. DP 7.4.1, $Z_{eq} = 2||(j3) = (j6)/(2 + j3)$

From Fig. 7.4.2, determine I_1 and I_2 and then V_{oc} by $V_{oc} = -2I_2 + 10\angle 0^0$.

DP 7.5 Use superposition to determine $v_C(t)$ **for the circuit shown in Fig. DP 7.5, if** $i_1 = 5\cos 2t\ A$, $i_2 = 2\cos t\ A$, **and** $v = 5\sin 2t\ V$

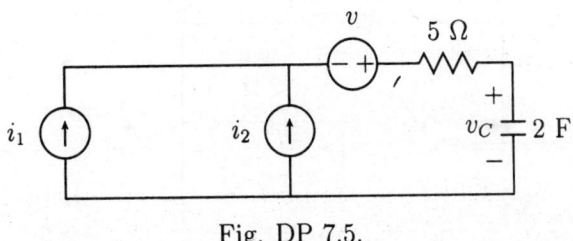

Fig. DP 7.5.

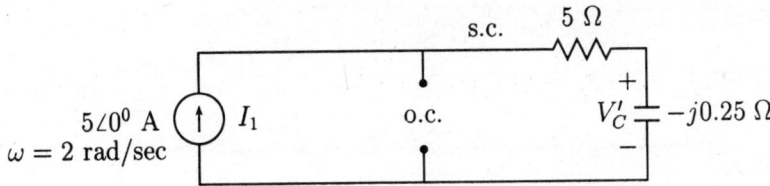

Fig. DP 7.5.1 : Phasor circuit with source I_1 alone.

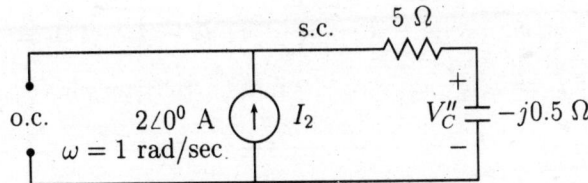

Fig. DP 7.5.2 : Phasor circuit with source I_2 alone.

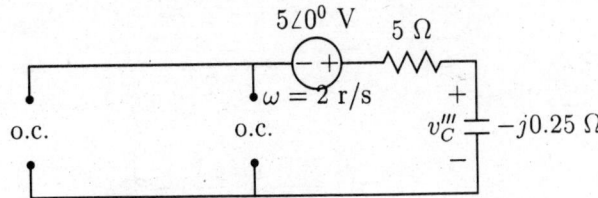

Fig. DP 7.5.3 : Phasor circuit with source V alone.

Hints : From Fig. DP 7.5.1,

$$V_C' = (-j0.25)(5\angle 0^0) = 1.25\angle -90^0$$

and $$v_C' = 1.25\cos(2t - 90^0)\ V$$

From Fig. DP 7.5.2,

$$V_C'' = (-j0.5)(2\angle 0^0) = 1\angle - 90^0$$

and $\qquad v_C'' = 1\cos(t - 90^0)\ V$

From Fig. 7.5.3, $V_C''' = 0 \qquad$ thus, by superposition

$$v_C = v_C' + v_C'' + v_C''' = 1.25\cos(2t - 90^0) + 1\cos(t - 90^0)$$

DP 7.6 Use mesh analysis to determine I for the circuit in Fig. DP 7.6.

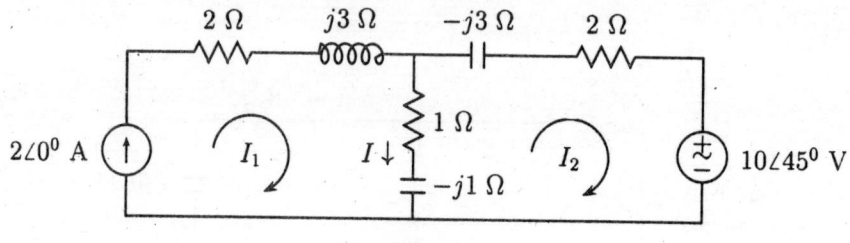

Fig. DP 7.6.

Hints : Mesh-equations are :

$$M_1 : \qquad I_1 = 2\angle 0^0$$
$$M_2 : \qquad [(2 - j3) + (1 - j1)]\,I_2 - (1 - j1)I_1 = -10\angle 45^0$$

Put the value of $I_1 = 2\angle 0^0$ into second equation to determine I_2 and so determine $I = I_1 - I_2$.

DP 7.7 Find the currents I_1 and I_2 by superposition for the circuit of Fig. DP 7.7.

Hints : From Fig. DP 7.7.1,

$$I_1' = -I_2' = \frac{5\angle 0^0}{2 + j3 + 2 - j3}$$

From Fig. DP 7.7.2,

$$I_1'' = -\frac{2 - j3}{2 + j3 + 2 - j3}(1\angle 0^0)$$
$$I_2'' = -\frac{2 + j3}{2 + j3 + 2 - j3}(1\angle 0^0)$$

From Fig. DP 7.7.3,

$$I_1''' = -I_2''' = \frac{5\angle - 90^0}{2 + j3 + 2 - j3}$$

Thus, by superposition $I_1 = I_1' + I_1'' + I_1'''$ and $I_2 = I_2' + I_2'' + I_2'''$

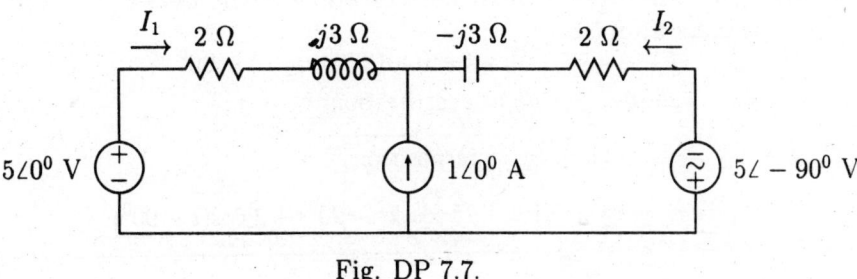

Fig. DP 7.7.

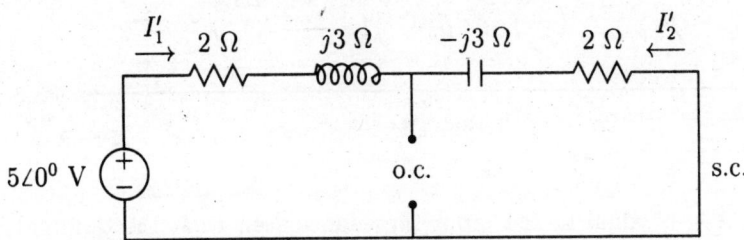

Fig. DP 7.7.1 : Circuit with $5\angle 0^0$V source alone.

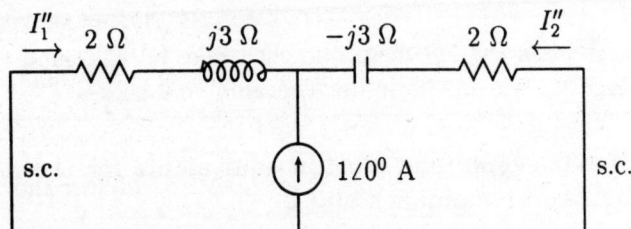

Fig. DP 7.7.2 : Circuit with $1\angle 0^0$ A source alone.

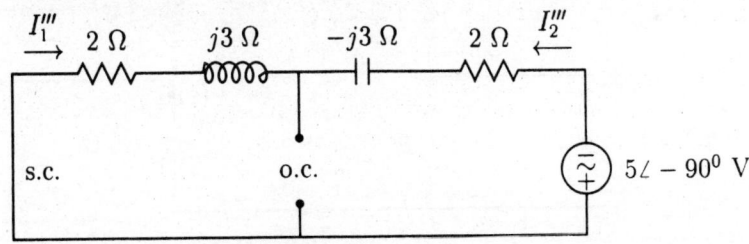

Fig. DP 7.7.3 : Circuit with $5\angle - 90^0$ V source alone.

DP 7.8 Find the Thevenin's equivalent circuit for the network in Fig. DP 7.8 between terminals a and b.

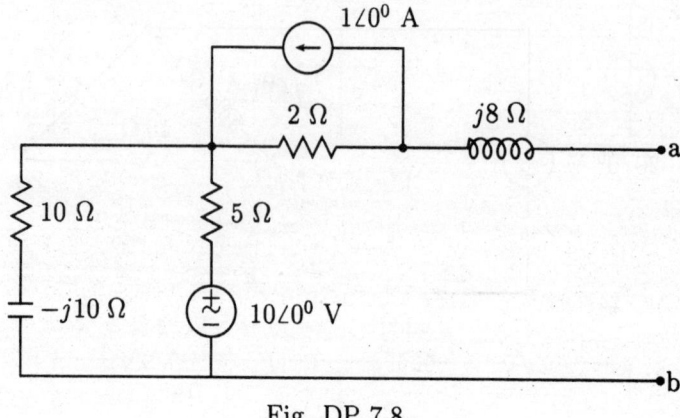

Fig. DP 7.8.

Hints: The Z_{Th} is equal to the input impedance seen from the terminals a and b when the current source is replaced by open circuit and voltage source by short circuit. Thus

$$Z_{Th} = Z_{in} = j8 + 2 + (5\|(10 - j10))$$

Determine the current in the left mesh and then take KVL around the loop containing open circuit voltage V_{ab} which will be the Thevenin voltage V_{Th}.

DP 7.9 Find the Thevenin and Norton equivalents for the network shown in Fig. DP 7.9, between terminals a and b.

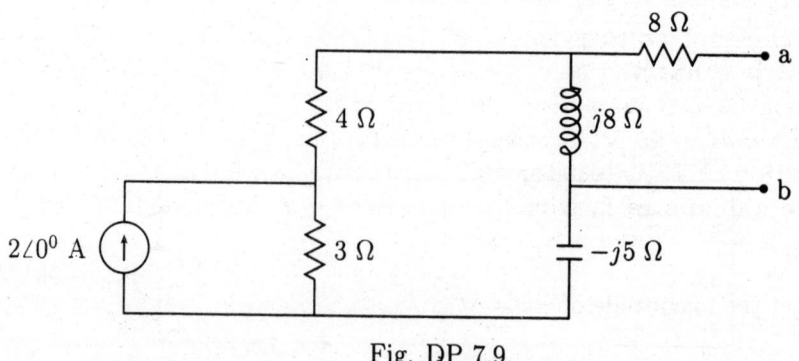

Fig. DP 7.9.

Hint: $V_{Th} = 8 + [j5\|(4 + 3 - j5)]$ and $V_{oc} = [3/(4 + 3)](2\angle 0^0)(j5)$

DP 7.10 Predict the current I in Fig. DP 7.10 in response to a voltage of $2\angle 0^0$ V. The impedance values are given in ohms. Use Thevenin's theorem.

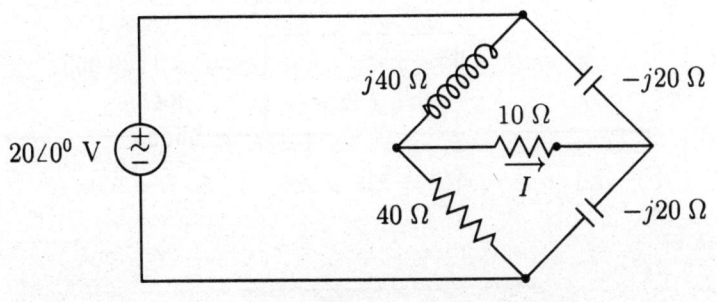

Fig. DP 7.10.

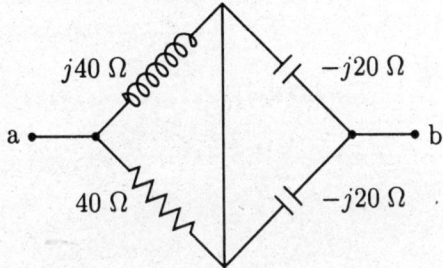

Fig. DP 7.10.1 Circuit for Z_{Th}.

GATE2000(EE)

Hints: The Z_{Th} can be determined from the Fig. DP 7.10.1 as:

$$Z_{Th} = (j40\|40) + ((-j20)\|(-j20))$$

DP 7.11 A constant voltage constant frequency sinusoidal voltage source of magnitude V_s is connected to a series circuit made of a resistor R and inductance L and a 50 μ F capacitor. The voltage across the 15 Ω resistor is 30 V, across the coil is 50 V, across the capacitor is 40 V. The voltage across the combination of 15 Ω resistor and coil together is 72.11 V. Determine the values of the inductance L, winding resistance R and source voltage V_s.

GATE-98 (EE)

Hints : Let the magnitude of current through the circuit be I. Then

$$15I = 30 \quad \Rightarrow I = 2 \text{ A}$$

$$I\sqrt{R^2 + \omega^2 L^2} = 50$$

or $$R^2 + \omega^2 L^2 = 625$$

$$\frac{1}{50 \times 10^{-6} \omega} I = 40 \qquad \Rightarrow \omega = 10^3 \text{ rad/sec.}$$

and $$I\sqrt{(R+15)^2 + \omega^2 L^2} = 72.11$$

or $$(R+15)^2 + \omega^2 L^2 = 1299.963$$

or $$R^2 + \omega^2 L^2 + 30R + 225 = 1299.963$$

or $$625 + 30R + 225 = 1299.963$$

Thus $$R = 15 \ \Omega$$

and $$\omega^2 L^2 = 625 - 225 \qquad \Rightarrow L = 0.02 \text{ H}$$

Then V_s can be

$$V_s = 2\sqrt{(15+15)^2 + \left(10^3 \times 0.02 - \frac{1}{10^3 \times 50 \times 10^{-6}}\right)^2}$$

****

Power in AC Circuits

SUMMARY OF THIS CHAPTER

8.1 INSTANTANEOUS POWER

The instantaneous power is given by $\quad p = vi \quad$ (8.1)

where v is the instantaneous voltage across a device and i is the instantaneous current through the device. The instantaneous power delivered to a device in the sinusoidal steady-state is

$$
\begin{aligned}
p(t) = vi &= (V_m \cos \omega t)\left[I_m \cos(\omega t + \phi)\right] \\
&= V_m I_m \cos(\omega t + \phi) \cos \omega t \\
&= \frac{V_m I_m}{2}\left[\cos(2\omega t + \phi) + \cos \phi\right] \\
&= \frac{V_m I_m}{2} \cos \phi + \frac{V_m I_m}{2} \cos(2\omega t + \phi)
\end{aligned}
$$

Thus the frequency of instantaneous power is twice the supply frequency.

8.2 AVERAGE POWER

The general formula for obtaining the average power is given by

$$
P_{av} = \frac{1}{T} \int_0^T p(t)dt \tag{8.2}
$$

where T is the time period of the periodic waveform, $p(t)$.

The average power in single phase ac circuits can be obtained as

$$
P_{av} = \frac{V_m I_m}{2} \cos \phi \tag{8.3}
$$

where ϕ is the phase angle between voltage and current and $\cos \phi$ is called power factor (pf).

This is the power absorbed by the resistive component of the load is also known as the **power** or **real power**. The product of the rms voltage value and rms current value $V_{rms}I_{rms}$ is called the **apparent power** and is measured in units of volt ampere.

The average power delivered to the resistor is

$$P_R = \frac{V_m I_m}{2} \cos 0 = \frac{1}{2} I_m^2 R = \frac{1}{2} \frac{V_m^2}{R}$$

The average power delivered to an inductor or a capacitor is zero, i.e.,

$$P_{av} = \frac{1}{2} V_m I_m \cos(\pm 90^0) = 0$$

If the current through a resistor of resistance R is

$$i(t) = I + I_{m1} \sin(\omega_1 t + \theta_1) + I_{m2} \sin(\omega_2 t + \theta_2) + \ldots\ldots + I_{mn} \sin(\omega_n t + \theta_n)$$

the average power will be

$$P_{av} = \frac{1}{2} \left[\left(\sqrt{2} I \right)^2 + I_{m1}^2 + I_{m2}^2 + \ldots\ldots + I_{mn}^2 \right] R$$

R. M. S. or Effective Values of Periodic Signals

The rms or effective value of a periodic signal, $x(t)$ is

$$X_{rms} = \sqrt{\frac{1}{T} \int_0^T x^2(t) dt} \tag{8.4}$$

where T is the time period of periodic signal $x(t)$.

The rms values of sinusoidal voltage and current are:

$$I_{rms} = \sqrt{\frac{1}{T} \int_0^T i^2(t) dt} = \frac{I_m}{\sqrt{2}} \tag{8.5}$$

$$V_{rms} = \sqrt{\frac{1}{T} \int_0^T v^2(t) dt} = \frac{V_m}{\sqrt{2}} \tag{8.6}$$

The ratio of the rms value of any periodic wave to its average value is called its **form factor**.

The average power in terms of rms values is

$$P_{av} = V_{rms} I_{rms} \cos \phi \tag{8.7}$$

where $V_{rms}I_{rms}$ is known to be **apparent power**.

The average power delivered to a resistor is

$$P_{av} = I_{rms}^2 R = \frac{V_{rms}^2}{R}$$

Power Factor

$$pf = \cos\phi = \frac{P_{av}}{V_{rms}I_{rms}}$$

When current lags the voltage, the power factor is considered lagging. When current leads the voltage, power factor is considered leading.

8.3 COMPLEX POWER

The complex power is

$$\mathbf{S} = P_{av} + jQ = \mathbf{V_{rms}I^*_{rms}} \tag{8.8}$$

where $\mathbf{S}$ is the complex power measured in volt-amperes, P_{av} is the average power measured in watts, Q is the reactive or quadrature power measured in volt-amperes reactive, and $\mathbf{I^*_{rms}}$ is the conjugate of $\mathbf{I_{rms}}$.

For any impedance $Z(j\omega) = R + jX$ $\qquad \mathbf{V_{rms}} = \mathbf{I_{rms}}Z$ Then

$$\mathbf{S} = \mathbf{I_{rms}}Z\mathbf{I^*_{rms}} = (\mathbf{I_{rms}I^*_{rms}})Z = I^2_{rms}Z$$

where bold letters stand for complex numbers.

Some other important formulae are :

$$P_{av} = \text{Re}\left[\mathbf{V_{rms}I^*_{rms}}\right] = \frac{1}{2}\text{Re}\left[\mathbf{V_mI^*_m}\right] = \frac{1}{2}\text{Re}\left[|\mathbf{I}_m|^2 Z\right] = \frac{1}{2}\text{Re}\left[|\mathbf{V}_m|^2 Y^*\right] \tag{8.9}$$

$$Q = \text{Im}\left[\mathbf{V_{rms}I^*_{rms}}\right] = \frac{1}{2}\text{Im}\left[\mathbf{V_mI^*_m}\right] = \frac{1}{2}\text{Im}\left[|\mathbf{I}_m|^2 Z\right] = \frac{1}{2}\text{Im}\left[|\mathbf{V}_m|^2 Y^*\right] \tag{8.10}$$

For a resistor R

$$\mathbf{S} = P_{av} + jQ = I^2_{rms}R$$

Hence $\qquad\qquad P_{av} = I^2_{rms}R \qquad \text{and} \qquad Q = 0$

Thus a resistor absorbs real power from the circuit.

For an inductor L

$$\mathbf{S} = P_{av} + jQ = I^2_{rms}(jX_L) = I^2_{rms}(j\omega L)$$

Hence $\qquad\qquad P_{av} = 0 \qquad \text{and} \qquad Q = I^2_{rms}X_L$

Thus an inductor absorbs reactive power from the circuit.

For a capacitor C

$$\mathbf{S} = P_{av} + jQ = I^2_{rms}\left(-j\frac{1}{\omega C}\right) = -jI^2_{rms}X_C$$

Hence $\qquad\qquad P_{av} = 0 \qquad \text{and} \qquad Q = -I^2_{rms}X_C$

Thus a capacitor generates reactive power in the circuit.

8.4   COMPLEX POWER CONSERVATION

From the conservation of energy, total real power supplied by the source is equal the sum of the real powers consumed by each load. At the same time, balance between the reactive power must be maintained. Thus the total complex power supplied by the source is the sum of the complex powers delivered to each load. With reference to Figure 8.1, we have

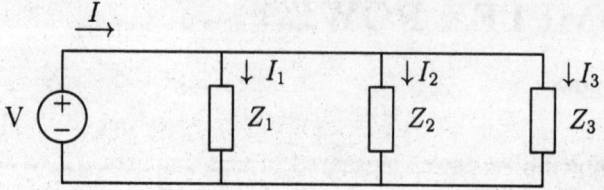

Figure 8.1: Showing three loads connected to a single voltage source

$$S = VI^* = VI_1{}^* + VI_2{}^* + VI_3{}^* \tag{8.11}$$

8.5   MAXIMUM POWER TRANSFER THEOREM

The maximum power is delivered from a source to a load when the load impedance is the conjugate of the source impedance. When the load impedance is the conjugate of the source impedance, the load and source are said to be matched.

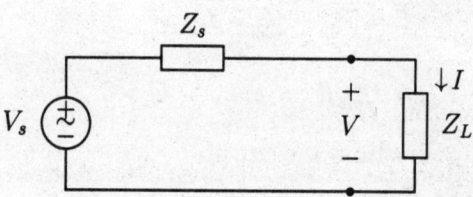

Figure 8.2: Practical phasor voltage source circuit model.

Let us consider a Thevenin's equivalent circuit as shown in Figure 8.2. This circuit is nothing but a practical voltage source circuit model with

V_s Phasor voltage source (peak)
$Z_s = R_s + jX_s$ Source impedance
$Z_L = R + jX$ Load impedance

The phasor current through the load is

$$I = \frac{V_s}{Z_s + Z_L} = \frac{V_s}{(R_s + R) + j(X_s + X)} \tag{8.12}$$

The power consumed (or average power) by the load is

$$P_{Load} = \frac{1}{2}|I|^2 R = \frac{1}{2}\frac{|V_s|^2 R}{(R_s + R)^2 + (X_s + X)^2} \tag{8.13}$$

For the maximum power transfer, following conditions must be satisfied:

$$\frac{\partial P_{Load}}{\partial X} = 0 \tag{8.14}$$

$$\frac{\partial P_{Load}}{\partial R} = 0 \tag{8.15}$$

From the eqn 8.14, we have

$$X_s + X = 0 \qquad \text{or} \qquad X = -X_s \tag{8.16}$$

From the eqn 8.15, we have

$$R = \sqrt{R_s^2 + (X_s + X)^2} \tag{8.17}$$

Setting the value for X, we have

$$R = R_s \tag{8.18}$$

Thus, for the maximum power transfer

$$Z_L = R + jX = R_s - jX_s = Z_s^* \tag{8.19}$$

and the maximum power transfer is

$$(P_{Load})_{max} = \frac{1}{2}\frac{|V_s|^2 R}{(R_s + R)^2 + (X_s + X)^2} = \frac{|V_s|^2}{8R_s} \tag{8.20}$$

From above explanation, we conclude that

1. When $Z_L = R + jX$ (both R and X are variables), the maximum power condition is

$$Z_L = R_s - jX_s = Z_s^*$$

2. When $Z_L = R + jX$ (only R is variable and X is fix), the maximum power transfer condition is

$$R = \sqrt{R_s^2 + (X_s + X)^2}$$

3. When $Z_L = R + jX$ (only X is variable and R is fix), the maximum power transfer condition is

$$X = -X_s$$

4. When $Z_L = R$ (only R is variable and X is zero), the maximum power transfer condition is

$$R = \sqrt{R_s^2 + X_s^2}$$

Superposition of Power in AC Circuits

The superposition is not applicable to instantaneous power, that is,

$$p \neq p' + p''$$

The superposition is also not applicable to average power if the sinusoidal sources have same frequencies, that is,

$$P_{av} \neq P'_{av} + P''_{av} \qquad \text{if} \qquad \omega' = \omega''$$

The superposition is applicable to average power if the sinusoidal sources have different frequencies, that is,

$$P_{av} = P'_{av} + P''_{av} \qquad \text{if} \qquad \omega' \neq \omega''$$

8.6   SOLVED PROBLEMS

SP 8.1 Find the average power absorbed by the impedance $Z_L = 6 - j8 \, \Omega$ due to current $I = 10\angle 45^0$.

SOLUTION: The average power is

$$P_{av} = \frac{1}{2}I_m^2 R = \frac{1}{2}10^2(6) = 300 \text{ W}$$

SP 8.2 Find the average power absorbed by the impedance $Z_L = 6 - j8 \, \Omega$ due to current $I = 3 + j4$ A.

SOLUTION: The magnitude of maximum current is

$$I_m = \sqrt{3^2 + 4^2} = 5 \text{ A} \qquad \text{and} \qquad P_{av} = \frac{1}{2}5^2(6) = 75 \text{ W}$$

SP 8.3 A current source i_s is connected through a $5 - \Omega$ resistor. Find the average power absorbed by resistor if i_s equals (a) $2\cos 2t$ A
   (b) $2\cos(2t + 45^0)$ A
   (c) $2\cos 2t - 3\cos 4t$ A
   (d) $2\cos 2t - 3\cos 2t$ A
   (e) $2\cos 2t - 3\cos(2t + 45^0)$ A
   (f) $2\cos 2t - 3\cos(2t + 45^0) + 4\sin t + 2$ A.

SOLUTION :

(a) $P_{av} = \frac{1}{2}I_m^2 R = \frac{1}{2}2^2(5) = 10$ W

(b) $P_{av} = \frac{1}{2}I_m^2 R = \frac{1}{2}2^2(5) = 10$ W

(c) $P_{av} = \frac{1}{2}(I_{m1}^2 + I_{m2}^2)R$

$$= \frac{1}{2}\left[2^2 + (-3)^2\right](5) = 32.5 \text{ W}$$

(d) Given $i_s = 2\cos 2t - 3\cos 2t = -\cos 2t$

thus $P_{av} = \frac{1}{2}(-1)^2(5) = 2.5$ W

(e) Given $i_s = 2\cos 2t - 3\cos(2t + 45^0)$

$$= 2\cos 2t - 3\cos 2t \cos 45^0 + 3\sin 2t \sin 45^0$$

$$= -0.12\cos 2t + 2.12\sin 2t = A\sin(2t + \phi)$$

where $A = \sqrt{(-0.12)^2 + (2.12)^2}$

thus $P_{av} = \frac{1}{2}\left[(-0.12)^2 + 2.12^2\right](5) = 11.27$ W

(f) Given $i_s = 2\cos 2t - 3\cos(2t + 45^0) + 4\sin t + 2$

$$= -0.12\cos 2t + 2.12\sin 2t + 4\sin t + 2$$

thus $P_{av} = \frac{1}{2}\left[(-0.12)^2 + (2.12)^2 + 4^2 + (2\sqrt{2})^2\right](5) = 71.27$ W

SP 8.4 Determine S, P_{av}, and Q delivered to a series combination of a 15 Ω resistor and a 1-H inductor if the voltage across the combination is $v = 5\sqrt{2}\cos(2t + 45^0)$.

SOLUTION

$$V_{rms} = 5\angle 45^0 \qquad Z = 15 + j2$$

then

$$I_{rms} = \frac{V_{rms}}{Z} = \frac{5\angle 45^0}{15 + j2} = \frac{5\angle 45^0}{15.13\angle 7.6^0} = 0.33\angle 37.4^0$$

$$S = V_{rms}I_{rms}^* = (5\angle 45^0)(0.33\angle -37.4^0) = 1.65\angle 7.6^0 = 1.64 + j0.22$$

thus $\quad P_{av} = \text{Re}[S] = 1.64$ W $\quad$ and $\quad Q = \text{Im}[S] = 0.22$ VAR

SP 8.5 In the circuit shown in **Figure 8.1**, $V = 230\angle 0^0$ V, $Z_1 = 10 + j0$ Ω, $Z_2 = 3 + j4$ Ω, and $Z_3 = 10 - j10$ Ω. Find the power consumed by each load and the total complex power.

SOLUTION: Individual currents are:

$$I_1 = \frac{230\angle 0^0}{10 + j0} = 23 + j0 \text{ A}$$

$$I_2 = \frac{230\angle 0^0}{3 + j4} = 27.6 - j36.8 \text{ A}$$

$$I_3 = \frac{230\angle 0^0}{10 - j10} = 11.5 - j11.5 \text{ A}$$

Individual complex powers are:

$$VI_1{}^* = 230\angle 0^0(23 - j0) = 5290 \text{ W} - j0 \text{ VAR}$$

$$VI_2{}^* = 230\angle 0^0(27.6 + j36.8) = 6348 \text{ W} + j8464 \text{ VAR}$$

$$VI_3{}^* = 230\angle 0^0(11.5 + j11.5) = 2645 \text{ W} - j2645 \text{ VAR}$$

The total complex power is

$$\begin{aligned}
S &= S_1 + S_2 + S_3 \\
&= 5290 \text{ W} - j0 \text{ VAR} + 6348 \text{ W} + j8464 \text{ VAR} + 2645 \text{ W} - j2645 \text{ VAR} \\
&= 14283 \text{ W} + j5819 \text{ VAR}
\end{aligned}$$

The complex powers may also be determined from the equation

$$S_1 = \frac{|V|^2}{Z_1^*} = 5290 \text{ W} - j0 \text{ VAR}$$

$$S_2 = \frac{|V|^2}{Z_2^*} = +6348 \text{ W} + j8464 \text{ VAR}$$

$$S_2 = \frac{|V|^2}{Z_3^*} = 2645 \text{ W} - j2645 \text{ VAR}$$

SP 8.6 For the circuit shown in Fig. SP 8.6, find the complex power absorbed by each element in the circuit.

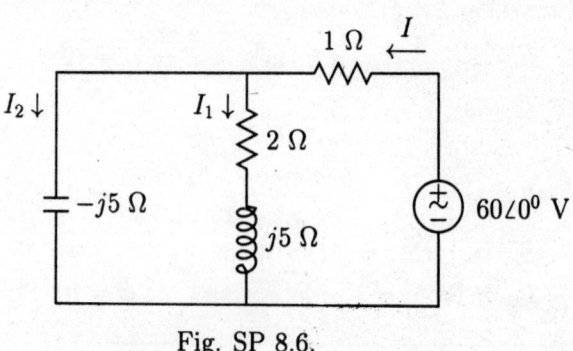

Fig. SP 8.6.

SOLUTION: Total impedance seen by the voltage source

$$\dot{Z}_{in} = 1 + (-j5)\|(2 + j5) = 1 + \frac{(-j5)(2 + j5)}{-j5 + 2 + j5} = 13.5 - j5 = 14.4\angle - 20.32^0 \text{ } \Omega$$

Source current

$$I = \frac{60\angle 0^0}{14.4\angle - 20.32^0} = 4.17\angle 20.32^0 \text{ A}$$

Current through $2 + j5$ Ω branch

$$I_1 = \frac{-j5}{-j5 + 2 + j5}(4.17\angle 20.32^0) = 10.43\angle - 69.68^0 \text{ A}$$

Current through $-j5$ Ω branch

$$I_2 = \frac{2 + j5}{-j5 + 2 + j5}(4.17\angle 20.32^0) = 11.24\angle 88.52^0 \text{ A}$$

Complex power absorbed the elements

$$S_{1\Omega} = \frac{1}{2}I^2 R = \frac{1}{2}(4.17)^2(1) = 8.7 + j0 \text{ VA}$$

$$S_{2\Omega} = \frac{1}{2}(10.43)^2(2) = 108.78 + j0 \text{ VA}$$

$$S_{(j5)\Omega} = \frac{1}{2}(10.43)^2(j5) = 0 + j271.2 \text{ VA}$$

$$S_{(-j5)\Omega} = \frac{1}{2}(11.24)^2(-j5) = 0 - j315.84 \text{ VA}$$

$$S_{60\angle 0^0 V} = -\frac{1}{2}(60\angle 0^0)(4.17\angle - 20.32^0) = -117.31 + j43.44 \text{ VA}$$

SP 8.7 Determine the average power generated by each source and average power delivered to each impedance in the circuit of Fig. SP 8.7.

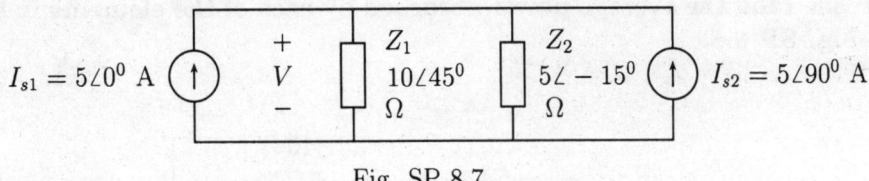

$I_{s1} = 5\angle 0^0 \text{ A}$ $+$ V $-$ Z_1 $10\angle 45^0$ Ω Z_2 $5\angle - 15^0$ Ω $I_{s2} = 5\angle 90^0 \text{ A}$

Fig. SP 8.7

SOLUTION : Given

$$Z_1 = 10\angle 45^0 \ \Omega \quad Y_1 = 0.1\angle - 45^0 = 0.07 - j0.07 \text{ S}$$
$$Z_2 = 5\angle - 15^0 \ \Omega \quad Y_2 = 0.2\angle 15^0 = 0.19 + j0.05 \text{ S}$$
$$Y = Y_1 + Y_2 = 0.26 - j0.02 = 0.26\angle - 4.4^0 \text{ S}$$

Voltage across parallel combination

$$V = \frac{5\angle 0^0 + 5\angle 90^0}{Y} = \frac{5 + j5}{0.26\angle - 4.4^0} = \frac{7.07\angle 45^0}{0.26\angle - 4.4^0} = 27.19\angle 49.4^0 \text{ V}$$

Average power generated by sources

$$(P_{av})_{5\,A} = \frac{1}{2}\text{Re}[VI_{s1}^*]$$

$$= \frac{1}{2}\text{Re}\left[(27.19\angle 49.4^0)(5\angle 0^0)\right]$$

$$= \frac{1}{2}\text{Re}\,[88.47 + j103.22] = 44.24 \text{ W}$$

$$(P_{av})_{j5\,A} = \frac{1}{2}\text{Re}[VI_{s2}^*]$$

$$= \frac{1}{2}\text{Re}\left[(27.19\angle 49.4^0)(5\angle - 90^0)\right]$$

$$= \frac{1}{2}\text{Re}\,[103.22 - j88.47] = 51.61 \text{ W}$$

Average power delivered to the impedances

$$(P_{av})_{Z_1} = \frac{1}{2}\text{Re}\left[|V|^2 Y_1^*\right]$$

$$= \frac{1}{2}\text{Re}\left[(27.19)^2(0.1\angle 45^0)\right]$$

$$= \frac{1}{2}\text{Re}\,[52.28 + j52.28] = 26.14 \text{ W}$$

$$(P_{av})_{Z_2} = \frac{1}{2}\text{Re}\left[|V|^2 Y_2^*\right]$$

$$= \frac{1}{2}\text{Re}\left[(27.19)^2(0.2\angle - 15^0)\right]$$

$$= \frac{1}{2}\text{Re}\,[142.82 - j38.27] = 71.41 \text{ W}$$

SP 8.8 Find the average power absorbed by each of the elements in the circuit of Fig. SP 8.8.

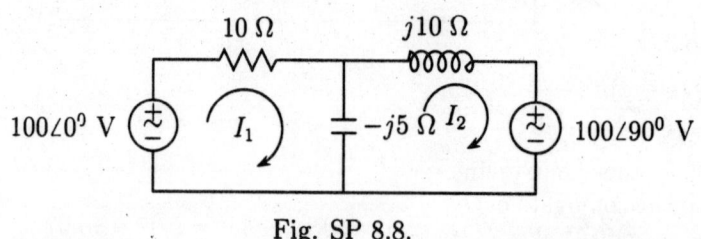

Fig. SP 8.8.

SOLUTION :
Mesh-current equations are :

$$(10 - j5)I_1 + j5I_2 = 100$$
$$j5I_1 + j5I_2 = -100\angle 90^0 = -j100$$

Solving for I_1 and I_2 by Cramer's rule

$$I_1 = \frac{\begin{vmatrix} 100 & j5 \\ -j100 & j5 \end{vmatrix}}{\begin{vmatrix} 10-j5 & j5 \\ j5 & j5 \end{vmatrix}} = 10\angle 90^0 \quad \text{and} \quad I_2 = \frac{\begin{vmatrix} 10-j5 & 100 \\ j5 & -j100 \end{vmatrix}}{\begin{vmatrix} 10-j5 & j5 \\ j5 & j5 \end{vmatrix}} = 22.36\angle -153.4^0$$

Average power absorbed by the elements are :

$$(P_{av})_{100V} = -\frac{1}{2}\text{Re}\,[V_1 I_1^*]$$

$$= -\frac{1}{2}\text{Re}\,\left[(100)(10\angle -90^0)\right] = -\frac{1}{2}\text{Re}\,[-j1000] = 0$$

$$(P_{av})_{10\Omega} = \frac{1}{2}\text{Re}\,\left[|I_1|^2(10)\right] = \frac{1}{2}(10^2)(10) = 500 \text{ W}$$

$$(P_{av})_{j10\Omega} = \frac{1}{2}\text{Re}\,\left[|I_2|^2(j10)\right] = 0$$

$$(P_{av})_{100\angle 90^0 V} = \frac{1}{2}\text{Re}\,[V_2 I_2^*]$$

$$= \frac{1}{2}\text{Re}\,\left[(100\angle 90^0)(22.36\angle 153.4^0)\right]$$

$$= \frac{1}{2}\text{Re}\,\left[2236\angle 243.4^0\right]$$

$$= \frac{1}{2}\text{Re}\,[-1000.1 - j1999.3] = -500 \text{ W}$$

$$(P_{av})_{(-j5)\Omega} = \frac{1}{2}\text{Re}\,\left[|I_1 - I_2|^2(-j5)\right] = 0$$

SP 8.9 With reference to Fig. SP 8.9, find the value of Z_L if it absorbs maximum average power. And what is the value of this maximum power.

SOLUTION:
From Fig. 7-26.1, using voltage division method

$$V_{oc} = \frac{-j5}{5+j10-j5}(100\angle 0^0) = \frac{500\angle -90^0}{7.07\angle 45^0} = 70.7\angle -135^0$$

From Fig. 7-26.2,

$$Z_{eq} = \frac{(5+j10)(-j5)}{5+j10-j5} = \frac{55.9\angle -26.57^0}{7.07\angle 45^0} = 7.91\angle -71.57^0 = 2.5 - j7.5 \ \Omega$$

From Fig. 7-26.3, under maximum average power transfer condition the load impedance will be equal to the conjugate of the source impedance, that is $Z_L = 2.5 + j7.5 \ \Omega$ and the load current is

$$I = \frac{70.7\angle -135^0}{2.5 + j7.5 + 2.5 - j7.5} = 14.14\angle -135^0 \text{ A}$$

Average power delivered to the load under matched condition

$$P_{av} = \frac{1}{2}|I|^2(2.5) = \frac{1}{2}(14.14)^2(2.5) = 249.9 \text{ W}$$

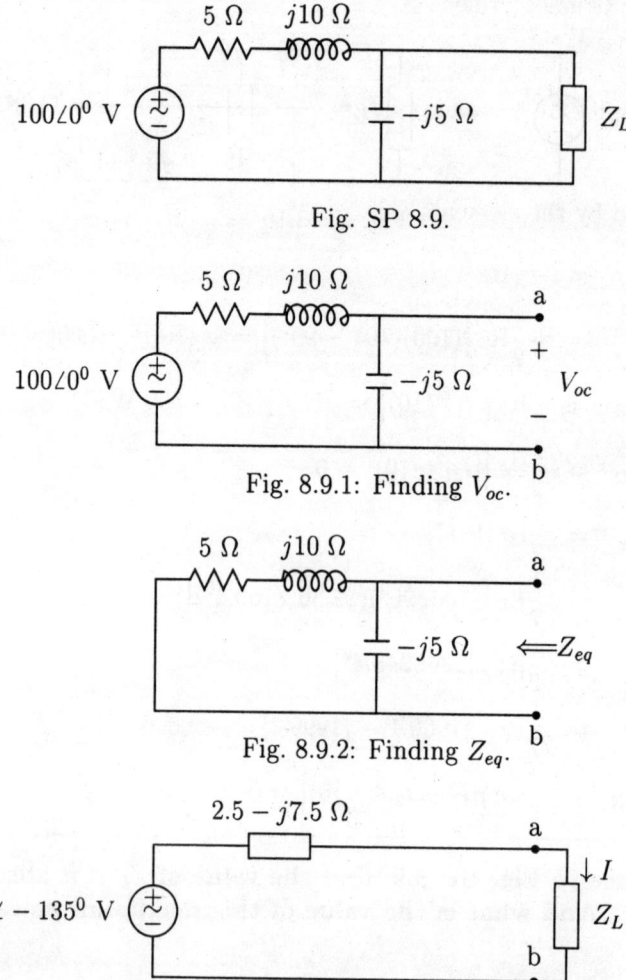

Fig. SP 8.9.

Fig. 8.9.1: Finding V_{oc}.

Fig. 8.9.2: Finding Z_{eq}.

Fig. 8.9.3: The Thevenin equivalent circuit.

SP 8.10 In the circuit of Fig. SP 8.10, $Z_1 = 10 + j0\ \Omega$, $Z_2 = 3 + j4\ \Omega$ and $V_s = 230$ V (rms) with 50 Hz.

(a) Find the total real power, total reactive power, total current and power factor at the sending end.

(b) Find the capacitance of the capacitor connected across the load to improve the power factor to unity power factor. This method of improving pf is known as the power factor correction.

SOLUTION: (a) From Fig. SP 8.10, the individual complex powers are:

$$S_1 = \frac{|V|^2}{Z_1^*} = 5290\ \text{W} - j0\ \text{VAR}$$

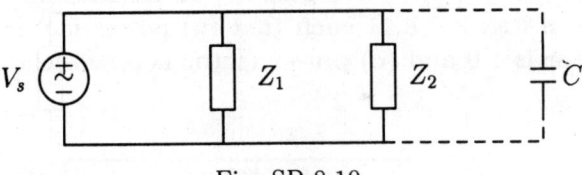

Fig. SP 8.10

$$S_2 = \frac{|V|^2}{Z_2^*} = +6348 \text{ W} + j8464 \text{ VAR}$$

The total complex power is

$$S = S_1 + S_2 = 11638 \text{ W} + j8464 \text{ VAR}$$

The total current will be

$$I = \frac{S^*}{V^*} = \frac{11638 - j8464}{230 - j0} = 50.6 - j36.8 = 62.57\angle - 36.03^0 \text{ A}$$

Thus, power factor at the sending end is

$$pf = \cos(36.03^0) = 0.8087 \text{ lagging}$$

(b) When a capacitor is connected across the load as shown in Fig. SP 8.10, the inductive impedance is

$$Z_C = \frac{1}{j2\pi f C} = -j0.0032/C$$

Thus, the complex power delivered to the capacitor is

$$S_C = \frac{|V|^2}{Z_C^*} = \frac{(230)^2}{j0.0032/C} = -j1.6531 \times 10^7 C$$

If the power factor is improved to unity power factor, the total real power is 11638 W at pf $= 1$. We can determine the total new complex power as

$$S_{new} = \frac{P}{pf} \angle \cos^{-1}(pf) = 11638\angle 0^0$$

From the complex power balance, we have

$$S_{new} = S_1 + S_2 + S_C$$

or

$$11638 + j0 = 5290 + j0 + 6348 + j8464 - j1.6531 \times 10^7 C$$

Thus

$$C = 0.5120 \ \mu F$$

SP 8.11 The average power delivered to an induction motor is 2 KW at power factor 0.707 (lagging) by a 200-V rms ac, 50 Hz source. Determine the value of capacitance in Fig. SP 8.11 such that (a) power factor is 0.866 lagging (b) the power factor is 1.0 and (c) power factor is 0.866 leading.

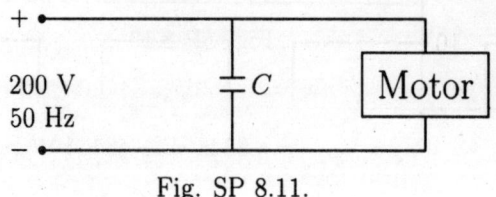

Fig. SP 8.11.

SOLUTION:

(a) The complex power drawn by motor is

$$S_1 = \frac{2000}{0.707}\angle\cos^{-1}(0.707) = 2828.85\angle45^0 = 2000 + j2000 \text{ VA}$$

In order to achieve a power factor of 0.866 lagging, the total complex power will be

$$S_T = \frac{2000}{0.866}\angle\cos^{-1}(0.866) = 2309.47\angle30^0 = 2000 + j1154.74 \text{ VA}$$

Thus, the complex power drawn by the capacitance is $S_C = -j845.26$ VA and

$$S_C = -j845.26 = |V|^2 Y_C^* = (200)^2[-j2\pi(50)C] \quad \Rightarrow C = \frac{845.26}{(200)^2(2\pi)(50)} = 67.3 \ \mu F$$

(b) In order to achieve upf, total complex power will be

$$S_T = \frac{2000}{1}\angle\cos^{-1}(1) = 2000 + j0 \text{ VA}$$

Thus the complex power drawn by capacitor is $S_C = -j2000$ VA and

$$S_C = -j2000 = |V|^2 Y_C^* = (200)^2[-j2\pi(50)C] \quad \Rightarrow C = \frac{2000}{(200)^2(2\pi)(50)} = 159.2 \ \mu F$$

(c) To get power factor 0.866 (leading), total complex power will be

$$S_T = \frac{2000}{0.866}\angle-\cos^{-1}(0.866) = 2309.47\angle-30^0 = 2000 - j1154.74 \text{ VA}$$

Thus, the capacitive plant will have complex power of $-j3154.74$ VA and

$$S_C = -j3154.74 = |V|^2 Y_C^* = (200)^2[-j2\pi(50)C] \quad \Rightarrow C = \frac{3154.74}{(200)^2(2\pi)(50)} = 251 \ \mu F$$

SP 8.12 Determine the average power delivered to a 10-Ω resistor by each of the following waveforms shown in Fig. SP 8.12.

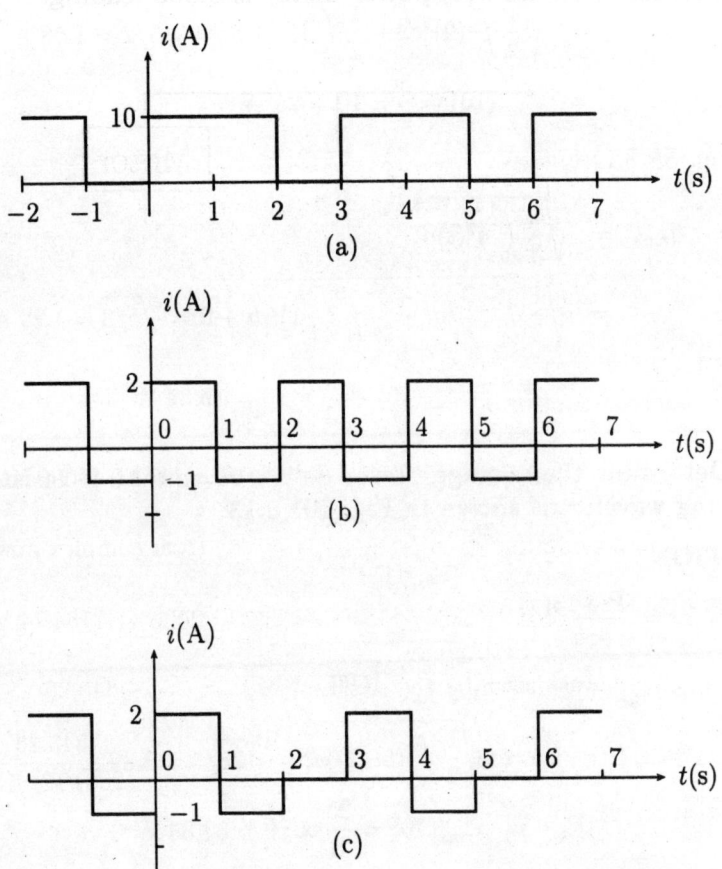

Fig. SP 8.12(a-c)

SOLUTION :

(a) From Fig. SP 8.12(a)

$$I_{rms} = \sqrt{\frac{1}{3}\int_0^3 i^2(t)dt}$$

$$= \sqrt{\frac{1}{3}\int_0^2 (10)^2 dt} = \sqrt{200/3} = 8.16 \text{ A}$$

$$P_{av} = I_{rms}^2(10) = \frac{200}{3} \times 10 = 666.67 \text{ W}$$

(b) From Fig. SP 8.12(b)

$$I_{rms} = \sqrt{\frac{1}{2}\int_0^2 i^2(t)dt}$$

$$= \sqrt{\frac{1}{2}\int_0^1 (2)^2 dt + \frac{1}{2}\int_1^2 (-1)^2 dt} = \sqrt{5/2} = 1.58 \text{ A}$$

$$P_{av} = I_{rms}^2(10) = \frac{5}{2} \times 10 = 25 \text{ W}$$

(c) From Fig. SP 8.12(c)

$$I_{rms} = \sqrt{\frac{1}{3}\int_0^3 i^2(t)dt}$$

$$= \sqrt{\frac{1}{3}\int_0^1 (2)^2 dt + \frac{1}{3}\int_1^2 (-1)^2 dt + 0} = \sqrt{5/3} = 1.29 \text{ A}$$

$$P_{av} = I_{rms}^2(10) = \frac{5}{3} \times 10 = 16.67 \text{ W}$$

SP 8.13 Determine the average power delivered to a 10-Ω resistor by each of the following waveforms shown in Fig. SP 8.13.

SOLUTION

(a) From Fig. SP 8.13(a)

$$I_{rms} = \sqrt{\frac{1}{2}\int_0^2 i^2(t)dt}$$

$$= \sqrt{\frac{1}{2}\int_0^2 (5t/2)^2 dt} = \sqrt{25/3} = 2.89 \text{ A}$$

$$P_{av} = I_{rms}^2(10) = \frac{25}{3} \times 10 = 83.33 \text{ W}$$

(b) From Fig. SP 8.13(b)

$$I_{rms} = \sqrt{\frac{1}{2}\int_0^2 i^2(t)dt}$$

$$= \sqrt{\frac{1}{2}\int_0^1 t^2 dt + \frac{1}{2}\int_1^2 [-(t-2)]^2 dt} = \sqrt{1/3} = 0.58 \text{ A}$$

$$P_{av} = I_{rms}^2(10) = \frac{1}{3} \times 10 = 3.33 \text{ W}$$

(c) From Fig. SP 8.13(c)

$$I_{rms} = \sqrt{\frac{1}{2}\int_0^2 i^2(t)dt}$$

$$= \sqrt{\frac{1}{2}\int_0^1 t^2 dt + \frac{1}{2}\int_1^2 (1)^2 dt} = \sqrt{2/3} = 0.82 \text{ A}$$

$$P_{av} = I_{rms}^2(10) = \frac{2}{3} \times 10 = 20/3 = 6.67 \text{ W}$$

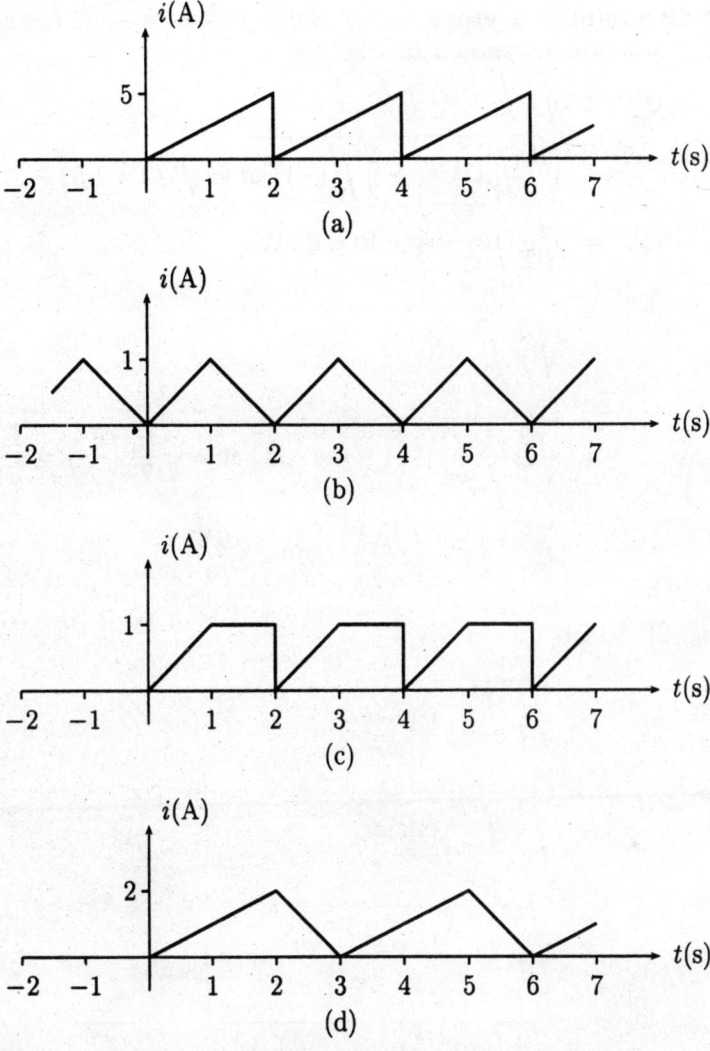

Fig. SP 8.13(a-d)

(d) From Fig. SP 8.13(d)

$$I_{rms} = \sqrt{\frac{1}{3} \int_0^3 i^2 dt}$$

$$= \sqrt{\frac{1}{3} \int_0^2 t^2 dt + \frac{1}{3} \int_2^3 [-2(t-3)]^2 dt} = \sqrt{4/3} = 1.15 \text{ A}$$

$$P_{av} = I_{rms}^2 (10) = \frac{4}{3} \times 10 = 40/3 = 13.33 \text{ W}$$

SP 8.14 Determine the average power delivered to a 10-Ω resistor by each of the following waveforms shown in Fig. SP 8.14.

SOLUTION

(a) From Fig. SP 8.14(a)

$$I_{rms} = \sqrt{\frac{1}{T}\int_{-T/2}^{T/2} i^2(t)dt}$$

$$= \sqrt{\frac{2}{T}\int_0^{T/2} I_m^2 \cos^2 \frac{\pi}{T}tdt}$$

$$= \sqrt{\frac{2}{T}\frac{I_m^2}{2}\int_0^{T/2}\left(1+\cos\frac{2\pi}{T}t\right)dt} = \sqrt{\frac{I_m^2}{T}\left(\frac{T}{2}\right)} = \frac{I_m}{\sqrt{2}}$$

$$P_{av} = I_{rms}^2(10) = \left(\frac{I_m}{\sqrt{2}}\right)^2(10) = 5I_m^2$$

(b) From Fig. SP 8.14(b)

$$I_{rms} = \sqrt{\frac{1}{T}\int_{-T/2}^{T/2} i^2(t)dt}$$

$$= \sqrt{\frac{1}{T}\int_{-T/4}^{T/4} i^2(t)dt}$$

$$= \sqrt{\frac{2}{T}\int_0^{T/4} I_m^2 \cos^2 \frac{2\pi}{T}tdt}$$

$$= \sqrt{\frac{2}{T}\frac{I_m^2}{2}\int_0^{T/4}\left(1+\cos\frac{4\pi}{T}t\right)dt} = \sqrt{\frac{I_m^2}{T}\left(\frac{T}{4}\right)} = \frac{I_m}{2}$$

$$P_{av} = I_{rms}^2(10) = \left(\frac{I_m}{2}\right)^2(10) = 2.5I_m^2$$

(c) From Fig. SP 8.14(c)

$$I_{rms} = \sqrt{\frac{1}{1}\int_0^1 i^2(t)dt}$$

$$= \sqrt{\int_0^1 \sin^2 \pi tdt}$$

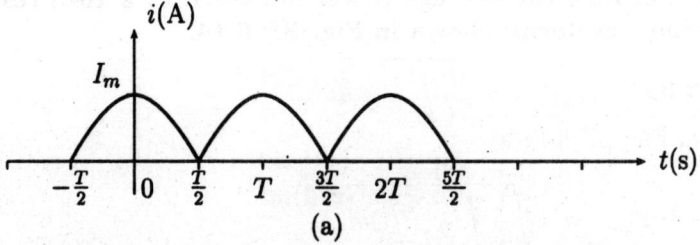

(a)

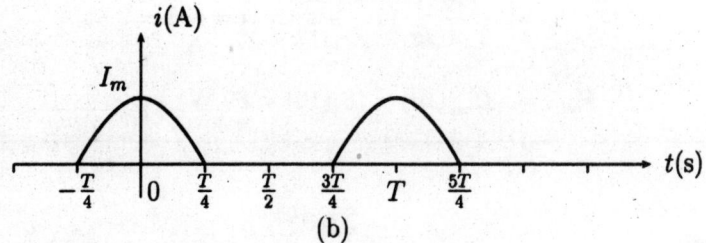

(b)

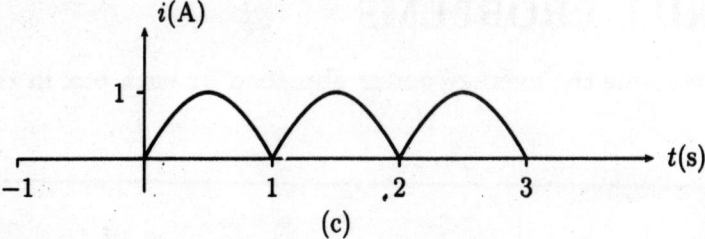

(c)

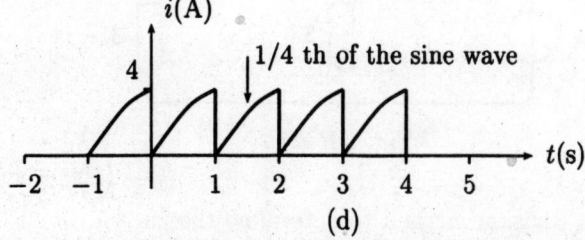

(d)

Fig. SP 8.14(a-d).

$$= \sqrt{\frac{1}{2} \int_0^1 (1 - \cos 2\pi t)\, dt} = \frac{1}{\sqrt{2}} \text{ A}$$

$$P_{av} = I_{rms}^2 (10) = \left(\frac{1}{\sqrt{2}}\right)^2 (10) = 5 \text{ W}$$

(d) From Fig. SP 8.14(d)

$$I_{rms} = \sqrt{\frac{1}{1}\int_0^1 i^2(t)dt}$$

$$= \sqrt{\int_0^1 4^2 \sin^2(\pi/2)t\, dt}$$

$$= \sqrt{\frac{16}{2}\int_0^1 (1 - \cos \pi t)\, dt} = \sqrt{8} = 2.83 \text{ A}$$

$$P_{av} = I_{rms}^2(10) = (8)(10) = 80 \text{ W}$$

8.7   DRILL PROBLEMS

DP 8.1 Determine the average power absorbed by each box in the network of **Fig. DP 8.1.**

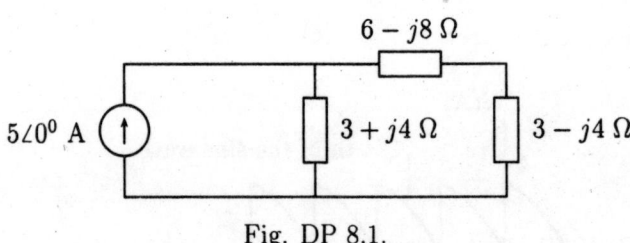

Fig. DP 8.1.

Hints : Use current division method to determine the current in each branch in the circuit of Fig. DP 7-8. Then find the average power absorbed by each box with the help of formula $P_{av} = \frac{1}{2}\text{Re}[|I|^2 Z]$.

DP 8.2 Find S, P_{av} and Q supplied by the current source shown in **Fig. DP 8.2.**

Hints : Find the total impedance seen by the current source in Fig. DP 8.2 as

$$Z_{in} = 5 + (5 - j0.5)||(10 + j2)$$
$$= 5 + \frac{(5 - j0.5)(10 + j2)}{5 - j0.5 + 10 + j2}$$

Use formula $S = \frac{1}{2}|I|^2(Z_{in})$ to determine complex power and so determine P_{av} and Q.

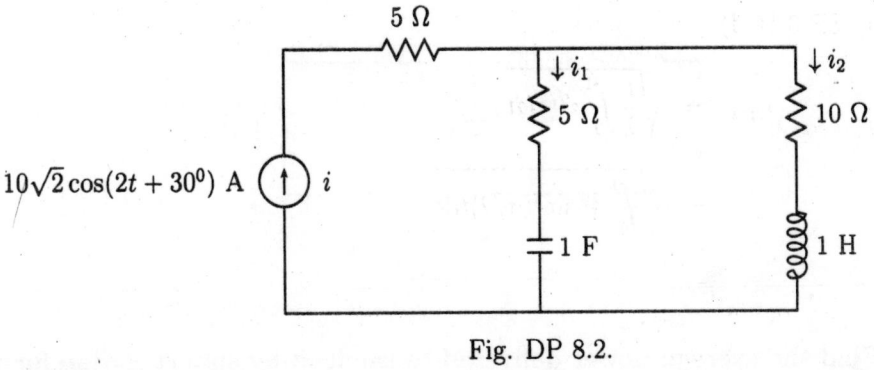

Fig. DP 8.2.

DP 8.3 Find the average power supplied by the dependent source in the circuit of Fig. DP 8.3.

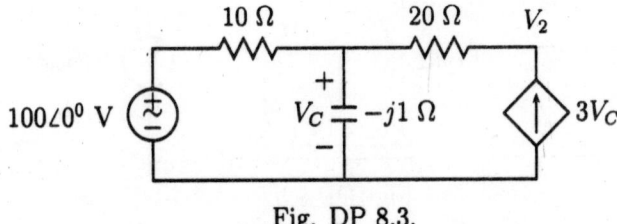

Fig. DP 8.3.

Hints : Obtain V_C and V_2 from the following nodal equations :

$$\frac{100\angle 0^0 - V_C}{10} = \frac{V_C}{-j1} + \frac{V_C - V_2}{20}$$

$$\frac{V_C - V_2}{20} = -3V_C$$

Now use $P_{av} = \frac{1}{2}\mathrm{Re}[V_2(3V_C)^*]$ to determine average power supplied by this dependent source.

DP 8.4 Find the value of C that will enable maximum power in the 1-Ω resistor of circuit in Fig. DP 8.4.

Hints : Find equivalent impedance seen by 1-Ω resistor as :

$$Z_{eq} = (1 + j4)\|(-j\frac{1}{2C})$$

$$= \frac{(1 + j4)(-j\frac{1}{2C})}{1 + j4 - j\frac{1}{2C}} = \frac{\left(\frac{2}{C} - j\frac{1}{2C}\right)}{1 + j\left(4 - \frac{1}{2C}\right)}$$

Express the last result in $Z_{eq} = x + jy$ form and solve the equation $\mathrm{Re}[Z_{eq}] = 1$(load resistance) to determine the value of C.

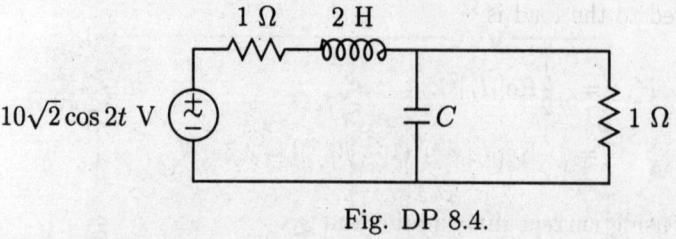

Fig. DP 8.4.

DP 8.5 Find the average power delivered to the load by superposition for the circuit shown in Fig. DP 8.5.

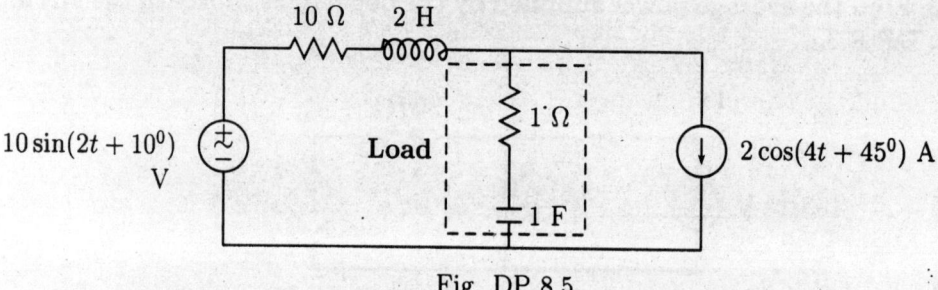

Fig. DP 8.5.

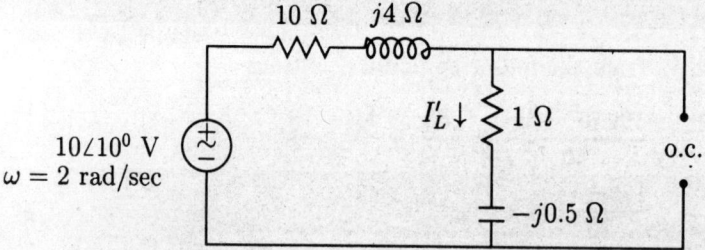

Fig. DP 8.5.1: Phasor circuit with voltage source alone.

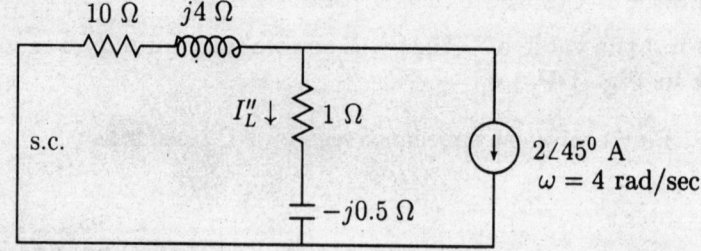

Fig. DP 8.5.2: Phasor circuit with current source alone

GATE-98 (EE)

Hints : From Fig. 7-12.1,

$$I'_L = \frac{10\angle 10^0}{10 + j4 + 1 - j0.5} = 0.87\angle -7.65^0$$

Average power delivered to the load is

$$P'_{av} = \frac{1}{2}\text{Re}[|I'_L|^2 Z_L]$$

$$= \frac{1}{2}\text{Re}[|(0.87)^2(1 - j0.5)] = 0.38 \text{ W}$$

From Fig. DP 7-12.2, using current division method

$$I''_L = -\frac{10 + j8}{10 + j8 + 1 - j0.25}(2\angle 45^0) = -1.9\angle 48.49^0$$

Hence $\quad P''_{av} = \frac{1}{2}\text{Re}[(-1.9)^2(1 - j0.25)] = 1.81 \text{ W}$

Thus by superposition $\quad P_{av} = P'_{av} + P''_{av} = 0.38 + 1.81 = 2.91\text{W}$

DP 8.6 **Find the complex power for each element in the circuit of Fig. DP 8.6.**

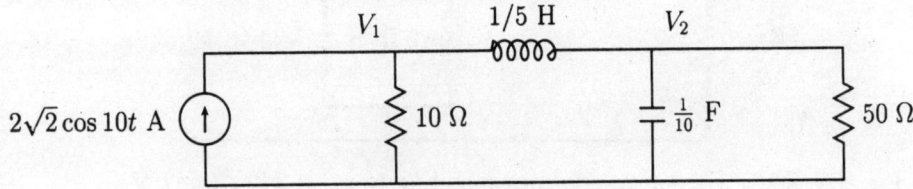

Fig. DP 8.6.

Hints : Apply nodal analysis to determine V_1 and V_2 in phasors. Use formula $S = \frac{1}{2}|V|^2 Y^*$ to determine complex power for each element of this circuit.

DP 8.7 **A circuit consisting of a single resistor R and an inductor L in series is driven by a 25 V rms, 50 Hz sinusoidal voltage source. a capacitor is to be placed in parallel with the source to improve the power factor. Given that the average power dissipated in the R is 100 W and that reactive power delivered to the L is 75 Var, what value of C will yield a 0.9 power factor lagging as seen by the source?**

Hints : Complex power in series RL branch is $S_1 = 100 + j75$ VA.
The total complex power in order to achieve power factor of 0.9 is

$$S_T = \frac{100}{0.9}\angle \cos^{-1}(0.9) = 100 + j48.43 \text{ VA}$$

Hence the complex power absorbed by the capacitor is $S_C = -j26.57$ VA. The required value of capacitance will be found by the equation :

$$S_C = -j26.57 = |V|^2 Y_C^* = 25^2(-j2\pi)(50)C$$

DP 8.8 Find the rms value of the voltage $v_{ab}(t)$ in the circuit of Fig. DP 8.8 if $v_s(t) = 240\sqrt{2}\sin t + 70\sqrt{2}\cos 3t$ volts.

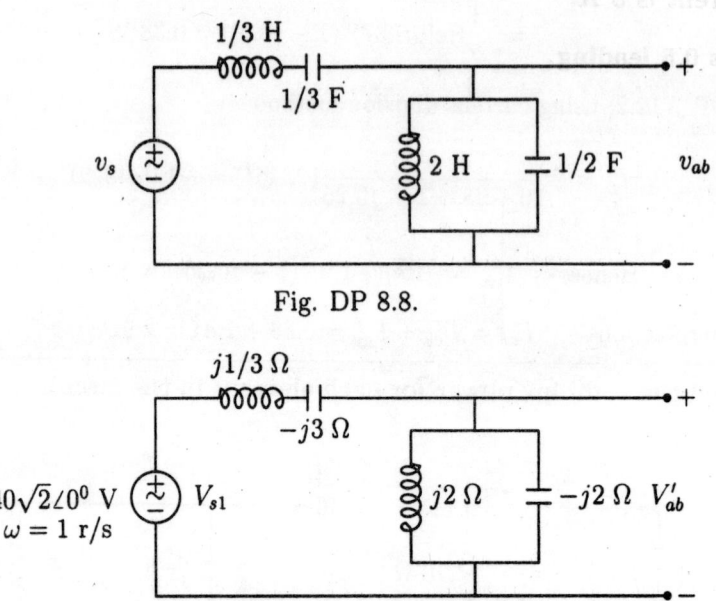

Fig. DP 8.8.

Fig. DP 8.8.1: Phasor circuit corresponding to $v_{s1} = 240\sqrt{2}\sin t$ V.

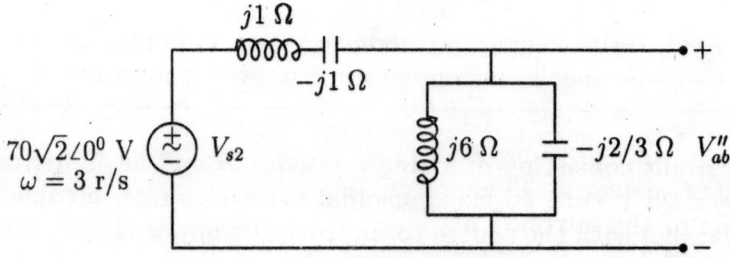

Fig. DP 8.8.2: Phasor circuit corresponding to $v_{s2} = 70\sqrt{2}\cos 3t$ V

GATE-94(EE)

Hints : It is noted from Fig. DP 8.8.1 and Fig. DP 8.8.2 that outputs are equal to their inputs. Thus, by superposition :

$$v_{ab} = v'_{ab} + v''_{ab} = 240\sqrt{2}\sin t + 70\sqrt{2}\cos 3t \text{ V}$$

and

$$(V_{ab})_{rms} = \sqrt{\left(\frac{240\sqrt{2}}{\sqrt{2}}\right)^2 + \left(\frac{70\sqrt{2}}{\sqrt{2}}\right)^2}$$

DP 8.9 For the circuit shown in Fig. DP 8.9, find the value of reactance when

(a) the power factor is 0.866 lagging.

(b) current is 5 A.

(c) pf is 0.5 leading.

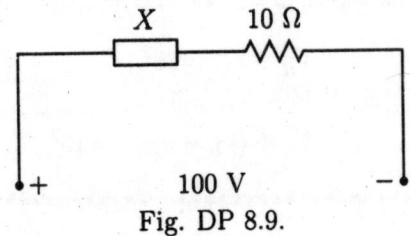

Fig. DP 8.9.

Hints:

(a)   pf $= 0.866 = 10/(X^2 + 10^2)$

(b) $100^2 = (5X)^2 + (10 \times 5)^2$

(c)   pf $= 0.5 = 10/[(-X)^2 + 10^2]$

DP 8.10 A voltage sinusoidal source, $v(t)$ is connected across the series connected RLC network. The circuit is operating in the steady-state. Given that rms value of the current, $i(t)$ is 1 A, rms value of voltage, $v(t)$ is 1 V, $\omega = 2$ rad/sec, and $L = 1$ H. Determine the rms value of voltage across the each of the elements if the current lags the voltage by 45^0.

Hints: Given that $V = 1\angle 0^0$ and $I = 1\angle -45^0$. Thus,

$$Z = \frac{V}{I} = 1\angle 45^0 = R + j2 - j\frac{1}{2C}$$

or

$$\frac{1}{\sqrt{2}} + j(\frac{1}{\sqrt{2}} - 2) = R - j\frac{1}{2C}$$

Comparing the real and imaginary parts on both sides of this equation, we have $R = 1/\sqrt{2}$ and $-1/2C = 1/\sqrt{2} - 2$ from which C can be determined.

DP 8.11 A voltage sinusoidal source, $v(t)$ is connected across the series connected RLC network. The circuit is operating in the steady-state. Given that rms value of the voltage across the combination of resistor and inductor is 20 V, rms value of the voltage across the combination of capacitor and inductor is 9 V, and rms value of the voltage across the combination of resistor, inductor and capacitor is 15 V. Determine the rms value of voltage across the each of the three elements.

Hints: We may find the rms values of V_R, V_L and V_C from the following equations as:

$$V_R^2 + V_L^2 = 20^2$$
$$V_L - V_C = 9$$
$$V_R^2 + (V_L - V_C)^2 = 15^2$$

****

Network Functions

SUMMARY OF THIS CHAPTER

9.1 ONE-PORT NETWORK

A two-terminal network is known as the one-port network as shown in Figure 9.1.

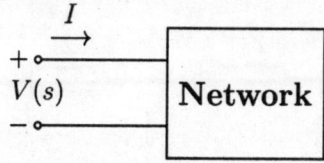

Figure 9.1: One-port network.

With reference to Figure 9.1, we have

$$\text{Driving point impedance} \qquad Z(s) = \frac{V(s)}{I(s)}$$

$$\text{Driving point admittance} \qquad Y(s) = \frac{I(s)}{V(s)} = \frac{1}{Z(s)}$$

9.2 TWO-PORT NETWORK

Figure 9.2 shows the two-port network. The left port is connected to the driving source (or the input), and the right port is connected a load.

With reference to Figure 9.2, we have

$$\text{Voltage transfer function} \qquad G_{12}(s) = \frac{V_2(s)}{V_1(s)}$$

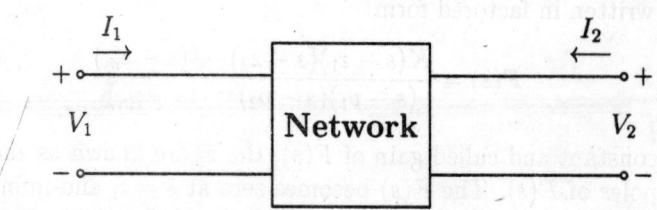

Figure 9.2: Two-port network with standard directions for port voltages and currents.

Current transfer function $\qquad \alpha_{12}(s) = \dfrac{I_2(s)}{I_1(s)}$

Transfer impedance function $\qquad Z_{12}(s) = \dfrac{V_2(s)}{I_1(s)}$

Transfer admittance function $\qquad Y_{12}(s) = \dfrac{I_2(s)}{V_1(s)}$

9.3   LADDER NETWORKS

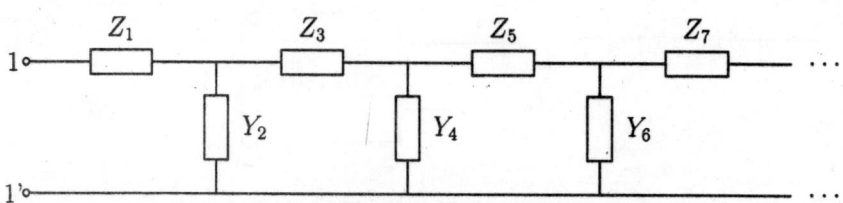

Figure 9.3: A ladder network.

With reference to Figure 9.3, the driving point immittance (a combination of impedance and admittance) function may be given by

$$Z = Z_1 + \cfrac{1}{Y_2 + \cfrac{1}{Z_3 + \cfrac{1}{Y_4 + \cfrac{1}{Z_5 + \cfrac{1}{Y_6 + \cdots}}}}} \qquad (9.1)$$

This equation is called continued fraction.

9.4   POLES AND ZEROS

In general, the network function is a ratio of polynomials in s as:

$$F(s) = \frac{b_0 s^m + b_1 s^{m-1} + \cdots + b_{m-1}s + b_m}{a_0 s^n + a_1 s^{n-1} + \cdots + a_{n-1}s + a_n} \qquad (9.2)$$

It also can be written in factored form

$$F(s) = \frac{K(s - z_1)(s - z_2)\cdots(s - z_m)}{(s - p_1)(s - p_2)\cdots(s - p_n)} \qquad (9.3)$$

where K is a constant and called gain of $F(s)$, the z_i are known as zeros of $F(s)$ and p_i are known as poles of $F(s)$. The $F(s)$ becomes zero at $s = z_i$ and infinity at $s = p_i$. The poles and zeros of a system are also called its critical frequencies. The location of poles and zeros of a network function on the complex plane is known as the pole-zero plot. The complex plane has an orthogonal pair of axes of which the horizontal one is the real and vertical one the imaginary axis. Since we have used the letter s for poles and zeros, we call this plane the s-plane.

For any rational network function, the total number of poles are equal to the total number of zeros. If $n > m$, the number of zeros at infinity are $n - m$. If $n < m$, the number of poles at infinity are $m - n$. The Figure 9.4 shows the pole-zero plot of a specific network function, $F(s)$.

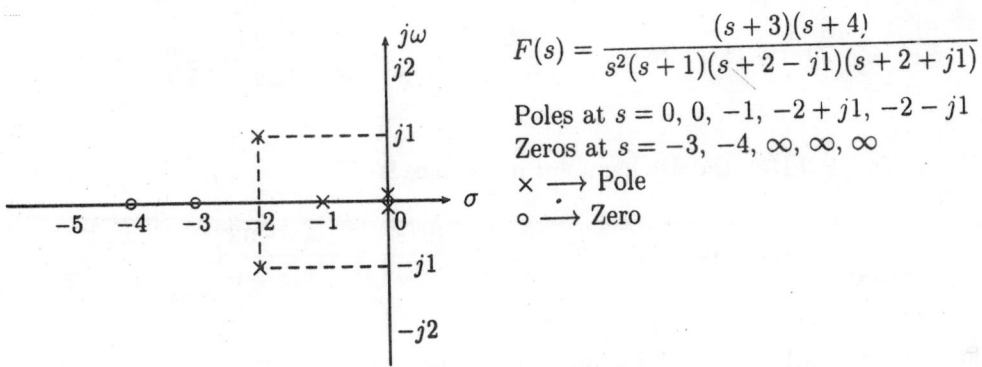

$$F(s) = \frac{(s + 3)(s + 4)}{s^2(s + 1)(s + 2 - j1)(s + 2 + j1)}$$

Poles at $s = 0, 0, -1, -2 + j1, -2 - j1$
Zeros at $s = -3, -4, \infty, \infty, \infty$
$\times \longrightarrow$ Pole
$\circ \longrightarrow$ Zero

Figure 9.4: Pole-zero plot of a specific network function $F(s)$.

9.5 SOLVED PROBLEMS

SP 9.1 Determine $Z(s)$ for the following networks shown in Fig. SP 9.1.

SOLUTION:
(a) From Fig. SP 9.1(a), the driving point impedance is

$$Z(s) = 5 + 2s + 1/s = \frac{2(s^2 + 2.5s + 0.5)}{s}$$

(b) From Fig. SP 9.1(b), the driving point impedance is

$$Z(s) = 5 + [2s\|(1/s)] = 5 + \frac{(2s)(1/s)}{2s + 1/s} = \frac{5(s^2 + 0.2s + 0.5)}{s^2 + 0.5}$$

(c) From Fig. SP 9.1(c), the driving point impedance is

$$Z(s) = 5 + [(1/s)\|(2s + 1/s)] = 5 + \frac{(1/s)(2s + 1/s)}{(2s + 1/s + 1/s)}$$

$$= \frac{5(s^3 + 0.2s^2 + s + 0.1)}{s(s^2 + 1)}$$

(d) From Fig. SP 9.1(d), the driving point impedance is

$$Z(s) = (1/s)\|(5 + 2s) = \frac{(5 + 2s)/s}{5 + 2s + 1/s} = \frac{s + 2.5}{s^2 + 2.5s + 0.5}$$

(e) From Fig. SP 9.1(e), the driving point impedance is

$$Z(s) = 10 + [(2/s)\|5] = 10 + \frac{10/s}{5 + 2/s} = \frac{10(s + 0.6)}{s + 0.4}$$

(f) From Fig. SP 9.1(f), the driving point impedance is

$$Z(s) = (2 + 4s)\|(1/s) = \frac{(2 + 4s)/s}{2 + 4s + 1/s} = \frac{s + 0.5}{s^2 + 0.5s + 0.25}$$

(g) From Fig. SP 9.1(g), the driving point admittance is

$$Y(s) = \frac{1}{s + 1} + s + \frac{1}{2}$$

$$= \frac{s^2 + 1.5s + 1.5}{s + 1}$$

Hence $$Z(s) = \frac{1}{Y(s)}$$

$$= \frac{s + 1}{s^2 + 1.5s + 1.5}$$

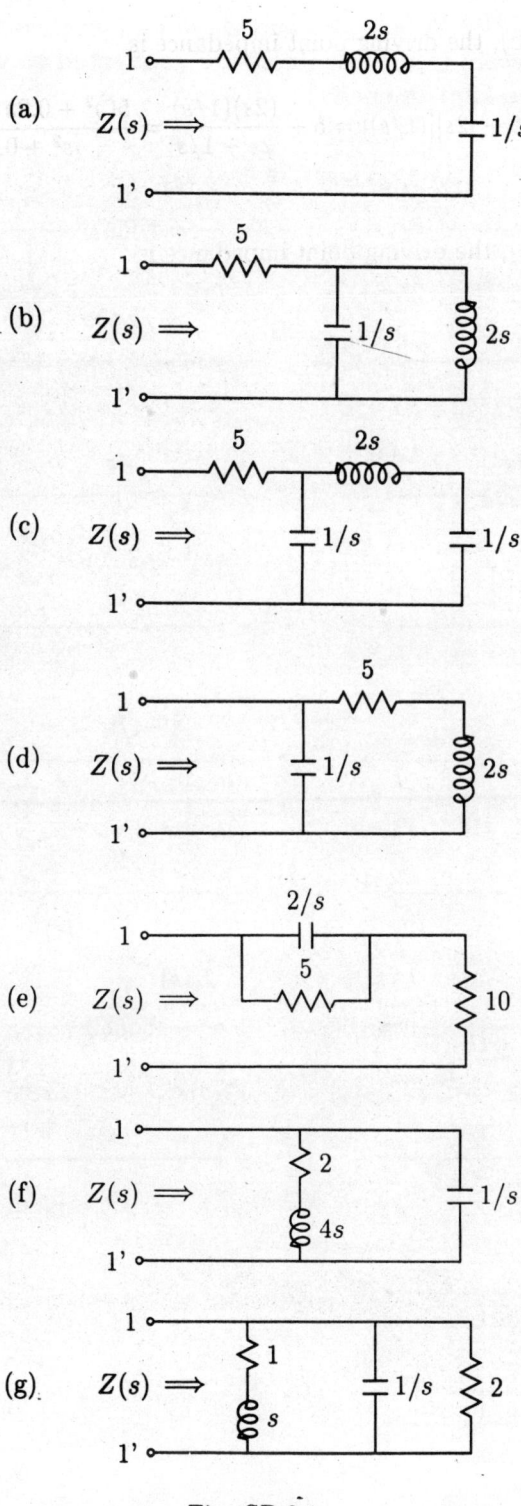

Fig. SP 9.1.

SP 9.2 Find $G_{12} = V_2/V_1$, $Y_{11} = I_1/V_1$, and $Z_{12} = V_2/I_1$ for the following networks in Fig. SP 9.2. Wherein input terminals are connected to voltage sources and output terminals are kept opened.

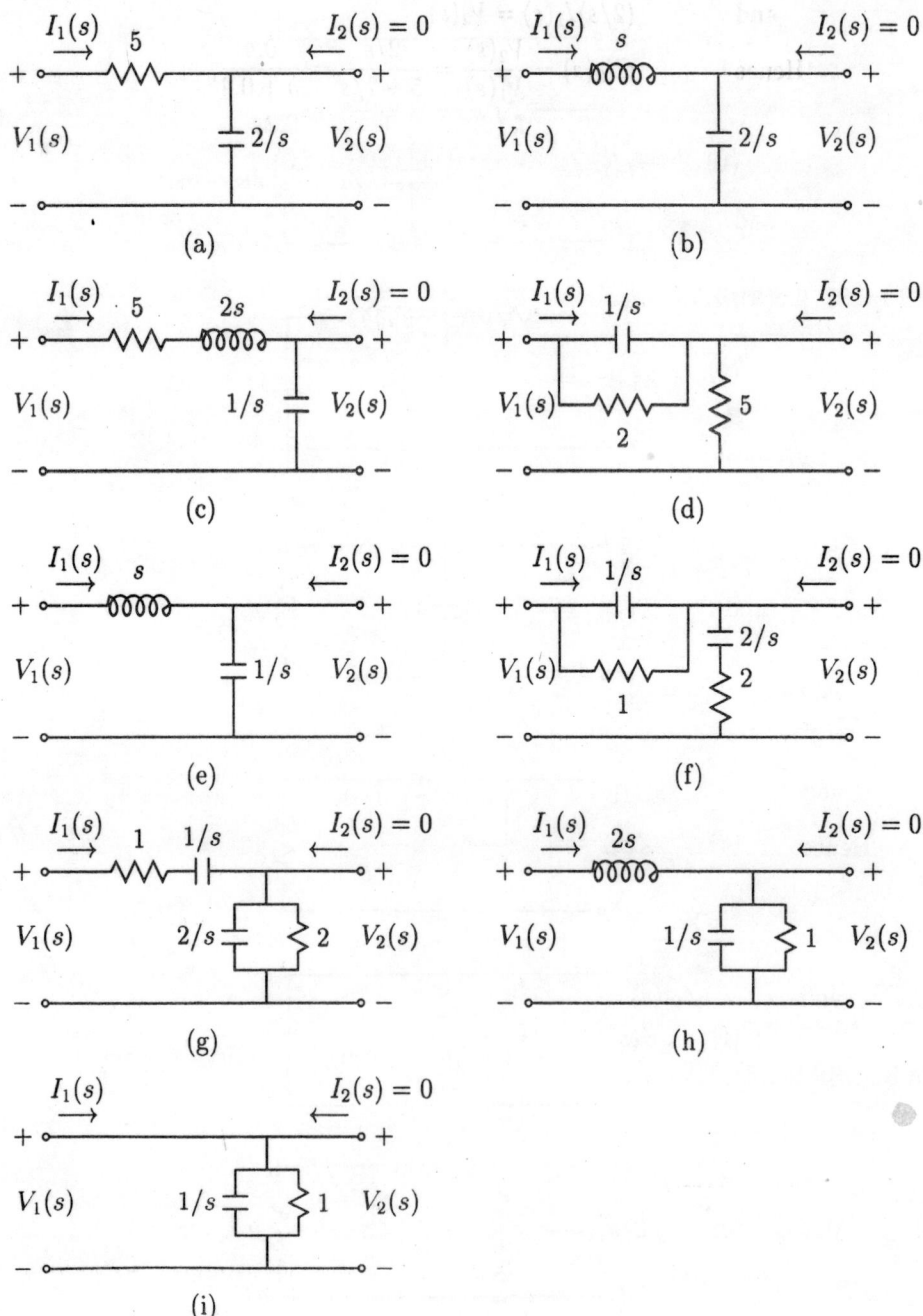

Fig. SP 9.2.

SOLUTION:
(a) From Fig. SP 9.2(a), KVL equations are:

$$(5 + 2/s)I_1(s) = V_1(s)$$

and $\qquad (2/s)I_1(s) = V_2(s)$

Hence $\qquad G_{12}(s) = \dfrac{V_2(s)}{V_1(s)} = \dfrac{2/s}{5 + 2/s} = \dfrac{0.4}{s + 0.4}$

$$Y_{11}(s) = \dfrac{I_1(s)}{V_1(s)} = \dfrac{1}{5 + 2/s} = \dfrac{0.2s}{s + 1}$$

and $\qquad Z_{12}(s) = \dfrac{V_2(s)}{I_1(s)} = \dfrac{2}{s}$

(b) From Fig. SP 9.2(b), KVL equations are:

$$(s + 2/s)I_1(s) = V_1(s)$$

and $\qquad (2/s)I_1(s) = V_2(s)$

Hence $\qquad G_{12}(s) = \dfrac{V_2(s)}{V_1(s)} = \dfrac{2/s}{s + 2/s} = \dfrac{2}{s^2 + 2}$

$$Y_{11}(s) = \dfrac{I_1(s)}{V_1(s)} = \dfrac{1}{s + 2/s} = \dfrac{s}{s^2 + 2}$$

and $\qquad Z_{12}(s) = \dfrac{V_2(s)}{I_1(s)} = \dfrac{2}{s}$

(c) From Fig. SP 9.2(c), KVL equations are:

$$(5 + 2s + 1/s)I_1(s) = V_1(s)$$

and $\qquad (1/s)I_1(s) = V_2(s)$

Hence $\qquad G_{12}(s) = \dfrac{V_2(s)}{V_1(s)} = \dfrac{1/s}{5 + 2s + 1/s} = \dfrac{0.5}{s^2 + 2.5s + 0.5}$

$$Y_{11}(s) = \dfrac{I_1(s)}{V_1(s)} = \dfrac{1}{5 + 2s + 1/s} = \dfrac{0.5s}{s^2 + 2.5s + 0.5}$$

and $\qquad Z_{12}(s) = \dfrac{V_2(s)}{I_1(s)} = \dfrac{1}{s}$

(d) From Fig. SP 9.2(d), KVL equations are:

$$[(2\|1/s) + 5]I_1(s) = V_1(s)$$

and $\qquad 5I_1(s) = V_2(s)$

Hence $\qquad G_{12}(s) = \dfrac{V_2(s)}{V_1(s)} = \dfrac{5}{(2\|1/s) + 5} = \dfrac{s + 0.5}{s + 0.7}$

$$Y_{11}(s) = \dfrac{I_1(s)}{V_1(s)} = \dfrac{1}{(2\|1/s) + 5} = \dfrac{0.2(s + 0.5)}{s + 0.7}$$

and $\qquad Z_{12}(s) = \dfrac{V_2(s)}{I_1(s)} = 5$

(e) From Fig. SP 9.2(e), KVL equations are:

$$(s + 1/s)I_1(s) = V_1(s)$$

and $\quad (1/s)I_1(s) = V_2(s)$

Hence $\quad G_{12}(s) = \dfrac{V_2(s)}{V_1(s)} = \dfrac{1/s}{s + 1/s} = \dfrac{1}{s^2 + 1}$

$$Y_{11}(s) = \dfrac{I_1(s)}{V_1(s)} = \dfrac{1}{s + 1/s} = \dfrac{s}{s^2 + 1}$$

and $\quad Z_{12}(s) = \dfrac{V_2(s)}{I_1(s)} = \dfrac{1}{s}$

(f) From Fig. SP 9.2(f), KVL equations are:

$$[(1/s||1) + 2 + 2/s]\, I_1(s) = V_1(s)$$

and $\quad (2 + 2/s)I_1(s) = V_2(s)$

Hence $\quad G_{12}(s) = \dfrac{V_2(s)}{V_1(s)} = \dfrac{2 + 2/s}{(1||1/s) + 2 + 2/s} = \dfrac{(s + 1)^2}{s^2 + 2.5s + 1}$

$$Y_{11}(s) = \dfrac{I_1(s)}{V_1(s)} = \dfrac{1}{(1/s||1) + 2 + 2/s} = \dfrac{0.5s(s + 1)}{s^2 + 2.5s + 1}$$

and $\quad Z_{12}(s) = \dfrac{V_2(s)}{I_1(s)} = \dfrac{2(s + 1)}{s}$

(g) From Fig. SP 9.2(g), KVL equations are:

$$[1 + 1/s + (2||2/s)]\, I_1(s) = V_1(s)$$

and $\quad (2||2/s)I_1(s) = V_2(s)$

Hence $\quad G_{12}(s) = \dfrac{V_2(s)}{V_1(s)} = \dfrac{(2||2/s)}{1 + 1/s + (2||2/s)} = \dfrac{2s}{s^2 + 4s + 1}$

$$Y_{11}(s) = \dfrac{I_1(s)}{V_1(s)} = \dfrac{1}{1 + 1/s + (2||2/s)} = \dfrac{s(s + 1)}{s^2 + 4s + 1}$$

and $\quad Z_{12}(s) = \dfrac{V_2(s)}{I_1(s)} = (2||2/s) = \dfrac{2}{s + 1}$

(h) From Fig. SP 9.2(h), KVL equations are:

$$[2s + (1||1/s)]\, I_1(s) = V_1(s)$$

and $\quad (1||1/s)I_1(s) = V_2(s)$

Hence $\quad G_{12}(s) = \dfrac{V_2(s)}{V_1(s)} = \dfrac{(1||1/s)}{2s + (1||1/s)} = \dfrac{0.5}{s^2 + s + 0.5}$

$$Y_{11}(s) = \dfrac{I_1(s)}{V_1(s)} = \dfrac{1}{2s + (1||1/s)} = \dfrac{0.5(s + 1)}{s^2 + s + 0.5}$$

and $\quad Z_{12}(s) = \dfrac{V_2(s)}{I_1(s)} = \dfrac{1/s}{1 + 1/s} = \dfrac{1}{s + 1}$

(i) From Fig. SP 9.2(i), KVL equations are:

$$V_1(s) = (1\|(1/s)) I_1 = V_2(s)$$

Hence
$$G_{12}(s) = \frac{V_2(s)}{V_1(s)} = 1$$

$$Y_{11}(s) = \frac{I_1(s)}{V_1(s)} = \frac{1}{(1\|1/s)} = s + 1$$

and
$$Z_{12}(s) = \frac{V_2(s)}{I_1(s)} = (1\|1/s) = \frac{1}{s+1}$$

SP 9.3 Compute $\alpha_{12} = I_2/I_1$ and $Z_{12} = V_2/I_1$ for the following planar networks (no two branches intersect at a point) shown in Fig. SP 9.3.

SOLUTION:

(a) Using current division formula in Fig. SP 9.3(a),

$$I_2(s) = \frac{R}{R + 1/sC} I_1(s)$$

Thus
$$\alpha_{12} = \frac{I_2(s)}{I_1(s)} = \frac{R}{R + 1/sC} = \frac{s}{s + 1/RC}$$

and
$$Z_{12}(s) = \frac{V_2(s)}{I_1(s)} = \frac{(1/sC)I_2(s)}{I_1(s)} = \frac{1/C}{s + 1/sC}$$

(b) Using current division formula in Fig. SP 9.3(b),

$$I_2(s) = \frac{4/s}{5 + 2/s + 4/s} I_1(s) = \frac{4}{5s + 6} I_1(s)$$

Hence
$$\alpha_{12}(s) = \frac{I_2(s)}{I_1(s)} = \frac{0.8}{s + 1.2}$$

and
$$Z_{12}(s) = \frac{V_2(s)}{I_1(s)} = \frac{(5 + 2/s)I_2(s)}{I_1(s)} = \frac{4(s + 0.4)}{s(s + 1.2)}$$

(c) Using current division formula in Fig. SP 9.3(c),

$$I_2(s) = \frac{2/s}{2/s + (2 + 4/s)} I_1(s) = \frac{1}{s + 3} I_1(s)$$

Thus
$$\alpha_{12}(s) = \frac{I_2(s)}{I_1(s)} = \frac{1}{s + 3}$$

and
$$Z_{12}(s) = \frac{V_2(s)}{I_1(s)} = \frac{(4/s)I_2(s)}{I_1(s)} = \frac{4}{s(s + 3)}$$

(d) Using current division formula in Fig. SP 9.3(d),

$$I_2(s) = \frac{1 + s}{(1 + s) + (1/s + 2)} I_1(s) = \frac{s^2 + s}{s^2 + 3s + 1} I_1(s)$$

Thus
$$\alpha_{12}(s) = \frac{I_2(s)}{I_1(s)} = \frac{s(s + 1)}{s^2 + 3s + 1}$$

and
$$Z_{12}(s) = \frac{2I_2(s)}{I_1(s)} = \frac{2s(s + 1)}{s^2 + 3s + 1}$$

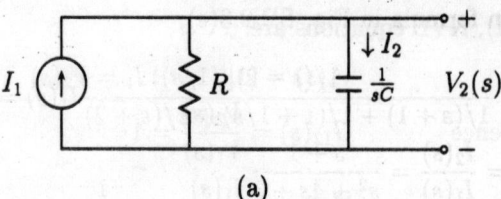

(a)

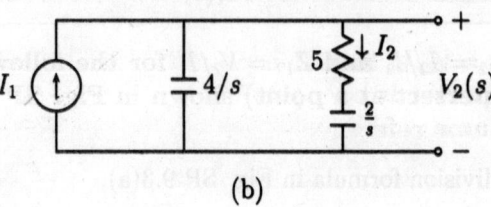

(b)

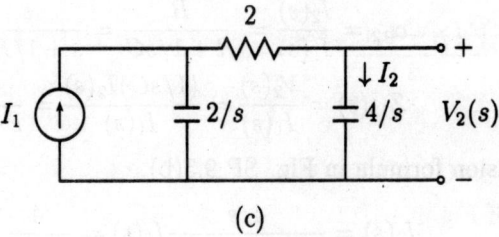

(c)

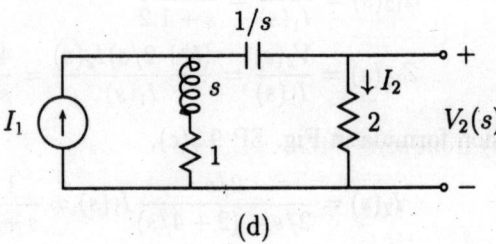

(d)

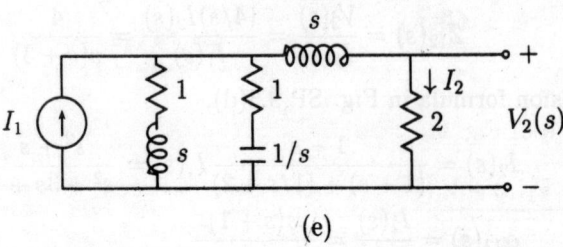

(e)

Fig. SP 9.3.

(e) Using current division formula in Fig. SP 9.3(e),

$$I_2(s) = \frac{1/(s+2)}{1/(s+1) + 1/(1+1/s) + 1/(s+2)} I_1(s) = \frac{s+1}{s^2 + 4s + 3} I_1(s)$$

Thus

$$\alpha_{12}(s) = \frac{I_2(s)}{I_1(s)} = \frac{s+1}{s^2 + 4s + 3}$$

and

$$Z_{12}(s) = \frac{V_2(s)}{I_1(s)} = \frac{2I_2(s)}{I_1(s)} = \frac{2(s+1)}{s^2 + 4s + 3}$$

SP 9.4 Determine $G_{12} = V_2/V_1$ for the each of the following networks shown in Fig. SP 9.4, wherein port 1 is connected to a voltage source V_1 and port 2 is designated by V_2. Assume reference polarities for the voltages.

SOLUTION:

(a)
By KCL at the node a of Fig. 9.4.1(a),

$$\frac{V_1(s) - V_a(s)}{2} = \frac{V_a(s)}{1/s} + \frac{V_a(s) - V_2(s)}{2}$$

Thus

$$V_1(s) = 2(s+1)V_a(s) - V_2(s) \qquad [9.4.1a]$$

By KCL at the node 2 of Fig. 9.4.1(a),

$$\frac{V_a(s) - V_2(s)}{2} = \frac{V_2(s)}{1/s}$$

Thus

$$V_a(s) = (1 + 2s)V_2(s) \qquad [9.4.2a]$$

Putting the value of $V_a(s)$ from eqn [9.4.2a] into eqn [9.4.1a],

$$V_1(s) = 2(s+1)(1+2s)V_2(s) - V_2(s)$$

or

$$\frac{V_2(s)}{V_1(s)} = \frac{1}{2(s+1)(1+2s) - 1}$$

Thus

$$G_{12}(s) = \frac{V_2(s)}{V_1} = \frac{0.25}{s^2 + 1.5s + 0.25}$$

(b)
By KCL at the node a of Fig. 9.4.1(b),

$$\frac{V_1(s) - V_a(s)}{1/s} = \frac{V_a(s)}{2} + \frac{V_a(s) - V_2(s)}{1/s}$$

Thus

$$sV_1(s) = (2s + 1/2)V_a(s) - sV_2(s) \qquad [9.4.1b]$$

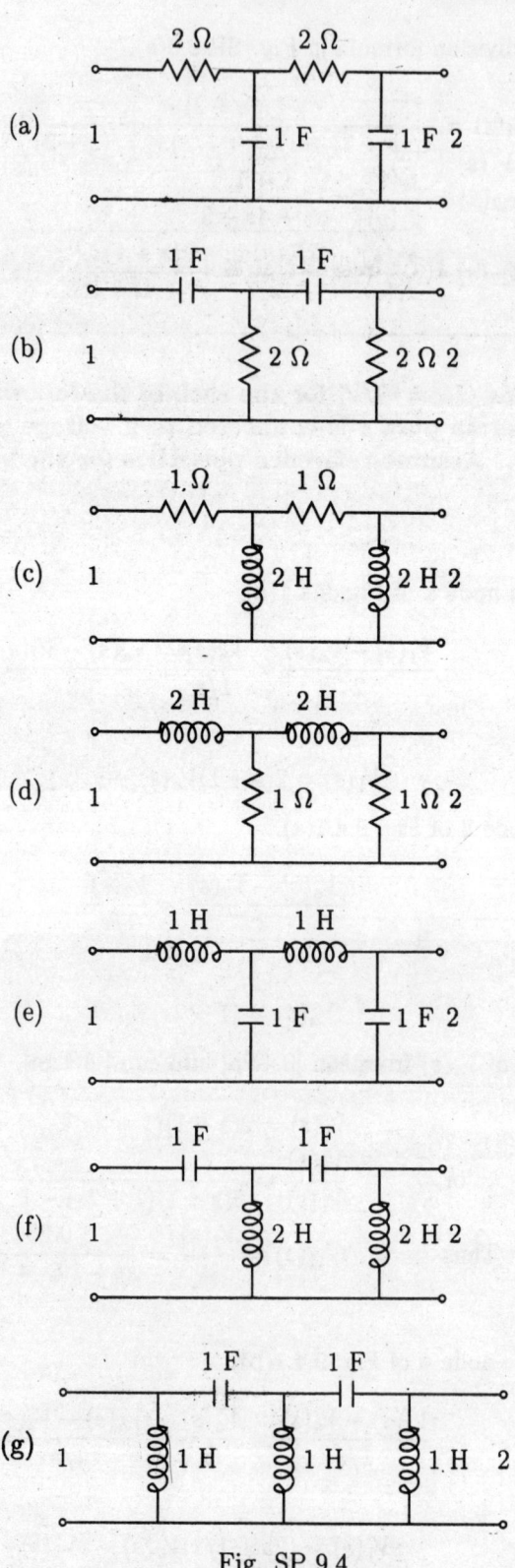

Fig. SP 9.4.

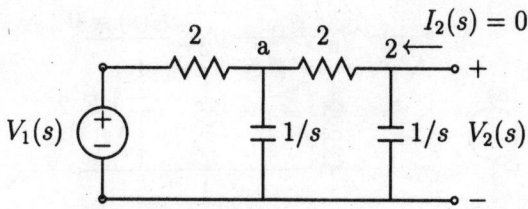

Fig. 9.4.1(a): Redrawn circuit of Fig. SP 9.4(a).

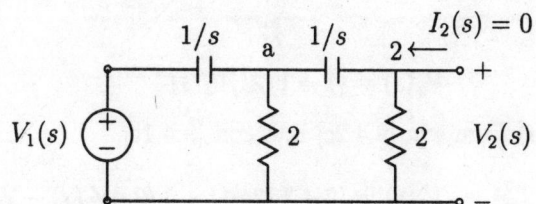

Fig. 9.4.1(b): Redrawn circuit of Fig. SP 9.4(b).

By KCL at the node **2** of Fig. 9.4.1(b),

$$\frac{V_a(s) - V_2(s)}{1/s} = \frac{V_2(s)}{2}$$

Thus

$$V_a(s) = \frac{s + 1/2}{s} V_2(s) \qquad [9.4.2b]$$

Putting the value of $V_a(s)$ from eqn [9.4.2b] into eqn [9.4.1b],

$$sV_1(s) = \frac{(2s + 1/2)(s + 1/2)}{s} V_2(s) - sV_2(s)$$

or

$$\frac{V_2(s)}{V_1(s)} = \frac{s^2}{(2s + 1/2)(s + 1/2) - s^2}$$

Thus

$$G_{12}(s) = \frac{V_2(s)}{V_1(s)} = \frac{s^2}{s^2 + 1.5s + 0.25}$$

(c)

By KCL at the node **a** of Fig. 9.4.1(c),

$$\frac{V_1(s) - V_a(s)}{1} = \frac{V_a(s)}{2s} + \frac{V_a(s) - V_2(s)}{1}$$

Thus

$$V_1(s) = (2 + 1/2s)V_a(s) - V_2(s) \qquad [9.4.1c]$$

By KCL at the node **2** of Fig. 9.4.1(c),

$$\frac{V_a(s) - V_2(s)}{1} = \frac{V_2(s)}{2s}$$

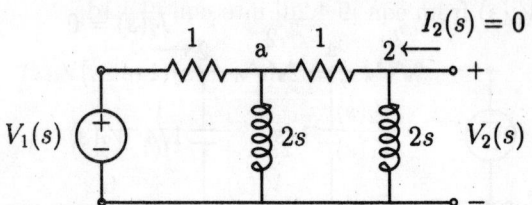

Fig. 9.4.1(c): Redrawn circuit of Fig. SP 9.4(c).

Thus

$$V_a(s) = (1 + 1/2s)V_2(s) \qquad [9.4.2c]$$

Putting the value of $V_a(s)$ from eqn [9.4.2c] into eqn [9.4.1c],

$$V_1(s) = (2 + 1/2s)(1 + 1/2s)V_2(s) - V_2(s)$$

or $\qquad \dfrac{V_2(s)}{V_1(s)} = \dfrac{1}{(2 + 1/2s)(1 + 1/2s) - 1}$

Therefore $\qquad G_{12}(s) = \dfrac{V_2(s)}{V_1(s)} = \dfrac{s^2}{s^2 + 1.5s + 0.25}$

(d)

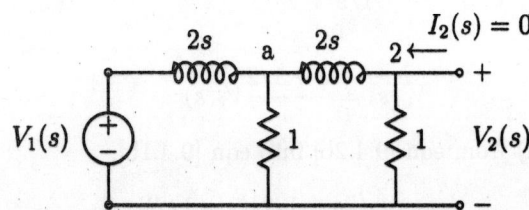

Fig. 9.4.1(d): Redrawn circuit of Fig. SP 9.4(d).

By KCL at the node a of Fig. 9.4.1(d),

$$\frac{V_1(s) - V_a(s)}{2s} = \frac{V_a(s)}{1} + \frac{V_a(s) - V_2(s)}{2s}$$

or

$$V_1(s) = 2s(1 + 1/s)V_a(s) - V_2(s) \qquad [9.4.1d]$$

By KCL at the node 2 of Fig. 9.4.1(d),

$$\frac{V_a(s) - V_2(s)}{2s} = \frac{V_2(s)}{1}$$

Thus

$$V_a(s) = (1 + 2s)V_2(s) \qquad [9.4.2d]$$

Putting the value of $V_a(s)$ from eqn [9.4.2d] into eqn [9.4.1d],

$$V_1(s) = 2s(1+1/s)(1+2s)V_2(s) - V_2(s)$$

or

$$\frac{V_2(s)}{V_1(s)} = \frac{1}{2s(1+1/s)(1+2s)-1}$$

Therefore

$$G_{12}(s) = \frac{V_2(s)}{V_1(s)} = \frac{0.25}{s^2 + 1.5s + 0.25}$$

(e)

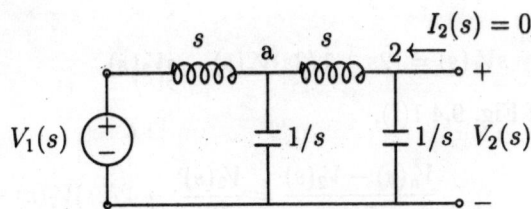

Fig. 9.4.1(e): Redrawn circuit of Fig. SP 9.4(e).

By KCL at the node **a** of Fig. 9.4.1(e),

$$\frac{V_1(s) - V_a(s)}{s} = \frac{V_a(s)}{1/s} + \frac{V_a(s) - V_2(s)}{s}$$

Thus

$$V_1(s) = (2 + s^2)V_a(s) - V_2(s) \qquad [9.4.1e]$$

By KCL at the node **2** of Fig. 9.4.1(e),

$$\frac{V_a(s) - V_2(s)}{s} = \frac{V_2(s)}{1/s}$$

Thus

$$V_a(s) = (1 + s^2)V_2(s) \qquad [9.4.2e]$$

Putting the value of $V_a(s)$ from eqn [9.4.2e] into eqn [9.4.1e],

$$V_1(s) = (2 + s^2)(1 + s^2)V_2(s) - V_2(s)$$

or

$$\frac{V_2(s)}{V_1(s)} = \frac{1}{(2 + s^2)(1 + s^2) - 1}$$

Therefore

$$G_{12}(s) = \frac{V_2(s)}{V_1(s)} = \frac{1}{s^4 + 3s^2 + 1}$$

(f)

By KCL at the node **a** of Fig. 9.4.1(f),

$$\frac{V_1(s) - V_a(s)}{1/s} = \frac{V_a(s)}{2s} + \frac{V_a(s) - V_2(s)}{1/s}$$

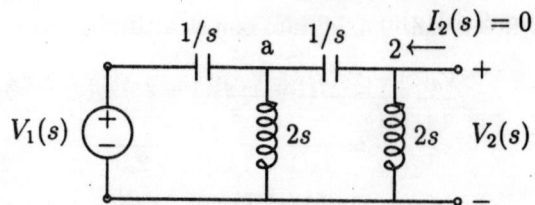

Fig. 9.4.1(f): Redrawn circuit of Fig. SP 9.4(f).

Thus

$$sV_1(s) = (2s + 1/2s)V_a(s) - sV_2(s) \qquad [9.4.1f]$$

By KCL at the node **2** of Fig. 9.4.1(f),

$$\frac{V_a(s) - V_2(s)}{1/s} = \frac{V_2(s)}{2s}$$

Thus

$$V_a(s) = (1 + 1/2s^2)V_2(s) \qquad [9.4.2f]$$

Putting the value of $V_a(s)$ from eqn [9.4.2f] into eqn [9.4.1f],

$$sV_1(s) = (2s + 1/2s)(1 + 1/2s^2)V_2(s) - sV_2(s)$$

or

$$\frac{V_2(s)}{V_1(s)} = \frac{s}{(2s + 1/2s)(1 + 1/2s^2) - s}$$

Thus

$$G_{12}(s) = \frac{V_2(s)}{V_1(s)} = \frac{s^4}{s^4 + 1.5s^2 + 0.25}$$

(g)

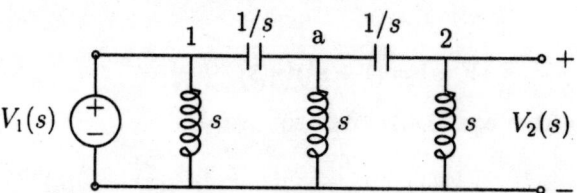

Fig. 9.4.1(g): Redrawn circuit of Fig. SP 9.4(g).

By KCL at the node **a** of Fig. 9.4.1(g),

$$\frac{V_1(s) - V_a(s)}{1/s} = \frac{V_a(s)}{s} + \frac{V_a(s) - V_2(s)}{1/s}$$

Thus

$$sV_1(s) = (2s + 1/s)V_a(s) - sV_2(s) \qquad [9.4.1g]$$

By KCL at the node **2** of Fig. 9.4.1(g),

$$\frac{V_a(s) - V_2(s)}{1/s} = \frac{V_2(s)}{s}$$

Thus

$$V_a(s) = (1 + 1/s^2)V_2(s) \qquad [9.4.2g]$$

Putting the value of $V_a(s)$ from eqn [9.4.2g] into eqn [9.4.1g],

$$sV_1(s) = (2s + 1/s)(1 + 1/s^2)V_2(s) - sV_2(s)$$

or $\qquad \dfrac{V_2(s)}{V_1(s)} = \dfrac{s}{(2s + 1/s)(1 + 1/s^2) - s}$

Therefore $\qquad G_{12}(s) = \dfrac{V_2(s)}{V_1(s)} = \dfrac{s^4}{s^4 + 3s^2 + 1}$

SP 9.5 **For each of the following networks shown in Fig. SP 9.5, compute** $G_{12} = V_2/V_1$, $Z_{12} = V_2/I_1$, $\alpha_{12} = I_2/I_1$, **and** $Z_{11} = V_1/I_1$.

(a) **SOLUTION**
Starting with the right end, we can have following equations for the network of Fig. 9.5 .1(a):

$$I_2 = -V_2 \qquad [9.5.1a]$$

$$I_b = sV_2 + V_2 = (s + 1)V_2 \qquad [9.5.2a]$$

$$V_a = sI_b + V_2 = [s(s + 1) + 1]V_2 \qquad [9.5.3a]$$

$$I_1 = sV_a + I_b = [s(s^2 + s + 1) + (s + 1)]V_2 \qquad [9.5.4a]$$

$$V_1 = sI_1 + V_a = [s(s^3 + s^2 + 2s + 1) + (s^2 + s + 1)]V_2 \qquad [9.5.5a]$$

Hence

$$G_{12} = \frac{V_2}{V_1} = \frac{V_2}{(s^4 + s^3 + 3s^2 + 2s + 1)V_2} = \frac{1}{s^4 + s^3 + 3s^2 + 2s + 1}$$

$$Z_{12} = \frac{V_2}{I_1} = \frac{V_2}{(s^3 + s^2 + 2s + 1)V_2} = \frac{1}{s^3 + s^2 + 2s + 1}$$

$$\alpha_{12} = \frac{I_2}{I_1} = \frac{-V_2}{(s^3 + s^2 + 2s + 1)V_2} = -\frac{1}{(s^3 + s^2 + 2s + 1)}$$

and $\qquad Z_{11} = \dfrac{V_1}{I_1} = \dfrac{(s^4 + s^3 + 3s^2 + 2s + 1)V_2}{(s^3 + s^2 + 2s + 1)V_2} = \dfrac{s^4 + s^3 + 3s^2 + 2s + 1}{s^3 + s^2 + 2s + 1}$

(b)

Starting with the right end, we can have following equations for the network of Fig. 9.5.1(b):

$$I_2 = -V_2 \qquad\qquad [9.5.1b]$$

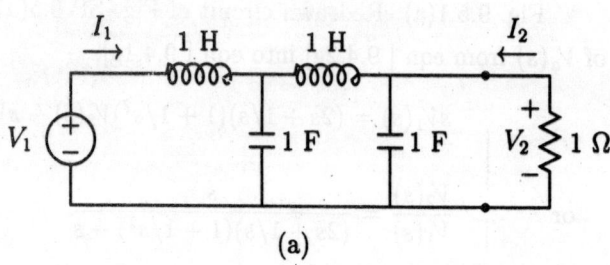

(a)

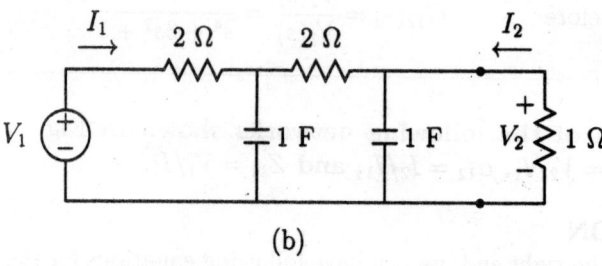

(b)

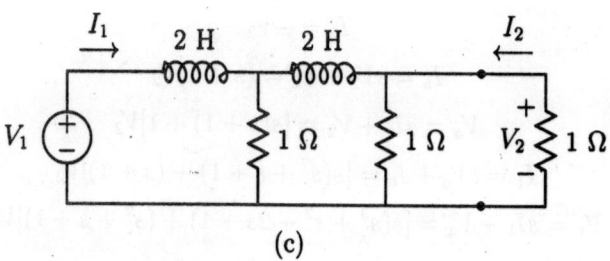

(c)

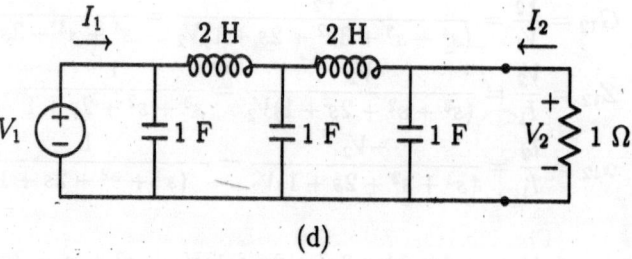

(d)

Fig. SP 9

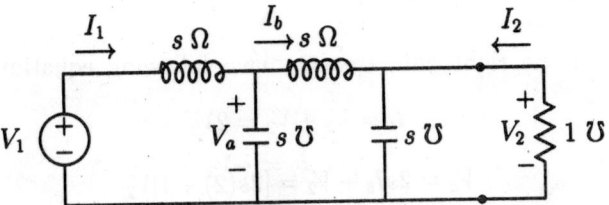

Fig. 9.5.1(a): Redrawn circuit of Fig. SP 9.5(a).

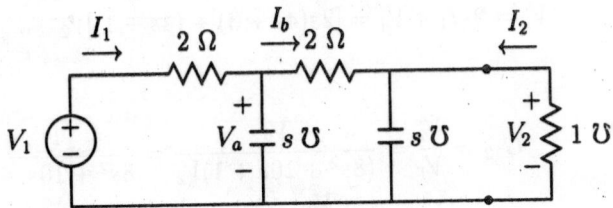

Fig. 9.5.1(b): Redrawn circuit of Fig. SP 9.5(b).

$$I_b = sV_2 + V_2 = (s+1)V_2 \qquad\qquad [9.5.2b]$$

$$V_a = 2I_b + V_2 = [2(s+1)+1]V_2 \qquad\qquad [9.5.3b]$$

$$I_1 = sV_a + I_b = [s(2s+3)+(s+1)]V_2 \qquad\qquad [9.5.4b]$$

$$V_1 = 2I_1 + V_a = [2(2s^2+4s+1)+(2s+3)]V_2 \qquad\qquad [9.5.5b]$$

Hence

$$G_{12} = \frac{V_2}{V_1} = \frac{V_2}{(4s^2+10s+5)V_2} = \frac{1}{4s^2+10s+5}$$

$$Z_{12} = \frac{V_2}{I_1} = \frac{V_2}{(2s^2+4s+1)V_2} = \frac{1}{2s^2+4s+1}$$

$$\alpha_{12} = \frac{I_2}{I_1} = \frac{-V_2}{(2s^2+4s+1)V_2} = -\frac{1}{2s^2+4s+1}$$

and $$\qquad Z_{11} = \frac{V_1}{I_1} = \frac{(4s^2+10s+5)V_2}{(2s^2+4s+1)V_2} = \frac{4s^2+10s+5}{2s^2+4s+1}$$

(c)

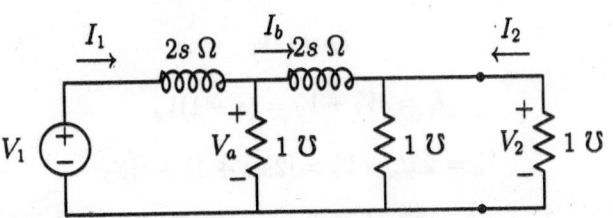

Fig. 9.5.1(c): Redrawn circuit of Fig. SP 9.5(c).

Starting with the right end, we can have following equations for the network of Fig. 9.5.1(c):

$$I_2 = -V_2 \qquad\qquad [9.5.1c]$$

$$I_b = V_2 + V_2 = 2V_2 \qquad\qquad [9.5.2c]$$

$$V_a = 2sI_b + V_2 = [2s(2) + 1]V_2 \qquad\qquad [9.5.3c]$$

$$I_1 = V_a + I_b = [(4s+1) + 2]V_2 \qquad\qquad [9.5.4c]$$

$$V_1 = 2sI_1 + V_a = [2s(4s+3) + (4s+1)]V_2 \qquad\qquad [9.5.5c]$$

Hence

$$G_{12} = \frac{V_2}{V_1} = \frac{V_2}{(8s^2 + 10s + 1)V_2} = \frac{1}{8s^2 + 10s + 1}$$

$$Z_{12} = \frac{V_2}{I_1} = \frac{V_2}{(4s+3)V_2} = \frac{1}{4s+3}$$

$$\alpha_{12} = \frac{I_2}{I_1} = \frac{-V_2}{(4s+3)V_2} = -\frac{1}{4s+3}$$

and $\qquad Z_{11} = \dfrac{V_1}{I_1} = \dfrac{(8s^2 + 10s + 1)V_2}{(4s+3)V_2} = \dfrac{8s^2 + 10s + 1}{4s+3}$

(d)

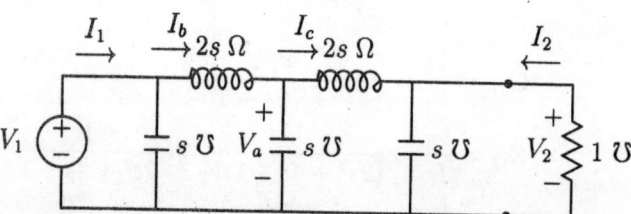

Fig. 9.5.1(d): Redrawn circuit of Fig. SP 9.5(d).

Starting with the right end, we can have following equations for the network of Fig. 9.5.1(d):

$$I_2 = -V_2 \qquad\qquad [9.5.1d]$$

$$I_c = sV_2 + V_2 = (s+1)V_2 \qquad\qquad [9.5.2d]$$

$$V_a = 2sI_c + V_2 = [2s(s+1) + 1]V_2 \qquad\qquad [9.5.3d]$$

$$I_b = sV_a + I_c = [s(2s^2 + 2s + 1) + (s+1)]V_2 \qquad\qquad [9.5.4d]$$

$$V_1 = 2sI_b + V_a = [2s(2s^3 + 2s^2 + 2s + 1) + (2s^2 + 2s + 1)]V_2 \qquad\qquad [9.5.5d]$$

$$I_1 = sV_1 + I_b = [s(4s^4 + 4s^3 + 6s^2 + 4s + 1) + (2s^3 + 2s^2 + 2s + 1)]V_2 \qquad\qquad [9.5.6d]$$

Hence

$$G_{12} = \frac{V_2}{V_1} = \frac{V_2}{(4s^4 + 4s^3 + 6s^2 + 4s + 1)V_2} = \frac{1}{4s^4 + 4s^3 + 6s^2 + 4s + 1}$$

$$Z_{12} = \frac{V_2}{I_1} = \frac{V_2}{(4s^5 + 4s^4 + 8s^3 + 6s^2 + 3s + 1)V_2} = \frac{1}{4s^5 + 4s^4 + 8s^3 + 6s^2 + 3s + 1}$$

$$\alpha_{12} = \frac{I_2}{I_1} = \frac{-V_2}{(4s^5 + 4s^4 + 8s^3 + 6s^2 + 3s + 1)V_2} = -\frac{1}{4s^5 + 4s^4 + 8s^3 + 6s^2 + 3s + 1}$$

$$Z_{11} = \frac{V_1}{I_1} = \frac{(4s^4 + 4s^3 + 6s^2 + 4s + 1)V_2}{(4s^5 + 4s^4 + 8s^3 + 6s^2 + 3s + 1)V_2} = \frac{4s^4 + 4s^3 + 6s^2 + 4s + 1}{4s^5 + 4s^4 + 8s^3 + 2s^2 + 3s + 1}$$

SP 9.6 Determine the Driving point impedance function for each of the following networks by continued fraction method.

SOLUTION:

(a) From Fig. SP 9.6(a), given $Z_1 = 2s\ \Omega$, $Y_2 = s\ \mho$, $Z_3 = 2s\ \Omega$, and $Y_4 = s\ \mho$. Hence

$$Z_{11} = 2s + \cfrac{1}{s + \cfrac{1}{2s + \frac{1}{s}}}$$

$$= \frac{4s^4 + 6s^2 + 1}{2s(s^2 + 1)}$$

(b) From Fig. SP 9.6(b), given $Z_1 = \frac{s(1/s)}{s + 1/s} = \frac{s}{s^2 + 1}\ \Omega$, $Y_2 = s\ \mho$, $Z_3 = \frac{s}{s^2 + 1}\ \Omega$, and $Y_4 = s\ \mho$.   Hence

$$Z_{11} = \frac{s}{s^2 + 1} + \cfrac{1}{s + \cfrac{1}{\frac{s}{s^2 + 1} + \frac{1}{s}}}$$

$$= \frac{5s^4 + 5s^2 + 1}{s(3s^4 + 5s^2 + 2)}$$

(c) From Fig. SP 9.6(c), given $Z_1 = 1\ \Omega$, $Y_2 = s\ \mho$, $Z_3 = 1\ \Omega$, and $Y_4 = s\ \mho$. Hence

$$Z_{11} = 1 + \cfrac{1}{s + \cfrac{1}{1 + \frac{1}{s}}}$$

$$= \frac{s^2 + 3s + 1}{s(s + 2)}$$

(d) From Fig. SP 9.6(d), given $Z_1 = \frac{1}{1 + 1/s + s} = \frac{s}{s^2 + s + 1}\ \Omega$, $Y_2 = s + 1\ \mho$, $Z_3 = \frac{1}{1 + 1/s} = \frac{s}{s + 1}\ \Omega$, and $Y_4 = s\ \mho$.   Hence

$$Z_{11} = \frac{s}{s^2 + s + 1} + \cfrac{1}{s + 1 + \cfrac{1}{\frac{s}{s + 1} + \frac{1}{s}}}$$

$$= \frac{2s^4 + 5s^3 + 6s^2 + 3s + 1}{s^5 + 4s^4 + 7s^3 + 7s^2 + 4s + 1)}$$

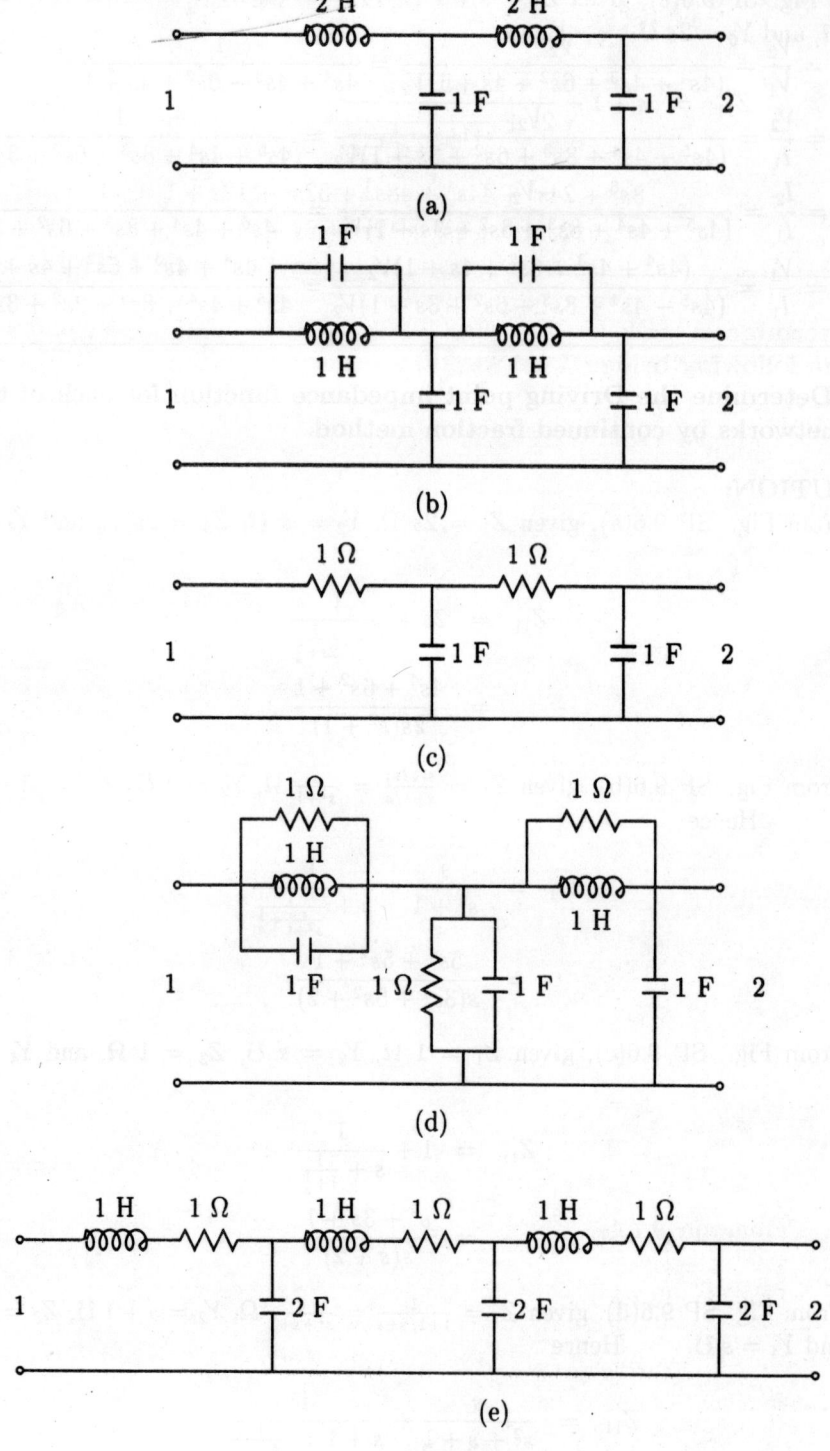

Fig. SP 9.6.

(e) From Fig. SP 9.6(e), given $Z_1 = s + 1\ \Omega$, $Y_2 = 2s\ \mho$, $Z_3 = s + 1\ \Omega$, $Y_4 = 2s\ \mho$, $Z_5 = s + 1\ \Omega$, and $Y_6 = 2s\ \mho$.   Hence

$$
Z_{11} = s + 1 + \cfrac{1}{2s + \cfrac{1}{s + 1 + \cfrac{1}{2s + \cfrac{1}{s + 1 + \frac{1}{2s}}}}}
$$

$$
= \frac{8s^6 + 24s^5 + 44s^4 + 48s^3 + 32s^2 + 12s + 1}{2s(4s^4 + 8s^3 + 12s^2 + 8s + 3)}
$$

SP 9.7 Determine $Y_{11} = V_1/I_1$, $Y_{12} = I_2/V_1$, $G_{12} = V_2/V_1$, $Z_{12} = V_2/I_1$ and $\alpha_{12} = I_2/I_1$ **for the following bridge T-networks.**

SOLUTION:

(a)

Mesh equations for the network of Fig. 9.7.1(a) are:

$$15I_1 + 5I_2 - 10I_3 = V_1 \qquad [9.7.1a]$$

$$5I_1 + 17I_2 + 10I_3 = 0 \qquad [9.7.2a]$$

and

$$-10I_1 + 10I_2 + 40I_3 = 0 \qquad [9.7.3a]$$

The determinant is

$$
\Delta = \begin{vmatrix} 15 & 5 & -10 \\ 5 & 17 & 10 \\ -10 & 10 & 40 \end{vmatrix} = 5000
$$

Using Cramer's rule,

$$
I_1 = \frac{1}{\Delta} \begin{vmatrix} V_1 & 5 & -10 \\ 0 & 17 & 10 \\ 0 & 10 & 40 \end{vmatrix} = \frac{580V_1}{\Delta} \qquad [9.7.4a]
$$

and

$$
I_2 = \frac{1}{\Delta} \begin{vmatrix} 15 & V_1 & -10 \\ 5 & 0 & 10 \\ -10 & 0 & 40 \end{vmatrix} = \frac{-300V_1}{\Delta} \qquad [9.7.5a]
$$

From eqn [9.7.4a] $Y_{11} = \dfrac{I_1}{V_1} = \dfrac{580}{\Delta} = \dfrac{580}{5000} = 29/250$

From eqn [9.7.5a] $Y_{12} = \dfrac{I_2}{V_1} = -\dfrac{300}{\Delta} = -3/50$

Further $G_{12} = \dfrac{V_2}{V_1} = \dfrac{-2I_2}{V_1} = 3/25$

$Z_{12} = \dfrac{V_2}{I_1} = \dfrac{-2I_2}{I_1} = 30/29$

$\alpha_{12} = \dfrac{I_2}{I_1} = -15/29.$

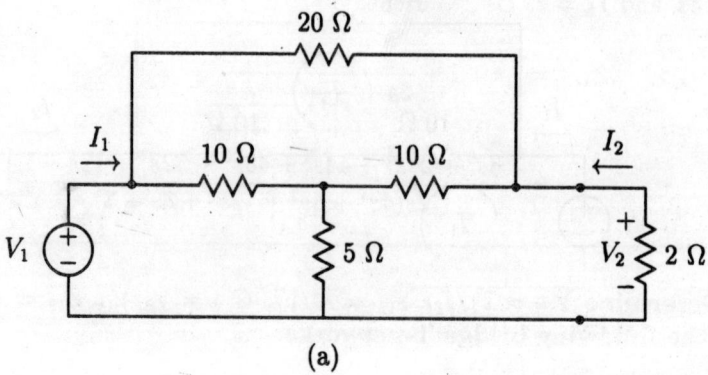

(a)

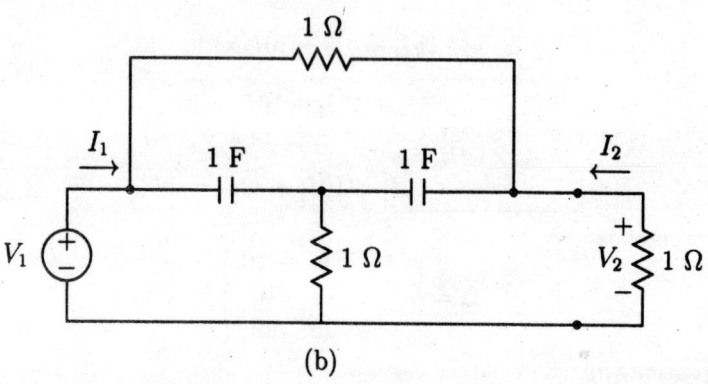

(b)

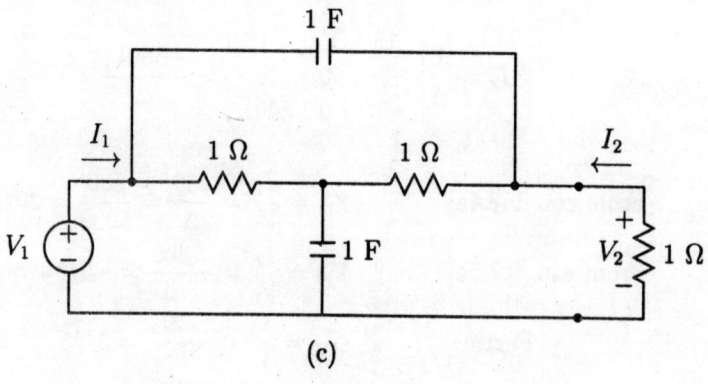

(c)

Fig. SP 9.7.

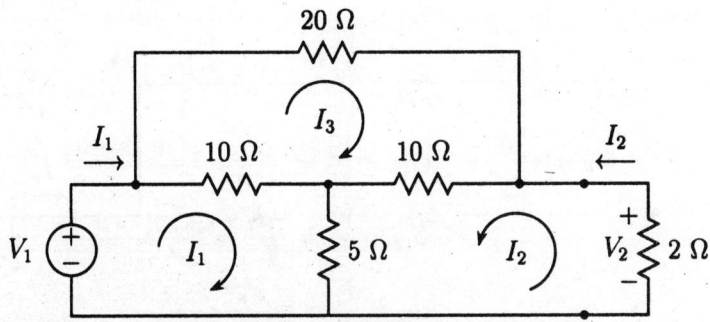

Fig. 9.7.1(a): Redrawn circuit of Fig. SP 9.7(a).

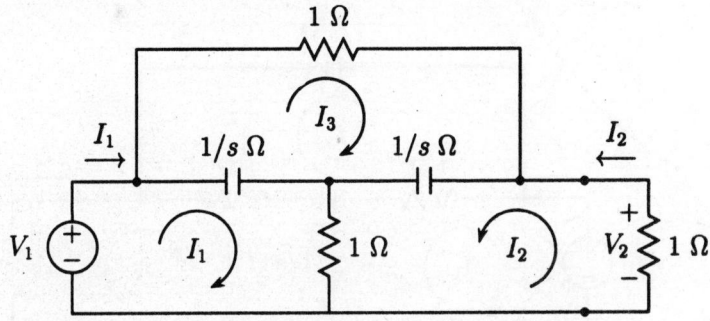

Fig. 9.7.1(b): Redrawn circuit of Fig. SP 9.7(b).

(b)

Mesh equations for the network of Fig. 9.7.1(b) are:

$$(1 + 1/s)I_1 + I_2 - (1/s)I_3 = V_1 \qquad [9.7.1b]$$

$$I_1 + (2 + 1/s)I_2 + (1/s)I_3 = 0 \qquad [9.7.2b]$$

and

$$-(1/s)I_1 + (1/s)I_2 + (1 + 2/s)I_3 = 0 \qquad [9.7.3b]$$

The determinant is

$$\Delta = \begin{vmatrix} 1+1/s & 1 & -1/s \\ 1 & 2+1/s & 1/s \\ -1/s & 1/s & 1+2/s \end{vmatrix} = 1 + 5/s + 2/s^2$$

Using Cramer's rule,

$$I_1 = \frac{1}{\Delta} \begin{vmatrix} V_1 & 1 & -1/s \\ 0 & 2+1/s & 1/s \\ 0 & 1/s & 1+2/s \end{vmatrix} = \frac{(2 + 5/s + 1/s^2)V_1}{\Delta} \qquad [9.7.4b]$$

and

$$I_2 = \frac{1}{\Delta} \begin{vmatrix} 1+1/s & V_1 & -1/s \\ 1 & 0 & 1/s \\ -1/s & 0 & 1+2/s \end{vmatrix} = \frac{-(1 + 2/s + 1/s^2)V_1}{\Delta} \qquad [9.7.5b]$$

From eqn [8 − 7.4b] $Y_{11} = \dfrac{I_1}{V_1} = \dfrac{2 + 5/s + 1/s^2}{\Delta} = \dfrac{2s^2 + 5s + 1}{s^2 + 5s + 2}$

From eqn [9.7.5b] $Y_{12} = \dfrac{I_2}{V_1} = -\dfrac{1 + 2/s + 1/s^2}{\Delta} = -\dfrac{s^2 + 2s + 1}{s^2 + 5s + 2}$

Further $G_{12} = \dfrac{V_2}{V_1} = \dfrac{-I_2}{V_1} = \dfrac{s^2 + 2s + 1}{s^2 + 5s + 2}$

$Z_{12} = \dfrac{V_2}{I_1} = \dfrac{-I_2}{I_1} = \dfrac{s^2 + 2s + 1}{2s^2 + 5s + 1}$

$\alpha_{12} = \dfrac{I_2}{I_1} = -\dfrac{s^2 + 2s + 1}{2s^2 + 5s + 1}$

(c)

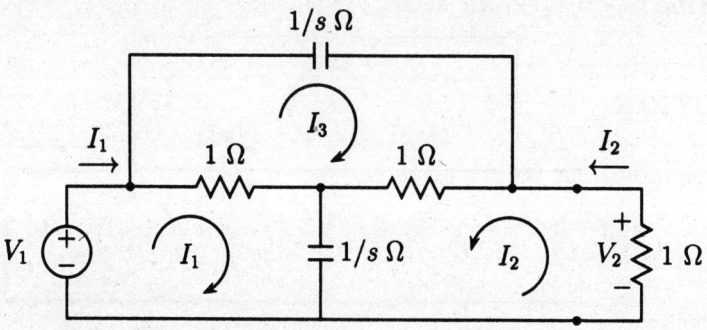

Fig. 9.7.1(c): Redrawn circuit of Fig. SP 9.7(c).

Mesh equations for the network of Fig. 9.7.1(c) are:

$$(1 + 1/s)I_1 + (1/s)I_2 - I_3 = V_1 \qquad [9.7.1c]$$

$$(1/s)I_1 + (2 + 1/s)I_2 + I_3 = 0 \qquad [9.7.2c]$$

and

$$-I_1 + I_2 + (2 + 1/s)I_3 = 0 \qquad [9.7.3c]$$

The determinant is

$$\Delta = \begin{vmatrix} 1 + 1/s & 1/s & -1 \\ 1/s & 2 + 1/s & 1 \\ -1 & 1 & 2 + 1/s \end{vmatrix} = 1 + 4/s + 3/s^2$$

Using Cramer's rule,

$$I_1 = \frac{1}{\Delta} \begin{vmatrix} V_1 & 1/s & -1 \\ 0 & 2 + 1/s & 1 \\ 0 & 1 & 2 + 1/s \end{vmatrix} = \frac{(3 + 4/s + 1/s^2)V_1}{\Delta} \qquad [9.7.4c]$$

and

$$I_2 = \frac{1}{\Delta} \begin{vmatrix} 1 + 1/s & V_1 & -1 \\ 1/s & 0 & 1 \\ -1 & 0 & 2 + 1/s \end{vmatrix} = \frac{-(1 + 2/s + 1/s^2)V_1}{\Delta} \qquad [9.7.5c]$$

From eqn [9.7.4c] $Y_{11} = \dfrac{I_1}{V_1} = \dfrac{3 + 4/s + 1/s^2}{\Delta} = \dfrac{3s^2 + 4s + 1}{s^2 + 4s + 3}$

From eqn [9.7.5c] $Y_{12} = \dfrac{I_2}{V_1} = -\dfrac{1 + 2/s + 1/s^2}{\Delta} = -\dfrac{s^2 + 2s + 1}{s^2 + 4s + 3}$

Further $G_{12} = \dfrac{V_2}{V_1} = \dfrac{-I_2}{V_1} = \dfrac{s^2 + 2s + 1}{s^2 + 4s + 3}$

$Z_{12} = \dfrac{V_2}{I_1} = \dfrac{-I_2}{I_1} = \dfrac{s^2 + 2s + 1}{3s^2 + 4s + 1}$

$\alpha_{12} = \dfrac{I_2}{I_1} = -\dfrac{s^2 + 2s + 1}{3s^2 + 4s + 1}$

SP 9.8 Find $G_{12} = V_2/V_1$ for each of the following bridge T-networks in Fig. 9.8.

SOLUTION:
(a)
Mesh equations for the circuit of Fig. 9.8.1(a) are:

$$3I_1 - 2I_3 = V_1 \qquad\qquad [9.8.1a]$$

$$I_1 + 2I_3 = V_2 \qquad\qquad [9.8.2a]$$

and

$$-2I_1 + 8I_3 = 0 \qquad\qquad [9.8.3a]$$

From eqn [9.8.3a]
$$2I_1 = 8I_3 \qquad \Rightarrow I_1 = 4I_3$$

Putting this value of I_1 into eqn [9.8.1a] and eqn [9.8.2a],

$$V_1 = 3 \times 4I_3 - 2I_3 = 10I_3$$
and $V_2 = 4I_3 + 2I_3 = 6I_3$
Hence $G_{12} = \dfrac{V_2}{V_1} = 6/10 = 3/5$

(b)
Mesh equations for the circuit of Fig. 9.8.1(b) are:

$$(1 + 1/s)I_1 - (1/s)I_3 = V_1 \qquad\qquad [9.8.1b]$$

$$I_1 + (1/s)I_3 = V_2 \qquad\qquad [9.8.2b]$$

and

$$-(1/s)I_1 + (1 + 2/s)I_3 = 0 \qquad\qquad [9.8.3b]$$

From eqn [9.8.3b]
$$I_1 = (2 + s)I_3$$

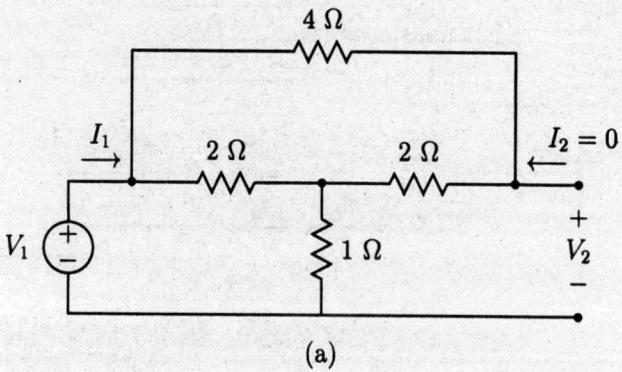

(a)

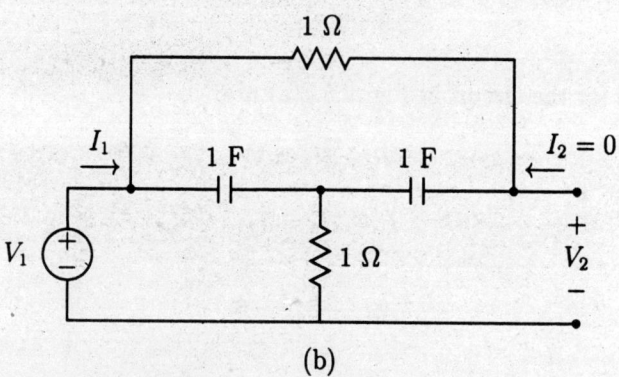

(b)

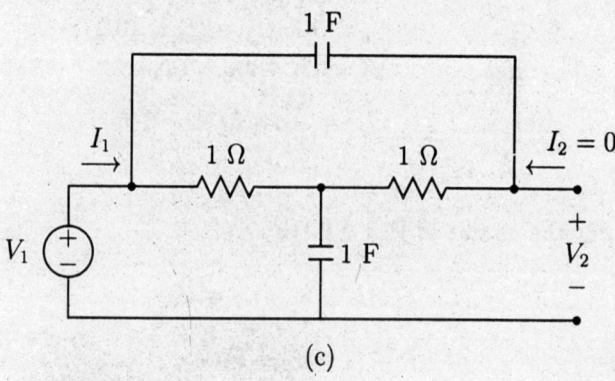

(c)

Fig. SP 9.8.

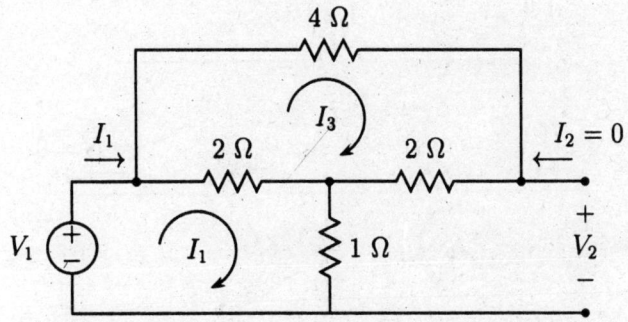

Fig. 9.8.1(a): Redrawn circuit of Fig. SP 9.8(a).

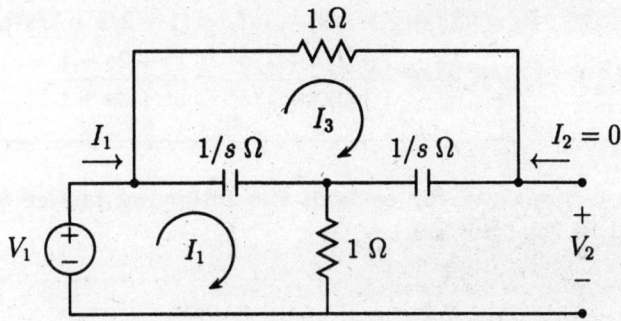

Fig. 9.8.1(b): Redrawn circuit of Fig. SP 9.8(b).

Putting this value of I_1 into eqn [9.8.1b] and eqn [9.8.2b],

$$V_1 = [(1 + 1/s)(2 + s) - 1/s]\,I_3 = (s + 3 + 1/s)I_3$$

and $\quad V_2 = [(2 + s) + 1/s]\,I_3$

Hence $\quad G_{12} = \dfrac{V_2}{V_1} = \dfrac{s + 2 + 1/s}{s + 3 + 1/s} = \dfrac{s^2 + 2s + 1}{s^2 + 3s + 1}$

(c)

Mesh equations for the circuit of Fig. 9.8.1(c) are:

$$(1 + 1/s)I_1 - I_3 = V_1 \qquad\qquad [9.8.1c]$$

$$(1/s)I_1 + I_3 = V_2 \qquad\qquad [9.8.2c]$$

and

$$-I_1 + (2 + 1/s)I_3 = 0 \qquad\qquad [9.8.3c]$$

From eqn [9.8.3c]

$$I_1 = (2 + 1/s)I_3$$

Putting this value of I_1 into eqn [9.8.1c] and eqn [9.8.2c],

$$V_1 = [(1 + 1/s)(2 + 1/s) - 1]\,I_3 = (1 + 3/s + 1/s^2)I_3$$

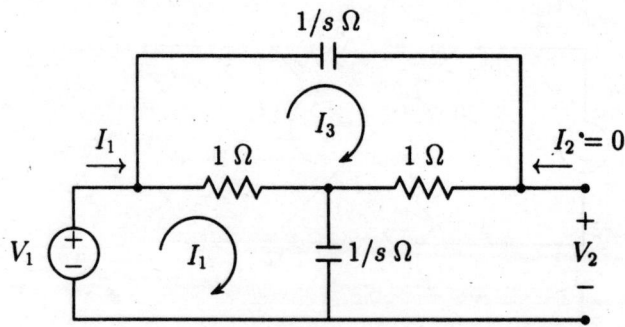

Fig. 9.8.1(c): Redrawn circuit of Fig. SP 9.8(c).

and $\qquad V_2 = [(1/s)(2 + 1/s) + 1]\, I_3 = (1 + 2/s + 1/s^2)I_3$

Hence $\qquad G_{12} = \dfrac{V_2}{V_1} = \dfrac{1 + 2/s + 1/s^2}{1 + 3/s + 1/s^2} = \dfrac{s^2 + 2s + 1}{s^2 + 3s + 1}$

SP 9.9 Find the $G_{12} = V_2/V_1$ for each of the following lattice networks (non-planar networks) in Fig. SP 9.9.

SOLUTION:
(a)
It is observed from Fig. 9.9.1(a) that

$$I_1' = \frac{V_1}{10 + 20} = \frac{V_1}{30} = I_1''$$

By KVL around the loop b-d-c-b,

$$V_2 = 20I_1' - 10I_1'' = 10I_1' = \frac{10}{30}V_1$$

Hence $\qquad G_{12} = \frac{V_2}{V_1} = 1/3$

(b)
It is observed from Fig. 9.9.1(b) that

$$I_1' = \frac{V_1}{1 + s + 1/s} = \frac{sV_1}{s^2 + s + 1} = I_1''$$

By KVL around the loop b-d-c-b,

$$V_2 = I_1' - (s + 1/s)I_1'' = (1 - s - 1/s)I_1' = -\frac{s^2 - s + 1}{s^2 + s + 1}V_1$$

Hence $\qquad G_{12} = \frac{V_2}{V_1} = -\frac{s^2 - s + 1}{s^2 + s + 1}$

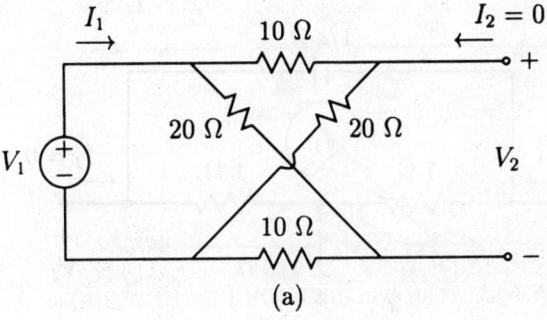

(a)

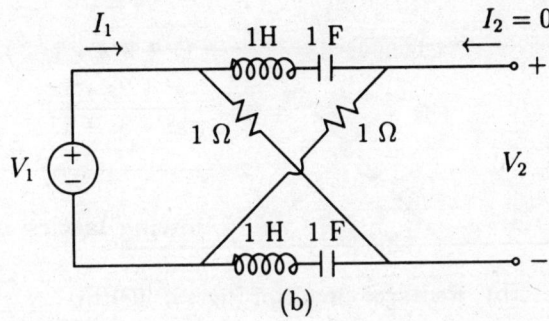

(b)

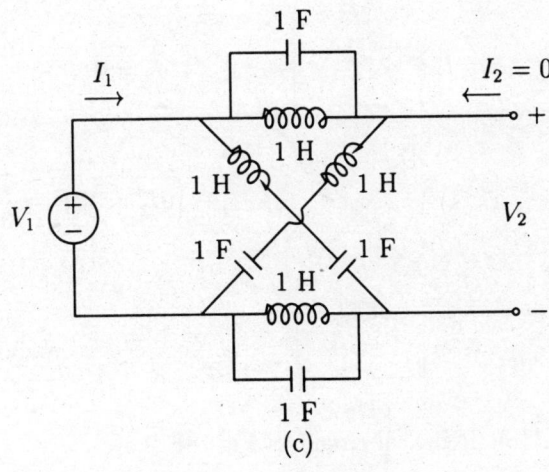

(c)

Fig. SP 9.9.

(c)

It is observed from Fig. 9.9.1(c) that

$$I_1' = \frac{V_1}{s + 1/s + (s\|1/s)} = \frac{s(s^2 + 1)V_1}{s^4 + 3s^2 + 1} = I_1''$$

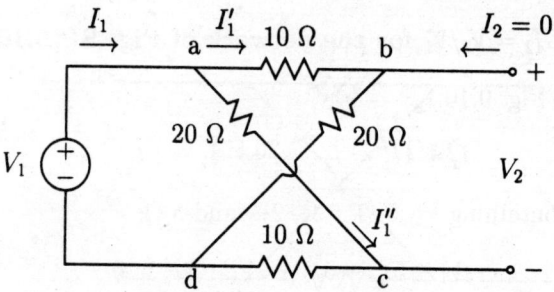

Fig. 9.9.1(a): Redrawn circuit of Fig. SP 9.9(a).

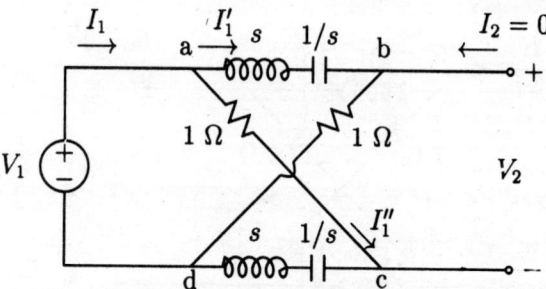

Fig. 9.9.1(b): Redrawn circuit of Fig. SP 9.9(b).

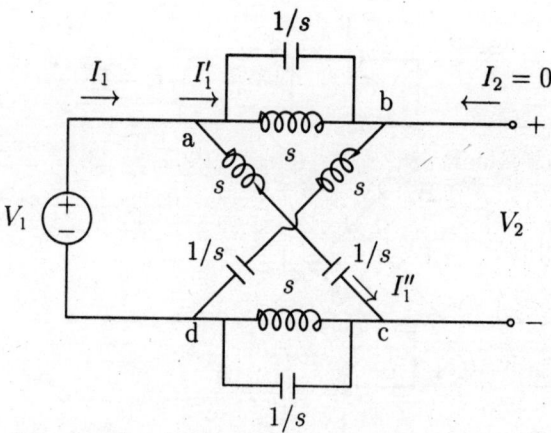

Fig. 9.9.1(c): Redrawn circuit of Fig. SP 9.9(c).

By KVL around the loop b-d-c-b,

$$V_2 = (s + 1/s)I_1' - (s||1/s)I_1'' = \frac{s^4 + s^2 + 1}{s^3 + s}I_1' = \frac{s^4 + s^2 + 1}{s(s^2 + 1)}\left[\frac{s(s^2 + 1)}{s^4 + 3s^2 + 1}V_1\right]$$

and $\quad G_{12} = \dfrac{V_2}{V_1} = \dfrac{s^4 + s^2 + 1}{s^4 + 3s^2 + 1}$

SP 9.10 Compute $G_{12} = V_2/V_1$ for the network of Fig. SP 9.10.

By KCL node **b** of Fig. 9.10.1,

$$I_x + 4I_x = I_2 \qquad \Rightarrow I_2 = 5I_x \qquad [9.10.1]$$

By KVL in the loop containing V_1, 5 Ω, 5 Ω, $2V_1$ and 5 Ω,

$$-V_1 + 5I_x + 5I_2 + 2V_1 + 5I_2 = 0$$

$$\text{Hence} \qquad V_1 + 5I_x + 10I_2 = 0 \qquad [9.10.2]$$

Putting the value I_2 from eqn [9.10.1] into eqn [9.10.2],

$$V_1 + 5I_x + 10 \times 5I_x = 0 \qquad \Rightarrow V_1 = -55I_x \qquad [9.10.3]$$

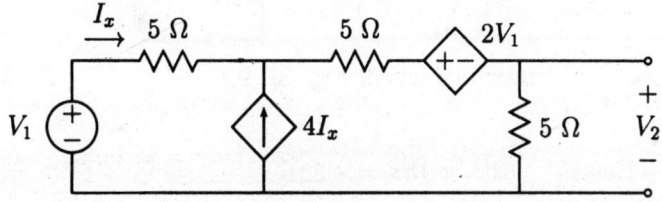

Fig. SP 9.10.

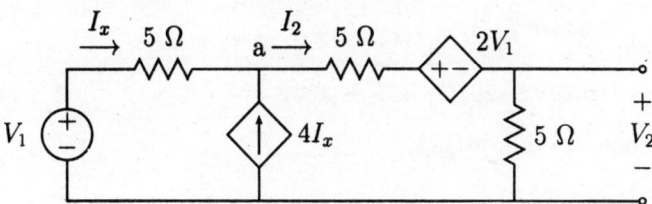

Fig. SP 9.10.1: Redrawn circuit of Fig. SP 9.10.

The output equation will be

$$V_2 = 5I_2 = 5 \times 5I_x = 25I_x \qquad [9.10.4]$$

From eqn [9.10.3] and eqn [9.10.4], we have

$$G_{12} = \frac{V_2}{V_1} = \frac{25I_x}{-55I_x} = -\frac{5}{11}$$

SP 9.11 Determine $\alpha_{12} = I_2/I_1$ for the network in Fig. SP 9.11.

SOLUTION:
By KVL around the second mesh from left of Fig. 9.11.1,

$$3I_x + 5\left(\frac{4}{3}I_1 + I_2 + I_x\right) + 5I_x = 0$$

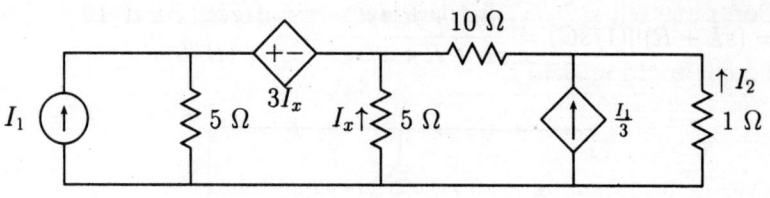

Fig. SP 9.11.

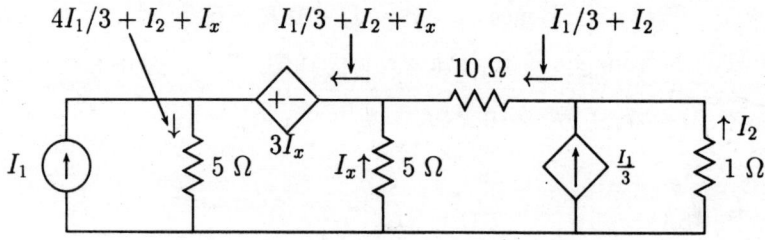

Fig. 9.11.1: Redrawn circuit of Fig. SP 9.11.

Hence $20I_1 + 15I_2 = -39I_x$ [9.11.1]

Again by KVL around the loop containing 1 Ω, 10 Ω, and 5 Ω, gives

$$10(I_2 + I_1/3) - 5I_x + I_2 = 0$$

Hence $10I_1 + 33I_2 = 15I_x$ [9.11.2]

On solving eqn [9.11.1] and eqn [9.11.2], we get

$$I_1 = -(252/85)I_x$$

and $I_2 = (23/17)I_x$

Therefore

$$\alpha_{12} = \frac{I_2}{I_1} = \frac{(23/17)I_x}{-(252/85)I_x} = -115/252$$

SP 9.12 **A one-port network has a coil of inductance L and resistance R shunted by a capacitance C. If the poles and zeros of driving point impedance $Z(s)$ of this network are:**
   poles at $-1 \pm j\sqrt{3}$
   and zeros at -2 with $Z(s)$ at $s = 0$ is 2,
   find $R, L,$ and C.

SOLUTION:
The driving point impedance $Z(s)$ of the network shown in Fig. 9.12 is

$$Z(s) = (sL + R)\|(1/sC) = \frac{(sL + R)/sC}{sL + R + 1/sC} = \frac{(1/C)(s + R/L)}{s^2 + (R/L)s + 1/LC}$$ [9.12.1]

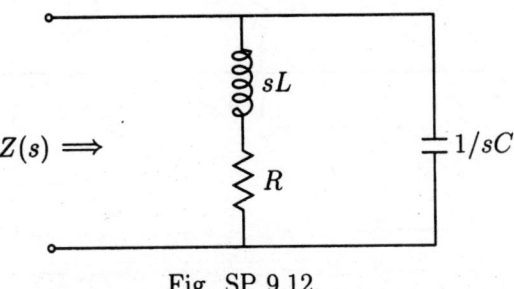

Fig. SP 9.12.

Therefore

$$Z(s) = \frac{(1/C)(s + R/L)}{\left[s + \frac{R}{2L} + j\sqrt{\frac{1}{LC} - (\frac{R}{2L})^2}\right]\left[s + \frac{R}{2L} - j\sqrt{\frac{1}{LC} - (\frac{R}{2L})^2}\right]}$$

Comparing this $Z(s)$ function with the given locations for the poles and zeros , we may write

$$\frac{R}{2L} = 1$$

$$\Rightarrow \frac{R}{L} = 2$$

and $$\sqrt{\frac{1}{LC} - \left(\frac{R}{2L}\right)^2} = \sqrt{3}$$

or $$\frac{1}{LC} - \left(\frac{R}{2L}\right)^2 = 3$$

Hence $$\frac{1}{LC} = 3 + 1^2 = 4$$ [9.12.3]

Also from eqn [9.12.1]

$$Z(s = 0) = 2 = \frac{(1/C)(R/L)}{1/LC} = R$$

Putting this value of R into eqn [9.12.2],

$$L = 1 \; H$$

Putting this value of L into eqn [9.12.3],

$$C = \frac{1}{4L} = 1/4 \; F$$

9.6   DRILL PROBLEMS

DP 9.1 Determine the transfer function $G_{12} = E_2/E_1$ for each of the following networks in Fig. DP 9.1.

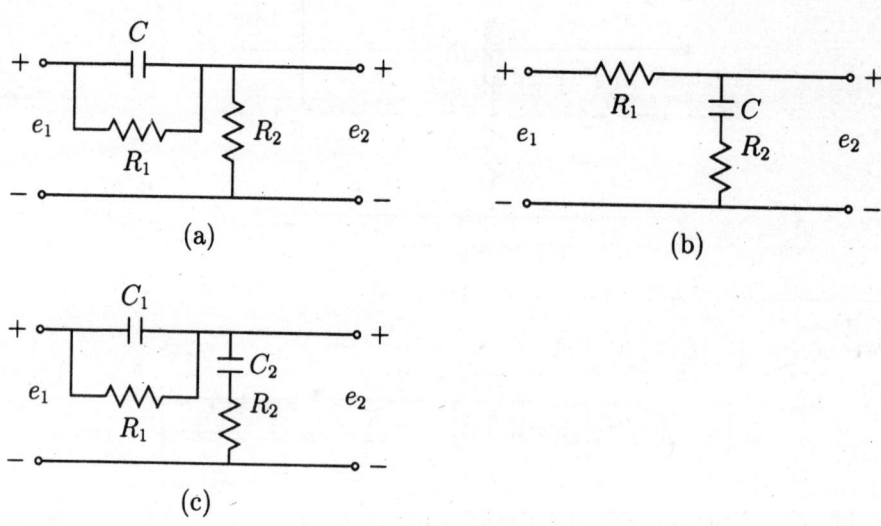

(a)

(b)

(c)

Fig. SP 9.1.

Hints:

(a) $$E_2/E_1 = \frac{R_2}{(R_1\|1/sC) + R_2}$$

(b) $$E_2/E_1 = \frac{R_2 + 1/sC}{R_1 + R_2 + 1/sC}$$

(a) $$E_2/E_1 = \frac{R_2 + 1/sC_2}{R_2 + 1/sC_2 + (R_1\|1/sC_1)}$$

DP 9.2 (a) Find the transfer function

$$G_{12} = \frac{V_2(s)}{V_1(s)}$$

for the network shown in Fig. DP 9.2.

(b) What is order of the system ?

(c) Now if the inductance value is changed to 2 H, what will be the order of the modified system.

GATE-91

Hints :

(a) $$V_2/V_1 = \frac{1 + 1/2s}{1 + 1/2s + [(1 + 1/s)\|(1 + s)]} = \frac{s + 1/2}{2(s + 1/4)}$$

(b) The order of the system is one.

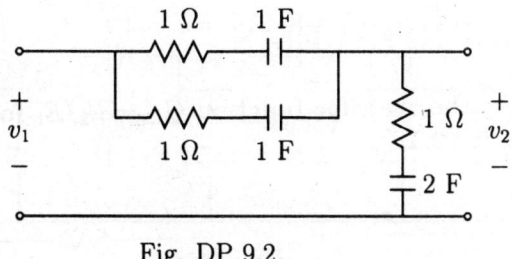

Fig. DP 9.2.

(c)
$$V_2/V_1 = \frac{1 + 1/2s}{1 + 1/2s + [(1 + 1/s)\|(1 + 2s)]}$$

DP 9.3 Determine (a) G_{12} (b) Z_{12} (c) Y_{12} and (d) α_{12} for the network in Fig. DP 9.3.

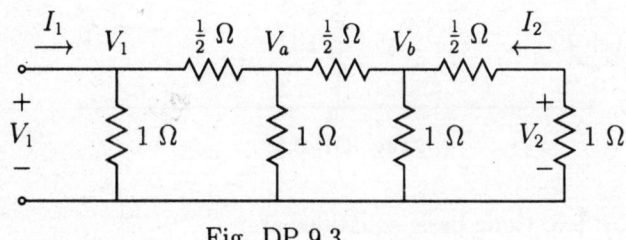

Fig. DP 9.3.

Hints: Node basis equations are:

$$I_1 = 3V_1 - 2V_a$$
$$5V_a - 2V_1 - 2V_b = 0$$
$$5V_b - 2V_a - 2V_2 = 0$$

and
$$3V_2 - 2V_b = 0$$

Express V_a, V_b, V_1, and I_1 in term of V_2 and then find the all.

DP 9.4 Determine $G_{12} = V_2/V_1$ for the network of Fig. DP 9.4.

Hint: Node equations at V_a and V_2 in Fig. DP 9.4 are:

$$\left(\frac{s^2+1}{s} + s + \frac{s^2+1}{s}\right)V_a - \frac{s^2+1}{s}V_1 - \frac{s^2+1}{s}V_2 = 0$$

and
$$\left(s + \frac{s^2+1}{s}\right)V_2 - \frac{s^2+1}{s}V_a = 0$$

Eliminate V_a to determine V_2/V_1.

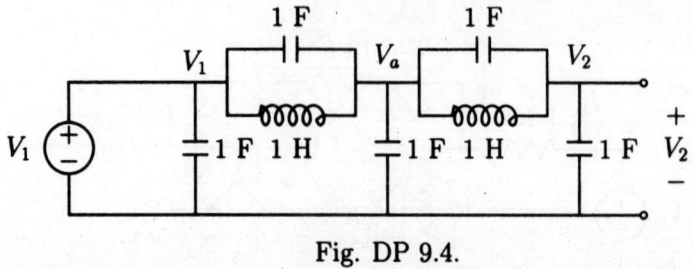

Fig. DP 9.4.

DP 9.5 Determine V_3/I_1 for the network in Fig. DP 9.5.

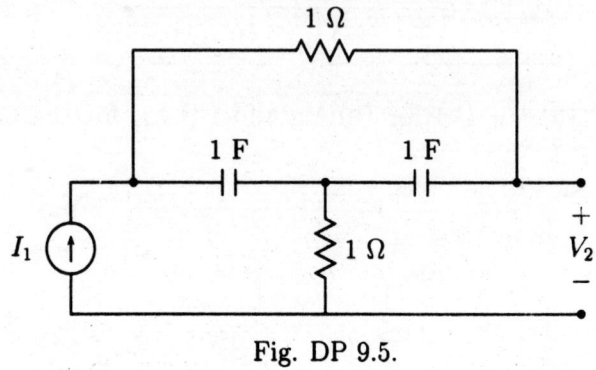

Fig. DP 9.5.

Hints : From Fig DP 9.5, Node basis equations are:

$$(s+1)V_1 - sV_2 - V_3 = I_1$$
$$- \quad sV_1 + (2s+1)V_2 - sV_3 = 0$$
$$\text{and} \quad - \quad V_1 - sV_2 + (s+1)V_3 = 0$$

Use Cramer's rule to find V_3 and so determine V_3/I_1.

DP 9.6 Determine V_2/V_i for the network of Fig. DP 9.6.

Hints : Node basis equations for the network of Fig. DP 9.6.1 are:

$$\left(\frac{1}{10^3} + \frac{1}{10^5} + \frac{1}{10^4}\right) V_1 - \frac{1}{10^4} V_2 = \frac{V_i}{10^3}$$

$$\text{and} \quad \left(\frac{1}{0.1} + \frac{1}{10^4}\right) V_2 - \frac{1}{10^4} V_1 = -10^6 V_1$$

$$\text{or} \quad \left(10^6 - \frac{1}{10^4}\right) V_1 + \left(\frac{1}{0.1} + \frac{1}{10^4}\right) V_2 = 0$$

Apply Cramer's rule to determine V_2 and so determine V_2/V_i

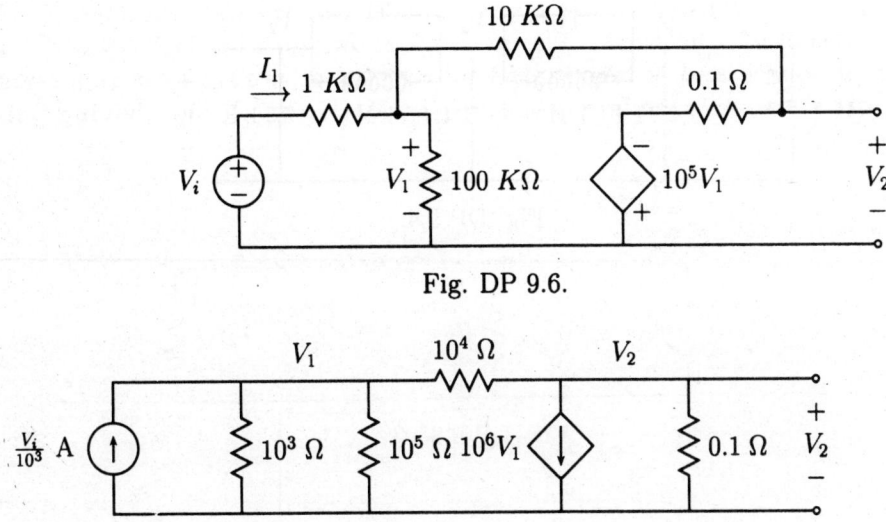

Fig. DP 9.6.

Fig. DP 9.6.1: Redrawn circuit of Fig. DP 9.6 with current sources.

DP 9.7 Determine the transfer function $G_{12} = V_2/V_1$ for the twin-T (or parallel T) network of Fig. DP 9.7.

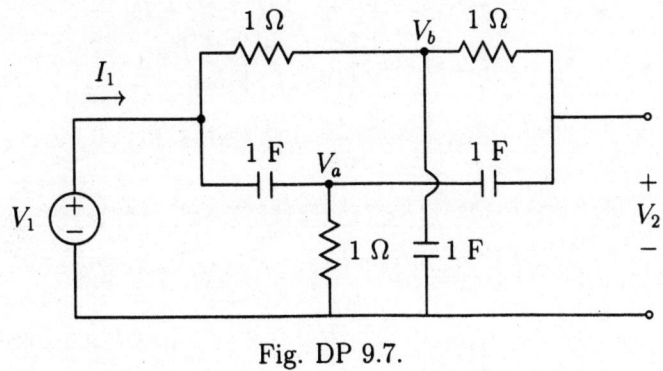

Fig. DP 9.7.

Hints : From Fig. DP 9.7, node basis equations at V_a, V_b, and V_2 are :

$$-sV_1 + (2s+1)V_a - sV_2 = 0$$
$$-V_1 + (2+s)V_b - V_2 = 0$$

and $\quad -sV_a - V_b + (s+1)V_2 = 0$

Express V_a in terms of V_1 and V_2 from last two equations and then substitute this value in the first equation to determine V_2/V_1

DP 9.8 Define the terms driving point impedance, transfer impedance, and transfer function of a network. The network shown in Fig. DP 9.8 is driven by a current source and is terminated by resistor at port 2. For this terminated 2-port network, calculate $G_{21}(s)$, $\alpha_{21}(s)$, $Z_{21}(s)$, $Y_{21}(s)$, and driving point impedance $Z_{11}(s)$.

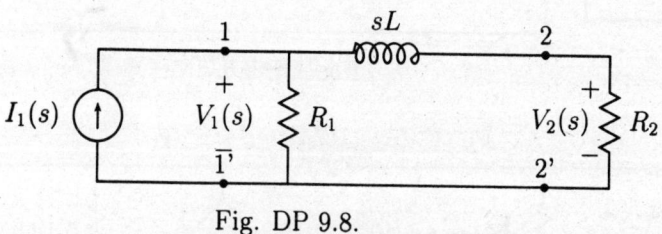

Fig. DP 9.8.

IES-94(EE)

Hints:

$$G_{21}(s) = \frac{V_2(s)}{V_1(s)} = \frac{R_2}{R_2 + sL}$$

$$\alpha_{21}(s) = \frac{I_2(s)}{I_1(s)} = \frac{-R_1}{sL + R_1 + R_2}$$

$$Z_{21}(s) = \frac{V_2(s)}{V_1(s)} = \frac{R_1 R_2}{sL + R_1 + R_2}$$

$$Y_{21}(s) = \frac{I_2(s)}{V_1(s)} = \frac{-V_2(s)/R_2}{V_1(s)}$$

$$Z_{11}(s) = \frac{V_1(s)}{I_1(s)} = R_1 \| (sL + R_2)$$

****

Two-Port Parameters

SUMMARY OF THIS CHAPTER

10.1 OPEN-CIRCUIT IMPEDANCE PARAMETERS

V_1 and V_2 are expressed in terms of I_1 and I_2 as:

$$V_1 = z_{11}I_1 + z_{12}I_2 \qquad (10.1)$$
$$V_2 = z_{21}I_1 + z_{22}I_2 \qquad (10.2)$$

In matrix form
$$\begin{bmatrix} V_1 \\ V_2 \end{bmatrix} = \begin{bmatrix} z_{11} & z_{12} \\ z_{21} & z_{22} \end{bmatrix} \begin{bmatrix} I_1 \\ I_2 \end{bmatrix} \qquad (10.3)$$

From equations 10.1 and 10.2
$$z_{11} = \left. \frac{V_1}{I_1} \right|_{I_2=0}$$

$$z_{21} = \left. \frac{V_2}{I_1} \right|_{I_2=0}$$

$$z_{12} = \left. \frac{V_1}{I_2} \right|_{I_1=0}$$

and $\qquad z_{22} = \left. \frac{V_2}{I_2} \right|_{I_1=0}$

The z_{11} and z_{21} are determined from Figure 10.1, and the remaining two impedances z_{12} and z_{22} are obtained from Figure 10.2 as:

10.2 SHORT-CIRCUIT ADMITTANCE PARAMETERS

I_1 and I_2 are expressed in terms of V_1 and V_2 as:

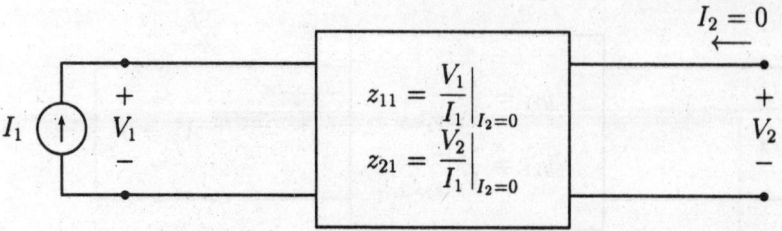

Figure 10.1: Determination of z_{11} and z_{21}.

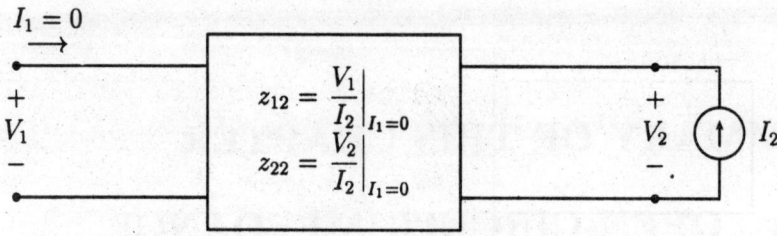

Figure 10.2: Determination of z_{12} and z_{22}.

$$I_1 = y_{11}V_1 + y_{12}V_2 \qquad (10.4)$$
$$I_2 = y_{21}V_1 + y_{22}V_2 \qquad (10.5)$$

In matrix form

$$\begin{bmatrix} I_1 \\ I_2 \end{bmatrix} = \begin{bmatrix} y_{11} & y_{12} \\ y_{21} & y_{22} \end{bmatrix} \begin{bmatrix} V_1 \\ V_2 \end{bmatrix} \qquad (10.6)$$

From equations 10.4 and 10.5

$$y_{11} = \left. \frac{I_1}{V_1} \right|_{V_2=0}$$

$$y_{21} = \left. \frac{I_2}{V_1} \right|_{V_2=0}$$

$$y_{12} = \left. \frac{I_1}{V_2} \right|_{V_1=0}$$

and $\qquad y_{22} = \left. \frac{I_2}{V_2} \right|_{V_1=0}$

The y_{11} and y_{21} are determined from Figure 10.3, and the remaining two admittances y_{12} and y_{22} are obtained from Figure 10.4 as:

10.3   TRANSMISSION PARAMETERS

The transmission parameters are also called the $ABCD$ parameters or chain parameters. V_1 and I_1 are expressed in terms of V_2 and I_2 as:

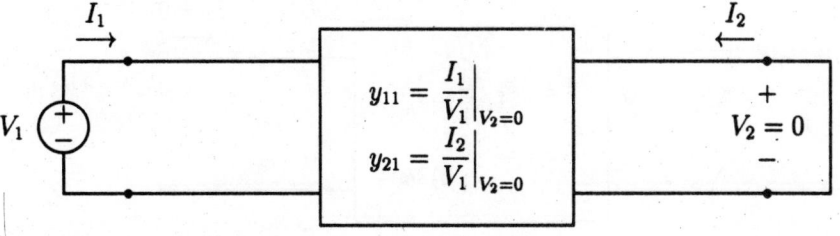

Figure 10.3: Determination of y_{11} and y_{21}.

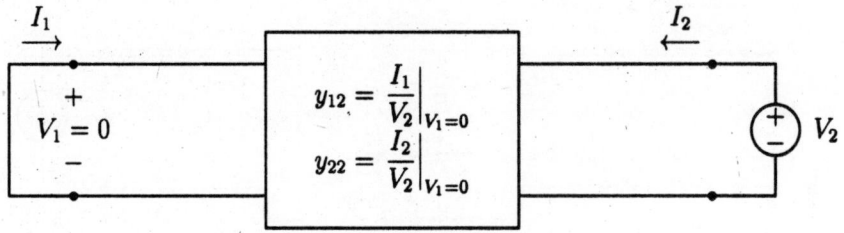

Figure 10.4: Determination of y_{12} and y_{22}.

$$V_1 = AV_2 - BI_2 \qquad (10.7)$$
$$I_1 = CV_2 - DI_2 \qquad (10.8)$$

In matrix form

$$\begin{bmatrix} V_1 \\ I_1 \end{bmatrix} = \begin{bmatrix} A & B \\ C & D \end{bmatrix} \begin{bmatrix} V_2 \\ -I_2 \end{bmatrix} \qquad (10.9)$$

From equations 10.7 and 10.8

$$\frac{1}{A} = \left. \frac{V_2}{V_1} \right|_{I_2=0}$$

$$\frac{1}{C} = \left. \frac{V_2}{I_1} \right|_{I_2=0}$$

$$-\frac{1}{B} = \left. \frac{I_2}{V_1} \right|_{V_2=0}$$

and

$$-\frac{1}{D} = \left. \frac{I_2}{I_1} \right|_{V_2=0}$$

The A and C are determined from Figure 10.5 and remaining two parameters B and D are obtained from Figure 10.6 as:

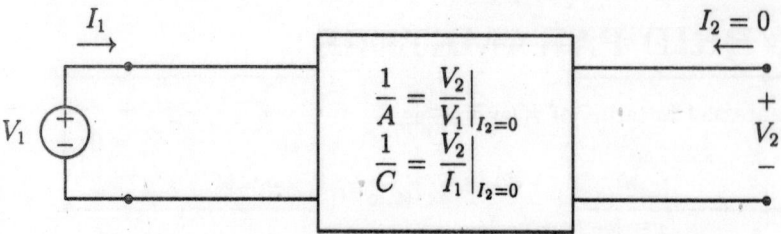

Figure 10.5: Determination of A and C.

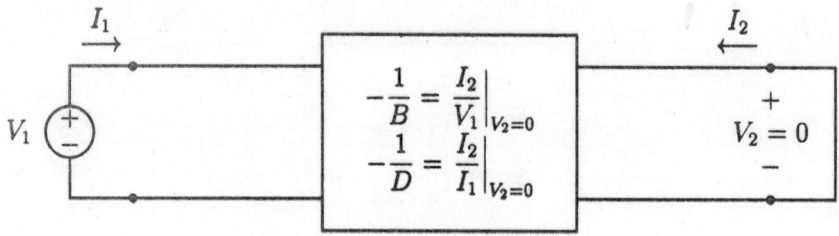

Figure 10.6: Determination of B and D.

10.4 INVERSE TRANSMISSION PARAMETERS

V_2 and I_2 are expressed in terms of V_1 and I_1 as:

$$V_2 = A'V_1 - B'I_1 \tag{10.10}$$
$$I_2 = C'V_1 - D'I_1 \tag{10.11}$$

In matrix form

$$\begin{bmatrix} V_2 \\ I_2 \end{bmatrix} = \begin{bmatrix} A' & B' \\ C' & D' \end{bmatrix} \begin{bmatrix} V_1 \\ -I_1 \end{bmatrix} \tag{10.12}$$

From equations 10.10 and 10.11

$$A' = \left. \frac{V_2}{V_1} \right|_{I_1=0}$$

$$C' = \left. \frac{I_2}{V_1} \right|_{I_1=0}$$

$$-B' = \left. \frac{V_2}{I_1} \right|_{V_1=0}$$

and

$$-D' = \left. \frac{I_2}{I_1} \right|_{V_1=0}$$

10.5 HYBRID PARAMETERS

V_1 and I_2 are expressed in terms of I_1 and V_2 as:

$$V_1 = h_{11}I_1 + h_{12}V_2 \qquad (10.13)$$
$$I_2 = h_{21}I_1 + h_{22}V_2 \qquad (10.14)$$

In matrix form

$$\begin{bmatrix} V_1 \\ I_2 \end{bmatrix} = \begin{bmatrix} h_{11} & h_{12} \\ h_{21} & h_{22} \end{bmatrix} \begin{bmatrix} I_1 \\ V_2 \end{bmatrix} \qquad (10.15)$$

From equations 10.13 and 10.14

$$h_{11} = \left.\frac{V_1}{I_1}\right|_{V_2=0} \qquad h_{21} = \left.\frac{I_2}{I_1}\right|_{V_2=0}$$

$$h_{12} = \left.\frac{V_1}{V_2}\right|_{I_1=0} \quad \text{and} \quad h_{22} = \left.\frac{I_2}{V_2}\right|_{I_1=0}$$

The h_{11} and h_{21} are determined from Figure 10.7, and the remaining two parameters h_{12} and h_{22} are obtained from Figure 10.8 as:

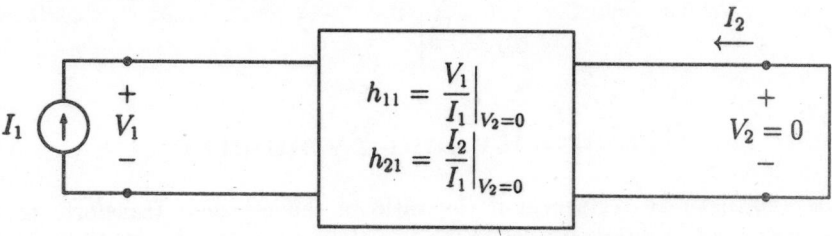

Figure 10.7: Determination of h_{12} and h_{22}.

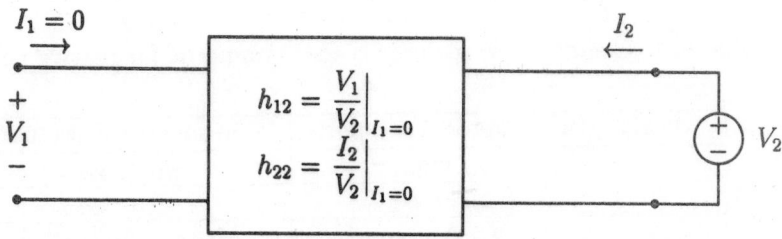

Figure 10.8: Determination of h_{11} and h_{21}.

10.6   INVERSE HYBRID PARAMETERS

I_1 and V_2 are expressed in terms of V_1 and I_2 as:

$$I_1 = g_{11}V_1 + g_{12}I_2 \tag{10.16}$$

$$V_2 = g_{21}V_1 + g_{22}I_2 \tag{10.17}$$

In matrix form

$$\begin{bmatrix} I_1 \\ V_2 \end{bmatrix} = \begin{bmatrix} g_{11} & g_{12} \\ g_{21} & g_{22} \end{bmatrix} \begin{bmatrix} V_1 \\ I_2 \end{bmatrix} \tag{10.18}$$

From equations 10.16 and 10.17

$$g_{11} = \left. \frac{I_1}{V_1} \right|_{I_2=0}$$

$$g_{21} = \left. \frac{V_2}{V_1} \right|_{I_2=0}$$

$$g_{12} = \left. \frac{I_1}{I_2} \right|_{V_1=0}$$

and

$$g_{22} = \left. \frac{V_2}{I_2} \right|_{V_1=0}$$

Condition for Reciprocity and Symmetry

A network is known to be reciprocal if the ratio of the response transform to input transform is invariant to an interchange of the position of the input and the response in the network.

A two-port network is known to be symmetric if the port can be interchanged without changing the port voltages and currents.

Table 10.1: Condition for reciprocity and symmetric for passive networks.

Matrix	for reciprocal two-ports	for symmetrical two-ports
Y	$y_{12} = y_{21}$	$y_{11} = y_{22}$
Z	$z_{12} = z_{21}$	$z_{11} = z_{22}$
h	$h_{12} = -h_{21}$	$h_{11}h_{22} - h_{12}h_{21} = 1$
g	$g_{12} = -g_{21}$	$g_{11}g_{22} - g_{12}g_{21} = 1$
$ABCD$	$AD - BC = 1$	$A = D$
$A'B'C'D'$	$A'D' - B'C' = 1$	$A' = D'$

10.7 SOLVED PROBLEMS

SP 10.1 Find the y and z parameters for each of the following networks in Fig. SP 10.1.

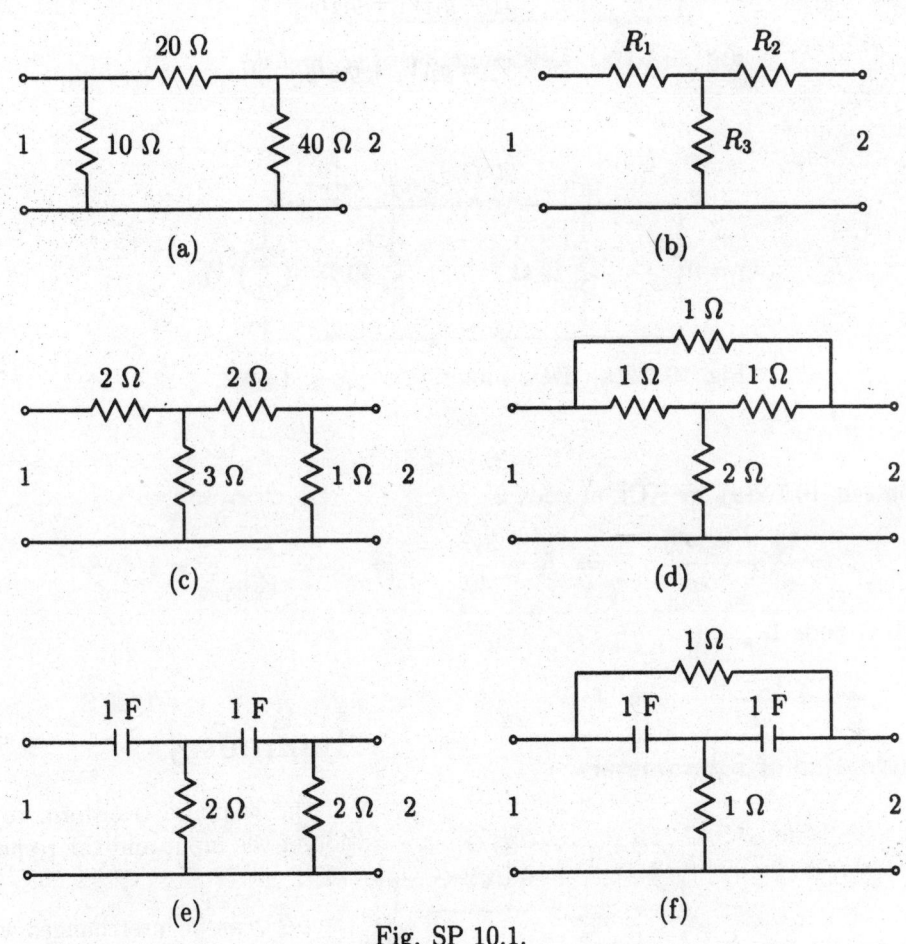

Fig. SP 10.1.

SOLUTION:

(a) Determination of y-parameters:

From Fig. 10.1.1(a), by KCL at node **1**

$$I_1 = \frac{V_1}{10} + \frac{V_1 - 0}{20} \quad \Rightarrow \quad I_1 = \frac{3V_1}{20} \quad \Rightarrow \quad y_{11} = \left.\frac{I_1}{V_1}\right|_{V_2=0} = 3/20 \text{ S}$$

By KCL at node **2**

$$-I_2 = \frac{V_1 - 0}{20} \quad \Rightarrow \quad y_{21} = \left.\frac{I_2}{V_1}\right|_{V_2=0} = -1/20 \text{ S}$$

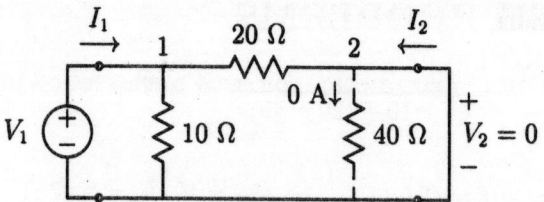

Fig. 10.1.1(a): Determination of y_{11} and y_{21}.

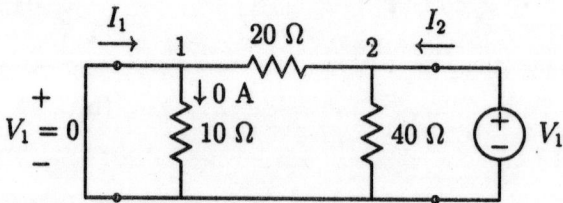

Fig. 10.1.2(a): Determination of y_{12} and y_{22}.

From Fig. 10.1.2(a), by KCL at node **2**

$$I_2 = \frac{V_2}{40} + \frac{V_2 - 0}{20} \qquad \Rightarrow I_2 = \frac{3V_2}{40} \qquad \Rightarrow y_{22} = \frac{I_2}{V_2}\bigg|_{V_1=0} = 3/40 \text{ S}$$

By KCL at node **1**

$$-I_1 = \frac{V_2 - 0}{20} \qquad \Rightarrow I_1 = -\frac{1}{20}V_2 \qquad \Rightarrow y_{12} = \frac{I_1}{V_2}\bigg|_{V_1=0} = -1/20 \text{ S}$$

Determination of z-parameters

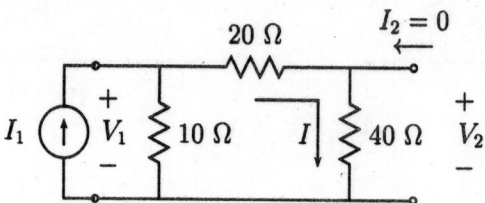

Fig. 10.1.3(a): Determination of z_{11} and z_{21}.

By KCL at node **1** of Fig. 9-1.3(a),

$$I_1 = \frac{V_1}{10} + \frac{V_1}{20 + 40} \qquad \Rightarrow I_1 = \frac{7V_1}{60} \qquad \Rightarrow z_{11} = \frac{V_1}{I_1}\bigg|_{I_2=0} = \frac{60}{7} \ \Omega$$

Alternately the z_{11} which is input impedance, can also be computed as:

$$z_{11} = 10\|(20 + 40) = \frac{10 \times 60}{10 + 60} = 60/7 \ \Omega$$

By current division formula,

$$I = \frac{10}{10 + (20 + 40)} I_1 = I_1/7$$

then $\qquad V_2 = 40I = 40I_1/7 \qquad \Rightarrow z_{21} = V_2/I_1 = 40/7 \; \Omega$

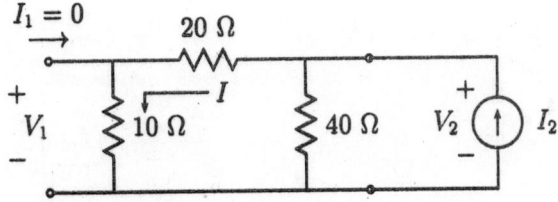

Fig. 10.1.4(a): Determination of z_{12} and z_{22}.

From Fig. 10.1.4(a), z_{22} can be determined as

$$z_{22} = 40||(20 + 10) = \frac{40 \times 30}{40 + 30} = 120/7 \; \Omega$$

By current division formula,

$$I = \frac{40}{40 + (20 + 10)} I_2 = \frac{4}{7} I_2$$

then $\qquad V_1 = 10I = 10 \times \frac{4}{7} I_2 \qquad \Rightarrow z_{12} = V_1/I_2 = 40/7 \; \Omega$

(b) **Determination of y-parameters:**

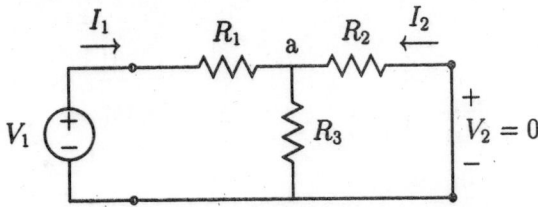

Fig. 10.1.1(b): Determination of y_{11} and y_{21}.

By KCL at node a of Fig. 10.1.1(b),

$$\frac{V_1 - V_a}{R_1} = \frac{V_a}{R_3} + \frac{V_a - 0}{R_2} \qquad \Rightarrow V_a = \frac{R_2 R_3}{R_1 R_2 + R_2 R_3 + R_1 R_3} V_1$$

I_1 can be expressed as

$$I_1 = \frac{V_a}{R_3} + \frac{V_a}{R_2}$$

Putting the value of V_a in the last equation, we get

$$I_1 = \frac{R_2 + R_3}{R_1 R_2 + R_2 R_3 + R_1 R_3} V_1 \qquad \Rightarrow y_{11} = \frac{I_1}{V_1} = \frac{R_2 + R_3}{R_1 R_2 + R_2 R_3 + R_1 R_3}$$

I_2 may be given by equation.

$$-I_2 = \frac{V_a}{R_2} = \frac{R_3}{R_1 R_2 + R_2 R_3 + R_1 R_3} V_1 \qquad \Rightarrow y_{21} = \frac{I_2}{V_1} = -\frac{R_3}{R_1 R_2 + R_2 R_3 + R_1 R_3}$$

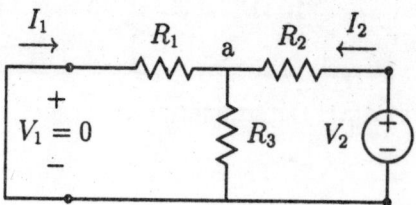

Fig. 10.1.2(b): Determination of y_{12} and y_{22}.

From Fig. 10.1.2(b), KCL at node **a** is

$$\frac{V_2 - V_a}{R_2} = \frac{V_a}{R_3} + \frac{V_a - 0}{R_1} \qquad \Rightarrow V_a = \frac{R_1 R_3}{R_1 R_2 + R_2 R_3 + R_1 R_3} V_2$$

I_2 will be

$$I_2 = \frac{V_a}{R_3} + \frac{V_a}{R_1} = \frac{R_1 + R_3}{R_1 R_3} V_a$$

Setting the value of V_a in the last equation and then simplifying, we have

$$I_2 = \frac{R_1 + R_3}{R_1 R_2 + R_2 R_3 + R_1 R_3} V_2 \qquad \Rightarrow y_{22} = I_2 / V_2 = \frac{R_1 + R_3}{R_1 R_2 + R_2 R_3 + R_1 R_3}$$

I_1 may be expressed as

$$-I_1 = \frac{V_a}{R_1} = \frac{R_3}{R_1 R_2 + R_2 R_3 + R_1 R_3} V_2 \qquad \Rightarrow Y_{12} = I_1 / V_2 = -\frac{R_3}{R_1 R_2 + R_2 R_3 + R_1 R_3}$$

Determination of z parameters:
    From Fig. 10.1.3(b),

$$z_{11} = \left. \frac{V_1}{I_1} \right|_{I_2=0} = \frac{(R_1 + R_3) I_1}{I_1} = R_1 + R_3$$

and

$$z_{21} = \left. \frac{V_2}{I_1} \right|_{I_2=0} = \frac{R_3 I_1}{I_1} = R_3$$

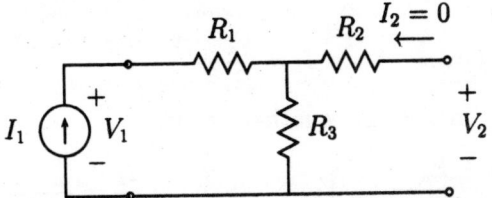

Fig. 10.1.3(b): Determination of z_{11} and z_{21}.

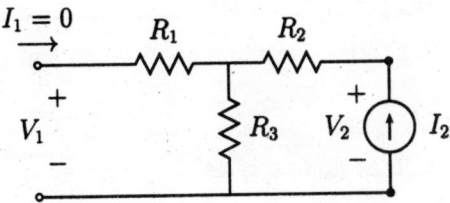

Fig. 10.1.4(b): Determination of z_{12} and z_{22}.

From Fig. 10.1.4(b),

$$z_{22} = \left.\frac{V_2}{I_2}\right|_{I_1=0} = \frac{(R_2 + R_3)I_2}{I_2} = R_2 + R_3$$

and

$$z_{12} = \left.\frac{V_1}{I_2}\right|_{I_1=0} = \frac{R_3 I_2}{I_2} = R_3$$

(c) **Determination of y parameters:**

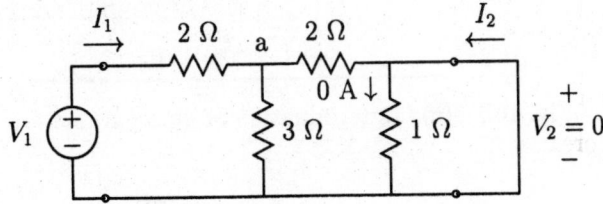

Fig. 10.1.1(c): Determination of y_{11} and y_{21}.

From Fig. 10.1.1(c), KCL at node **a** gives

$$\frac{V_1 - V_a}{2} = \frac{V_a}{3} + \frac{V_a - 0}{2} \qquad \Rightarrow V_a = \frac{3}{8}V_1$$

Now

$$y_{11} = \left.\frac{I_1}{V_1}\right|_{V_2=0} = \frac{(V_1 - V_a)/2}{V_1} = \frac{V_1 - (3/8)V_1}{2V_1} = 5/16 \text{ S}$$

and $\qquad y_{21} = \dfrac{I_2}{V_1}\bigg|_{V_2=0} = \dfrac{-V_a/2}{V_1} = \dfrac{-(1/2)(3/8)V_1}{V_1} = -3/16 \text{ S}$

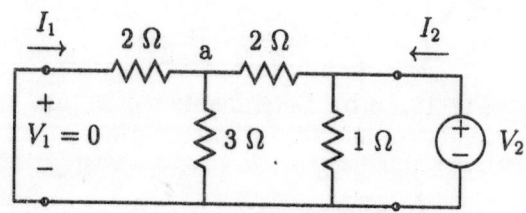

Fig. 10.1.2(c): Determination of y_{12} and y_{22}.

From Fig. 10.1.2(c), KCL at node a gives

$$\frac{V_2 - V_a}{2} = \frac{V_a}{3} + \frac{V_a - 0}{2} \qquad \Rightarrow V_a = \frac{3}{8}V_2$$

Now

$$y_{12} = \frac{I_1}{V_2}\bigg|_{V_1=0} = \frac{-V_a/2}{V_2} = \frac{-(1/2)(3/8)V_2}{V_2} = -3/16 \text{ S}$$

and $\qquad y_{22} = \dfrac{I_2}{V_2}\bigg|_{V_1=0} = \dfrac{V_2/1 + (V_2 - V_a)/2}{V_2} = \dfrac{(3/2)V_2 - (1/2)(3/8)V_2}{V_2} = 21/16 \text{ S}$

Determination of z parameters:

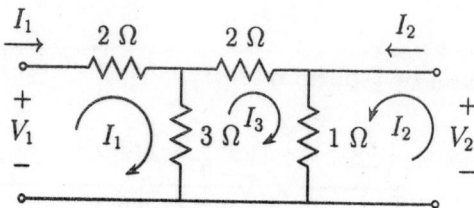

Fig. 10.1.3(c): Determination of all z parameters.

Mesh-current equations for the network of Fig. 10.1.3(c) are:
Mesh 1:

$$5I_1 - 3I_3 = V_1 \qquad\qquad\qquad [10.1.1c]$$

Mesh 2:

$$I_2 + I_3 = V_2 \qquad\qquad\qquad [10.1.2c]$$

Mesh 3:

$$-3I_1 + I_2 + 6I_3 = 0 \qquad\qquad\qquad [10.1.3c]$$

From eqn [10.1.3c]

$$I_3 = I_1/2 - I_2/6$$

Putting this value of I_3 into equations [10.1.1c] and [10.1.2c],

$$V_1 = 5I_1 - 3(I_1/2 - I_2/6)$$

Hence

$$V_1 = \frac{7}{2}I_1 + \frac{1}{2}I_2 \tag{10.1.4c}$$

and

$$V_2 = I_2 + \frac{1}{2}I_1 - \frac{1}{6}I_2$$

Hence

$$V_2 = \frac{1}{2}I_1 + \frac{5}{6}I_2 \tag{10.1.5c}$$

From equations [10.1.4c] and [10.1.5c], we can directly write

$$z_{11} = 7/2 \ \Omega, \ z_{12} = 1/2 \ \Omega, \ z_{21} = 1/2 \ \Omega, \ \text{and} \ z_{22} = 5/6 \ \Omega$$

(d) **Determination of y parameters:**

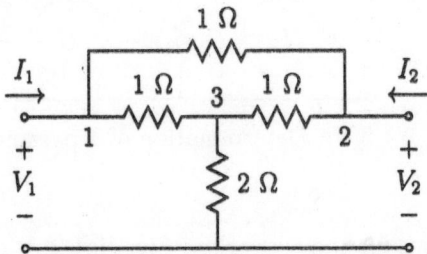

Fig. 10.1.1(d): Determination of all y parameters.

Node basis equations for the circuit of Fig. 10.1.1(d) are:

Node 1:

$$2V_1 - V_2 - V_3 = I_1 \tag{10.1.1d}$$

Node 2:

$$-V_1 + 2V_2 - V_3 = I_2 \tag{10.1.2d}$$

Node 3:

$$-V_1 - V_2 + (5/2)V_3 = 0 \tag{10.1.3d}$$

From eqn [10.1.3d]

$$V_3 = \frac{2}{5}(V_1 + V_2)$$

Putting this value of V_3 into equations [10.1.1d] and [10.1.2d]

$$2V_1 - V_2 - \frac{2}{5}(V_1 + V_2) = I_1$$

Thus

$$\frac{8}{5}V_1 - \frac{7}{5}V_2 = I_1 \tag{10.1.4d}$$

Next
$$-V_1 + 2V_2 - \frac{2}{5}(V_1 + V_2) = I_2$$

Thus
$$-\frac{7}{5}V_1 + \frac{8}{5}V_2 = I_2 \qquad\qquad [10.1.5d]$$

From equations [10.1.4d] and [10.1.5d], we can directly write

$$y_{11} = 8/5 \text{ S}, \ y_{12} = -7/5 \text{ S}, \ y_{21} = -7/5 \text{ S}, \text{ and } y_{22} = 8/5 \text{ S}$$

Determination of z parameters:

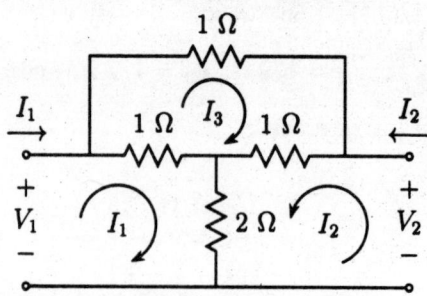

Fig. 10.1.2(d): Determination of z parameters.

Mesh-current equations for the network of Fig. 10.1.2(d) are:

Mesh 1:
$$3I_1 + 2I_2 - I_3 = V_1 \qquad\qquad [10.1.6d]$$

Mesh 2:
$$2I_1 + 3I_2 + I_3 = V_2 \qquad\qquad [10.1.7d]$$

Mesh 3:
$$-I_1 + I_2 + 3I_3 = 0 \qquad\qquad [10.1.8d]$$

From eqn [10.1.8d],
$$I_3 = (I_1 - I_2)/3$$

Putting this value of I_3 into equations [10.1.6d] and [10.1.7d],

$$3I_1 + 2I_2 - (I_1 - I_2)/3 = V_1$$

Thus
$$\frac{8}{3}I_1 + \frac{7}{3}I_2 = V_1 \qquad\qquad [10.1.9d]$$

and
$$2I_1 + 3I_2 + (I_1 - I_2)/3 = V_2$$

Thus
$$\frac{7}{3}I_1 + \frac{8}{3}I_2 = V_2 \qquad\qquad [10.1.10d]$$

From equations [10.1.9d] and [10.1.10d], we can directly write

$$z_{11} = 8/3 \; \Omega, \; z_{12} = 7/3 \; \Omega, \; z_{21} = 7/3 \; \Omega \text{ and } z_{22} = 8/3 \; \Omega$$

(e) Determination of y parameters:

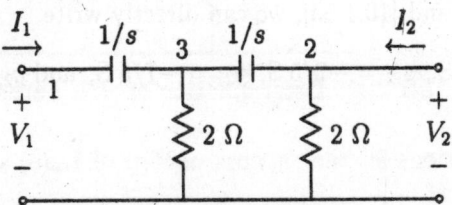

Fig. 10.1.1(e): Determination of all y parameters.

Node basis equations for the network of Fig. 10.1.1(e) are:

Node 1:
$$sV_1 - sV_3 = I_1 \qquad\qquad [10.1.1e]$$

Node 2:
$$(s + 1/2)V_2 - sV_3 = I_2 \qquad\qquad [10.1.2e]$$

Node 3:
$$-sV_1 - sV_2 + (2s + 1/2)V_3 = 0 \qquad\qquad [10.1.3e]$$

From eqn [10.1.3e]
$$V_3 = \frac{s(V_1 - V_2)}{2s + 1/2}$$

Putting this value of V_3 into equations [10.1.1e] and [10.1.2e],

$$sV_1 - s\left[\frac{s(V_1 + V_2)}{2s + 1/2}\right] = I_1$$

Thus
$$\frac{2s^2 + s}{4s + 1}V_1 - \frac{2s^2}{4s + 1}V_2 = I_1 \qquad\qquad [10.1.4e]$$

and
$$(s + 1/2)V_2 - s\left[\frac{s(V_1 + V_2)}{2s + 1/2}\right] = I_2$$

Thus
$$-\frac{2s^2}{4s + 1}V_1 + \frac{4s^2 + 6s + 1}{8s + 2}V_2 = I_2 \qquad\qquad [10.1.5e]$$

Hence, from equations [10.1.4e] and [10.1.5e], we have

$$y_{11} = \frac{2s^2 + s}{4s + 1}, \; y_{12} = -\frac{2s^2}{4s + 1} \; y_{21} = -\frac{2s^2}{4s + 1} \text{ and } y_{22} = \frac{4s^2 + 6s + 1}{8s + 2}$$

Determination of z parameters:

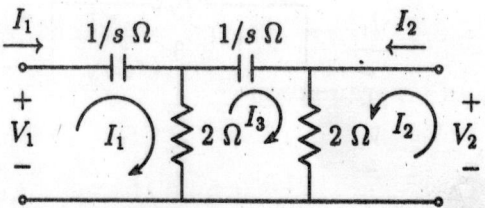

Fig. 10.1.2(e): Determination of all z parameters.

Mesh-current equations for the network of Fig. 10.1.2(e) are:
Mesh 1:
$$(2 + 1/s)I_1 - 2I_3 = V_1 \qquad\qquad [10.1.6e]$$

Mesh 2:
$$2I_2 + 2I_3 = V_2 \qquad\qquad [10.1.7e]$$

Mesh 3:
$$4 - 2I_1 + 2I_2 + (4 + 1/s)I_3 = 0 \qquad\qquad [10.1.8e]$$

From eqn [10.1.8e]
$$I_3 = \frac{2(I_1 - I_2)}{4 + 1/s}$$

Putting this value of I_3 into equations [10.1.6e] and [10.1.7e],

$$(2 + 1/s)I_1 - 2\left[\frac{2(I_1 - I_2)}{4 + 1/s}\right] = V_1$$

Thus
$$\frac{4s^2 + 6s + 1}{4s^2 + s}I_1 + \frac{4s}{4s + 1}I_2 = V_1 \qquad\qquad [10.1.9e]$$

Next
$$2I_2 + 2\left[\frac{2(I_1 - I_2)}{4 + 1/s}\right] = V_2$$

Thus
$$\frac{4s}{4s + 1}I_1 + \frac{4s + 2}{4s + 1}I_2 = V_2 \qquad\qquad [10.1.10e]$$

From equations [10.1.9e] and [10.1.10e], we have

$$z_{11} = \frac{4s^2 + 6s + 1}{4s^2 + s}, \; z_{12} = \frac{4s}{4s + 1}, \; z_{21} = \frac{4s}{4s + 1} \text{ and } z_{22} = \frac{4s + 2}{4s + 1}$$

(f) Determination of y parameters:

Node basis equations for the network of Fig.10.1.1(f) are:
Node 1:
$$(s + 1)V_1 - V_2 - sV_3 = I_1 \qquad\qquad [10.1.1f]$$

Node 2:
$$-V_1 + (s + 1)V_2 - sV_3 = I_2 \qquad\qquad [10.1.2f]$$

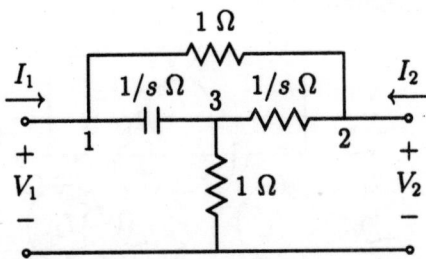

Fig. 10.1.1(f): Determination of all y parameters.

Node 3:

$$-sV_1 - sV_2 + (1 + 2s)V_3 = 0 \qquad [10.1.3f]$$

From eqn [10.1.3f]

$$V_3 = \frac{s(V_1 + V_2)}{1 + 2s}$$

Putting this value of V_3 into equations [10.1.1f] and [10.1.2f],

$$(s + 1)V_1 - V_2 - s\left[\frac{s(V_1 + V_2)}{1 + 2s}\right] = I_1$$

Thus

$$\frac{s^2 + 3s + 1}{1 + 2s}V_1 - \frac{s^2 + 2s + 1}{1 + 2s}V_2 = I_1 \qquad [10.1.4f]$$

Next

$$-V_1 + (s + 1)V_2 - s\left[\frac{s(V_1 + V_2)}{1 + 2s}\right] = I_2$$

Thus

$$-\frac{s^2 + 2s + 1}{1 + 2s}V_1 + \frac{s^2 + 3s + 1}{1 + 2s}V_2 = I_2 \qquad [10.1.5f]$$

From equations [10.1.4f] and [10.1.5f], we have

$$y_{11} = \frac{s^2 + 3s + 1}{2s + 1}, \ y_{12} = -\frac{s^2 + 2s + 1}{1 + 2s}, \ y_{21} = -\frac{s^2 + 2s + 1}{1 + 2s}, \text{ and } y_{22} = \frac{s^2 + 3s + 1}{1 + 2s}$$

Determination of z parameters:

Mesh-current equations for the network of Fig. 10.1.2(f) are:

Mesh 1:

$$(1 + 1/s)I_1 + I_2 - 1/s\ I_3 = V_1 \qquad [10.1.6f]$$

Mesh 2:

$$I_1 + (1 + 1/s)I_2 + 1/s\ I_3 = V_2 \qquad [10.1.7f]$$

Mesh 3:

$$-1/s\ I_1 + 1/s\ I_2 + (2/s + 1)I_3 = 0 \qquad [10.1.8f]$$

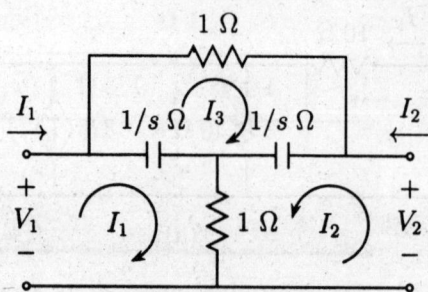

Fig. 10.1.2(f): Determination of all z parameters.

From eqn [10.1.8f]

$$I_3 = \frac{(I_1 - I_2)/s}{2/s + 1}$$

Putting this value of I_3 into equations [10.1.6f] and [10.1.7f],

$$(1 + 1/s)I_1 + I_2 - 1/s\left[\frac{(I_1 - I_2)/s}{2/s + 1}\right] = V_1$$

Thus

$$\frac{s^2 + 3s + 1}{s^2 + 2s}I_1 + \frac{s^2 + 2s + 1}{s^2 + 2s}I_2 = V_1 \qquad [10.1.9f]$$

Next

$$I_1 + (1 + 1/s)I_2 + 1/s\left[\frac{(I_1 - I_2)/s}{2/s + 1}\right] = V_2$$

Thus

$$\frac{s^2 + 2s + 1}{s^2 + 2s}I_1 + \frac{s^2 + 3s + 1}{s^2 + 2s}I_2 = V_2 \qquad [10.1.10f]$$

From equations [10.1.9f] and [10.1.10f], we have

$$z_{11} = \frac{s^2 + 3s + 1}{s^2 + 2s}, \ z_{12} = \frac{s^2 + 2s + 1}{S^2 + 2s}, \ z_{21} = \frac{s^2 + 2s + 1}{s^2 + 2s} \text{ and } z_{22} = \frac{s^2 + 3s + 1}{s^2 + 2s}$$

SP 10.2 Find the y and z parameters for each of the following networks in Fig. SP 10.2.

SOLUTION:

(a) **Determination of y parameters:**
Nodal equations for the circuit of Fig. 10.2.1(a) are:
Node 1:

$$\frac{1}{10}V_1 - \frac{1}{10}V_2 = I_1 \qquad [10.2.1a]$$

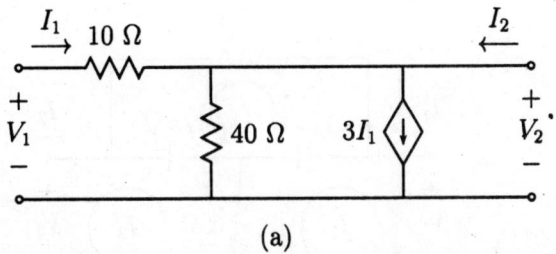

(a)

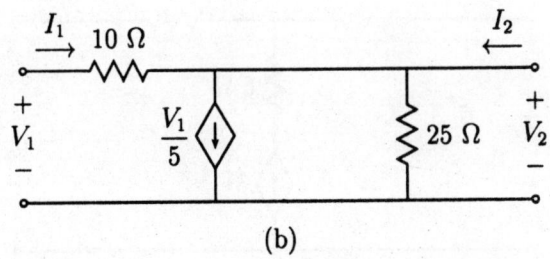

(b)

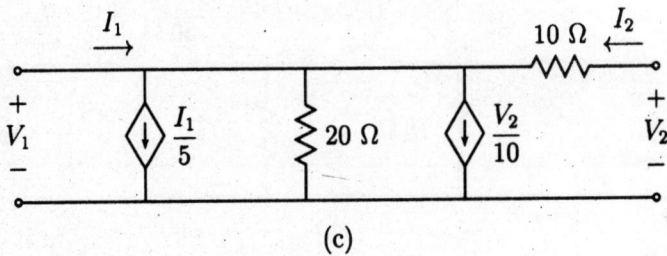

(c)

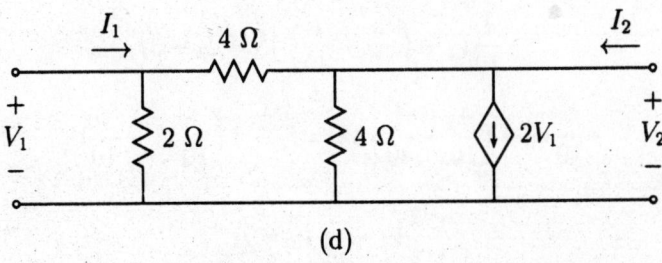

(d)

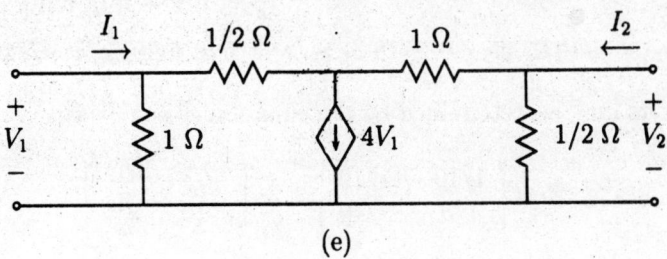

(e)

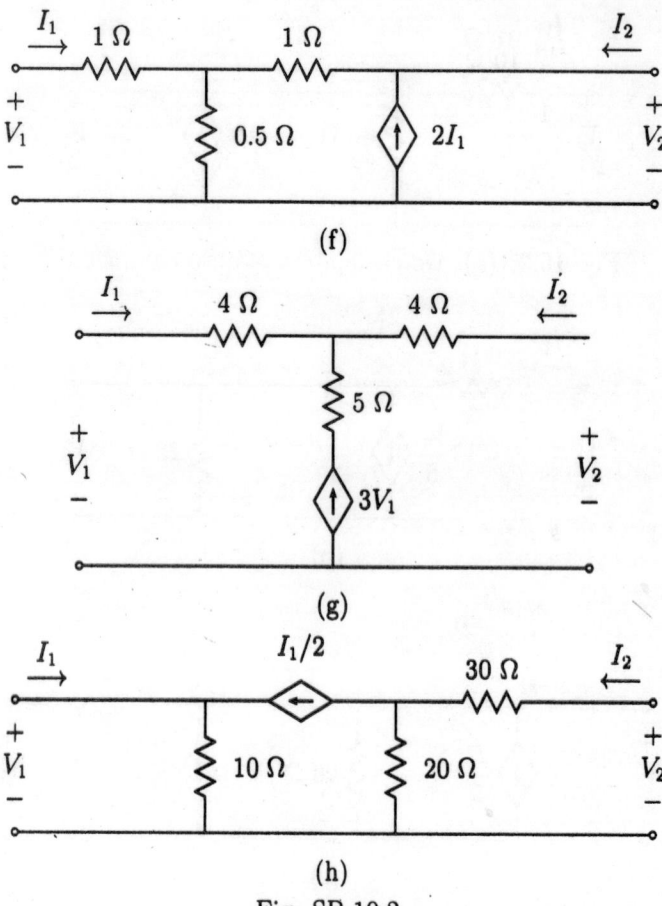

(f)

(g)

(h)

Fig. SP 10.2.

Node 2:

$$-\frac{1}{10}V_1 + \left(\frac{1}{10} + \frac{1}{40}\right)V_2 = I_2 - 3I_1$$

or

$$-\frac{1}{10}V_1 + \left(\frac{1}{10} + \frac{1}{40}\right)V_2 = I_2 - 3\left(\frac{1}{10}V_1 - \frac{1}{10}V_2\right)$$

Hence

$$\frac{1}{5}V_1 - \frac{7}{40}V_2 = I_2 \qquad\qquad [10.2.2a]$$

The y parameters can be written directly from equations [10.2.1a] and [10.2.2a] as,

$$y_{11} = 1/10 \text{ S}, \ y_{12} = -1/10 \text{ S}, \ y_{21} = 1/5 \text{ S}, \text{ and } y_{22} = -7/40 \text{ S}$$

The z parameters may be determined by the equation $Z = Y^{-1}$ as:

$$Z = \begin{bmatrix} 1/10 & -1/10 \\ 1/5 & -7/40 \end{bmatrix}^{-1} = \begin{bmatrix} \frac{-7/40}{\Delta} & \frac{1/10}{\Delta} \\ \frac{-1/5}{\Delta} & \frac{1/10}{\Delta} \end{bmatrix}$$

where $\Delta = -7/400 + 1/50 = 1/400$, hence

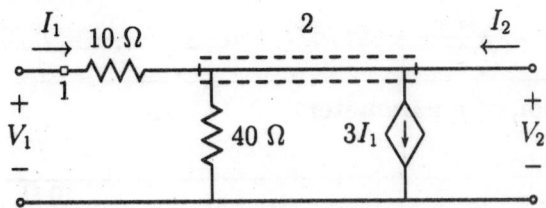

Fig. 10.2.1(a): Determination of all y parameters:

$$z_{11} = \frac{-7/40}{\Delta} = -70 \ \Omega, \ z_{12} = \frac{1/10}{\Delta} = 40 \ \Omega,$$

$$z_{21} = \frac{-1/5}{\Delta} = -80 \ \Omega, \text{ and } z_{22} = \frac{1/10}{\Delta} = 40 \ \Omega.$$

(b) Determination of y parameters:

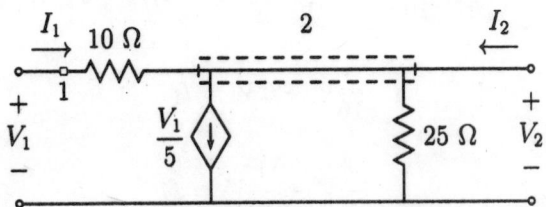

Fig. 10.2.1(b): Determination of all y parameters.

Nodal equations for the circuit of Fig. 10.2.1(b) are:
Node 1:

$$\frac{1}{10}V_1 - \frac{1}{10}V_2 = I_1 \qquad\qquad [10.2.1b]$$

Node 2:

$$-\frac{1}{10}V_1 + \left(\frac{1}{10} + \frac{1}{25}\right)V_2 = I_2 - \frac{V_1}{5}$$

Hence

$$\frac{1}{10}V_1 + \frac{7}{50}V_2 = I_2 \qquad\qquad [10.2.2b]$$

The y parameters may be written directly from equations [10.2.1b] and [10.2.2b] as,

$$y_{11} = 1/10 \text{ S}, \ y_{12} = -1/10 \text{ S}, \ y_{21} = 1/10 \text{ S}, \text{ and } y_{22} = 7/50 \text{ S}$$

The z parameters may be determined by

$$Z = Y^{-1} = \begin{bmatrix} 1/10 & -1/10 \\ 1/10 & 7/50 \end{bmatrix}^{-1} = \begin{bmatrix} \frac{7/50}{\Delta} & \frac{1/10}{\Delta} \\ \frac{-1/10}{\Delta} & \frac{1/10}{\Delta} \end{bmatrix}$$

where $\Delta = 7/500 + 1/100 = 3/125$, hence

$$y_{11} = \frac{7/50}{\Delta} = 35/6 \ \Omega, \ z_{12} = \frac{1/10}{\Delta} = 25/6 \ \Omega,$$

$$z_{21} = \frac{-1/10}{\Delta} = -25/6 \ \Omega, \text{ and } z_{22} = \frac{1/10}{\Delta} = 25/6 \ \Omega$$

(c) **Determination of y parameters:**

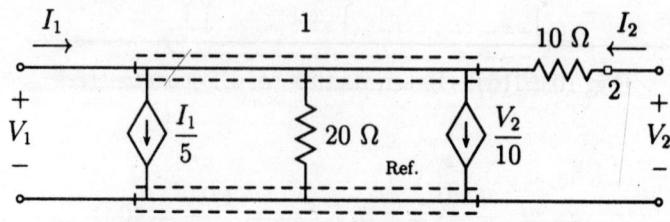

Fig. 10.2.1(c): Determination of all y parameters.

Nodal equations for the circuit of Fig. 10.2.1(c) are:

Node 1:

$$\left(\frac{1}{20} + \frac{1}{10}\right) V_1 - \frac{1}{10} V_2 = I_1 - I_1/5 - V_2/10$$

Thus

$$\frac{3}{16} V_1 + 0 \ V_2 = I_1 \qquad\qquad [10.2.1c]$$

Node 2:

$$\frac{-1}{10} V_1 + \frac{1}{10} V_2 = I_2 \qquad\qquad [10.2.2c]$$

The y parameters may be written directly from equations [10.2.1c] and [10.2.2c] as:

$$y_{11} = 3/16 \text{ S, } y_{12} = 0 \text{ , } y_{21} = -1/10 \text{ S, and } y_{22} = 1/10 \text{ S}$$

The z parameters may be determined by

$$Z = Y^{-1} = \begin{bmatrix} 3/16 & 0 \\ -1/10 & 1/10 \end{bmatrix}^{-1} = \begin{bmatrix} \frac{1/10}{\Delta} & 0 \\ \frac{1/10}{\Delta} & \frac{3/16}{\Delta} \end{bmatrix}$$

where $\Delta = 3/160$, hence

$$z_{11} = \frac{1/10}{\Delta} = 16/3 \ \Omega, \ z_{12} = 0,$$

$$z_{21} = \frac{1/10}{\Delta} = 16/3 \ \Omega, \text{ and } z_{22} = \frac{3/16}{\Delta} = 10 \ \Omega$$

(d)

Nodal equations for the network of Fig. 10.2.1(d) are:

Node 1:

$$\left(\frac{1}{2} + \frac{1}{4}\right) V_1 - \frac{1}{4} V_2 = I_1 \qquad \Rightarrow \frac{3}{4} V_1 - \frac{1}{4} V_2 = I_1 \qquad [10.2.1d]$$

Node 2:

$$\left(\frac{1}{4} + \frac{1}{4}\right) V_2 - \frac{1}{4} V_1 = I_2 - 2V_1 \qquad \Rightarrow \frac{7}{4} V_1 + \frac{1}{2} V_2 = I_2 \qquad [10.2.2d]$$

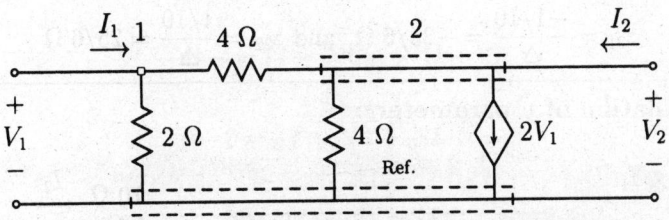

Fig. 10.2.1(d): Determination of all y parameters.

The y parameters may be written directly from equations [10.2.1d] and [10.2.2d] as:

$$y_{11} = 3/4 \text{ S}, \ y_{12} = -1/4 \text{ S}, \ y_{21} = 7/4 \text{ S, and } y_{22} = 1/2 \text{ S}$$

The z parameters may be determined by

$$Z = Y^{-1} = \begin{bmatrix} 3/4 & -1/4 \\ 7/4 & 1/2 \end{bmatrix}^{-1} = \begin{bmatrix} \frac{1/2}{\Delta} & \frac{1/4}{\Delta} \\ \frac{-7/4}{\Delta} & \frac{3/4}{\Delta} \end{bmatrix}$$

where $\Delta = 3/8 + 7/16 = 13/16$, hence

$$z_{11} = \frac{1/2}{\Delta} = 8/13 \ \Omega, \ z_{12} = \frac{1/4}{\Delta} = 4/13 \ \Omega,$$

$$z_{21} = \frac{-7/4}{\Delta} = -28/13 \ \Omega, \text{ and } z_{22} = \frac{3/4}{\Delta} = 12/13 \ \Omega$$

(e)

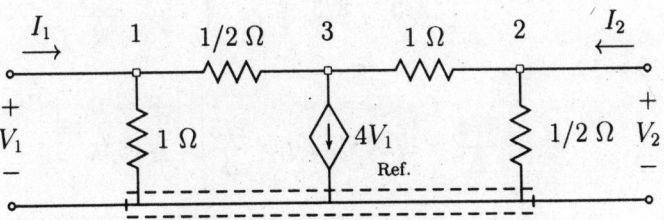

Fig. 10.2.1(e): Determination of all y parameters.

Nodal equations for circuit of Fig. 10.2.1(e) are:

Node 1:

$$(1 + 2)V_1 - 2V_3 = I_1$$

or

$$3V_1 - 2V_3 = I_1 \qquad [10.2.1e]$$

Node 2:

$$(2 + 1)V_2 - V_3 = I_2$$

or

$$3V_2 - V_3 = I_2 \qquad [10.2.2e]$$

Node 3:

$$(2+1)V_3 - 2V_1 - V_2 = -4V_1$$

or

$$3V_3 + 2V_1 - V_2 = 0 \qquad [10.2.3e]$$

From eqn [10.2.3e]

$$V_3 = (V_2 - 2V_1)/3$$

Putting this value of V_3 into equations [10.2.1e] and [10.2.2e],

$$3V_1 - 2\left[\frac{1}{3}(V_2 - 2V_1)\right] = I_1$$

Thus

$$\frac{13}{3}V_1 - \frac{2}{3}V_2 = I_2 \qquad [10.2.4e]$$

Next

$$3V_2 - \frac{1}{3}(V_2 - 2V_1) = I_2$$

Thus

$$\frac{2}{3}V_1 + \frac{8}{3}V_2 = I_2 \qquad [10.2.5e]$$

The y parameters can be given from equations [10.2.4e] and [10.2.5e] as:

$$y_{11} = 13/3 \text{ S},\ y_{12} = -2/3 \text{ S},\ y_{21} = 2/3 \text{ S, and } y_{22} = 8/3 \text{ S}$$

The z parameters may be determined by

$$Z = Y^{-1} = \begin{bmatrix} 13/3 & -2/3 \\ 2/3 & 8/3 \end{bmatrix}^{-1} = \begin{bmatrix} \frac{8/3}{\Delta} & \frac{2/3}{\Delta} \\ \frac{-2/3}{\Delta} & \frac{13/3}{\Delta} \end{bmatrix}$$

where $\Delta = 108/9 = 12$, hence

$$z_{11} = \frac{1/2}{\Delta} = 1/24\ \Omega,\ z_{12} = \frac{2/3}{\Delta} = 1/18\ \Omega,$$

$$z_{21} = \frac{-2/3}{\Delta} = -1/18\ \Omega,\text{ and } z_{22} = \frac{13/3}{\Delta} = 13/36\ \Omega$$

(f)

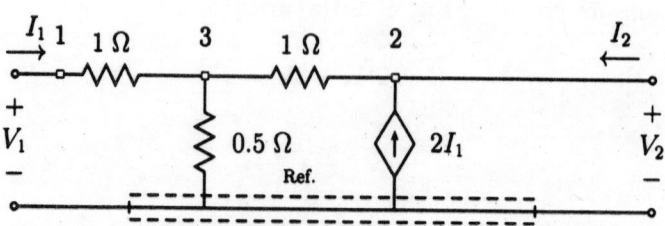

Fig. 10.2.1(f): Determination of all y parameters.

Nodal equations for circuit of Fig. 10.2.1(f) are:

Node 1:
$$V_1 - V_3 = I_1 \qquad\qquad [10.2.1f]$$

Node 2:
$$V_2 - V_3 = 2I_1 + I_2 \qquad\qquad [10.2.2f]$$

Node 3:
$$4V_3 - V_1 - V_2 = 0 \qquad\qquad [10.2.3f]$$

From eqn [10.2.3f]
$$V_3 = (V_1 + V_2)/4 \qquad\qquad [10.2.4f]$$

Putting this value of V_3 into eqn [10.2.1f],

$$V_1 - (V_1 + V_2)/4 = I_1 \qquad \Rightarrow \quad \frac{3}{4}V_1 - \frac{1}{4}V_2 = I_1 \qquad [10.2.5f]$$

Putting the value of V_3 from eqn [10.2.4f] and I_1 from eqn [10.2.5f] into eqn [10.2.2f],

$$V_2 - \frac{V_1 + V_2}{4} = 2\left[\frac{3}{4}V_1 - \frac{1}{4}V_2\right] + I_2 \qquad \Rightarrow \quad -\frac{7}{4}V_1 + \frac{5}{4}V_2 = I_2 \qquad [10.2.6f]$$

Thus y parameters may be written directly from equations [10.2.5f] and [10.2.6f] as:

$$y_{11} = 3/4 \text{ S}, \ y_{12} = -1/4 \text{ S}, \ y_{21} = -7/4 \text{ S}, \text{ and } y_{22} = 5/4 \text{ S}$$

The z parameters may be determined by

$$Z = Y^{-1} = \begin{bmatrix} 3/4 & -1/4 \\ -7/4 & 5/4 \end{bmatrix}^{-1} = \begin{bmatrix} \frac{5/4}{\Delta} & \frac{1/4}{\Delta} \\ \frac{7/4}{\Delta} & \frac{3/4}{\Delta} \end{bmatrix}$$

where $\Delta = 1/2$, hence

$$z_{11} = \frac{5/4}{\Delta} = 5/2 \ \Omega, \ z_{12} = \frac{1/4}{\Delta} = 1/2 \ \Omega,$$

$$z_{21} = \frac{7/4}{\Delta} = 7/2 \ \Omega, \text{ and } z_{22} = \frac{3/4}{\Delta} = 3/2 \ \Omega$$

(g)
   Nodal equations for the circuit of Fig. 10.2.1(g) are:

Node 1:
$$\frac{1}{4}V_1 - \frac{1}{4}V_3 = I_1 \qquad\qquad [10.2.1g]$$

Node 2:
$$\frac{1}{4}V_2 - \frac{1}{4}V_3 = I_2 \qquad\qquad [10.2.2g]$$

Node 3:
$$\left(\frac{1}{4} + \frac{1}{4} + \frac{1}{5}\right) V_3 - \frac{1}{4}V_1 - \frac{1}{4}V_2 - \frac{1}{5}V_4 = 0$$

Hence
$$-\frac{1}{4}V_1 - \frac{1}{4}V_2 + \frac{7}{10}V_3 - \frac{1}{5}V_4 = 0 \qquad\qquad [10.2.3g]$$

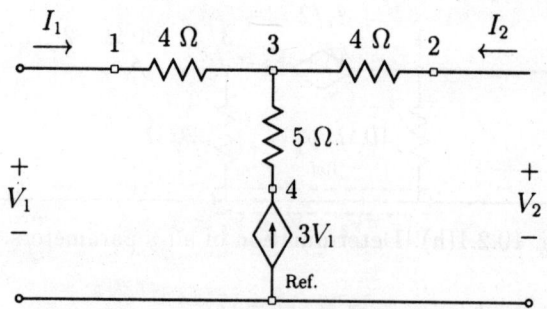

Fig. 10.2.1(g): Determination of all y parameters.

Node 4:

$$\frac{1}{5}V_4 - \frac{1}{5}V_3 = 3V_1 \qquad\qquad [10.2.4g]$$

From eqn [10.2.4g]

$$V_4 = 15V_1 + V_3 \qquad\qquad [10.2.5g]$$

Putting this value of V_4 into eqn [10.2.3g],

$$-\frac{1}{4}V_1 - \frac{1}{4}V_2 + \frac{7}{10}V_3 - \frac{1}{5}(15V_1 + V_3) = 0 \quad\Rightarrow\quad V_3 = \frac{13}{2}V_1 + \frac{1}{2}V_2 \qquad [10.2.6g]$$

Putting this value of V_3 into equations [10.2.1g] and [10.2.2g] as

$$\frac{1}{4}V_1 - \frac{1}{4}\left(\frac{13}{2}V_1 + \frac{1}{2}V_2\right) = I_1 \quad\Rightarrow\quad -\frac{11}{8}V_1 - \frac{1}{8}V_2 = I_1 \qquad [10.2.7g]$$

and

$$\frac{1}{4}V_2 - \frac{1}{4}\left(\frac{13}{2}V_1 + \frac{1}{2}V_2\right) = I_2 \quad\Rightarrow\quad -\frac{13}{8}V_1 + \frac{1}{8}V_2 = I_2 \qquad [10.2.8g]$$

Thus y parameters may be written directly from equations [10.2.7g] and [10.2.8g] as:

$$y_{11} = -11/8 \text{ S}, \ y_{12} = -1/8 \text{ S}, \ y_{21} = -13/8 \text{ S}, \text{ and } y_{22} = 1/8 \text{ S}$$

The z parameters may be determined by

$$Z = Y^{-1} = \begin{bmatrix} -11/8 & -1/8 \\ -13/8 & 1/8 \end{bmatrix}^{-1} = \begin{bmatrix} \frac{1/8}{\Delta} & \frac{1/8}{\Delta} \\ \frac{13/8}{\Delta} & \frac{-11/8}{\Delta} \end{bmatrix}$$

where $\Delta = -3/8$, hence

$$z_{11} = \frac{1/8}{\Delta} = -1/3 \ \Omega, \ z_{12} = \frac{1/8}{\Delta} = -1/3 \ \Omega,$$

$$z_{21} = \frac{13/8}{\Delta} = -13/3 \ \Omega, \text{ and } z_{22} = \frac{-11/8}{\Delta} = 11/3 \ \Omega$$

(h)

Nodal equations for the circuit of Fig. 10.2.1(h) are:

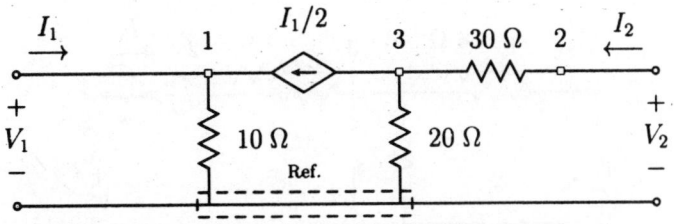

Fig. 10.2.1(h): Determination of all y parameters.

Node 1:

$$\frac{1}{10}V_1 = I_1 + I_1/2 \qquad \Rightarrow \quad \frac{1}{15}V_1 = I_1 \qquad\qquad [10.2.1h]$$

Node 2:

$$\frac{1}{30}V_2 - \frac{1}{30}V_3 = I_2 \qquad\qquad [10.2.2h]$$

Node 3:

$$\left(\frac{1}{30} + \frac{1}{20}\right)V_3 - \frac{1}{30}V_2 = -I_1/2$$

Setting for I_1 into last eqn.,

$$\left(\frac{1}{30} + \frac{1}{20}\right)V_3 - \frac{1}{30}V_2 = -\frac{1}{2}\left[\frac{1}{15}V_1\right] \qquad \Rightarrow \quad V_3 = \frac{2}{5}(V_2 - V_1) \qquad [10.2.3h]$$

Putting this value of V_3 into eqn [10.2.2h],

$$\frac{1}{30}V_2 - \frac{1}{30}\left[\frac{2}{5}(V_2 - V_1)\right] = I_2 \qquad \Rightarrow \quad \frac{1}{75}V_1 + \frac{1}{50}V_2 = I_2 \qquad [10.2.4h]$$

Thus, y parameters will be written directly from equations [10.2.1h] and [10.2.4h] as:

$$y_{11} = 1/15 \text{ S}, \ y_{12} = 0, \ y_{21} = 1/75 \text{ S}, \text{ and } y_{22} = 1/50 \text{ S}$$

SP 10.3 Find the z and y parameters for the networks in Fig. SP 10.3.

SOLUTION:
(a)
Mesh-current equations for the network of Fig. 10.3.1(a) are:
Mesh 1:

$$(Z_1 + Z_3)I_1 + Z_3 I_2 = V_1 \qquad\qquad [10.3.1a]$$

Mesh 2:

$$(kZ_2 + Z_3)I_1 + (Z_2 + Z_3)I_2 = V_2 \qquad\qquad [10.3.2a]$$

where $\Delta = Z_1(Z_2 + Z_3) + Z_2 Z_3(1 - k)$.
Hence z parameters may be written directly from equations [10.3.1a] and [10.3.2a] as:

$$z_{11} = Z_1 + Z_3, \ z_{12} = Z_3, \ z_{21} = kZ_2 + Z_3, \text{ and } z_{22} = Z_2 + Z_3$$

The y parameters may be determined by

$$Y = Z^{-1} = \begin{bmatrix} Z_1 + Z_3 & Z_3 \\ kZ_2 + Z_3 & Z_2 + Z_3 \end{bmatrix}^{-1} = \begin{bmatrix} \frac{Z_2+Z_3}{\Delta} & \frac{-Z_3}{\Delta} \\ \frac{-(kZ_2+Z_3)}{\Delta} & \frac{Z_1+Z_3}{\Delta} \end{bmatrix}$$

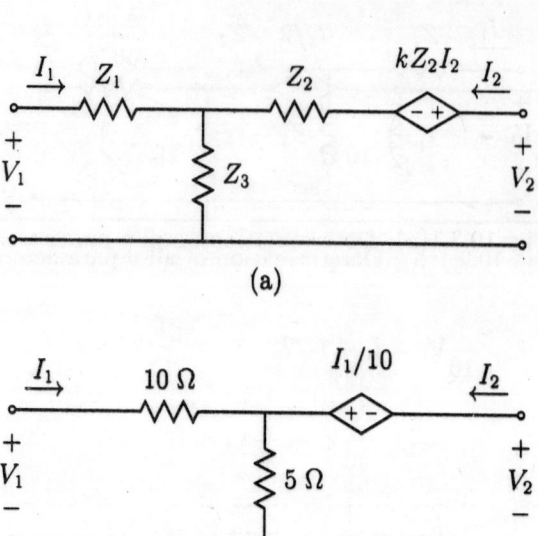

(a)

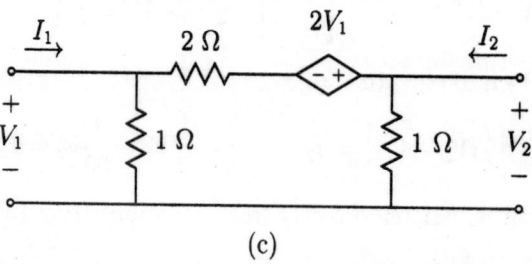

(b)

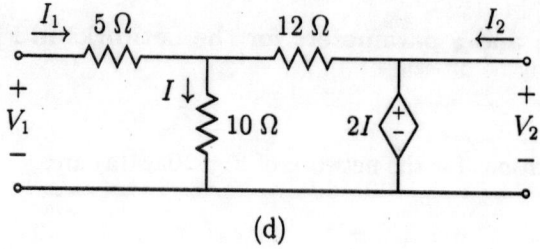

(c)

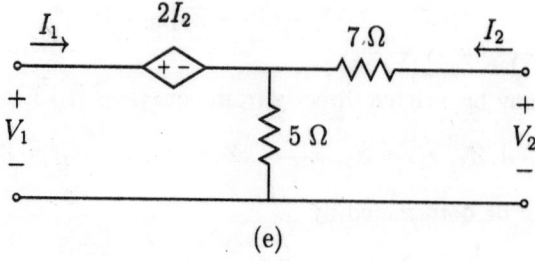

(d)

(e)

Fig. SP 10.3.

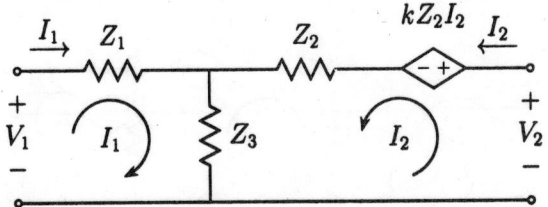

Fig. 10.3.1(a): Determination of all z parameters.

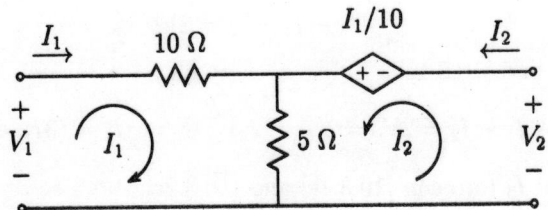

Fig. 10.3.1(b): Determination of all z parameters.

(b)

Mesh-current equations for the network of Fig. 10.3.1(b) are:

Mesh 1:

$$15I_1 + 5I_2 = V_1 \qquad\qquad [10.3.1b]$$

Mesh 2:

$$5I_1 - \frac{1}{10}I_1 + 5I_2 = V_2 \qquad \Rightarrow 4.9I_1 + 5I_2 = V_2 \qquad [10.3.2b]$$

Hence, z parameters may be written directly from equations [10.3.1b] and [10.3.2b] as:

$$z_{11} = 15 \ \Omega, \ z_{12} = 5 \ \Omega, \ z_{21} = 4.9 \ \Omega, \text{ and } z_{22} = 5 \ \Omega$$

The y parameters may be determined by

$$Y = Z^{-1} = \begin{bmatrix} 15 & 5 \\ 4.9 & 5 \end{bmatrix}^{-1} = \begin{bmatrix} 5/\Delta & -5/\Delta \\ -4.9/\Delta & 15/\Delta \end{bmatrix}$$

where $\Delta = 50.5$, hence

$$y_{11} = 5/\Delta = 10/101 \ \mho, \ y_{12} = -5/\Delta = -10/101 \ \mho,$$

$$y_{21} = -4.9/\Delta = 49/505 \ \mho \text{ and } y_{22} = 15/\Delta = 30/101 \ \mho$$

(c)

Mesh-current equations for the network of Fig. 10.3.1(c) are:

Mesh 1:

$$I_1 - I_3 = V_1 \qquad\qquad [10.3.1c]$$

Mesh 2:

$$I_2 + I_3 = V_2 \qquad\qquad [10.3.2c]$$

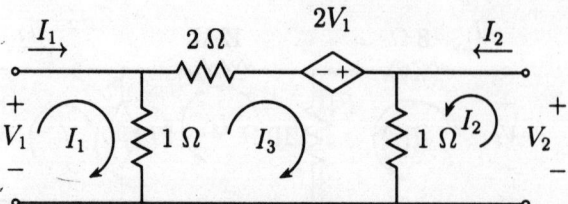

Fig. 10.3.1(c): Determination of z parameters.

Mesh 3:

$$4I_3 - I_1 + I_2 = 2V_1 \qquad [10.3.3c]$$

From eqns [10.3.3c] and [10.3.1c]

$$4I_3 - I_1 + I_2 = 2V_1 = 2(I_1 - I_3) \qquad \Rightarrow I_3 = (3I_1 - I_2)/6$$

Putting this value of I_3 into eqn [10.3.1c] and [10.3.2c],

$$I_1 - (3I_1 - I_2)/6 = V_1 \qquad \Rightarrow \frac{1}{2}I_1 + \frac{1}{6}I_2 = V_1 \qquad [10.3.4c]$$

and

$$I_2 + (3I_1 - I_2)/6 = V_2 \qquad \Rightarrow \frac{1}{2}I_1 + \frac{5}{6}I_2 = V_2 \qquad [10.3.5c]$$

Hence, z parameters may be written directly from equations [10.3.4c] and [10.3.5c] as:

$$z_{11} = 1/2 \ \Omega, \ z_{12} = 1/6 \ \Omega, \ z_{21} = 1/2 \ \Omega, \ \text{and} \ z_{22} = 5/6 \ \Omega$$

The y parameters may be determined by

$$Y = Z^{-1} = \begin{bmatrix} 1/2 & 1/6 \\ 1/2 & 5/6 \end{bmatrix}^{-1} = \begin{bmatrix} \frac{5/6}{\Delta} & \frac{-1/6}{\Delta} \\ \frac{-1/2}{\Delta} & \frac{1/2}{\Delta} \end{bmatrix}$$

where $\Delta = 1/3$, hence

$$y_{11} = \frac{5/6}{\Delta} = 5/2 \ \text{S}, \ y_{12} = -\frac{1/6}{\Delta} = -1/2 \ \text{S}$$

$$y_{21} = \frac{-1/2}{\Delta} = -3/2 \ \text{S}, \ \text{and} \ y_{22} = \frac{1/2}{\Delta} = 3/2 \ \text{S}$$

(d)

Mesh-current equations for the network of Fig. 10.3.1(d) are:
Mesh 1:

$$15I_1 - 10I_3 = V_1 \qquad [10.3.1d]$$

Mesh 2:

$$2I = V_2 \qquad \Rightarrow 2(I_1 - I_3) = V_2 \qquad [10.3.2d]$$

Mesh 3:

$$22I_3 - 10I_1 + 2I = 0 \quad \Rightarrow 22I_3 - 10I_1 + 2(I_1 - I_3) = 0 \qquad \Rightarrow I_3 = 2I_1/5 \quad [10.3.3d]$$

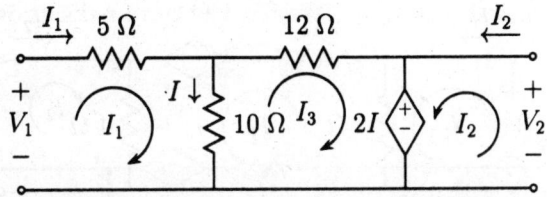

Fig. 10.3.1(d): Determination of all z parameters.

Putting this value of I_3 into eqns [10.3.1d] and [10.3.2d],

$$15I_1 - 10 \times \frac{2}{5}I_1 = V_1 \qquad \Rightarrow \quad 11I_1 = V_1 \qquad\qquad [10.3.4d]$$

And

$$2I_1 - 2 \times \frac{2}{5}I_1 = V_2 \qquad \Rightarrow \quad \frac{6}{5}I_1 = V_2 \qquad\qquad [10.3.5d]$$

Hence, z parameters may be written directly from equations [10.3.4d] and [10.3.5d] as:

$$z_{11} = 11\ \Omega,\ z_{12} = 0,\ z_{21} = 6/5\ \Omega,\ \text{and}\ z_{22} = 0$$

This network has no y parameters.

(e)

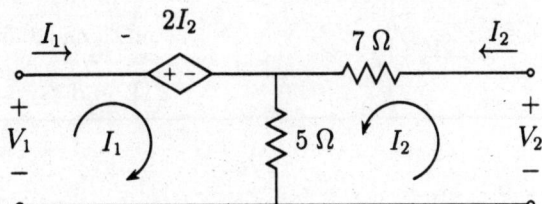

Fig. 10.3.1(e): Determination of all z parameters.

Mesh-current equations for the network of Fig. 10.3.1(e) are:

$$\text{Mesh 1}: \qquad 5I_1 + 7I_2 = V_1 \qquad\qquad [10.3.1e]$$

$$\text{Mesh 2}: \qquad 5I_1 + 12I_2 = V_2 \qquad\qquad [10.3.2e]$$

Hence, z parameters may be written directly from equations [10.3.1e] and [10.3.2e] as:

$$z_{11} = 5\ \Omega,\ z_{12} = 7\ \Omega,\ z_{21} = 5\ \Omega,\ \text{and}\ z_{22} = 12\ \Omega$$

The y parameters may be determined from equation $Y = Z^{-1}$ i.e,

$$Y = \begin{bmatrix} 5 & 7 \\ 5 & 12 \end{bmatrix}^{-1} = \begin{bmatrix} \frac{12}{\Delta} & -\frac{7}{\Delta} \\ \frac{-5}{\Delta} & \frac{5}{\Delta} \end{bmatrix}$$

where $\Delta = 25$, hence

$$y_{11} = \frac{12}{\Delta} = 12/25\ \text{S},\ y_{12} = -\frac{7}{\Delta} = -7/25\ \text{S},\ y_{21} = \frac{-5}{\Delta} = -1/5\ \text{S},\ \text{and}\ y_{22} = \frac{5}{\Delta} = 1/5\ \text{S}$$

SP 10.4 Find the y and z parameters for the networks shown in Fig. SP 10.4.

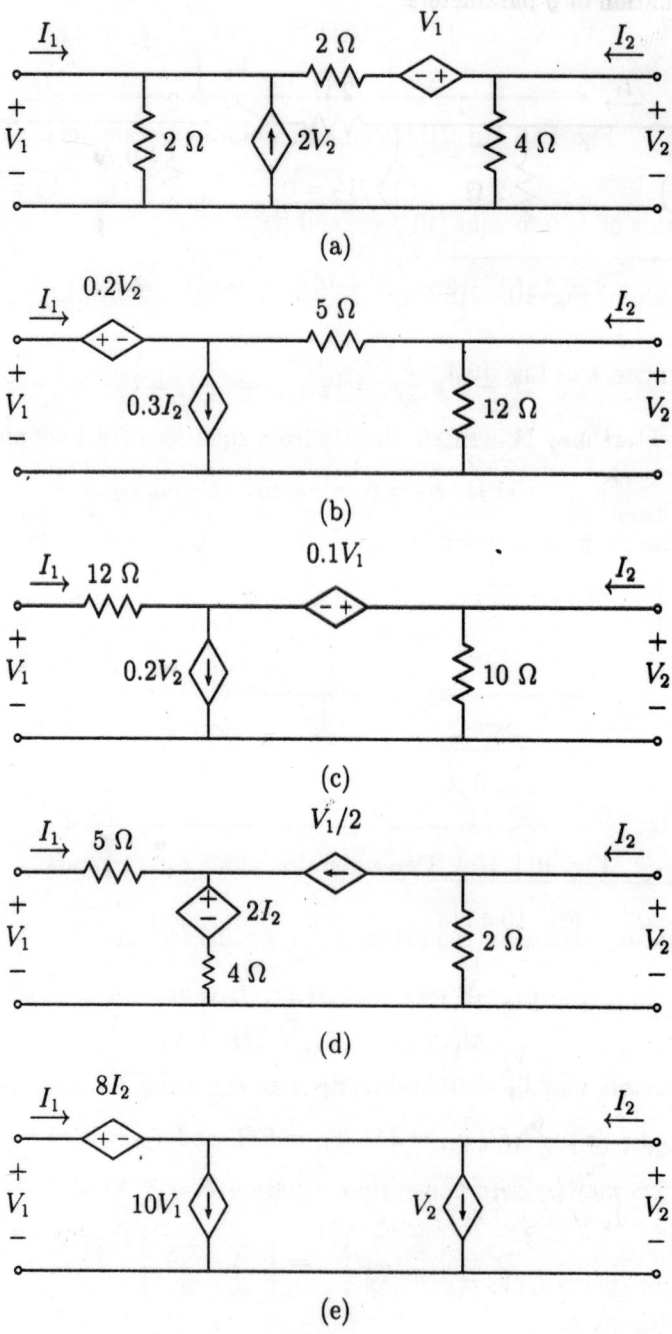

(a)

(b)

(c)

(d)

(e)

Fig. SP 10.4.

10.7. SOLVED PROBLEMS

SOLUTION:

(a) Determination of y parameters:

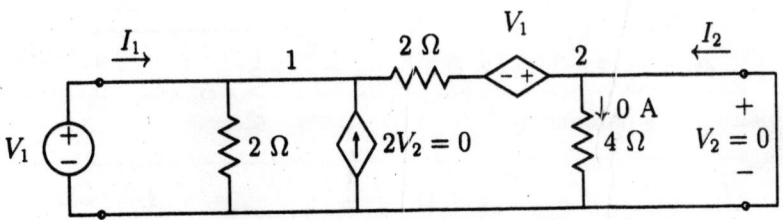

Fig. 10.4.1(a): Determination of y_{11} and y_{21}.

By KCL at node **1** of Fig. 10.4.1(a),

$$I_1 = \frac{V_1}{2} + \frac{V_1 - (-V_1)}{2} = \frac{3}{2}V_1 \qquad \Rightarrow \ y_{11} = I_1/V_1 = 3/2 \text{ S}$$

I_2 may be expressed as,

$$-I_2 = \frac{V_1 - (-V_1)}{2} = V_1 \qquad \Rightarrow \ y_{21} = I_2/V_1 = -1 \text{ S}$$

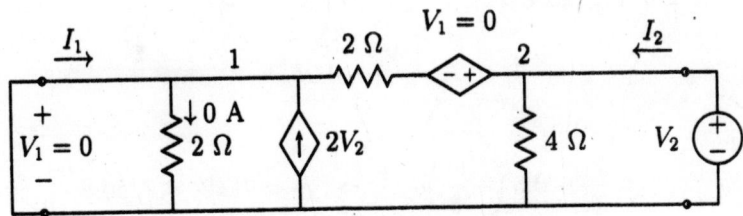

Fig. 10.4.2(a): Determination of y_{12} and y_{22}.

By KCL at node **2** of Fig. 10.4.2(a),

$$I_2 = \frac{V_2}{4} + \frac{V_2 - 0}{2} \qquad \Rightarrow \ y_{22} = I_2/V_2 = 3/4 \text{ S}$$

By KCL at node **1** of Fig. 10.4.2(a),

$$I_1 + \frac{V_2 - 0}{2} + 2V_2 = 0$$

$$\Rightarrow \ y_{12} = I_1/V_2 = -5/2 \text{ S}$$

The z parameters may be obtained from equation $Z = Y^{-1}$ i.e.,

$$Z = \begin{bmatrix} 3/2 & -5/2 \\ -1 & 3/4 \end{bmatrix}^{-1} = \begin{bmatrix} \frac{3/4}{\Delta} & \frac{5/2}{\Delta} \\ \frac{1}{\Delta} & \frac{3/2}{\Delta} \end{bmatrix}$$

where $\Delta = \frac{3}{2} \times \frac{3}{4} - \frac{5}{2} = -11/8$, hence

$$z_{11} = \frac{3/4}{\Delta} = -6/11 \ \Omega, \ z_{12} = \frac{5/2}{\Delta} = -20/11 \ \Omega,$$

$$z_{21} = \frac{1}{\Delta} = -8/11 \ \Omega, \text{ and } z_{22} = \frac{3/2}{\Delta} = -12/11 \ \Omega$$

(b) Determination of y parameters:

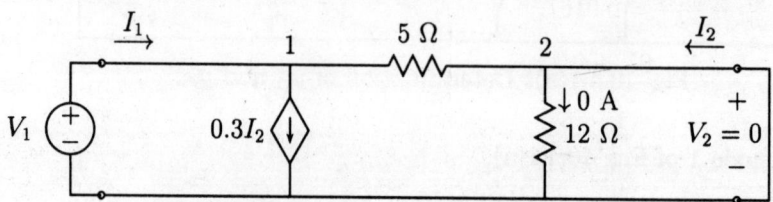

Fig. 10.4.1(b): Determination of y_{11} and y_{21}.

By KCL at node **2** of Fig. 10.4.1(b),

$$-I_2 = \frac{V_1 - 0}{5} \qquad \Rightarrow \ y_{21} = I_2/V_1 = -1/5 \ \text{S}$$

By KCL at node **1** of Fig. 10.4.2(b),

$$I_1 = \frac{V_1}{5} + 0.3I_2$$

Setting for I_2,

$$I_1 = \frac{V_1}{5} + 0.3 \left(-\frac{V_1}{5}\right) \qquad \Rightarrow \ y_{11} = I_1/V_1 = 7/50 \ \text{S}$$

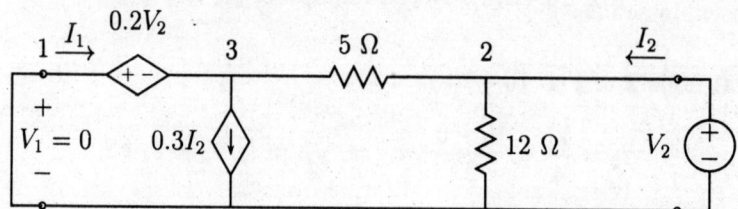

Fig. 10.4.2(b): Determination of y_{12} and y_{22}.

By KCL at node **2** of Fig. 10.4.2(b),

$$I_2 = \frac{V_2}{12} + \frac{V_2 - (-0.2V_2)}{5} = \frac{97V_2}{300} \qquad \Rightarrow \ y_{22} = I_2/V_2 = 97/300 \ \text{S}$$

By KCL at node **3** of Fig. 10.4.2(b),

$$I_1 + \frac{V_2 - (-0.2V_2)}{5} = 0.3I_2$$

Setting for I_2,

$$I_1 + \frac{1.2}{5}V_2 = 0.3\left(\frac{97V_2}{300}\right) \qquad \Rightarrow y_{12} = I_1/V_2 = -0.143 \text{ S}$$

The z parameters may be obtained from equation $Z = Y^{-1}$ i.e.,

$$Z = \begin{bmatrix} 7/50 & -0.143 \\ -1/5 & 97/300 \end{bmatrix}^{-1} = \begin{bmatrix} \frac{97/300}{\Delta} & \frac{0.143}{\Delta} \\ \frac{1/5}{\Delta} & \frac{7/50}{\Delta} \end{bmatrix}$$

where $\Delta = \frac{7}{50} \times \frac{97}{300} - \frac{0.143}{5} = 0.0167$, hence

$$z_{11} = \frac{97/300}{\Delta} = 19.36 \ \Omega, \ z_{12} = \frac{0.143}{\Delta} = 8.56 \ \Omega,$$

$$z_{21} = \frac{1/5}{\Delta} = 11.98 \ \Omega, \text{ and } z_{22} = \frac{7/50}{\Delta} = 8.38 \ \Omega$$

(c) **Determination of y parameters:**

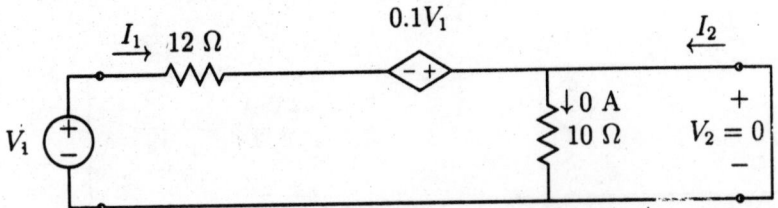

Fig. 10.4.1(c): Determination of y_{11} and y_{21}.

With reference to Fig. 10.4.1(c), I_1 may be expressed as:

$$I_1 = \frac{V_1 - (-0.1V_1)}{12} = \frac{1.1V_1}{12} \qquad \Rightarrow y_{11} = I_1/V_1 = 11/120 \text{ S}$$

I_2 may also be expressed as:

$$-I_2 = \frac{V_1 - (-0.1V_1)}{12} = \frac{1.1V_1}{12} \qquad \Rightarrow y_{21} = I_2/V_1 = -11/120 \text{ S}$$

Fig. 10.4.2(c): Determination of y_{12} and y_{22}.

By KCL at node **2** of Fig. 10.4.2(c),

$$I_2 = \frac{V_2}{10} + 0.2V_2 + \frac{V_2 - 0}{12} = 0.383V_2 \qquad \Rightarrow y_{22} = I_2/V_2 = 0.383 \text{ S}$$

I_1 may be expressed as:

$$-I_1 = \frac{V_2 - 0}{12} = 0.083V_2 \qquad \Rightarrow y_{12} = I_1/V_2 = -0.083 \text{ S}$$

The z parameters may be obtained from equation $Z = Y^{-1}$ i.e.,

$$Z = \begin{bmatrix} 0.092 & -0.083 \\ -0.092 & 0.383 \end{bmatrix}^{-1} = \begin{bmatrix} \frac{0.383}{\Delta} & \frac{0.083}{\Delta} \\ \frac{0.092}{\Delta} & \frac{0.092}{\Delta} \end{bmatrix}$$

where $\Delta = 0.092 \times 0.383 - 0.083 \times 0.092 = 0.0276$, hence

$$z_{11} = \frac{0.383}{\Delta} = 13.88 \ \Omega, \ z_{12} = \frac{0.083}{\Delta} = 3 \ \Omega,$$

$$z_{21} = \frac{0.092}{\Delta} = 3.33 \ \Omega, \text{ and } z_{22} = \frac{0.092}{\Delta} = 3.33 \ \Omega$$

(d) **Determination of y parameters:**

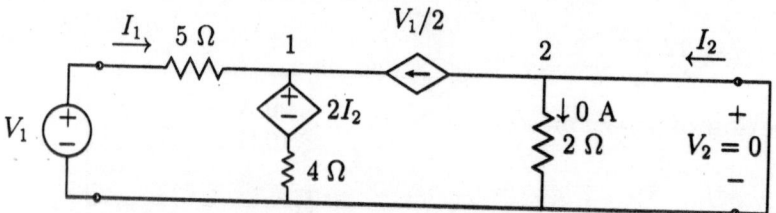

Fig. 10.4.1(d): Determination of y_{11} and y_{21}.

With reference to Fig. 10.4.1(d), I_2 may be expressed as:

$$I_2 = V_1/2 \qquad \Rightarrow y_{21} = I_2/V_1 = 1/2 \text{ S}$$

By KVL in the left mesh,

$$-V_1 + 5I_1 + 2I_2 + 4(I_1 + V_1/2) = 0$$

Setting for I_2,

$$-V_1 + 5I_1 + 2(V_1/2) + 4I_1 + 2V_1 = 0$$

$$\Rightarrow 2V_1 + 9I_1 = 0 \qquad \Rightarrow y_{11} = I_1/V_1 = -2/9 \text{ S}$$

With reference to Fig. 10.4.2(d), I_1 may be expressed as:

$$-I_1 = \frac{V_2 - 0}{5} \qquad \Rightarrow y_{12} = I_1/V_2 = -1/5 \text{ S}$$

By KVL in the left mesh,

$$5I_1 + 2I_2 + 4(I_1 + I_2 - V_2/2) = 0 \qquad \text{or} \qquad 9I_1 + 6I_2 - 2V_2 = 0$$

Setting I_1,

$$9(-V_1 + 6I_2 = 2V_2 \qquad \Rightarrow y_{22} = I_2/V_2 = 19/30 \text{ S}$$

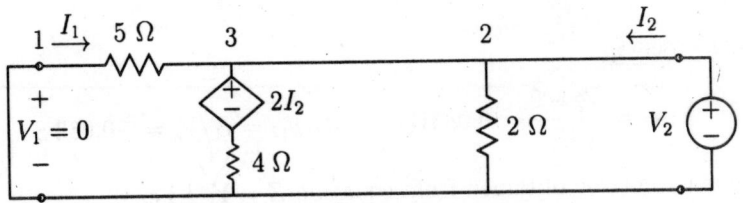

Fig. 10.4.2(d): Determination of y_{12} and y_{22}.

The z parameters may be obtained from equation $Z = Y^{-1}$ i.e.,

$$Z = \begin{bmatrix} -2/9 & -1/5 \\ 1/2 & 19/30 \end{bmatrix}^{-1} = \begin{bmatrix} \frac{19/30}{\Delta} & \frac{1/5}{\Delta} \\ \frac{-1/2}{\Delta} & \frac{-2/9}{\Delta} \end{bmatrix}$$

where $\Delta = -0.0407$, hence

$$z_{11} = \frac{19/30}{\Delta} = -15.56 \ \Omega, \ z_{12} = \frac{1/5}{\Delta} = -4.9 \ \Omega,$$

$$z_{21} = \frac{-1/2}{\Delta} = 12.29 \ \Omega, \text{ and } z_{22} = \frac{-2/9}{\Delta} = 5.46 \ \Omega$$

(e) **Determination of y parameters:**

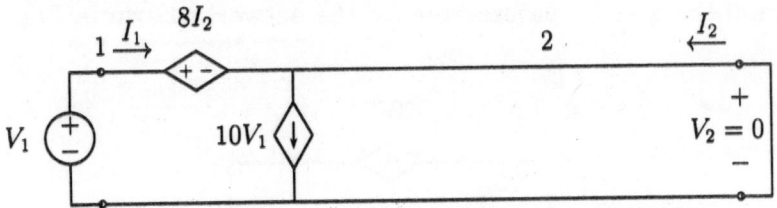

Fig. 10.4.1(e): Determination of y_{11} and y_{21}.

With reference to Fig. 10.4.1(e), KVL in the left mesh gives

$$V_1 = 8I_2 \quad \Rightarrow y_{21} = I_2/V_1 = 1/8 \text{ S}$$

and by KCL at node **2**,

$$I_1 = 10V_1 - I_2 \quad \Rightarrow I_1 = 10V_1 - V_1/8 = 79V_1/8 \quad \Rightarrow y_{11} = I_1/V_1 = 79/8 \text{ S}$$

With reference to Fig. 10.4.2(e), KVL around the loop consisting of source V_2 and dependent source $8I_2$ gives

$$V_2 + 8I_2 = 0 \quad \Rightarrow y_{22} = I_2/V_2 = -1/8 \text{ S}$$

And by KCL at node **2**,

$$I_1 = V_2 - I_2 \quad \text{or} \quad I_1 = V_2 - (-V_2/8) = 9V_2/8 \quad \text{or} \quad y_{12} = I_1/V_2 - 9/8 \text{ S}$$

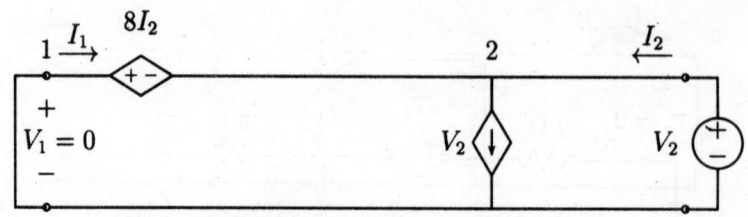

Fig. 10.4.2(e): Determination of y_{12} and y_{22}.

The z parameters may be obtained from equation $Z = Y^{-1}$ i.e.,

$$Z = \begin{bmatrix} 79/8 & 9/8 \\ 1/8 & -1/8 \end{bmatrix}^{-1} = \begin{bmatrix} \frac{-1/8}{\Delta} & \frac{-9/8}{\Delta} \\ \frac{-1/8}{\Delta} & \frac{79/8}{\Delta} \end{bmatrix}$$

where $\Delta = -\frac{79}{8} \times \frac{1}{8} - \frac{1}{8} \times \frac{9}{8} = -88/64 = -11/8$, hence

$$z_{11} = \frac{-1/8}{\Delta} = 1/11 \ \Omega, \ z_{12} = \frac{-9/8}{\Delta} = 9/11 \ \Omega,$$

$$z_{21} = \frac{-1/8}{\Delta} = 1/11 \ \Omega, \text{ and } z_{22} = \frac{79/8}{\Delta} = -79/11 \ \Omega$$

SP 10.5 Find the y and z parameters for the network shown in Fig. SP 10.5.

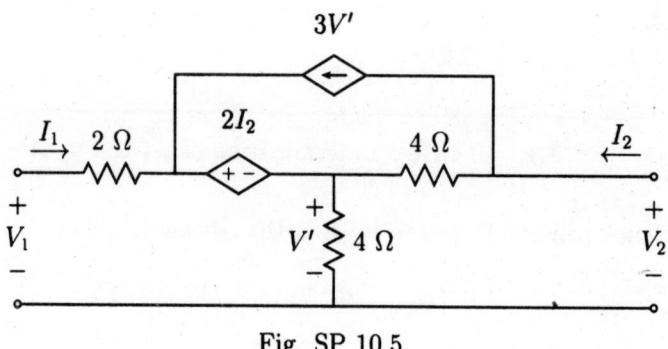

Fig. SP 10.5.

SOLUTION:

Mesh-current equations for the circuit of Fig. 10.5.1 are:
Mesh 1:

$$6I_1 + 4I_2 + 2I_2 = V_1$$

Thus

$$6I_1 + 6I_2 = V_1 \hspace{3cm} [10.5.1]$$

Mesh 2:

$$4I_1 + 8I_2 - 4I_3 = 0 \hspace{3cm} [10.5.2]$$

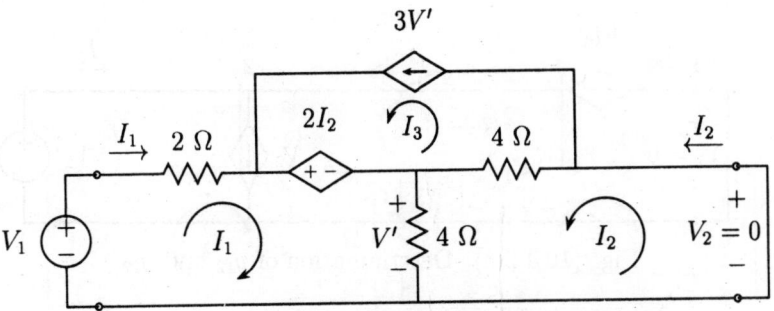

Fig. 10.5.1: Determination of y_{11} and y_{21}.

Mesh 3:

$$I_3 = 3V' = 3[4(I_1 + I_2)]$$

Thus

$$12I_1 + 12I_2 - I_3 = 0 \qquad [10.5.3]$$

$$\Delta = \begin{vmatrix} 6 & 6 & 0 \\ 4 & 8 & -4 \\ 12 & 12 & -1 \end{vmatrix} = -24$$

Hence, by Cramer's rule

$$I_1 = \frac{1}{\Delta} \begin{vmatrix} V_1 & 6 & 0 \\ 0 & 8 & -4 \\ 0 & 12 & -1 \end{vmatrix} = 40V_1/\Delta = -5V_1/3$$

Hence

$$y_{11} = I_1/V_1 = -5/3 \text{ S}$$

$$I_2 = \frac{1}{\Delta} \begin{vmatrix} 6 & V_1 & 0 \\ 4 & 0 & -4 \\ 12 & 0 & -1 \end{vmatrix} = 44V_1/\Delta = -11V_1/6$$

Hence

$$y_{21} = I_2/V_1 = -11/6 \text{ S}$$

Mesh-current equations for the circuit of Fig. 10.5.2 are:

Mesh 1:

$$2I_1 + 2I_2 + 4I_1 + 4I_2 = 0$$

Thus

$$6I_1 + 6I_2 = 0 \qquad [10.5.4]$$

Mesh 2:

$$4I_1 + 8I_2 - 4I_3 = V_2 \qquad [10.5.5]$$

Mesh 3:

$$I_3 = 3V' = 3[4(I_1 + I_2)]$$

Thus

$$12I_1 + 12I_2 - I_3 = 0 \qquad [10.5.6]$$

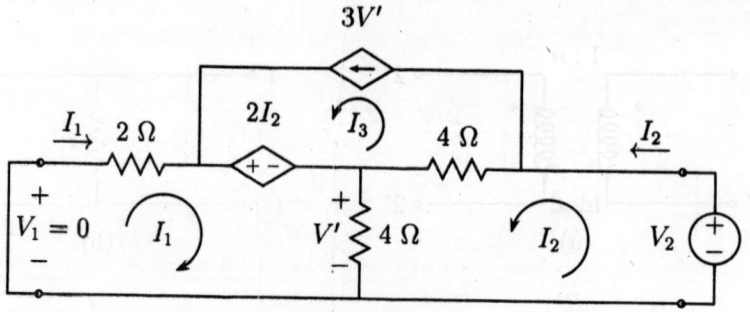

Fig. 10.5.2: Determination of y_{12} and y_{22}.

$$\Delta = \begin{vmatrix} 6 & 6 & 0 \\ 4 & 8 & -4 \\ 12 & 12 & -1 \end{vmatrix} = -24$$

Hence, by Cramer's rule

$$I_1 = \frac{1}{\Delta} \begin{vmatrix} 0 & 6 & 0 \\ V_2 & 8 & -4 \\ 0 & 12 & -1 \end{vmatrix} = 6V_2/\Delta = -V_2/4$$

Hence

$$y_{12} = I_1/V_2 = -1/4 \text{ S}$$

$$I_2 = \frac{1}{\Delta} \begin{vmatrix} 6 & 0 & 0 \\ 4 & V_2 & -4 \\ 12 & 0 & -1 \end{vmatrix} = -6V_2/\Delta = V_2/4$$

Hence

$$y_{22} = I_2/V_2 = 1/4 \text{ S}$$

SP 10.6 Find the transmission parameters for each of the following networks in Fig. SP 10.6.

SOLUTION:

(a)

From Fig. 10.6.1(a),

$$V_1 = V_2/n \qquad\qquad [10.6.1a]$$

$$I_1 = -nI_2 \qquad\qquad [10.6.2a]$$

In matrix form,

$$\begin{bmatrix} V_1 \\ I_1 \end{bmatrix} = \begin{bmatrix} 1/n & 0 \\ 0 & n \end{bmatrix} \begin{bmatrix} V_2 \\ -I_2 \end{bmatrix}$$

Hence

$$A = 1/n, \ B = 0, \ C = 0, \text{ and } D = n$$

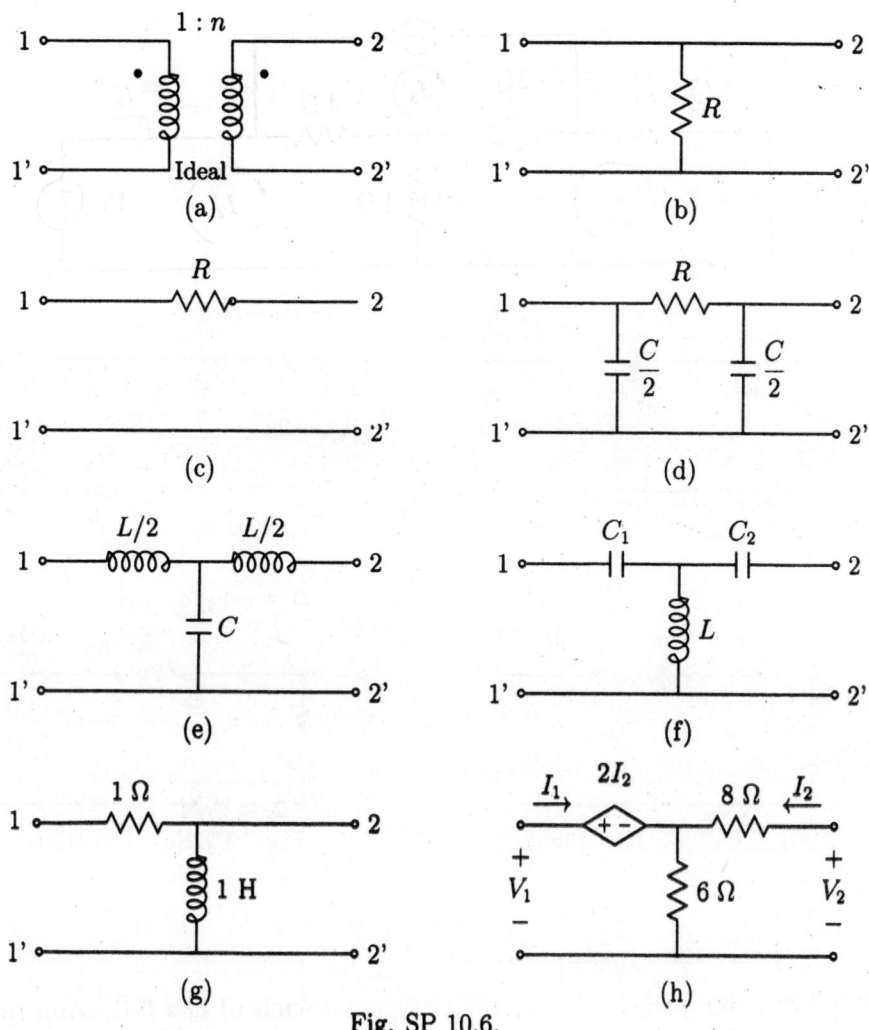

Fig. SP 10.6.

(b)

With reference to Fig. 10.6.1(b), we have

$$V_1 = V_2 \quad \Rightarrow A = V_1/V_2 = 1 \quad \text{and} \quad I_1 = V_1/R \quad \Rightarrow C = I_1/V_1 = 1/R$$

From the Fig. 10.6.2(b), we have

$$I_1 = -I_2 \quad \Rightarrow D = I_1/(-I_2) = 1 \quad \text{and} \quad B = V_1/(-I_2) = V_1/I_1 = 0$$

Fig. 10.6.1(a): Determination of transmission parameters.

(c)

From Fig. 10.6.1(c), we have

$$V_1 = V_2 \quad \Rightarrow A = V_1/V_2 = 1 \quad \text{and} \quad I_1 = I_2 = 0 \quad \Rightarrow C = I_1/V_2 = 0$$

From Fig. 10.6.2(c), we have

$$I_1 = -I_2 \quad \Rightarrow D = I_1/(-I_2) = 1 \quad \text{and} \quad -I_2 = V_1/R \quad \Rightarrow B = V_1/(-I_2) = R$$

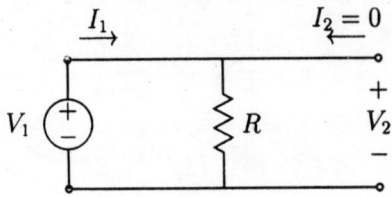

Fig. 10.6.1(b): For A and C

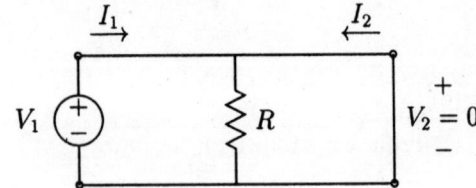

Fig. 10.6.2(b): For B and D

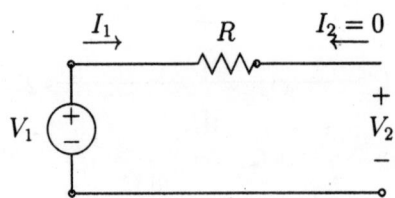

Fig. 10.6.1(c): For A and C

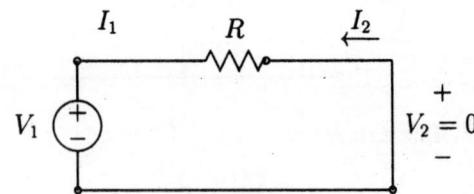

Fig. 10.6.2(c): For B and D

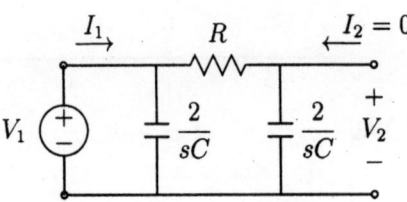

Fig. 10.6.1(d): For A and C.

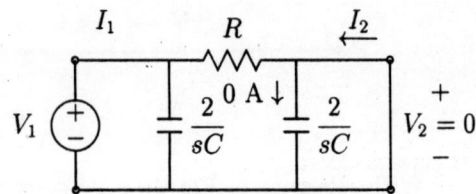

Fig. 10.6.2(d): For B and D.

(d)

With reference to Fig. 10.6.1(d), voltage division formula yields

$$V_2 = \frac{2/sC}{R + 2/sC} V_1 \quad \Rightarrow A = V_1/V_2 = \frac{R + 2/sC}{2/sC} = \frac{RCs + 2}{2}$$

And I_1 may be expressed as:

$$I_1 = \frac{V_1}{(2/sC)\|(R + 2/sC)} = \frac{Cs(RCs + 4)}{2(RCs + 2)} V_1$$

Setting for V_1,

$$I_1 = \frac{Cs(RCs+4)}{2(RCs+2)}\left[\frac{RCs+2}{2}V_2\right] = \frac{Cs(RCs+4)}{4}V_2 \quad \Rightarrow C = I_1/V_2 = \frac{Cs(RCs+4)}{4}$$

From Fig. 10.6.2(d), we have

$$-I_2 = \frac{V_1-0}{R} \quad \Rightarrow B = V_1/(-I_2) = R$$

By current division formula,

$$-I_2 = \frac{2/sC}{R+2/sC}I_1 \quad \Rightarrow D = I_1/(-I_2) = \frac{RCs+2}{2}$$

(e)

From Fig. 10.6.1(e), we have

$$A = \frac{V_1}{V_2} = \frac{(sL/2+1/sC)I_1}{(1/sC)I_1} = s^2LC/2+1 \quad \text{and} \quad C = I_1/V_2 = sC$$

From Fig. 10.6.2(e), KCL at node V_3 gives

$$\frac{V_1-V_3}{sL/2} = \frac{V_3}{1/sC} + \frac{V_3-0}{sL/2} \quad \Rightarrow V_3 = \frac{2}{s^2LC+4}V_1$$

The expression for I_2 will be

$$-I_2 = \frac{V_3}{sL/2} = \frac{2}{sL}\left(\frac{2}{s^2LC+4}V_1\right) \quad \Rightarrow B = V_1/(-I_2) = \frac{sL(s^2LC+4)}{4}$$

$$D = I_1/(-I_2) = \frac{(V_1-V_3)/(sL/2)}{V_3/(sL/2)} = \frac{V_1-V_3}{V_3}$$

$$\Rightarrow D = \frac{V_1}{V_3} - 1 = \frac{s^2LC+4}{2} - 1 = \frac{s^2LC+2}{2}$$

(f)

From Fig. 10.6.1(f), we have

$$A = \frac{V_1}{V_2} = \frac{(1/sC_1+sL)I_1}{sLI_1} = \frac{s^2LC_1+1}{s^2LC_1} \quad \text{and} \quad C = I_1/V_2 = \frac{I_1}{(sL)I_1} = 1/sL$$

From Fig. 10.6.2(f), KCL at node V_3 gives

$$\frac{V_1-V_3}{1/sC_1} = \frac{V_3}{sL} + \frac{V_3}{1/sC_2} \quad \Rightarrow V_3 = \frac{s^2LC_1}{s^2L(C_1+C_2)+1}V_1$$

The I_2 may be expressed as:

$$-I_2 = \frac{V_3}{1/sC_2} = sC_2\frac{s^2LC_1}{s^2L(C_1+C_2)+1}V_1 \quad \Rightarrow B = V_1/(-I_2) = \frac{s^2L(C_1+C_2)+1}{s^3LC_1C_2}$$

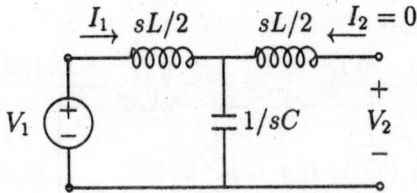

Fig. 10.6.1(e): For A and C.

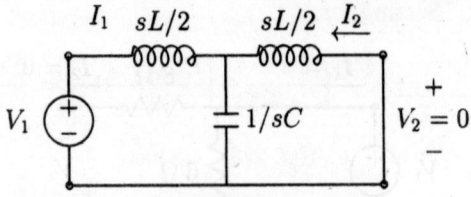

Fig. 10.6.2(e): For B and D.

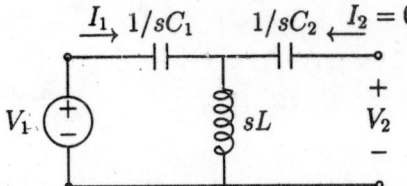

Fig. 10.6.1(f): For A and C.

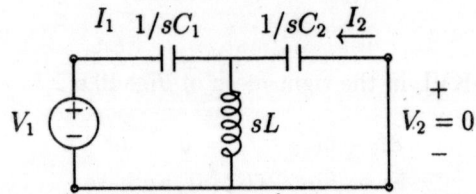

Fig. 10.6.2(f): For B and D.

and
$$D = I_1/(-I_2) = \frac{(V_1 - V_3)/(1/sC_1)}{V_3/(1/sC_2)} = \frac{C_1}{C_2}\left(\frac{V_1}{V_3} - 1\right)$$

$$\Rightarrow D = \frac{C_1}{C_2}\left(\frac{s^2L(C_1 + C_2) + 1}{s^2LC_1} - 1\right)$$

(g)

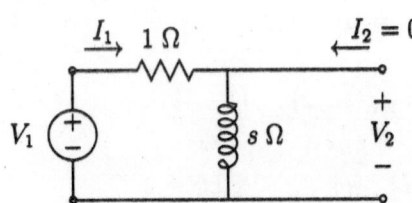

Fig. 10.6.1(g): For A and C.

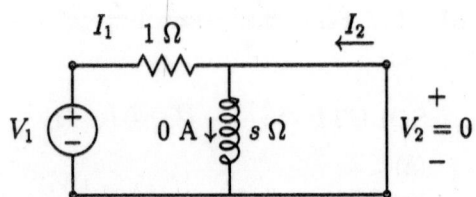

Fig. 10.6.2(g): For B and D.

With reference to Fig. 10.6.1(g), we have

$$A = V_1/V_2 = \frac{(s+1)I_1}{sI_1} = \frac{s+1}{s} \quad \text{and} \quad C = I_1/V_2 = \frac{I_1}{sI_1} = \frac{1}{s}$$

With reference to Fig. 10.6.2(g), we have

$$B = V_1/(-I_2) = V_1/I_1 = 1 \quad \text{and} \quad D = I_1/(-I_2) = I_1/I_1 = 1$$

(h)

From Fig. 10.6.1(h), we have

$$A = V_1/V_2 = V_1/V_1 = 1 \quad \text{and} \quad C = I_1/V_2 = I_1/6I_1 = 1/6$$

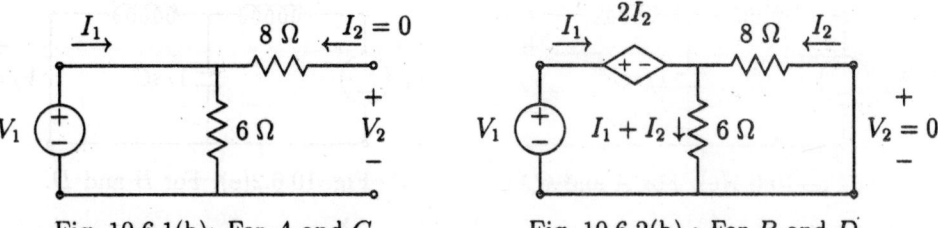

Fig. 10.6.1(h): For A and C.   Fig. 10.6.2(h) : For B and D.

KVL in the right mesh of Fig. 10.6.2(h) gives

$$8I_2 + 6(I_1 + I_2) = 0 \quad \text{or} \quad 14I_2 + 6I_1 = 0 \quad \text{or} \quad D = I_1/(-I_2) = 7/3$$

KVL in the left mesh of Fig. 10.6.2(h) gives

$$V_1 = 2I_2 + 6(I_1 + I_2) \quad \text{or} \quad V_1 = 6I_1 + 8I_2$$

Hence $\qquad B = V_1/(-I_2) = \dfrac{6I_1 + 8I_2}{-I_2} = 6\left(\dfrac{I_1}{-I_2}\right) - 8 = 6\,(7/3) - 8 = 6$

SP 10.7 Find the h parameters for each of the following networks in Fig. 10.7.

SOLUTION:

(a)
From Fig. 10.7.1(a),

$V_1 = 10I_1 \qquad \Rightarrow h_{11} = V_1/I_1 = 10 \qquad \text{and} \qquad I_1 = -I_2 \qquad \Rightarrow h_{21} = I_2/I_1 = -1$

From Fig. 10.7.2(a),

$$h_{12} = V_1/V_2 = V_2/V_2 = 1 \qquad \text{and} \qquad h_{22} = I_2/V_2 = \dfrac{V_2/20}{V_2} = 1/20$$

(b)
From Fig. 10.7.1(b), we have

$$h_{11} = V_1/I_1 = 5\|20 = 4\Omega \qquad \text{and} \qquad h_{21} = I_2/I_1 = \left[-\dfrac{20}{5+20}I_1\right]/I_1 = -4/5$$

From Fig. 10.7.2(b), we have

$$h_{12} = V_1/V_2 = \dfrac{20I_2}{(5+20)I_2} = 4/5 \qquad \text{and} \qquad h_{22} = I_2/V_2 = \dfrac{I_2}{(5+20)I_2} = 1/25$$

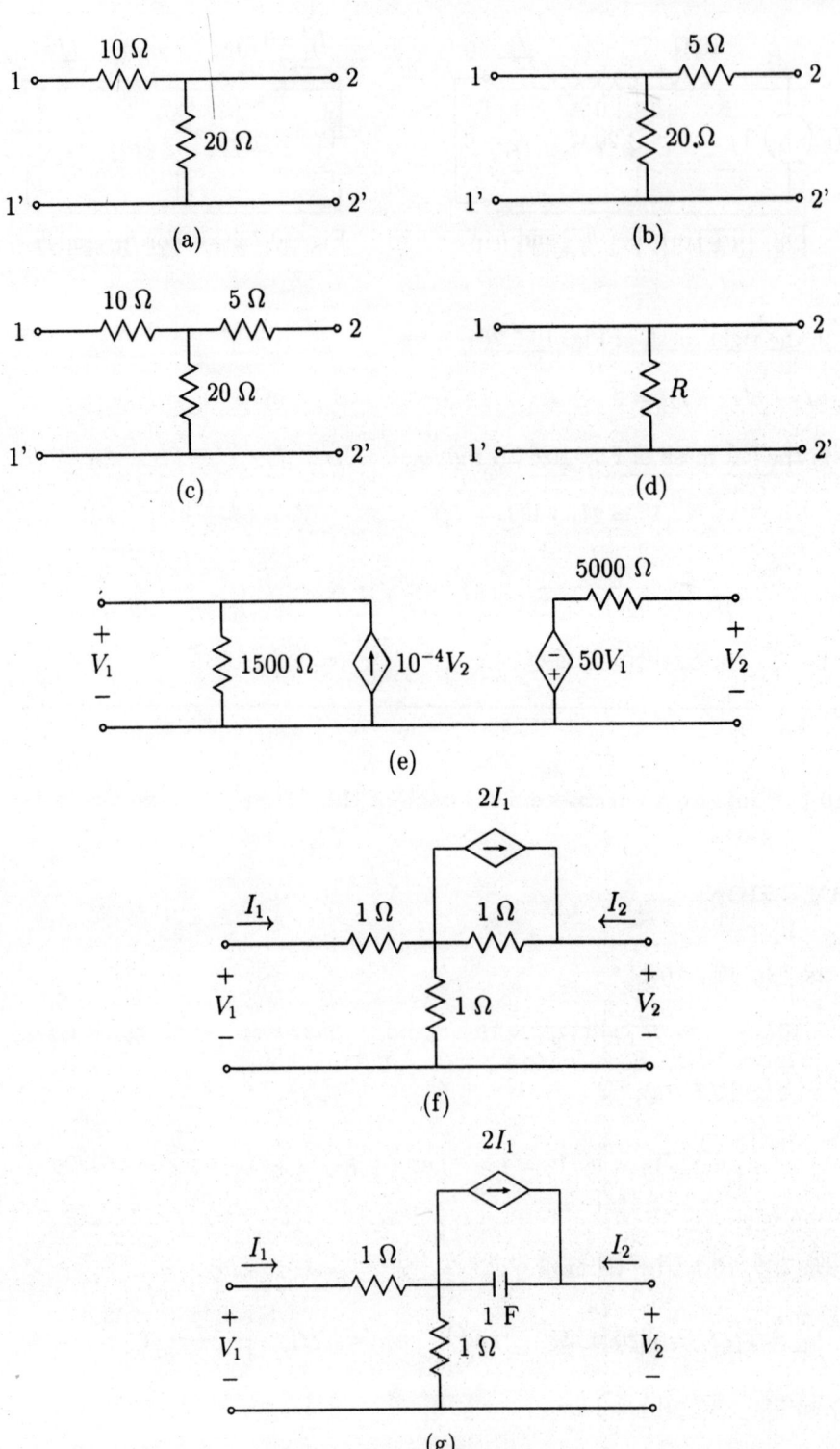

Fig. SP 10.7.

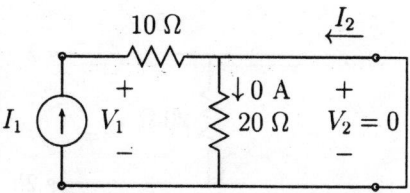

Fig. 10.7.1(a): For h_{11} and h_{21}.

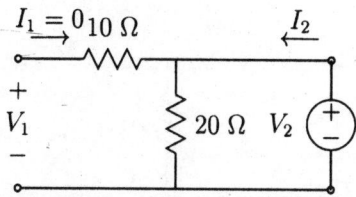

Fig. 10.7.2(a): For h_{12} and h_{22}.

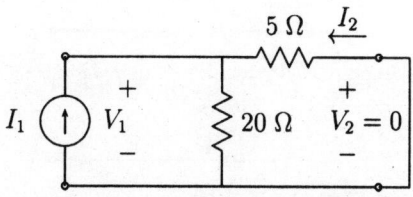

Fig. 10.7.1(b): For h_{11} and h_{21}.

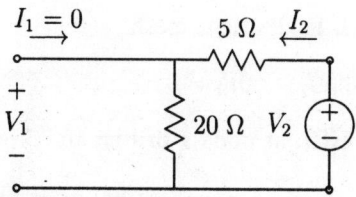

Fig. 10.7.2(b): For h_{12} and h_{22}.

(c)

From Fig. 10.7.1(c), we have

$$h_{11} = V_1/I_1 = 10 + (5||20) = 14\ \Omega \quad \text{and} \quad h_{21} = I_2/I_1 = \left(-\frac{20}{5+20}I_1\right)/I_1 = -4/5$$

From Fig. 10.7.2(c), we have

$$h_{12} = V_1/V_2 = \frac{20I_2}{(5+20)I_2} = 4/5 \quad \text{and} \quad h_{22} = I_2/V_2 = \frac{I_2}{(5+20)I_2} = 1/25$$

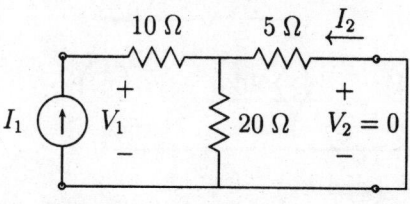

Fig. 10.7.1(c): For h_{11} and h_{21}.

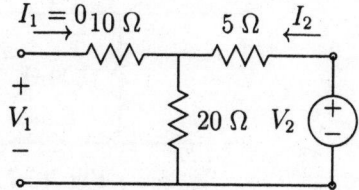

Fig. 10.7.2(c): For h_{12} and h_{22}.

(d)

From Fig. 10.7.1(d), we have

$$h_{11} = V_1/I_1 = 0 \quad \text{and} \quad h_{21} = I_2/I_1 = -I_1/I_1 = -1$$

From Fig. 10.7.2(d), we have

$$h_{12} = V_1/V_2 = V_2/V_2 = 1 \quad \text{and} \quad h_{22} = I_2/V_2 = \frac{V_2/R}{V_2} = 1/R$$

(e)

By KCL at node **1** of Fig. 10.7.1(e),

$$I_1 = \frac{V_1}{1500} \quad \Rightarrow \quad h_{11} = V_1/I_1 = 1500\Omega$$

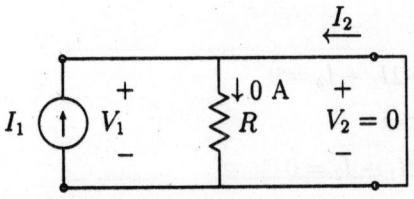

Fig. 10.7.1(d): For h_{11} and h_{21}.

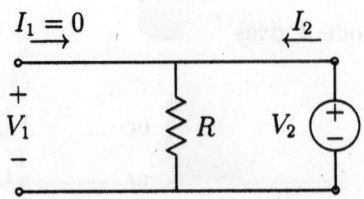

Fig. 10.7.2(d): For h_{12} and h_{22}.

By KVL in the right mesh,

$$5000I_2 - 50V_1 = 0 \quad \text{or} \quad 5000I_2 = 50 \times 1500I_1 \quad \text{or} \quad h_{21} = I_2/I_1 = 15$$

By KCL at node **1** of Fig. 10.7.2(e),

$$V_1/1500 = 10^{-4}V_2 \quad \Rightarrow \quad h_{12} = V_1/V_2 = 0.15$$

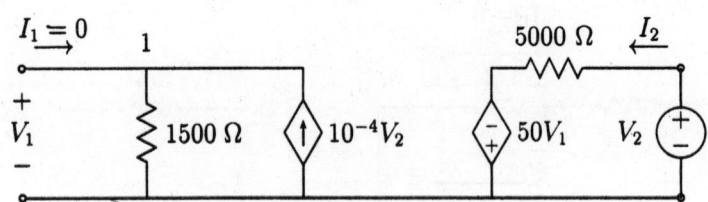

Fig. 10.7.1(e): For h_{11} and h_{21}.

Fig. 10.7.2(e): For h_{12} and h_{22}.

By KVL in the right mesh,

$$V_2 = 5000I_2 - 50V_1 \quad \text{or} \quad V_2 = 5000I_2 - 50 \times (0.15V_2) \quad \text{or} \quad 8.5V_2 = 5000I_2$$

Hence $$h_{22} = I_2/V_2 = 8.5/5000 = 1.7 \times 10^{-3} \text{ S}$$

(f)

From Fig. 10.7.1(f), KCL at node **a** gives

$$\frac{V_1 - V_a}{1} = \frac{V_a}{1} + \frac{V_a - 0}{1} \quad \Rightarrow \quad V_1 = 3V_a$$

KVL in the left mesh gives

$$V_1 = I_1 + V_a \quad \text{or} \quad V_1 = I_1 + V_1/3 \quad \text{or} \quad 2V_1 = 3I_1$$

Hence
$$h_{11} = V_1/I_1 = 3/2$$

KCL at node **2** gives

$$(V_a - 0)/1 + 2I_1 + I_2 = 0$$

or $\qquad V_1/3 + 2I_1 + I_2 = 0$

or $\qquad \dfrac{1}{3}(3I_1/2) + 2I_1 + I_2 = 0$

or $\qquad 5I_1/2 + I_2 = 0$

hence $\qquad h_{21} = I_2/I_1 = -5/2$

From Fig. 10.7.2(f), we have

$$h_{12} = V_1/V_2 = \frac{(1)I_2}{(1+1)I_2} = 1/2 \quad \text{and} \quad h_{22} = I_2/V_2 = \frac{I_2}{(1+1)I_2} = 1/2$$

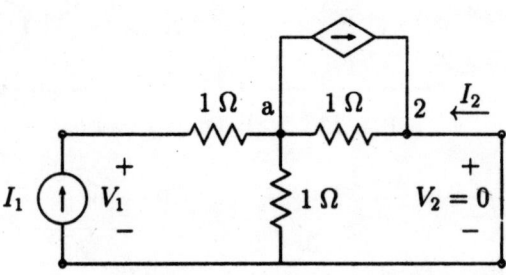

Fig. 10.7.1(f): For h_{11} and h_{21}.

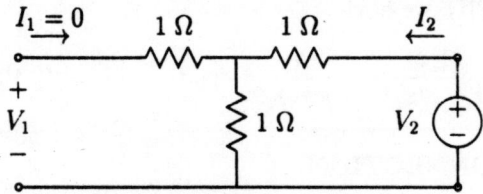

Fig. 10.7.2(f): For h_{12} and h_{22}.

(g)

From Fig. 10.7.1(g), KCL at node **a** gives

$$\frac{V_1 - V_a}{1} = \frac{V_a}{1} + \frac{V_a - 0}{1/s} \qquad \Rightarrow V_a = \frac{V_1}{2+s}$$

KVL in the left mesh gives

$$V_1 = I_1 + V_a \quad \text{or} \quad V_1 = I_1 + V_1/(2+s) \quad \text{or} \quad V_1 - \frac{1}{s+2}V_1 = I_1$$

hence $\qquad h_{11} = V_1/I_1 = \dfrac{s+2}{s+1}$

KCL at node **2** gives

$$sV_a + 2I_1 + I_2 = 0 \quad \text{or} \quad s\frac{V_1}{s+2} + 2I_1 + I_2 = 0$$

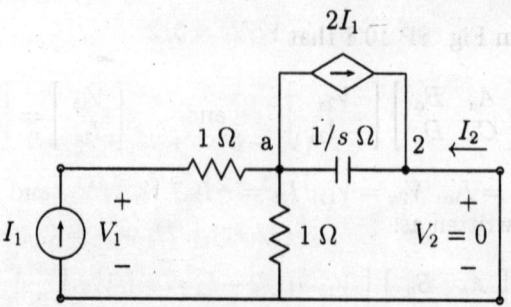

Fig. 10.7.1(g): For h_{11} and h_{21}.

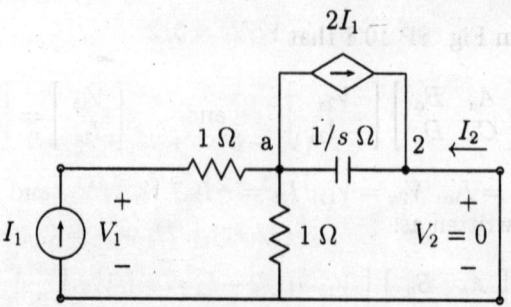

Fig. 10.7.2(g): For h_{12} and h_{22}.

or $\qquad \dfrac{s}{s+2} \times \dfrac{s+2}{s+1} I_1 + 2I_1 + I_2 = 0 \qquad$ or $\qquad \dfrac{3s+2}{s+1} I_1 + I_2 = 0$

hence $\qquad h_{21} = I_2/I_1 = -\dfrac{3s+2}{s+1}$

From Fig. 10.7.2(f), we have

$$h_{12} = V_1/V_2 = \frac{(1)I_2}{(1+1/s)I_2} = \frac{s}{s+1} \qquad \text{and} \qquad h_{22} = I_2/V_2 = \frac{I_2}{(1+1/s)I_2} = \frac{s}{s+1}$$

CASCADE CONNECTION:

The transmission parameters are useful in describing two-port networks which are connected in cascade.

SP 10.8 Fig. SP 10.8 shows a cascade connection of two-ports. Find the transmission parameters of the resulting two-port in terms of the ABCD parameters of the individual two-ports.

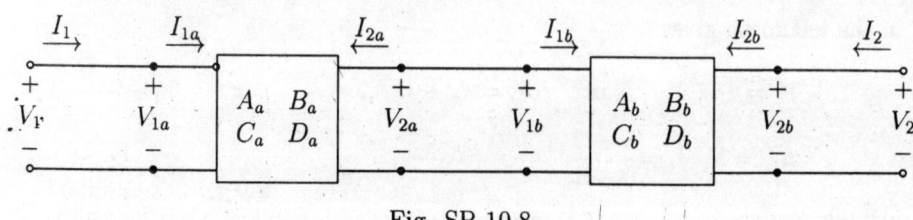

Fig. SP 10.8.

It is noted from Fig. SP 10.8 that

$$\begin{bmatrix} V_{1a} \\ I_{1a} \end{bmatrix} = \begin{bmatrix} A_a & B_a \\ C_a & D_a \end{bmatrix} \begin{bmatrix} V_{2a} \\ -I_{2a} \end{bmatrix} \quad \text{and} \quad \begin{bmatrix} V_{1b} \\ I_{1b} \end{bmatrix} = \begin{bmatrix} A_b & B_b \\ C_b & D_b \end{bmatrix} \begin{bmatrix} V_{2b} \\ -I_{2b} \end{bmatrix}$$

Since $V_1 = V_{1a}$, $I_1 = I_{1a}$, $V_{2a} = V_{1b}$, $I_{1b} = -I_{2a}$, $V_{2b} = V_2$, and $I_{2b} = I_2$, the above matrix equations can be written as:

$$\begin{bmatrix} V_1 \\ I_1 \end{bmatrix} = \begin{bmatrix} A_a & B_a \\ C_a & D_a \end{bmatrix} \begin{bmatrix} V_{1b} \\ I_{1b} \end{bmatrix} \quad \text{and} \quad \begin{bmatrix} V_{1b} \\ I_{1b} \end{bmatrix} = \begin{bmatrix} A_b & B_b \\ C_b & D_b \end{bmatrix} \begin{bmatrix} V_2 \\ -I_2 \end{bmatrix}$$

Combining we get

$$\begin{bmatrix} V_1 \\ I_1 \end{bmatrix} = \begin{bmatrix} A_a & B_a \\ C_a & D_a \end{bmatrix} \begin{bmatrix} A_b & B_b \\ C_b & D_b \end{bmatrix} \begin{bmatrix} V_2 \\ -I_2 \end{bmatrix}$$

Thus, we get

$$\begin{bmatrix} A & B \\ C & D \end{bmatrix} = \begin{bmatrix} A_a & B_a \\ C_a & D_a \end{bmatrix} \begin{bmatrix} A_b & B_b \\ C_b & D_b \end{bmatrix}$$

SP 10.9 Find the transmission for each of the following networks by cascade connection shown in Fig. SP 10.9.

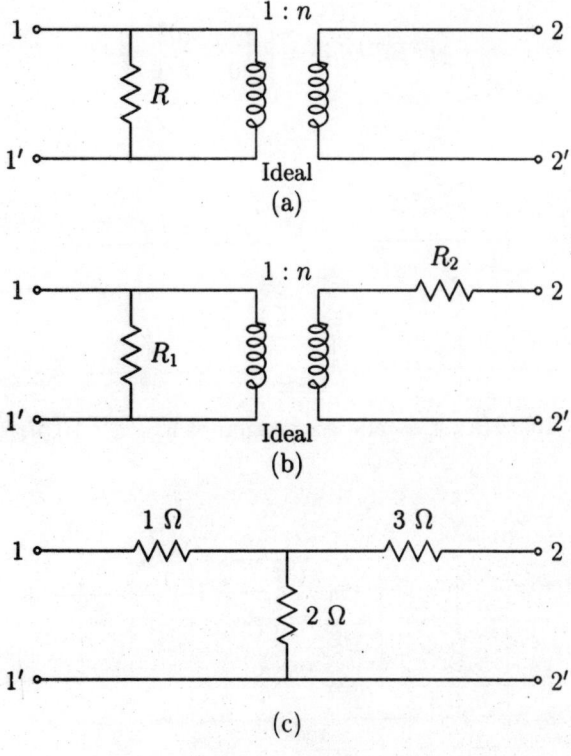

Fig. SP 10.9.

SOLUTION:

(a)

For the network N_1,

$$[ABCD]_1 = \begin{bmatrix} 1 & 0 \\ 1/R & 1 \end{bmatrix} \qquad \text{see SP 10.6(b)}$$

For the network N_2,

$$V_1 = V_2/n \qquad \text{and} \qquad I_1 = -nI_2$$

Hence

$$[ABCD]_2 = \begin{bmatrix} 1/n & 0 \\ 0 & n \end{bmatrix}$$

On combining, we obtain

$$\begin{bmatrix} A & B \\ C & D \end{bmatrix} = \begin{bmatrix} 1 & 0 \\ 1/R & 1 \end{bmatrix} \begin{bmatrix} 1/n & 0 \\ 0 & n \end{bmatrix} = \begin{bmatrix} 1/n & 0 \\ 1/nR & n \end{bmatrix}$$

(b) The cascade connection for the network of Fig. 10.9(b) is given in Fig. 10.9.1(b).

For the network N_1, the transmission matrix is

$$[ABCD]_{(1)} = \begin{bmatrix} 1 & 0 \\ 1/R_1 & 1 \end{bmatrix} \qquad \text{see SP 10.6(b)}$$

For the network N_2, the transmission matrix is

$$[ABCD]_{(2)} = \begin{bmatrix} 1/n & 0 \\ 0 & n \end{bmatrix} \qquad \text{see SP 10.5(a)}$$

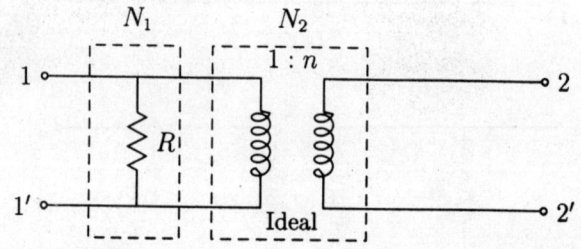

Fig. 10.9.1(a): Cascade connection of Fig. SP 10.9(a).

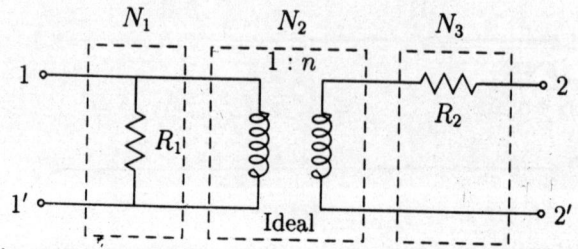

Fig. 10.9.1(b): Cascade connection of Fig. SP 10.9(b).

And for the network N_3, the transmission matrix is

$$[ABCD]_{(3)} = \begin{bmatrix} 1 & R_2 \\ 0 & 1 \end{bmatrix} \qquad \text{see SP 10.6(c)}$$

Thus, on combining, we obtain

$$\begin{bmatrix} A & B \\ C & D \end{bmatrix} = \begin{bmatrix} 1 & 0 \\ 1/R_1 & 1 \end{bmatrix} \begin{bmatrix} 1/n & 0 \\ 0 & n \end{bmatrix} \begin{bmatrix} 1 & R_2 \\ 0 & 1 \end{bmatrix} = \begin{bmatrix} 1/n & R_2/n \\ 1/nR_1 & R_2/nR_1 + n \end{bmatrix}$$

(c) The cascade connection for the network of Fig. SP 10.9(c) is given in Fig. 10.9.1(c).

For the network N_1, the transmission matrix is

$$[ABCD]_{(1)} = \begin{bmatrix} 1 & 1 \\ 0 & 1 \end{bmatrix} \qquad \text{see SP 10.6(c)}$$

For the network N_2, the transmission matrix is

$$[ABCD]_{(2)} = \begin{bmatrix} 1 & 0 \\ 1/2 & 1 \end{bmatrix} \qquad \text{see SP 10.6(b)}$$

And for the network N_3, the transmission matrix is

$$[ABCD]_{(3)} = \begin{bmatrix} 1 & 3 \\ 0 & 1 \end{bmatrix} \qquad \text{see SP 10.6(c)}$$

Thus, on combining, we obtain

$$\begin{bmatrix} A & B \\ C & D \end{bmatrix} = \begin{bmatrix} 1 & 1 \\ 0 & 1 \end{bmatrix} \begin{bmatrix} 1 & 0 \\ 1/2 & 1 \end{bmatrix} \begin{bmatrix} 1 & 3 \\ 0 & 1 \end{bmatrix} = \begin{bmatrix} 3/2 & 11/2 \\ 1/2 & 5/2 \end{bmatrix}$$

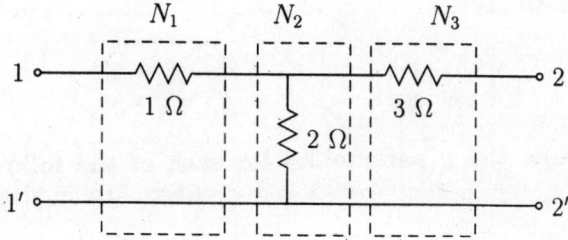

Fig. 10.9.1(c): Cascade connection of Fig. SP 10.9(c).

PARALLEL CONNECTION OF TWO-PORT NETWORKS:

The y parameters are useful in describing two-port networks which are connected in parallel .

SP 10.10 Fig. SP 10.10 shows a parallel connection of two-ports. Find the y parameters of the resulting two-port in terms of y parameters of the individual two-ports.

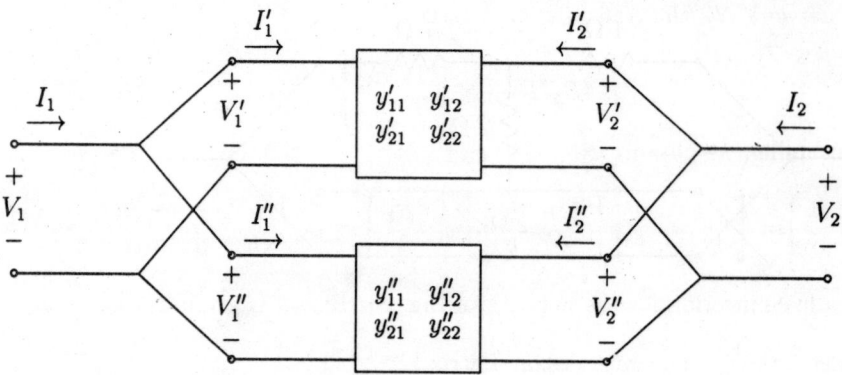

Fig. SP 10.10: Parallel connection of two two-port networks.

It is noted from Fig. SP 10.10 that

$$\begin{bmatrix} I_1' \\ I_2' \end{bmatrix} = \begin{bmatrix} y_{11}' & y_{12}' \\ y_{21}' & y_{22}' \end{bmatrix} \begin{bmatrix} V_1' \\ V_2' \end{bmatrix} \qquad \text{and} \qquad \begin{bmatrix} I_1'' \\ I_2'' \end{bmatrix} = \begin{bmatrix} y_{11}'' & y_{12}'' \\ y_{21}'' & y_{22}'' \end{bmatrix} \begin{bmatrix} V_1'' \\ V_2'' \end{bmatrix}$$

Since $V_1 = V_1' = V_1''$, $I_1 = I_1' + I_1''$, $V_2 = V_2' = V_2''$, and $I_2 = I_2' + I_2''$, the above matrix equations can be written as:

$$\begin{bmatrix} I_1' \\ I_2' \end{bmatrix} = \begin{bmatrix} y_{11}' & y_{12}' \\ y_{21}' & y_{22}' \end{bmatrix} \begin{bmatrix} V_1 \\ V_2 \end{bmatrix} \qquad \text{and} \qquad \begin{bmatrix} I_1'' \\ I_2'' \end{bmatrix} = \begin{bmatrix} y_{11}'' & y_{12}'' \\ y_{21}'' & y_{22}'' \end{bmatrix} \begin{bmatrix} V_1 \\ V_2 \end{bmatrix}$$

Adding these two matrix equations, we get

$$\begin{bmatrix} I_1 \\ I_2 \end{bmatrix} = \begin{bmatrix} I_1' + I_1'' \\ I_2' + I_2'' \end{bmatrix} = \begin{bmatrix} y_{11}' + y_{11}'' & y_{12}' + y_{12}'' \\ y_{21}' + y_{21}'' & y_{22}' + y_{22}'' \end{bmatrix} \begin{bmatrix} V_1 \\ V_2 \end{bmatrix}$$

Thus, resultant y matrix is

$$\begin{bmatrix} y_{11} & y_{12} \\ y_{21} & y_{22} \end{bmatrix} = \begin{bmatrix} y_{11}' + y_{11}'' & y_{12}' + y_{12}'' \\ y_{21}' + y_{21}'' & y_{22}' + y_{22}'' \end{bmatrix}$$

SP 10.11 Determine the y parameters for each of the following networks of Fig. SP 10.11.

SOLUTION:

(a) From network of Fig. SP 10.11(a), for the upper network, the Y matrix is

$$\begin{bmatrix} y_{11}' & y_{12}' \\ y_{21}' & y_{22}' \end{bmatrix} = \begin{bmatrix} 3/5 & -2/5 \\ -2/5 & 3/5 \end{bmatrix} \qquad\qquad \text{see SP 10.1(b)}$$

and for the lower network, the Y matrix is

$$\begin{bmatrix} y_{11}'' & y_{12}'' \\ y_{21}'' & y_{22}'' \end{bmatrix} = \begin{bmatrix} 3/2 & -1/2 \\ -1/2 & 5/6 \end{bmatrix} \qquad\qquad \text{see SP 10.1(a)}$$

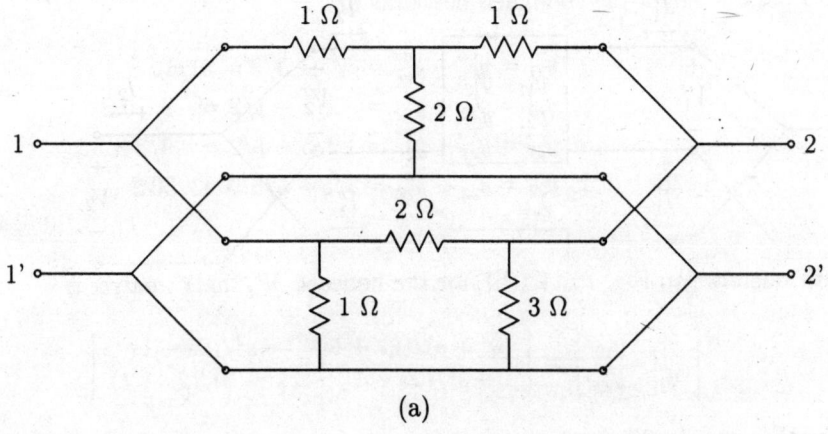

(a)

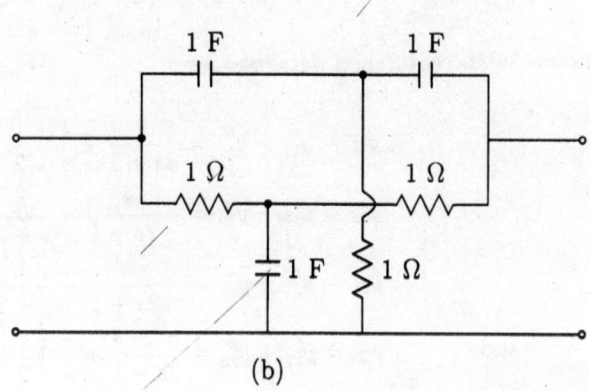

(b)

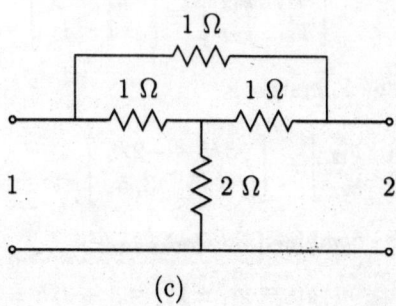

(c)

Fig. SP 10.11.

Hence, y parameters of the combined networks are:

$$y_{11} = y'_{11} + y''_{11} = 3/5 + 3/2 = 21/10 \text{ S}$$
$$y_{12} = y'_{12} + y''_{12} = -2/5 - 1/2 = -9/10 \text{ S}$$
$$y_{21} = y'_{21} + y''_{21} = -2/5 - 1/2 = -9/2 \text{ S}$$

and
$$y_{22} = y'_{22} + y''_{22} = 3/5 + 5/6 = 43/30 \text{ S}$$

(b)

With reference to Fig. 10.11.1(b), for the network N', the Y matrix is

$$\begin{bmatrix} y'_{11} & y'_{12} \\ y'_{21} & y'_{22} \end{bmatrix} = \begin{bmatrix} (s^2 + s)/(2s + 1) & -s^2/(2s + 1) \\ -s^2/(2s + 1) & (s^2 + s)/(2s + 1) \end{bmatrix}$$

and for the network N'', the Y matrix is

$$\begin{bmatrix} y''_{11} & y''_{12} \\ y''_{21} & y''_{22} \end{bmatrix} = \begin{bmatrix} (s + 1)/(s + 2) & -1/(s + 2) \\ -1/(s + 2) & (s + 1)/(s + 2) \end{bmatrix}$$

Hence, y parameters of the combined networks are:

$$y_{11} = y'_{11} + y''_{11} = \frac{s^2 + s}{2s + 1} + \frac{s + 1}{s + 2}$$

$$y_{12} = y'_{12} + y''_{12} = -\frac{s^2}{2s + 1} - \frac{1}{s + 2}$$

$$y_{21} = y'_{21} + y''_{21} = -\frac{s^2}{2s + 1} - \frac{1}{s + 2}$$

and
$$y_{22} = y'_{22} + y''_{22} = \frac{s^2 + s}{2s + 1} + \frac{s + 1}{s + 2}$$

(c)

With reference to the Fig. 10.11.1(c), for the network N', the Y matrix is

$$\begin{bmatrix} y'_{11} & y'_{12} \\ y'_{21} & y'_{22} \end{bmatrix} = \begin{bmatrix} 1 & -1 \\ -1 & 1 \end{bmatrix} \qquad \text{see DP 10.1(b)}$$

and for the network N'', the Y matrix is

$$\begin{bmatrix} y''_{11} & y''_{12} \\ y''_{21} & y''_{22} \end{bmatrix} = \begin{bmatrix} 3/5 & -2/5 \\ -2/5 & 3/5 \end{bmatrix} \qquad \text{see SP 10.1(b)}$$

Hence, y parameters of the combined networks are:

$$y_{11} = y'_{11} + y''_{11} = 1 + 3/5 = 8/5 \text{ S}$$
$$y_{12} = y'_{12} + y''_{12} = -1 - 2/5 = -7/5 \text{ S}$$
$$y_{21} = y'_{21} + y''_{21} = -1 - 2/5 = -7/5 \text{ S}$$

and
$$y_{22} = y'_{22} + y''_{22} = 1 + 3/5 = 8/5 \text{ S}$$

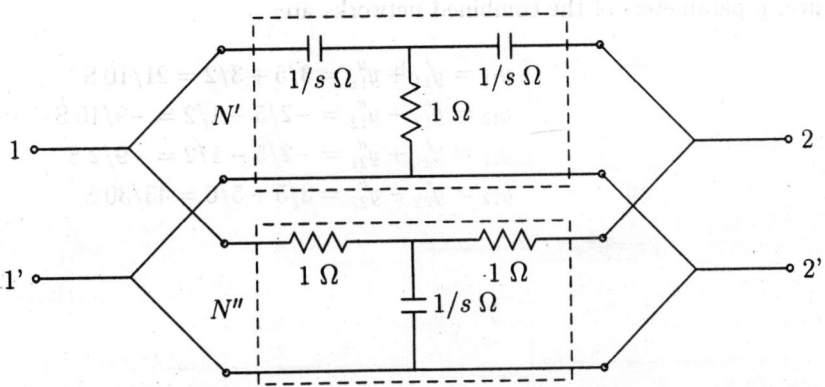

Fig. 10.11.1(b): Redrawn network from the network of Fig. SP 10.11(b).

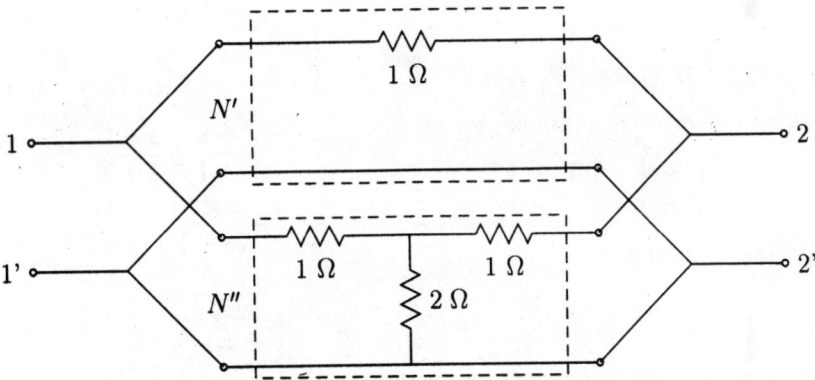

Fig. 10.11.1(c): Redrawn network from the network of Fig. SP 10.11(c).

SERIES CONNECTION OF TWO-PORT NETWORKS:

The z parameters are useful in describing two-port networks which are connected in series.

SP 10.12 Fig. SP 10.12 shows a series connection of two-ports. Find the z parameters of the resulting two-port in terms of z parameters of the individual two-ports.

It is noted from Fig. SP 10.12 that the individual matrix equations are:

$$\begin{bmatrix} V_1' \\ V_2' \end{bmatrix} = \begin{bmatrix} z_{11}' & z_{12}' \\ z_{21}' & z_{22}' \end{bmatrix} \begin{bmatrix} I_1' \\ I_2' \end{bmatrix} \quad \text{and} \quad \begin{bmatrix} V_1'' \\ V_2'' \end{bmatrix} = \begin{bmatrix} z_{11}'' & z_{12}'' \\ z_{21}'' & z_{22}'' \end{bmatrix} \begin{bmatrix} I_1'' \\ I_2'' \end{bmatrix}$$

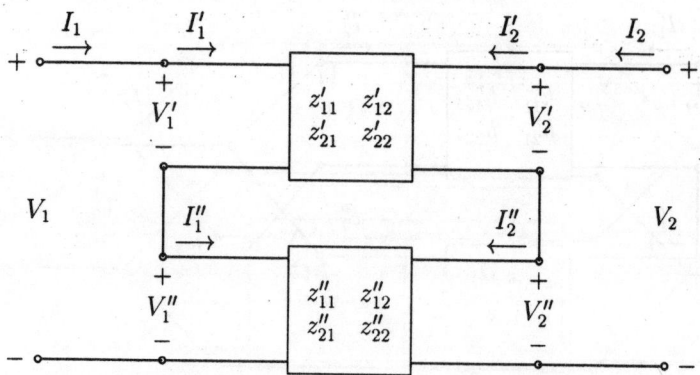

Fig. SP 10.12: Series connection of two two-port networks.

Since $V_1 = V_1' + V_1''$, $I_1 = I_1' = I_1''$, $V_2 = V_2' + V_2''$, and $I_2 = I_2' = I_2''$, the above matrix equations can be written as:

$$\begin{bmatrix} V_1' \\ V_2' \end{bmatrix} = \begin{bmatrix} z_{11}' & z_{12}' \\ z_{21}' & z_{22}' \end{bmatrix} \begin{bmatrix} I_1 \\ I_2 \end{bmatrix} \quad \text{and} \quad \begin{bmatrix} V_1'' \\ V_2'' \end{bmatrix} = \begin{bmatrix} z_{11}'' & z_{12}'' \\ z_{21}'' & z_{22}'' \end{bmatrix} \begin{bmatrix} I_1 \\ I_2 \end{bmatrix}$$

Adding these two matrix equations, we get

$$\begin{bmatrix} V_1 \\ V_2 \end{bmatrix} = \begin{bmatrix} V_1' + V_1'' \\ V_2' + V_2'' \end{bmatrix} = \begin{bmatrix} z_{11}' + z_{11}'' & z_{12}' + z_{12}'' \\ z_{21}' + z_{21}'' & z_{22}' + z_{22}'' \end{bmatrix} \begin{bmatrix} I_1 \\ I_2 \end{bmatrix}$$

Thus, resultant z matrix is

$$\begin{bmatrix} z_{11} & z_{12} \\ z_{21} & z_{22} \end{bmatrix} = \begin{bmatrix} z_{11}' + z_{11}'' & z_{12}' + z_{12}'' \\ z_{21}' + z_{21}'' & z_{22}' + z_{22}'' \end{bmatrix}$$

SERIES-PARALLEL CONNECTION OF TWO-PORT NETWORKS:

The h parameters are useful in describing two-port networks which are connected in series-parallel.

SP 10.13 Fig. SP 10.13 shows a series-parallel connection of two-ports. Find the h parameters of the resulting two-port in terms of h parameters of the individual two-ports.

It is noted from Fig. SP 10.13 that the individual matrix equations are:

$$\begin{bmatrix} V_1' \\ I_2' \end{bmatrix} = \begin{bmatrix} h_{11}' & h_{12}' \\ h_{21}' & h_{22}' \end{bmatrix} \begin{bmatrix} I_1' \\ V_2' \end{bmatrix} \quad \text{and} \quad \begin{bmatrix} V_1'' \\ I_2'' \end{bmatrix} = \begin{bmatrix} h_{11}'' & h_{12}'' \\ h_{21}'' & h_{22}'' \end{bmatrix} \begin{bmatrix} I_1'' \\ V_2'' \end{bmatrix}$$

Since $V_1 = V_1' + V_1''$, $I_1 = I_1' = I_1''$, $V_2 = V_2' = V_2''$, and $I_2 = I_2' + I_2''$, the above matrix equations can be written as:

$$\begin{bmatrix} V_1' I_2' \end{bmatrix} = \begin{bmatrix} h_{11}' & h_{12}' \\ h_{21}' & h_{22}' \end{bmatrix} \begin{bmatrix} I_1 \\ V_2 \end{bmatrix} \quad \text{and} \quad \begin{bmatrix} V_1'' \\ I_2'' \end{bmatrix} = \begin{bmatrix} h_{11}'' & h_{12}'' \\ h_{21}'' & h_{22}'' \end{bmatrix} \begin{bmatrix} I_1 \\ V_2 \end{bmatrix}$$

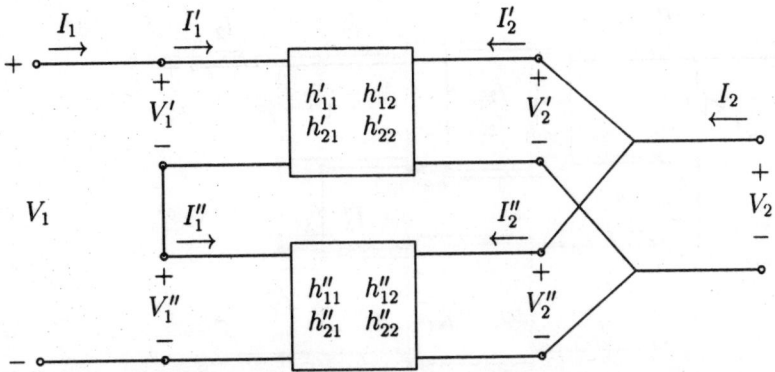

Fig. SP 10.13: Series-parallel connection of two two-port networks.

Adding these two matrix equations, we get

$$\left[\begin{array}{c} V_1 \\ I_2 \end{array}\right] = \left[\begin{array}{c} V_1' + V_1'' \\ I_2' + I_2'' \end{array}\right] = \left[\begin{array}{cc} h_{11}' + h_{11}'' & h_{12}' + h_{12}'' \\ h_{21}' + h_{21}'' & h_{22}' + h_{22}'' \end{array}\right] \left[\begin{array}{c} I_1 \\ V_2 \end{array}\right]$$

Thus, resultant h matrix is

$$\left[\begin{array}{cc} h_{11} & h_{12} \\ h_{21} & h_{22} \end{array}\right] = \left[\begin{array}{cc} h_{11}' + h_{11}'' & h_{12}' + h_{12}'' \\ h_{21}' + h_{21}'' & h_{22}' + h_{22}'' \end{array}\right]$$

PARALLEL-SERIES CONNECTION OF TWO-PORT NETWORKS:

The g parameters are useful in describing two-port networks which are connected in parallel-series.

SP 10.14 Fig. SP 10.14 shows a parallel-series connection of two-ports. Find the g parameters of the resulting two-port in terms of g parameters of the individual two-ports.

It is noted from Fig. SP 10.14 that the individual matrix equations are:

$$\left[\begin{array}{c} I_1' \\ V_2' \end{array}\right] = \left[\begin{array}{cc} g_{11}' & g_{12}' \\ g_{21}' & g_{22}' \end{array}\right] \left[\begin{array}{c} V_1' \\ I_2' \end{array}\right] \quad \text{and} \quad \left[\begin{array}{c} I_1'' \\ V_2'' \end{array}\right] = \left[\begin{array}{cc} g_{11}'' & g_{12}'' \\ g_{21}'' & g_{22}'' \end{array}\right] \left[\begin{array}{c} V_1'' \\ I_2'' \end{array}\right]$$

Since $V_1 = V_1' = V_1''$, $I_1 = I_1' + I_1''$, $V_2 = V_2' + V_2''$, and $I_2 = I_2' = I_2''$, the above matrix equations can be written as:

$$\left[\begin{array}{c} I_1' V_2' \end{array}\right] = \left[\begin{array}{cc} g_{11}' & g_{12}' \\ g_{21}' & g_{22}' \end{array}\right] \left[\begin{array}{c} V_1 \\ I_2 \end{array}\right] \quad \text{and} \quad \left[\begin{array}{c} I_1'' \\ V_2'' \end{array}\right] = \left[\begin{array}{cc} g_{11}'' & g_{12}'' \\ g_{21}'' & g_{22}'' \end{array}\right] \left[\begin{array}{c} V_1 \\ I_2 \end{array}\right]$$

Adding these two matrix equations, we get

$$\left[\begin{array}{c} I_1 \\ V_2 \end{array}\right] = \left[\begin{array}{c} I_1' + I_1'' \\ V_2' + V_2'' \end{array}\right] = \left[\begin{array}{cc} g_{11}' + g_{11}'' & g_{12}' + g_{12}'' \\ g_{21}' + g_{21}'' & g_{22}' + g_{22}'' \end{array}\right] \left[\begin{array}{c} V_1 \\ I_2 \end{array}\right]$$

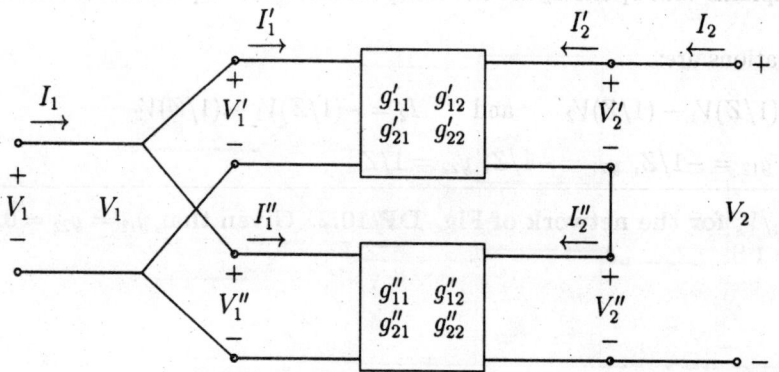

Fig. SP 10.14: Parallel-series connection of two two-port networks.

Thus, resultant g matrix is

$$\begin{bmatrix} g_{11} & g_{12} \\ g_{21} & g_{22} \end{bmatrix} = \begin{bmatrix} g'_{11} + g''_{11} & g'_{12} + g''_{12} \\ g'_{21} + g''_{21} & g'_{22} + g''_{22} \end{bmatrix}$$

10.8   DRILL PROBLEMS

DP 10.1 Find the y and z parameters for each of the following networks shown in Fig. 10.1.

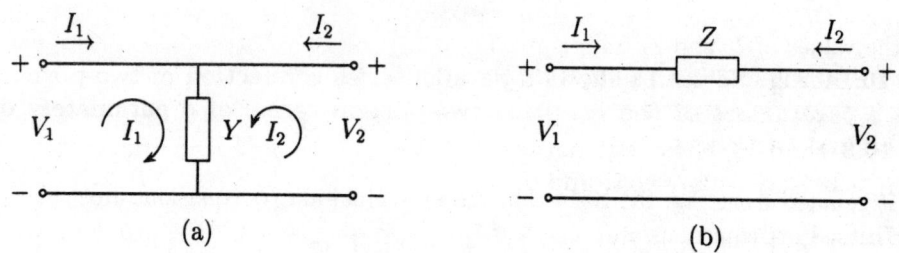

Fig. DP 10.1.

Hints: (a) Loop basis equations are:

$$V_1 = (1/Y)I_1 + (1/Y)I_2 \quad \text{and} \quad V_2 = (1/Y)I_1 + (1/Y)I_2$$

Hence, $z_{11} = 1/Y$, $z_{12} = 1/Y$, $z_{21} = 1/Y$, $z_{22} = 1/Y$

$V_1 = V_2$ explains that V_1 and V_2 are not independent, hence y parameters can not be determined.

(b) $I_1 = -I_2$ explains that I_1 and I_2 are not independent, hence z parameters can not be determined.

Node basis equations are:

$$I_1 = (1/Z)V_1 - (1/Z)V_2 \quad \text{and} \quad I_2 = -(1/Z)V_1 + (1/Z)V_2$$

Hence, $y_{11} = 1/Z$, $y_{12} = -1/Z$, $y_{21} = -1/Z$, $y_{22} = 1/Z$

DP 10.2 Find V_2/V_s for the network of Fig. DP 10.2. Given that $y_{11} = y_{22} = 0.5$ S and $y_{21} = y_{12} = 1$ S

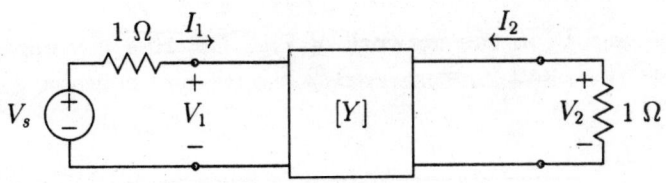

Fig. DP 10.2.

Hints: Port-equations are:

$$I_1 = 0.5V_1 + V_2 \qquad \text{[DP 10.2.1]}$$
$$I_2 = V_1 + 0.5V_2 \qquad \text{[DP 10.2.2]}$$

Input current

$$I_1 = (V_s - V_1)/1 \qquad \text{[DP 10.2.3]}$$

Ohm's law at output

$$-I_2 = V_2 \qquad \text{[DP 10.2.4]}$$

Set the values of I_1 and I_2 from eqns [DP 10.2.3] and [DP 10.2.4] into eqns [DP 10.2.1] and [DP 10.2.2] to determine V_2/V_s.

DP 10.3 Find V_2/V_s for the network of Fig. DP 10.3. Given that
$z_{11} = 18 \ \Omega$, $z_{12} = z_{21} = 6 \ \Omega$ and $z_{22} = 9 \ \Omega$.

Hints: Port-equations are:

$$V_1 = 18I_1 + 6I_2 \qquad \text{[DP 10.3.1]}$$
$$V_2 = 6I_1 + 9I_2 \qquad \text{[DP 10.3.2]}$$

Input current

$$I_1 = (V_s - V_1)/4 \qquad \text{[DP 10.3.3]}$$

Ohm's law at output

$$I_2 = -V_2/10 \qquad \text{[DP 10.3.4]}$$

Eliminate I_1 and I_2 from equations [DP 10.3.1] and [DP 10.3.2] with the help of equations [DP 10.3.3] and [DP 10.3.4] to determine V_2/V_s

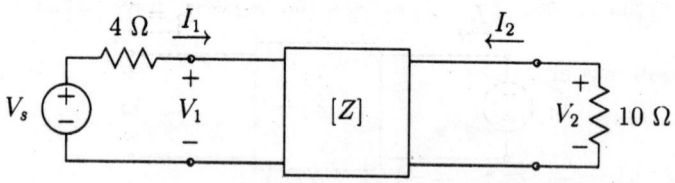

Fig. DP 10.3.

___

DP 10.4 Find V_1 and V_2 in the network of Fig. DP 10.4 if y parameters are: $y_{11} = 3/2$ S, $y_{22} = 5/6$ S, and $y_{12} = y_{21} = -1/2$ S

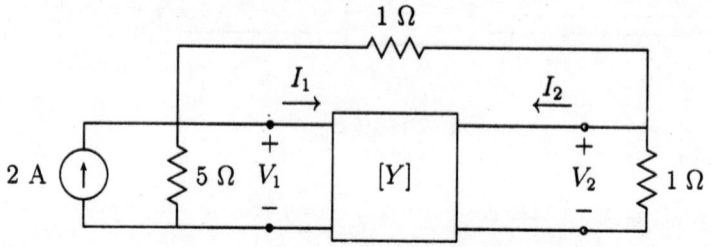

Fig. DP 10.4.

Hints : Port-equations are:

$$I_1 = \frac{3}{2}V_1 - \frac{1}{2}V_2 \qquad\qquad \text{[DP 10.4.1]}$$

$$I_2 = \frac{-1}{2}V_1 + \frac{5}{6}V_2 \qquad\qquad \text{[DP 10.4.2]}$$

KCL at input node

$$2 = V_1/5 + (V_1 - V_2)/1 + I_1 \quad \Rightarrow \quad I_1 = 2 - 6V_1/5 + V_2 \qquad \text{[DP 10.4.3]}$$

KCL at the output node

$$(V_1 - V_2)/1 = V_2/1 + I_2 \quad \Rightarrow \quad I_2 = V_1 - 2V_2 \qquad \text{[DP 10.4.4]}$$

Eliminate I_1 and I_2 from equations [DP 10.4.1] and [DP 10.4.2] with the help of equations [DP 10.4.3] and [DP 10.4.4] to determine V_1 and V_2.

___

DP 10.5 Determine $G_{12} = V_2/V_1$ for the network of Fig. DP 10.5.

Hints : Port-equations are:

$$V_1 = z_{11}I_1 + z_{12}I_2 \qquad\qquad \text{[DP 10.5.1]}$$

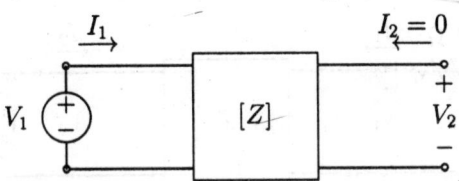

Fig. DP 10.5.

$$V_2 = z_{21}I_1 + z_{22}I_2 \qquad\qquad [DP\ 10.5.2]$$

Setting $V_1 = V_1$ and $I_2 = 0$, gives

$$V_1 = z_{11}I_1 \quad \text{and} \quad V_2 = z_{21}I_1 \quad \text{hence} \quad G_{12} = V_2/V_1 = z_{21}/z_{11}$$

DP 10.6 Find $G_{12} = V_2/V_1$ in terms of y parameters for the network of Fig. DP 10.6.

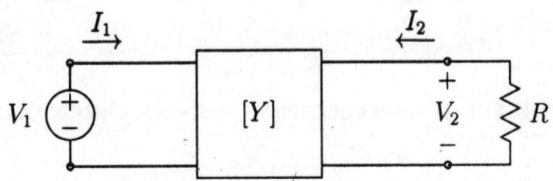

Fig. DP 10.6.

Hints: Port-equations are:

$$I_1 = y_{11}V_1 + y_{12}V_2 \qquad\qquad [DP9-6.1]$$

$$I_2 = y_{21}V_1 + y_{22}V_2 \qquad\qquad [DP9-6.2]$$

Set $I_2 = -V_2/R$ into eqn [DP 9-6.2] to determine $V_2/V_1 = -y_{21}/(y_{22} + 1/R)$

DP 10.7 Find $Z_{12} = V_2/I_1$ in terms of z parameters for network of Fig. DP 10.7.

Hints : Port-equations are:

$$V_1 = z_{11}I_1 + z_{12}I_2 \qquad\qquad [DP\ 10.7.1]$$

$$V_2 = z_{21}I_1 + z_{22}I_2 \qquad\qquad [DP\ 10.7.2]$$

Set $I_2 = -V_2/R$ into eqn [DP 10.7.2] to determine $Z_{12} = V_2/I_1 = z_{21}R/(R + z_{22})$

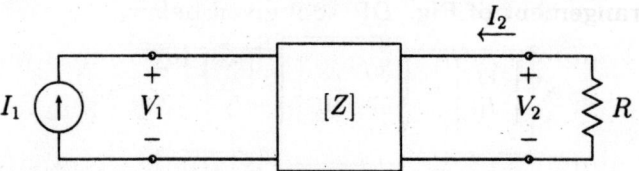

Fig. DP 10.7.

DP 10.8 Synthesize a T-network for which z parameters are:
$z_{11}(s) = 10 + 2s$, $z_{12}(s) = z_{21} = 2s$ and $z_{22}(s) = 1/2s + 2s$

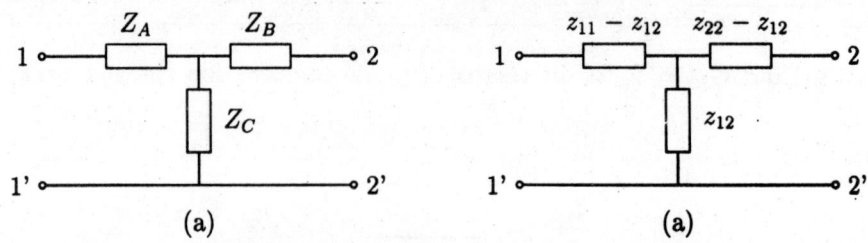

Fig. DP 10.8.1: (a) A T-network; (b) T-network representing z-parameters.

Hints:

With reference to Fig. DP 10.8.1,

$$Z_B = z_{12} = 2s$$

$$Z_A = z_{11} - z_{12} = 10 + 2s - 2s = 10$$

$$Z_C = z_{22} - z_{12} = 1/2s + 2s - 2s = 1/2s$$

Required circuit is shown in Fig. DP 10.8.2.

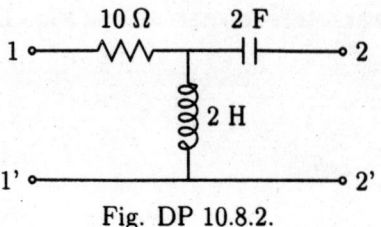

Fig. DP 10.8.2.

DP 10.9 In an arrangement of Fig. DP 10.9 given below,

$$\begin{bmatrix} V_1 \\ I_1 \end{bmatrix} = \begin{bmatrix} 30 & 23 \\ 13 & 10 \end{bmatrix} \begin{bmatrix} V_2 \\ -I_2 \end{bmatrix}$$

Find the A, B, C, D **parameters of** N'

GATE-94(EE)

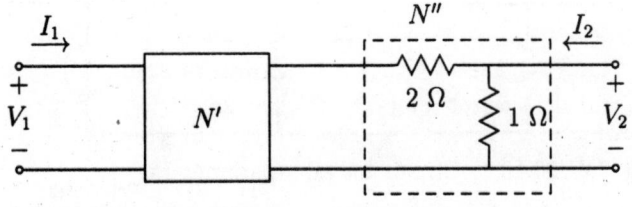

Fig. DP 10.9.

Hints : The transmission matrix for the network N'' is

$$[a''] = \begin{bmatrix} 3 & 2 \\ 1 & 1 \end{bmatrix}$$

Let the transmission matrix for the network N' be

$$[a'] = \begin{bmatrix} A' & B' \\ C' & D' \end{bmatrix}$$

Determine the transmission parameters for the network N' from the following matrix equation:

$$\begin{bmatrix} 30 & 23 \\ 13 & 10 \end{bmatrix} = \begin{bmatrix} A' & B' \\ C' & D' \end{bmatrix} \begin{bmatrix} 3 & 2 \\ 1 & 1 \end{bmatrix}$$

DP 10.10 Determine z-parameters using nodal analysis for the network shown in Fig. DP 10.10.

Hints: Nodal equations from Fig. DP 10.10.1 are:

$$\frac{V_1 - V_3}{1} = I_1$$

$$\frac{V_3}{2} + I = I_1$$

$$\frac{V_3}{3} = I + I_2$$

$$V_3 - V_2 = V_2$$

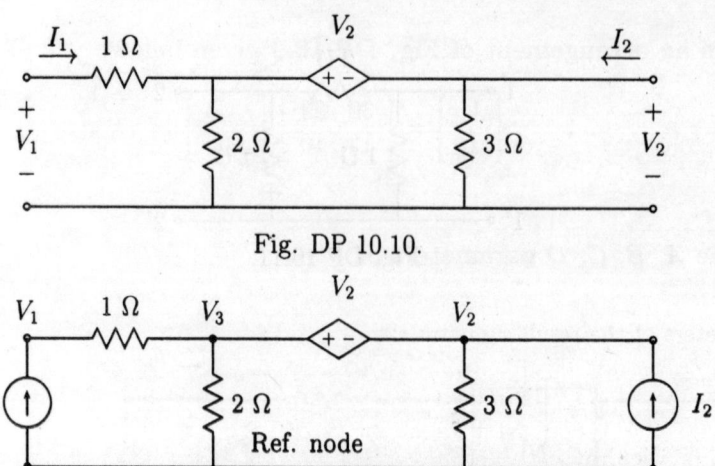

Fig. DP 10.10.

Fig. DP 10.10.1: Circuit for all z-parameters.

Second and third equations are manipulated to eliminate I with I_2 in in the right hand side. Thus the three equations are:

$$V_1 - V_3 = I_1$$
$$-V_1 + \frac{V_2}{3} + \frac{3}{2}V_3 = I_2$$
$$2V_2 - V_3 = 0$$

From them $\Delta = -4/3$. The z-parameters are given by

$$z_{11} = \frac{\Delta_{11}}{\Delta} = \frac{-10/3}{-4/3} = 5/2$$

$$z_{12} = \frac{\Delta_{12}}{\Delta} = 3/4$$

$$z_{21} = \frac{\Delta_{2i}}{\Delta} = 3/2$$

$$z_{21} = \frac{\Delta_{22}}{\Delta} = 1/4$$

DP 10.11 Two identical sections of the π-network shown in Fig. DP 10.11 are connected in parallel. Obtain the y-parameters of the resulting network.

Hints: y-parameters from the π-network are:

$$y_{11} = 2 + 1 = 3\,\mho$$
$$y_{22} = 2 + 1 = 3\,\mho$$
$$y_{12} = y_{21} = -2\,\mho$$

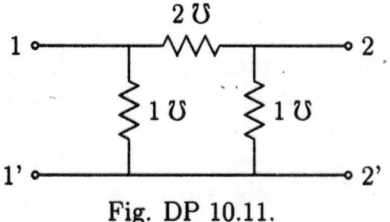

Fig. DP 10.11.

The y-parameters of the resulting network will be

$$Y = \begin{bmatrix} y_{11} + y_{11} & y_{12} + y_{12} \\ y_{21} + y_{21} & y_{22} + y_{22} \end{bmatrix} = \begin{bmatrix} 6 & -4 \\ -4 & 6 \end{bmatrix}$$

DP 10.12 Find the y-parameters using mesh analysis for the two-port network shown in Fig. DP 10.12.

IES-98(EC)

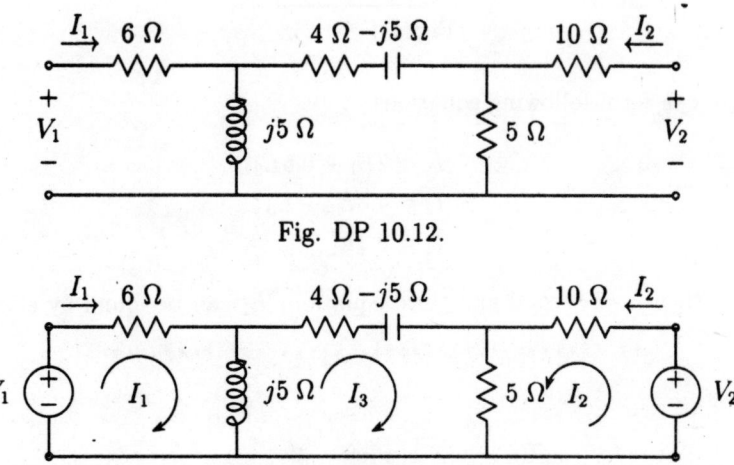

Fig. DP 10.12.

Fig. DP 10.12.1: Circuit for all y-parameters.

Hints: Mesh equations are:

$$(6 + j5)I_1 - j5I_3 = V_1$$
$$15I_2 + 5I_3 = V_2$$
$$-j5I_1 + 5I_2 + 9I_3 = 0$$

Make use of formulae $y_{11} = \frac{\Delta_{11}}{\Delta}$, $y_{12} = \frac{\Delta_{12}}{\Delta}$, $y_{21} = \frac{\Delta_{21}}{\Delta}$, and $y_{22} = \frac{\Delta_{22}}{\Delta}$ to determine all y-parameters.

DP 10.13 The admittance parameters of a 2-port network shown in Fig. DP 10.13 are given by $y_{11} = 2$ ℧, $y_{12} = -0.5$ ℧, $y_{21} = 4.8$ ℧, $y_{22} = 1$ ℧. The output port is terminated with a load admittance $Y_L = 0.2$ ℧. Find E_2 for each of the following conditions:

(a) $E_1 = 10\angle0^0$ V

(b) $I_1 = 10\angle0^0$ A

(c) A source $10\angle0^0$ V in series with a 0.25 Ω resistor is connected to input port.

GATE-2001(EC)

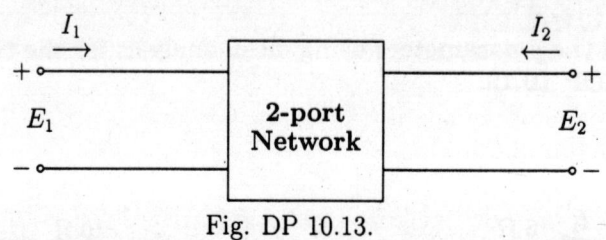

Fig. DP 10.13.

Hints: We can form following equations

$$I_1 = 2E_1 - 0.5E_2$$
$$I_2 = 4.8E_1 + E_2$$
$$I_2 = 0.2E_2$$

If $E_1 = 10\angle0^0$, then from second and third equations E_2 can be found by eliminating I_2.

****

Coupled-Inductors

A.1   INDUCTOR

A time-varying current i passing through a coil of N turns produces the time-varying flux, ϕ. The direction of the flux is given by right hand rule. If the right hand fingers are wrapped around in the direction of current, the thumb points in the direction of the magnetic flux, ϕ.

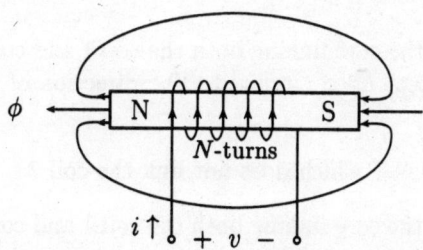

Figure A.1: Showing magnetic flux.

With reference to Figure A.1, by Faraday's law of electromagnetic induction, the electromotive force (emf)

$$
\begin{aligned}
v &= N\frac{d\phi}{dt} \\
&= N\frac{d\phi}{di} \times \frac{di}{dt} \\
&= L\frac{di}{dt}
\end{aligned}
$$

where $L = N\frac{d\phi}{di} = \frac{d(N\phi)}{di}$ is the self inductance of the coil. For the linear magnetic medium, the inductance will be given by

$$
L = \frac{N\phi}{i} \tag{A.1}
$$

or

$$Li = N\phi = \text{flux linkage with } N \text{ turns} \qquad (A.2)$$

A.2 COUPLED-INDUCTORS

Consider two coils are wound on a magnetic core as shown in Figure A.2. The coil 1 has N_1 turns and coil 2 has N_2 turns. The coil 1 is excited by the current i_1 while coil 2 is excited by the current i_2. The v_1 and v_2 are the induced voltages across coil 1 and coil 2 respectively.

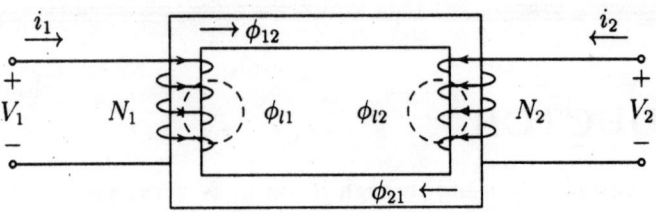

Figure A.2: Showing coupled-inductors.

The variables are defined as:

ϕ_{12} is the flux confined to the core linking both the coil 1 and coil 2 when coil 1 is excited by i_1 while coil 2 is kept open circuited. The direction of ϕ is determined by right hand rule.

ϕ_{l1} is the leakage flux of coil 1 which does not link the coil 2.

ϕ_{21} is the flux confined to the core linking both the coil 1 and coil 2 when coil 2 is excited by i_2 while coil 1 is kept open circuited.

ϕ_{l2} is the leakage flux of coil 2 which does not link the coil 1.

It is clear that $\phi_1 = \phi_{12} + \phi_{l1}$ and $\phi_2 = \phi_{21} + \phi_{l2}$.

Now when coil 1 is excited by i_1 and coil 2 is kept open circuited, the induced voltages across the coil 1 and coil 2 are:

$$v_1' = N_1 \frac{d}{dt}(\phi_{12} + \phi_{l2}) = N_1 \frac{d\phi_1}{dt}$$

$$\text{and} \quad v_2' = N_2 \frac{d\phi_{12}}{dt}$$

Similarly when coil 2 is excited by i_2 and coil 1 is kept open circuited, the induced voltages across the coil 1 and coil 2 are:

$$v_1'' = N_1 \frac{d\phi_{21}}{dt}$$

$$\text{and} \quad v_2'' = N_2 \frac{d}{dt}(\phi_{21} + \phi_{l2}) = N_2 \frac{d\phi_2}{dt}$$

Thus by superposition

$$
\begin{aligned}
v_1 &= v_1' + v_1'' \\
&= N_1 \frac{d\phi_1}{dt} + N_1 \frac{d\phi_{21}}{dt} \\
&= \frac{d}{dt}(N_1\phi_1) + \frac{d}{dt}(N_1\phi_{21})
\end{aligned}
$$

By letting (on the basis of eqn A2)

$$N_1\phi_1 = L_{11}i_1 \qquad (A.3)$$

and $\qquad N_1\phi_{21} = L_{12}i_2 \qquad (A.4)$

Thus

$$v_1 = L_{11}\frac{di_1}{dt} + L_{12}\frac{di_2}{dt} \qquad (A.5)$$

and

$$
\begin{aligned}
v_2 &= v_2' + v_2'' \\
&= N_2\frac{d\phi_{12}}{dt} + N_2\frac{d}{dt}(\phi_{21} + \phi_{l2}) \\
&= \frac{d}{dt}(N_2\phi_{12}) + \frac{d}{dt}(N_2\phi_2)
\end{aligned}
$$

Again by letting

$$N_2\phi_{12} = L_{21}i_1 \qquad (A.6)$$

and $\qquad N_2\phi_2 = L_{22}i_2 \qquad (A.7)$

Thus

$$v_2 = L_{21}\frac{di_1}{dt} + L_{22}\frac{di_2}{dt} \qquad (A.8)$$

The common mutual flux passes through the same reluctance, thus

$$L_{12} = L_{21} = M$$

where M is called mutual inductance.

Now from eqn (A.5) and eqn (A.8), we can write finally

$$v_1 = L_1\frac{di_1}{dt} + M\frac{di_2}{dt} \qquad (A.9)$$

$$v_2 = M\frac{di_1}{dt} + L_2\frac{di_2}{dt} \qquad (A.10)$$

Here $L_1 = L_{11}$ and $L_2 = L_{22}$ are always positive, but M may positive or negative.

A.3 COEFFICIENT OF COUPLING

The coefficient of coupling k is defined as the ratio of common flux to total flux produced by either coil. The value of k depends on the closeness and orientation of the two coils mutually coupled.

Mathematically,

$$k = \frac{\phi_{12}}{\phi_{12} + \phi_{l1}} = \frac{\phi_{12}}{\phi_1}$$

or

$$\phi_{12} = k\phi_1 \tag{A.11}$$

and

$$k = \frac{\phi_{21}}{\phi_{21} + \phi_{l2}} = \frac{\phi_{21}}{\phi_2}$$

or

$$\phi_{21} = k\phi_2 \tag{A.12}$$

Equations (A.3) and (A.4) now will become

$$N_1\phi_1 = L_1 i_1$$
$$N_2\phi_{12} = M i_1$$

Thus

$$\frac{M}{L_1} = \frac{N_2\phi_{12}}{N_1\phi_1} \tag{A.13}$$

Similarly, equations (A.7) and (A.8) will be

$$N_2\phi_2 = L_2 i_2$$
$$N_2\phi_{21} = M i_2$$

Thus

$$\frac{M}{L_2} = \frac{N_1\phi_{21}}{N_2\phi_2} \tag{A.14}$$

From equations (A.13) and (A.14), we have

$$\frac{M}{L_1} \times \frac{M}{L_2} = k^2$$

and

$$k = \frac{|M|}{\sqrt{L_1 L_2}} \tag{A.15}$$

where $0 \leq k \leq 1$.

The maximum value of k is unity. When $k = 1$, the two inductors are perfectly coupled. When $k = 0$, there is no coupling. The sign of M between two coils may easily be determined as:

If directions of ϕ_1 and ϕ_2 are same, M will be positive.

If directions of ϕ_1 and ϕ_2 are opposite to each other, M will be negative.

SP A.1 Find the sign of M for the circuit shown in Fig. SP A.1.

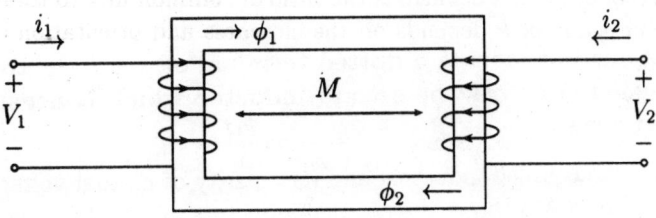

Fig. SP A.1.

In Fig. SP A.1, by right hand rule, the directions of ϕ_1 and ϕ_2 are same (aiding). Thus M will be positive and terminal-equations are:

$$v_1 = L_1 \frac{di_1}{dt} + M \frac{di_2}{dt}$$
$$v_2 = M \frac{di_1}{dt} + L_2 \frac{di_2}{dt}$$

SP A.2 Find the sign of M for the circuit shown in Fig. SP A.2.

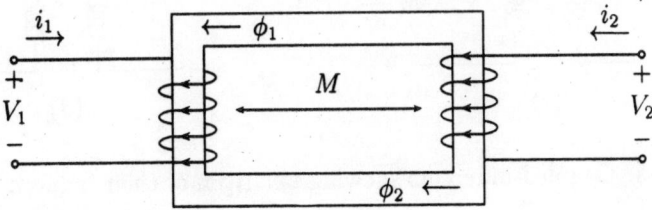

Fig. SP A.2.

In Fig. SP A.2, by right hand rule, the directions of ϕ_1 and ϕ_2 are opposing each other. Thus M will be negative and terminal-equations are:

$$v_1 = L_1 \frac{di_1}{dt} - M \frac{di_2}{dt}$$
$$v_2 = -M \frac{di_1}{dt} + L_2 \frac{di_2}{dt}$$

A.4 DOT CONVENTION

In case of symbolic representation of couple-inductors, dots are used to indicate the polarity of the induced voltage. According to dot conventions:

A current entering a dotted terminal of one inductor produces an induced voltage across the second inductor which is positive (+) at the dotted terminal.

Alternately a current leaving a dotted terminal of one inductor produces an induced voltage across the second inductor which is negative (-) at the dotted terminal.

Figure A.3 shows the coupled-inductors and its polarity of mutual voltage with different dot-placements, Figure A.4 shows the coupled-inductors and its terminal equations with different dot-placements, and Figure A.5 shows the coupled-inductors and its equivalent circuit with different dot-placements.

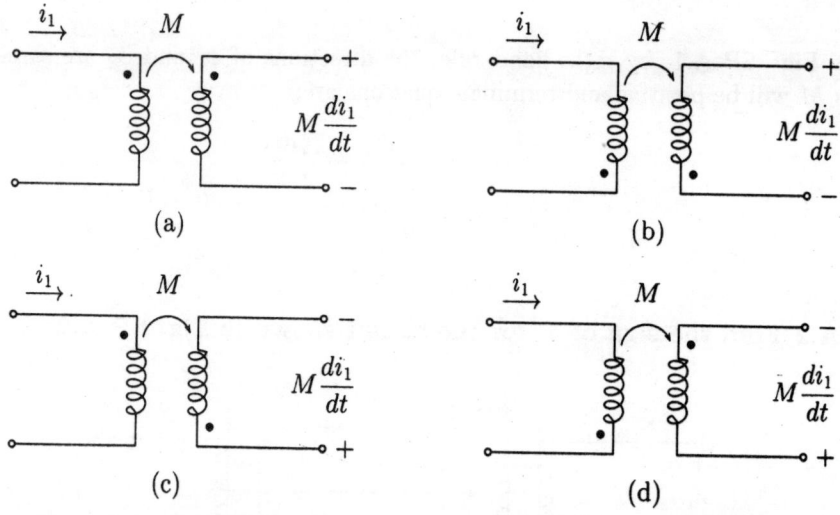

Figure A.3: Coupled-inductors showing polarities of their induced voltages.

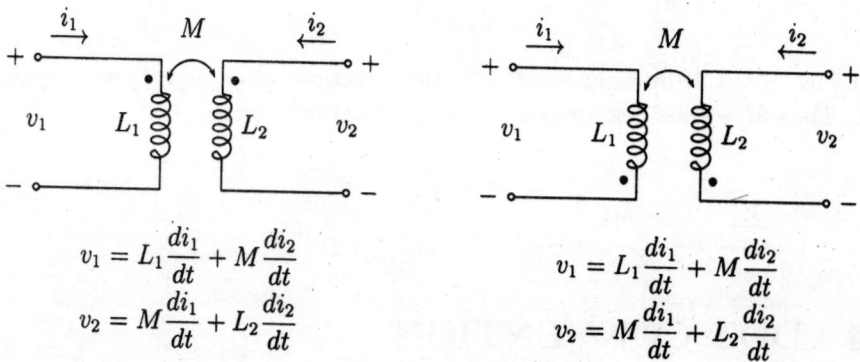

Figure A.4: Some Coupled-Inductors with their Terminal Equations. (contd.)

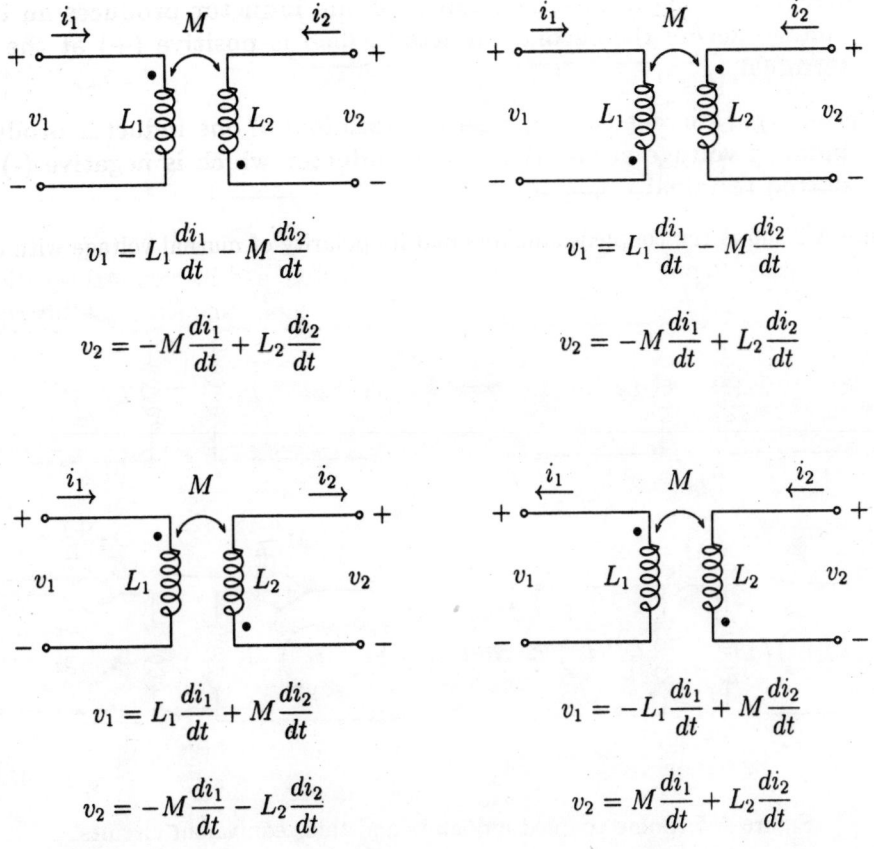

$$v_1 = L_1 \frac{di_1}{dt} - M \frac{di_2}{dt}$$

$$v_2 = -M \frac{di_1}{dt} + L_2 \frac{di_2}{dt}$$

$$v_1 = L_1 \frac{di_1}{dt} - M \frac{di_2}{dt}$$

$$v_2 = -M \frac{di_1}{dt} + L_2 \frac{di_2}{dt}$$

$$v_1 = L_1 \frac{di_1}{dt} + M \frac{di_2}{dt}$$

$$v_2 = -M \frac{di_1}{dt} - L_2 \frac{di_2}{dt}$$

$$v_1 = -L_1 \frac{di_1}{dt} + M \frac{di_2}{dt}$$

$$v_2 = M \frac{di_1}{dt} + L_2 \frac{di_2}{dt}$$

Figure A.4: Some Coupled-Inductors with their Terminal Equations

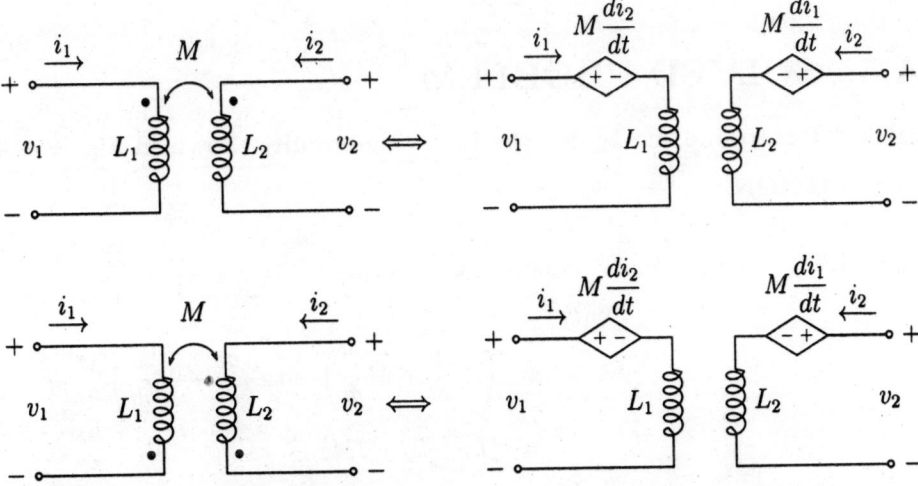

Figure A.5: Some coupled-inductors and their equivalent circuits. (contd.)

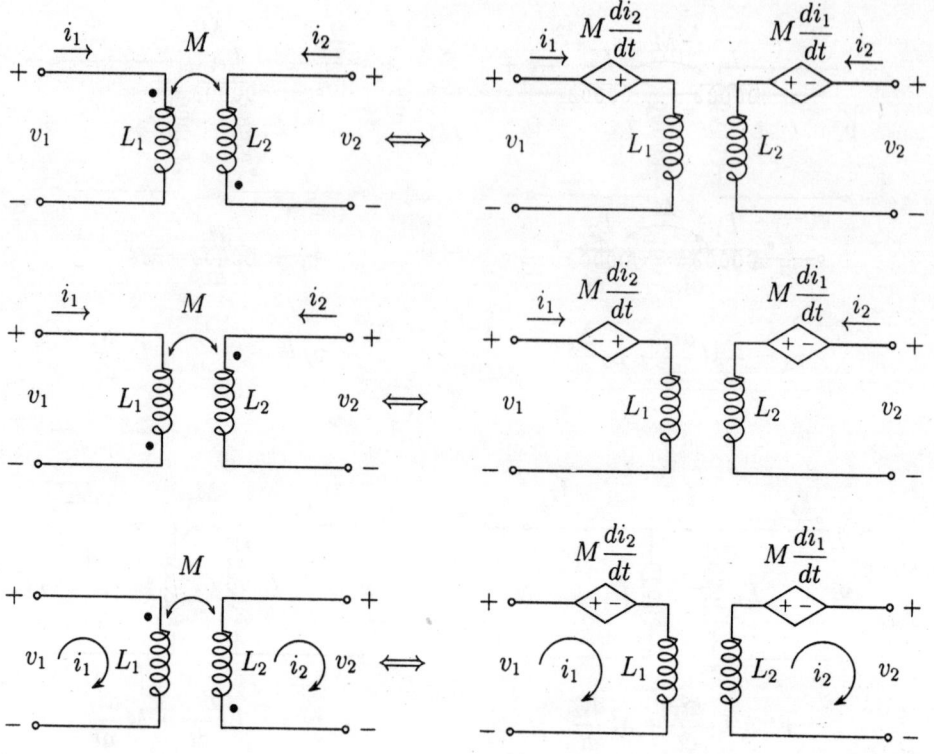

Figure A.5: Some coupled-inductors and their equivalent circuits.

A.5   SOLVED PROBLEMS

SP A.3 Determine the L_{eq} for the following circuits shown in Fig. SP A.3.

SOLUTION:

(a)

By KVL in the above left circuit of Fig. A.3.1(a)

$$v(t) = v_1 + v_2$$
$$= \left(L_1 \frac{di}{dt} + M \frac{di}{dt}\right) + \left(L_2 \frac{di}{dt} + M \frac{di}{dt}\right)$$
$$= (L_1 + L_2 + 2M) \frac{di}{dt}$$

Thus $v(t)/\dfrac{di}{dt} = L_1 + L_2 + 2M$

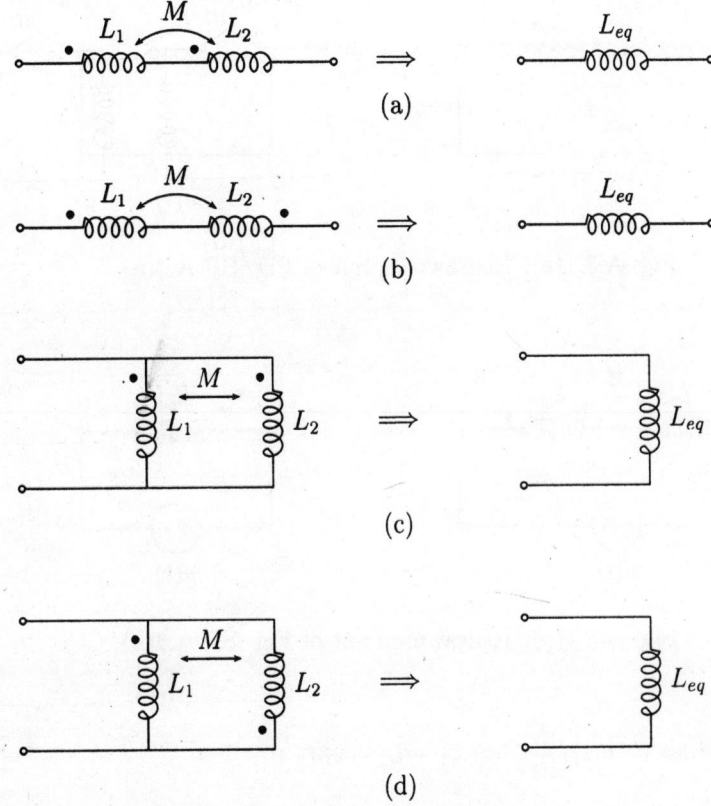

Fig. SP A.3.

From the equivalent circuit

$$v(t) = L_{eq}\frac{di}{dt}$$

or $\qquad v(t)/\dfrac{di}{dt} = L_{eq}$

Therefore

$$L_{eq} = L_1 + L_2 + 2M$$

(b)

By KVL in the above left circuit of Fig. A.3.1(b)

$$\begin{aligned}
v(t) &= v_1 + v_2 \\
&= \left(L_1\frac{di}{dt} - M\frac{di}{dt}\right) + \left(L_2\frac{di}{dt} - M\frac{di}{dt}\right) \\
&= (L_1 + L_2 - 2M)\frac{di}{dt}
\end{aligned}$$

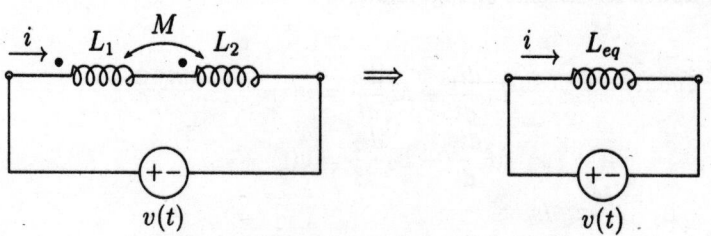

Fig. A.3.1(a): Redrawn circuit of Fig. SP A.3(a).

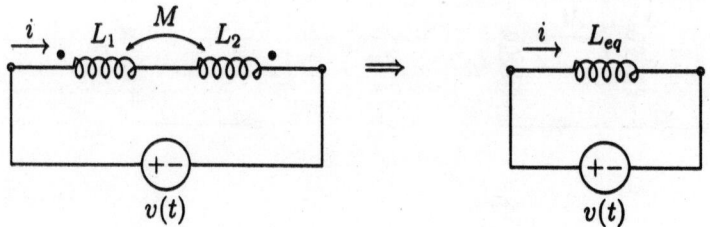

Fig. A.3.1(b): Redrawn circuit of Fig. SP A.3(b).

Thus $\quad v(t)\Big/\dfrac{di}{dt} = L_1 + L_2 - 2M$

From the equivalent circuit

$$v(t) = L_{eq}\frac{di}{dt}$$

or $\qquad v(t)\Big/\dfrac{di}{dt} = L_{eq}$

Therefore

$$L_{eq} = L_1 + L_2 - 2M$$

(c)

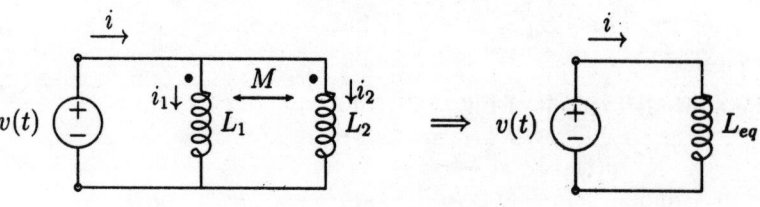

Fig. A.3.1(c): Redrawn circuit of Fig. SP A.3(c).

By KVL in the above left circuit of Fig. A.3.1(c)

$$L_1 \frac{di_1}{dt} + M \frac{di_2}{dt} = v(t)$$
$$M \frac{di_1}{dt} + L_2 \frac{di_2}{dt} = v(t)$$

In matrix form

$$\begin{bmatrix} L_1 & M \\ M & L_2 \end{bmatrix} \begin{bmatrix} di_1/dt \\ di_2/dt \end{bmatrix} = \begin{bmatrix} 1 \\ 1 \end{bmatrix} v(t)$$

Thus
$$\begin{bmatrix} di_1/dt \\ di_2/dt \end{bmatrix} = \begin{bmatrix} L_1 & M \\ M & L_2 \end{bmatrix}^{-1} \begin{bmatrix} 1 \\ 1 \end{bmatrix} v(t)$$

$$= \frac{1}{L_1 L_2 - M^2} \begin{bmatrix} L_2 & -M \\ -M & L_1 \end{bmatrix} \begin{bmatrix} 1 \\ 1 \end{bmatrix} v(t)$$

$$= \begin{bmatrix} \frac{L_2 - M}{L_1 L_2 - M^2} v(t) \\ \frac{L_1 - M}{L_1 L_2 - M^2} v(t) \end{bmatrix}$$

We find

$$\frac{di_1}{dt} = \frac{L_2 - M}{L_1 L_2 - M^2} v(t)$$

and
$$\frac{di_2}{dt} = \frac{L_1 - M}{L_1 L_2 - M^2} v(t)$$

We have

$$i = i_1 + i_2$$
$$\text{or} \quad \frac{di}{dt} = \frac{di_1}{dt} + \frac{di_2}{dt}$$
$$= \frac{L_2 - M}{L_1 L_2 - M^2} v(t) + \frac{L_1 - M}{L_1 L_2 - M^2} v(t)$$
$$= \frac{L_1 + L_2 - 2M}{L_1 L_2 - M^2} v(t)$$

Thus
$$v(t) / \frac{di}{dt} = \frac{L_1 L_2 - M^2}{L_1 + L_2 - 2M}$$

From the equivalent circuit

$$v(t) = L_{eq} \frac{di}{dt}$$

and
$$v(t) / \frac{di}{dt} = L_{eq}$$

Thus
$$L_{eq} = \frac{L_1 L_2 - M^2}{L_1 + L_2 - 2M}$$

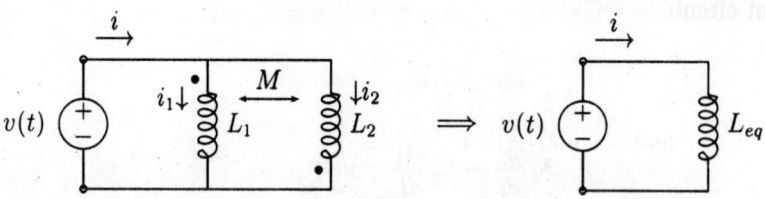

Fig. A.3.1(d): Redrawn circuit of Fig. SP A.3(d).

(d)

By KVL in the above left circuit of Fig. A.3.1(d)

$$L_1 \frac{di_1}{dt} - M \frac{di_2}{dt} = v(t)$$

$$-M \frac{di_1}{dt} + L_2 \frac{di_2}{dt} = v(t)$$

In matrix form

$$\begin{bmatrix} L_1 & -M \\ -M & L_2 \end{bmatrix} \begin{bmatrix} di_1/dt \\ di_2/dt \end{bmatrix} = \begin{bmatrix} 1 \\ 1 \end{bmatrix} v(t)$$

Thus
$$\begin{bmatrix} di_1/dt \\ di_2/dt \end{bmatrix} = \begin{bmatrix} L_1 & -M \\ -M & L_2 \end{bmatrix}^{-1} \begin{bmatrix} 1 \\ 1 \end{bmatrix} v(t)$$

$$= \frac{1}{L_1 L_2 - M^2} \begin{bmatrix} L_2 & M \\ M & L_1 \end{bmatrix} \begin{bmatrix} 1 \\ 1 \end{bmatrix} v(t)$$

$$= \begin{bmatrix} \frac{L_2+M}{L_1 L_2 - M^2} v(t) \\ \frac{L_1+M}{L_1 L_2 - M^2} v(t) \end{bmatrix}$$

We find

$$\frac{di_1}{dt} = \frac{L_2 + M}{L_1 L_2 - M^2} v(t)$$

and
$$\frac{di_2}{dt} = \frac{L_1 + M}{L_1 L_2 - M^2} v(t)$$

We have

$$i = i_1 + i_2$$

or
$$\frac{di}{dt} = \frac{di_1}{dt} + \frac{di_2}{dt}$$

$$= \frac{L_2 + M}{L_1 L_2 - M^2} v(t) + \frac{L_1 + M}{L_1 L_2 - M^2} v(t)$$

$$= \frac{L_1 + L_2 + 2M}{L_1 L_2 - M^2} v(t)$$

Thus
$$v(t) / \frac{di}{dt} = \frac{L_1 L_2 - M^2}{L_1 + L_2 + 2M}$$

From the equivalent circuit

$$v(t) = L_{eq}\frac{di}{dt}$$

and $\quad v(t)/\frac{di}{dt} = L_{eq}$

Thus $\quad L_{eq} = \dfrac{L_1 L_2 - M^2}{L_1 + L_2 + 2M}$

SP A.4 Write the mesh equations for the coupled networks of Fig. SP A.4.

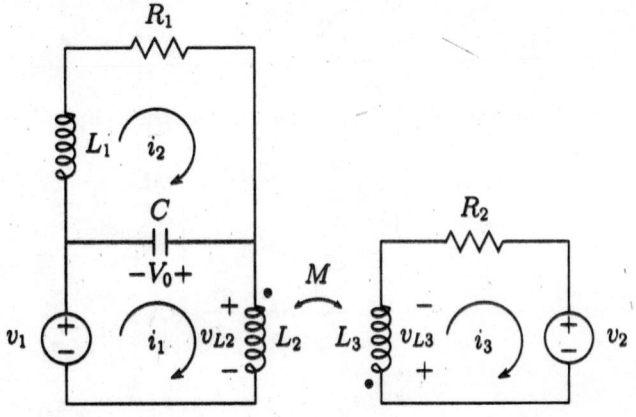

Fig. SP A.4.

SOLUTION: Let voltages across L_2 and L_3 be v_{L2} and v_{L3} respectively in the circuit of Fig. SP A.4. They are given as:

$$v_{L2} = L_2\frac{di_1}{dt} + M\frac{di_3}{dt}$$

$$v_{L3} = L_3\frac{di_3}{dt} + M\frac{di_1}{dt}$$

Mesh equations are:

Mesh 1 : $\quad -V_0 + \dfrac{1}{C}\displaystyle\int (i_1 - i_2)dt + v_{L2} = v_1(t)$

Mesh 2 : $\quad L_1\dfrac{di_2}{dt} + R_1 i_2 + V_0 + \dfrac{1}{C}\displaystyle\int (i_2 - i_1)dt = 0$

Mesh 3 : $\quad R_2 i_3 + v_2(t) + v_{L3} = 0$

Setting the values for v_{L2} and v_{L3} in the above equations, we will get the required equations.

SP A.5 Write the equations by KVL in the meshes of Fig. SP A.5.

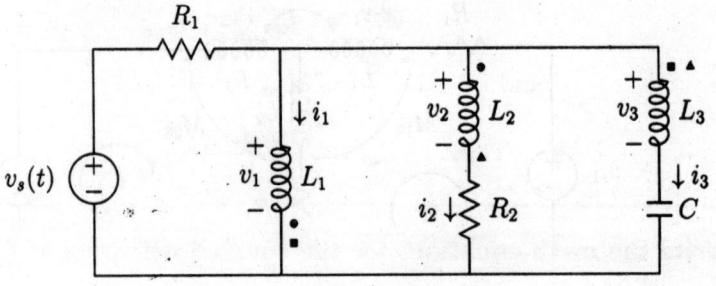

Fig. SP A.5.

SOLUTION: From Fig. SP A.5, voltages across the coupled inductors are:

$$v_1 = L_1 \frac{di_1}{dt} - M_{12} \frac{di_2}{dt} - M_{13} \frac{di_3}{dt}$$

$$v_2 = L_2 \frac{di_2}{dt} - M_{21} \frac{di_1}{dt} - M_{23} \frac{di_3}{dt}$$

$$v_3 = L_3 \frac{di_3}{dt} - M_{31} \frac{di_1}{dt} - M_{32} \frac{di_2}{dt}$$

By KVL

Mesh 1 : $R_1(i_1 + i_2 + i_3) + v_1 = v_s(t)$

Mesh 2 : $v_2 + R_2 i_2 - v_1 = 0$

Mesh 3 : $v_3 + \dfrac{1}{C} \int i_3 dt - R_2 i_2 - v_2 = 0$

Putting the values of v_1, v_2, and v_3 in above equations will give required equations.

SP A.6 Write the mesh equations for the coupled networks of Fig. SP A.6.

From Fig. SP A.6, the voltage across the coupled inductors are:

$$v_1 = L_1 \frac{di_1}{dt} + M_{12} \frac{d(i_1 - i_2)}{dt} - M_{13} \frac{di_2}{dt}$$

$$v_2 = L_2 \frac{d(i_1 - i_2)}{dt} + M_{21} \frac{di_1}{dt} - M_{23} \frac{di_2}{dt}$$

$$v_3 = L_3 \frac{di_2}{dt} - M_{31} \frac{di_1}{dt} - M_{32} \frac{d(i_1 - i_2)}{dt}$$

By KVL

Mesh 1 : $R_1 i_1 + v_1 + v_2 + R_2(i_1 - i_2) = v_{s1}$

Mesh 2 : $v_3 + R_2(i_2 - i_1) - v_2 = -v_{s2}$

Putting the values of v_1, v_2, and v_3 in above equations will give required equations.

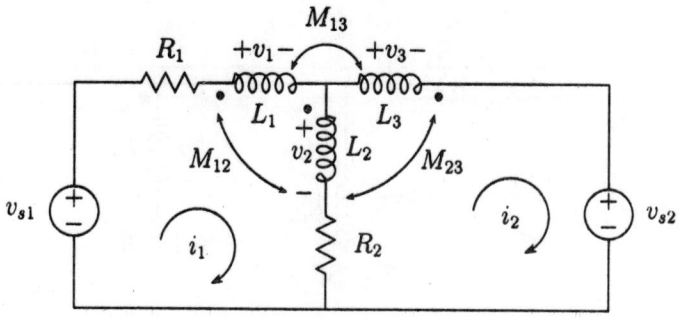

Fig. SP A.6.

SP A.7 Write the terminal equations for the three coupled inductors shown in Fig. SP A.7.

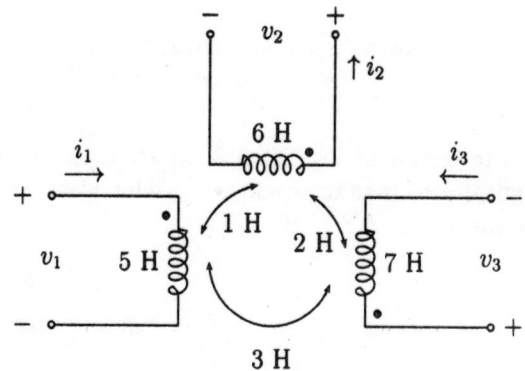

Fig. SP A.7.

From Fig. SP A.7, the terminal equations are:

$$v_1 = 5\frac{di_1}{dt} - 1\frac{di_2}{dt} - 3\frac{di_3}{dt}$$

$$v_2 = -6\frac{di_2}{dt} + 1\frac{di_1}{dt} - 2\frac{di_3}{dt}$$

$$v_3 = -7\frac{di_3}{dt} + 3\frac{di_1}{dt} - 2\frac{di_2}{dt}$$

SP A.8 From the circuit of Fig. SP A.8, (i) draw the dotted equivalent circuit; (ii) write KVL equations for the dotted equivalent circuit.

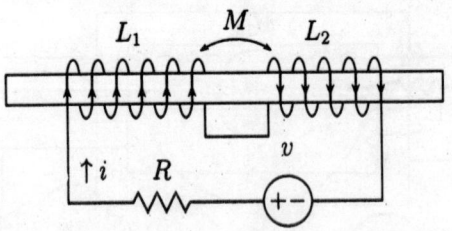

Fig. SP A.8.

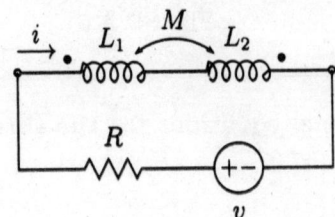

Fig. A.8.1: Dotted equivalent circuit of Fig SP A.8.

SOLUTION: For the circuit of Fig SP A.8, by right hand rule, we find that the flux in winding 1 is opposite to that in winding '2'. This gives negative M and dotted equivalent circuit is shown in Fig. A.8.1. By KVL

$$Ri + \left(L-1\frac{di}{dt} - M\frac{di}{dt}\right) + \left(L-1\frac{di}{dt} - M\frac{di}{dt}\right) = v$$

$$Ri + (L_1 + L_2 - 2M)\frac{di}{dt} = v$$

$$Ri + L_{eq}\frac{di}{dt} = v$$

where $L_{eq} = L_1 + L_2 - 2M$

SP A.9 From the circuit of **Fig. SP A.9**, (i) draw the dotted equivalent circuit; (ii) write mesh equations for the dotted equivalent circuit.

SOLUTION: (i) From Fig. SP A.9, by right hand rule, we find that the flux in winding 1 is opposite to that in winding 2. This gives negative M and dotted equivalent circuit is shown in Fig. A.9.1.

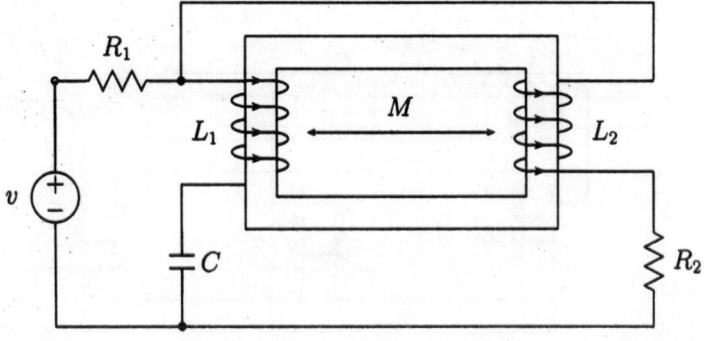

Fig. SP A.9.

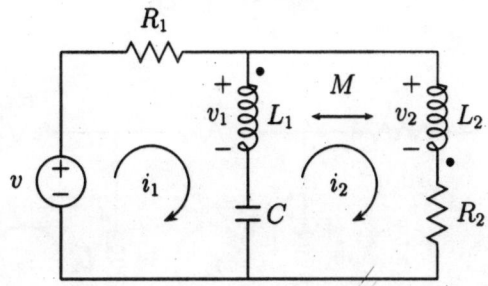

Fig. SP A.9.1: Dotted equivalent circuit of Fig. SP A.9.

(ii) From Fig. A.9.1, the induced voltages are:

$$v_1 = L_1 \frac{d(i_1 - i_2)}{dt} - M \frac{di_2}{dt}$$

$$v_2 = L_2 \frac{di_2}{dt} - M \frac{d(i_1 - i_2)}{dt}$$

Mesh equations are:

Mesh 1 : $R_1 i_1 + v_1 + \dfrac{1}{C} \displaystyle\int (i_1 - i_2) dt = v$

Mesh 2 : $v_2 + R_2 i_2 + \dfrac{1}{C} \displaystyle\int (i_2 - i_1) dt - v_1 = 0$

On putting the values of v_1 and v_2 in the above equations will give the required equations.

SP A.10 From the circuit of Fig. SP A.10,
   (i) draw the dotted equivalent circuit;
   (ii) write mesh equations for the dotted equivalent circuit.

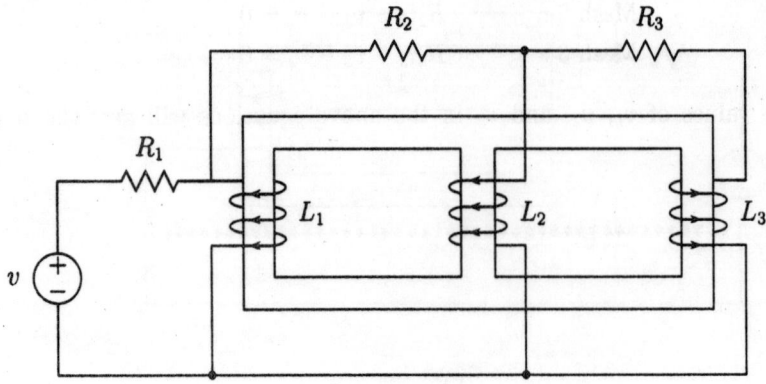

Fig. SP A.10.

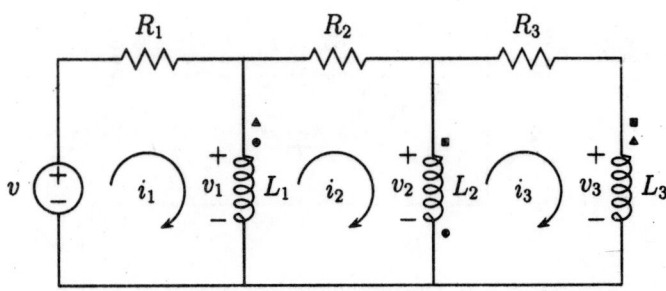

Fig. SP A.10.1: Dotted equivalent circuit of Fig. SP A.10.

SOLUTION: (i) From Fig. SP A.10, by right hand rule, we find that flux is downward in limb 1, downward in limb 2, and upward in limb 3 due to their currents flowing inward at their top terminals. Winding 1 and winding 2 do not produce aiding flux, winding 1 and winding 3 produce aiding flux, and winding 2 and winding 3 also produce aiding flux. Thus M_{12} or M_{21} is negative M_{13} or M_{31} is positive, and M_{23} or M_{32} is positive. The dotted equivalent circuit will be as shown in Fig. A.10.1

(ii) The induced voltages are:

$$v_1 = L_1 \frac{d(i_1 - i_2)}{dt} - M_{12}\frac{d(i_2 - i_3)}{dt} + M_{13}\frac{di_3}{dt}$$

$$v_2 = L_2 \frac{d(i_2 - i_3)}{dt} - M_{21}\frac{d(i_1 - i_2)}{dt} + M_{23}\frac{di_3}{dt}$$

$$v_3 = L_3 \frac{di_3}{dt} + M_{31}\frac{d(i_1 - i_2)}{dt} + M_{32}\frac{d(i_2 - i_3)}{dt}$$

Mesh equations are:

Mesh 1 : $R_1 i_1 + v_1 = v$

$$\text{Mesh 2}: \qquad R_2 i_2 + v_2 - v_1 = 0$$
$$\text{Mesh 3}: \qquad R_3 i_3 + v_3 - v_2 = 0$$

On putting the values of v_1, v_2, and v_3 in the above equations will give the required equations.

****

Cramer's Rule

Consider a set of simultaneous, linear algebraic equations written as:

$$a_{11}x_1 + a_{12}x_2 + \cdots + a_{1n}x_n = b_1$$
$$a_{21}x_1 + a_{22}x_2 + \cdots + a_{2n}x_n = b_2$$
$$\vdots$$
$$a_{n1}x_1 + a_{n2}x_2 + \cdots + a_{nn}x_n = b_n$$

where a_{ij} and b_j are known quantities and x_i are unknowns.

Writing these equations in matrix form, $AX = B$

$$\begin{bmatrix} a_{11} & a_{12} & \cdots & a_{1n} \\ a_{21} & a_{22} & \cdots & a_{2n} \\ \vdots & \vdots & \ddots & \vdots \\ a_{n1} & a_{n2} & \cdots & a_{nn} \end{bmatrix} \begin{bmatrix} x_1 \\ x_2 \\ \vdots \\ x_n \end{bmatrix} = \begin{bmatrix} b_1 \\ b_2 \\ \vdots \\ b_n \end{bmatrix}$$

If Δ is the system matrix and is given by

$$\Delta = \det(A) = \begin{vmatrix} a_{11} & a_{12} & \cdots & a_{1n} \\ a_{21} & a_{22} & \cdots & a_{2n} \\ \vdots & \vdots & \ddots & \vdots \\ a_{n1} & a_{n2} & \cdots & a_{nn} \end{vmatrix}$$

then according to the Cramer's rule

$$x_1 = \frac{\Delta_1}{\Delta} = \frac{1}{\Delta} \begin{vmatrix} b_1 & a_{12} & \cdots & a_{1n} \\ b_2 & a_{22} & \cdots & a_{2n} \\ \vdots & \vdots & \ddots & \vdots \\ b_n & a_{n2} & \cdots & a_{nn} \end{vmatrix}$$

$$x_2 = \frac{\Delta_2}{\Delta} = \frac{1}{\Delta} \begin{vmatrix} a_{11} & b_1 & \cdots & a_{1n} \\ a_{21} & b_2 & \cdots & a_{2n} \\ \vdots & \vdots & \ddots & \vdots \\ a_{n1} & b_n & \cdots & a_{nn} \end{vmatrix}$$

$$\vdots$$

$$x_n = \frac{\Delta_n}{\Delta} = \frac{1}{\Delta} \begin{vmatrix} a_{11} & a_{12} & \cdots & b_1 \\ a_{21} & a_{22} & \cdots & b_2 \\ \vdots & \vdots & \ddots & \vdots \\ a_{n1} & a_{n2} & \cdots & b_n \end{vmatrix}$$

where Δ_i, $i = 1, 2, \cdots, n$, is the determinant obtained from the Δ by replacing its i-th column with the column forming B.

The above solution is possible if $\Delta \neq 0$. If $\Delta = 0$ and the system is not homogeneous, it has, in general no solution. If $\Delta = 0$ and the system is homogeneous, it has only the trivial solution, i.e.; $x_1 = x_2 = \cdots = x_n = 0$.

****

Summary of Symbols and Units

Quantity	Symbol	Unit	Related Equation
Charge	Q, q	coulomb (C)	-
Current	I, i	ampere (A)	$i = \dfrac{dq}{dt}$
Magnetic flux	ϕ	weber (Wb)	-
Electromotive force	e, v	volt (V)	$e = N\dfrac{d\phi}{dt}$
Power	P, p	watt (W)	$p = vi$
Energy	W, w	joule (J)	$w = \int p\,dt$
Voltage	V, v	volt (V)	$v = \dfrac{dw}{dq}$
Capacitance	C	farad (F)	$C = \dfrac{\epsilon A}{d}$
Inductance	L	henry (H)	$L = \dfrac{\mu_0 N^2 A}{l}$
Mutual inductance	M	henry (H)	$M = k\sqrt{L_1 L_2}$
Resistance	R	ohm (Ω)	$R = \rho\dfrac{l}{A}$
Conductance	G	mho ($\mho$)	$G = 1/R$
Frequency	f	hertz (Hz)	-
Angular frequency	ω	radian/second (r/s)	$\omega = 2\pi f$

Multiple Choice Questions

First Order Circuits

1. An initially relaxed RC-series network with $R = 2\ M\Omega$ and $C = 1\ \mu F$ is switched on to a 10 V step input. The voltage across the capacitor after 2 seconds will be

   (a) zero   (b) 3.68 V   (c) 6.32 V   (d) 10 V

   IES-98(EE)

   Hints : $v_C(t) = 10(1 - e^{-t/\tau})$ where $\tau = RC = 2 \times 10^6 \times 1 \times 10^{-6} = 2$ sec.

   Therefore $v_C(2\ sec.) = 10(1 - e^{-1}) = 6.32$ V

2. An initially relaxed 100 mH inductor is switched ON at t = 1 sec. to an 2 A dc current source. The voltage across the inductor would be

   (a) zero   (b) $0.2\delta(t)$ V   (c) $0.2\delta(t - 1)$ V   (d) $0.2tu(t - 1)$V

   IES-97(EE)

   Hints : $sL = 0.1s$ and $i(t) = 2u(t - 1)$, thus $I(s) = 2e^{-s}/s$.

   The voltage across the inductor is $V_L(s) = (0.1s)(2e^{-s}/s) = 0.2e^{-s}$

        $\Rightarrow v_L(t) = \mathcal{L}^{-1}[V_L(s)] = 0.2\delta(t - 1)$

3. A system is represented by $\frac{dy}{dt} + 2y = 4tu(t)$

   The ramp component in the force response will be

   (a) $tu(t)$ (b) $2tu(t)$ (c) $3tu(t)$ (d) $4tu(t)$ IES-97(EE)

   Hints : Select $y_p = k_0 + k_1t$, and put it into differential equation as:

   $\frac{d}{dt}(k_0 + k_1t) + (k_0 + k_1t) = 4t$ $\Rightarrow (k_0 + k_0) + k_1t = 4t$

   Equating the coefficient of like terms, we get $k_1 = 4$

   Thus the ramp component of force response is $k_1tu(t) = 4tu(t)$

4. If the unit step response of a network is $(1 - e^{-\alpha t})$, then its unit impulse response will be

(a) $\alpha e^{-\alpha t}$ (b) $\frac{1}{\alpha}e^{-t/\alpha}$ (c) $\frac{1}{\alpha}e^{-\alpha t}$ (d) $(1-\alpha)e^{-\alpha t}$

IES-97(EE)

Hints : The unit impulse response is the derivative of unit step response,

thus unit impulse response $= \frac{d}{dt}(1-e^{-\alpha t}) = \alpha e^{-\alpha t}$

5. If the step response of an initially relaxed circuit is known then the ramp response can be obtained by

   (a) integrating the step response.

   (b) differentiating the step response.

   (c) integrating the step response twice.

   (d) differentiating the step response twice.

IES-93(EE)

Hint : Ans (a)

6. The circuit shown the figure, the switch S has been open for a long time. It is closed at $t = 0$. For $t > 0$ the current flowing through the inductor is

   (a) $i_L = 1.2 + 0.8e^{-2t}$

   (b) $i_L = 0.8 + 1.2e^{-2t}$

   (c) $i_L = 1.2 - 0.8e^{-2t}$

   (d) $i_L = 0.8 - 1.2e^{-2t}$

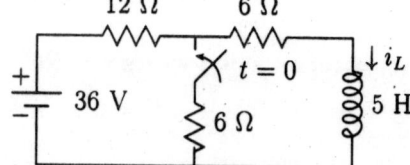

IES-93(EE)

Hints : $i_L(0^-) = \frac{36}{12+6} = i_L(0^+)$

$R_{Th} = (12\|6) + 6 = 10 \ \Omega$

$i_{L,ss} = (1/2)(36/15) = 6/5 = 1.2 \ A$

Hence $i_L = (i_L(0^+) - i_{L,ss})e^{-R_{Th}t/L} = 0.8e^{2t} + 1.2$

7. When the input to a system was withdrawn at t = 0, its output was found to decrease exponentially from 1000 units to 500 units in 1.386 secs. The time constant of system is

   (a) 0.500   (b) 0.693   (c) 1.386   (d) 2.000

IES-98(EE)

Hints : $f(t) = Ae^{-t/\tau}$ Thus $f(0) = 1000 = A$ and $f(1.386) = 500 = Ae^{-1.386/\tau}$
Thus $\frac{1000}{500} = \frac{A}{Ae^{-1.386/\tau}} = e^{1.386/\tau}$ or $\ln 2 = 1.386/\tau$ or, $\tau = 2.000$

8. The impulse response of a system is $5e^{-10t}$, its step response is equal to

   (a) $0.5e^{-10t}$ (b) $5(1 - e^{-10t})$ (c) $0.5(1 - e^{-10t})$ (d) $10(1 - e^{-10t})$

IES-96(EE)

Hints : step response $= \int_0^t 5e^{-10t}dt = 0.5(1 - e^{-10t})$

9. The equation for current $i(t)$ for $t > 0$ if $v_C(0) = 6$ V, for the given circuit is

(a) $i(t) = 8e^{-t}$ A

(b) $i(t) = e^{-t}$ A

(c) $i(t) = 6e^{-t}$ A

(d) $i(t) = 2e^{-t}$A

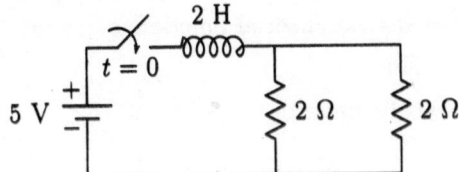

Hints : The value of current in RC-circuit is $i(t) = Ke^{-t/RC}$ where $RC = 2 \times (1/2) = 1$ and $K = i(0^+) = \frac{8-6}{2} = 1$ Thus $i(t) = e^{-t}$

10. The time constant of the circuit shown is

(a) 1 sec

(b) 1/2 sec

(c) 2 sec

(d) none of the above.

Hints : Thevenin resistance seen by the inductor terminal is $R_{Th} = 2||2 = 1\Omega$. Thus time constant $\tau = L/R_{Th} = 2/1 = 2$ sec.

11. If an RL circuit having impedance angle ϕ is switched in when the applied voltage wave is passing through an angle θ, there will be no switching transient if

(a) $\theta - \phi = 0$ (b) $\theta + \phi = 0$ (c) $\theta - \phi = 90^0$ (d) $\theta + \phi = 90^0$

IES-94(EE)

Hints : $i(t) = \frac{V_m}{\sqrt{R^2+\omega^2 L^2}} \left[\sin(\omega t - \tan^{-1} \frac{\omega L}{R}) - \sin(\omega t_0 - \tan^{-1} \frac{\omega L}{R})e^{-(t-t_0)/\tau} \right]$

or $i(t) = \frac{V_m}{Z} \left[\sin(\omega t - \phi) - \sin(\theta - \phi)e^{-(t-t_0)/\tau} \right]$

For no transient $\theta = \phi$ or $\theta - \phi = 0$

12. At a certain current, the energy stored in an iron coil is 1000 J and its copper loss is 2000 W. The time constant of the coil is

(a) 0.25   (b) 0.5   (c) 1.0   (d) 2.0

IES-93(EE)

Hints : $\tau = \frac{L}{R} = \frac{I^2 L}{I^2 R} = \frac{2(1/2)LI^2}{I^2 R} = \frac{2 \times 1000}{2000} = 1$ sec.

13. After closing the switch S at $t = 0$, the current $i(t)$ at any instant of time t in the network shown in the given figure will be

(a) $10 + 10e^{100t}$

(b) $10 - 10e^{100t}$

(c) $10 + 10e^{-100t}$

(d) $10 - 10e^{-100t}$

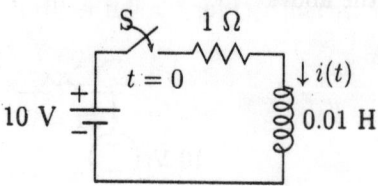

IES-97(EE)

Hints : $i(t) = \frac{V}{R}(1 - e^{-Rt/L}) = \frac{10}{1}(1 - e^{1t/0.01}) = 10 - 10e^{-100t}$

14. The impulse response of an RL circuit is a

(a) rising exponential function.

(b) decaying exponential function.

(c) step function.

(d) parabolic function.

IES-97(EE)

Hint : Ans (b)

15. The time constant of the network shown the given figure, is

(A) 2 RC

(B) 3RC

(C) RC/2

(D) 2RC/3

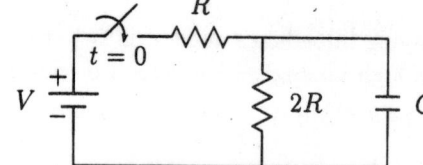

GATE-92(EE)

Hints : Equivalent resistance seen by the capacitor is

$R_{eq} = R||2R = \frac{R \times 2R}{R+2R} = 2R/3$

Time constant $\tau = R_{eq}C = 2RC/3$

16. Consider a DC voltage source connected to a series R-C circuit. When steady-state reaches, the ratio of energy stored in the capacitor to the total energy supplied by the voltage source, is equal to

(A) 0.362   (B) 0.500   (C) 0.632   (D) 1.000

Hints : Ans (B).

17. In the circuit of given figure, $v_C(0) = 6$ V. The energy absorbed by the 4 ohm resistor in the time interval $(0, \infty)$ is

(A) 36 J

(B) 16 J

(C) 256 J

(D) None of the above

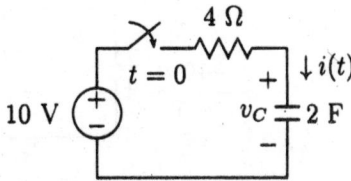

GATE-97(EC)

Hints :

$i(t) = Ke^{-t/RC} = \frac{10-6}{4}e^{-t/8} = e^{-t/8}$ A

$E = \int_0^\infty i^2(t)(R)dt = \int_0^\infty (e^{-t/8})^2(4)dt = -16e^{-t/4}|_0^\infty = 16$ J

18. A first order linear system is initially relaxed. For a unit step signal u(t), the response is $v_1(t) = (1 - e^{-3t})$ for $t > 0$. If a signal $3u(t) + \delta(t)$ is applied to the same initially relaxed system, the response will be

(a) $(3 - 6e^{-3t})u(t)$ (b) $(3 - 3e^{-3t})u(t)$

(c) $3u(t)$ (D) $(3 + 3e^{-3t})u(t)$

IES-95(EE)

Hints: Response to $3u(t) + \delta(t)$ is

$3v_1(t) + \frac{dv_1}{dt} = 3(1 - e^{-3t}) + 3e^{-3t} = 3u(t)$

Initial Conditions In The Networks

19. The circuit shown in figure is in the steady-state with switch 'S' open. The switch is closed at t = 0. The value of $v_C(0^+)$ and $v_C(\infty)$ will be respectively

(A) 2V, 0 V

(B) 0 V, 2 V

(C) 2 V, 2 V

(D) 0 V, 0 V

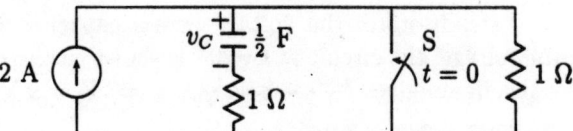

IES- 97(EE)

Hints : $v_C(0^-) = (2 A)(1 \, \Omega) = 2$ V $= v_C(0^+)$ and $v(\infty) = 0$

20. The value of $i_2(0^+)$ in the circuit of given figure is

(a) $i_2(0^+) = V/R_2$

(b) $i_2(0^+) = V/(R_1 + R_2)$

(c) $i_2(0^+) = 0$

(d) none

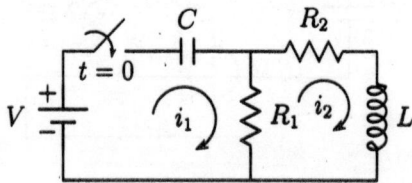

Hints : The inductor behaves as open circuit at t = 0^+, thus $i_2(0^+) = 0$

21. In the circuit shown the figure is open for a long time and steady-state is reached. S is closed at t = 0. The current I at t = 0^+ is

(a) 4 A

(b) 3 A

(c) 2 A

(d) 1 A

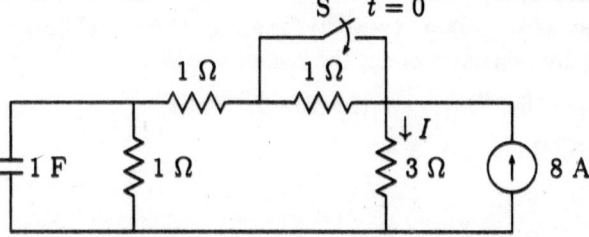

IES-95(EE)

Hints :

For $t < 0$ steady-state, the voltage across capacitor is 4 V. With this value of capacitor voltage, the circuit at t = 0^+ is shown in the figure. By KCL at the right node of this figure gives $\frac{4-V_1}{1} + 8 = \frac{V_1}{3}$ $\Rightarrow V_1 = 9$ V

Thus $I = V_1/3 = 9/3 = 3$ A

22. In the network shown in the figure, the circuit was initially in the steady-state condition with switch closed. At the instant when switch is opened, find the rate of decay of current through the inductance.

(a) zero

(b) 0.5 A/s

(c) 1 A/s

(d) 2 A/s

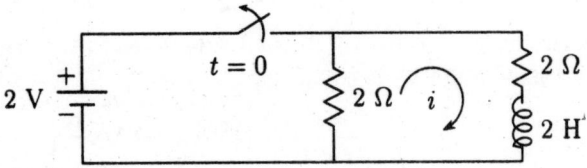

IES-93(EE)

Hints :

$i(t) = i(0^+)e^{-R_{Th}t/L} = 1e^{-(2+2)t/2} = e^{-2t}$ where $i(0^-) = i(0^+) = 2/2 = 1$ V

Hence $\frac{di}{dt}(0^+) = -2e^{-2\times0} = -2$ A/s Thus rate of decay is 2 A/s.

23. In the series RC circuit shown in figure, the voltage across C starts increasing when the dc source is switched on. The rate of increase of voltage across C at the instant just after the switch is closed (ie at t = 0$^+$), will be

    (a) zero

    (b) infinity

    (c) RC

    (d) 1/RC

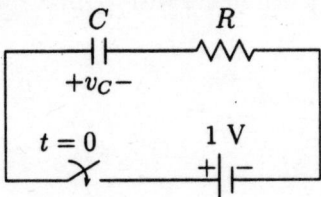

GATE-96(EE)

Hints : $v_C(t) = V(1 - e^{-t/RC})$ or, $\frac{dv_C}{dt} = \frac{e^{-t/RC}}{RC}$ or, $\frac{dv_C}{dt}(0^+) = 1/RC$

24. In the circuit shown in the figure, the switch is closed at t = 0. The induced voltage v_2 will have a maximum value of

    (a) 0.6 V   (b) 1 V   (c) 3.78 V   (d) 6 V

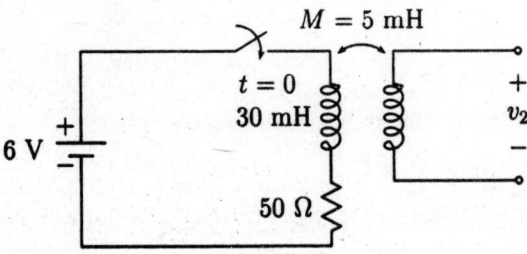

Hints: $i(t) = \frac{6}{50}(1 - e^{-t/\tau})$ where $\tau = \frac{30\times10^{-3}}{50}$

$\frac{di}{dt} = -\frac{6}{50}\left(-\frac{1}{\tau}e^{-t/\tau}\right)$

$\frac{di}{dt}\big|_{max} = \frac{di}{dt}(0) = \frac{6}{50\tau}$

Hence the maximum value of is $v_2 = M\frac{di}{dt}(0) = \frac{5\times10^{-3}\times6\times50}{50\times30\times10^{-3}} = 1$ V

25. In the circuit of given figure, the switch is closed at t =0 with $i_l(0) = 0$ and $v_C(0) = 0$. In the steady-state v_C is equal to

   (A) 200 V   (B) 300 V   (C) zero   (D) 100 V

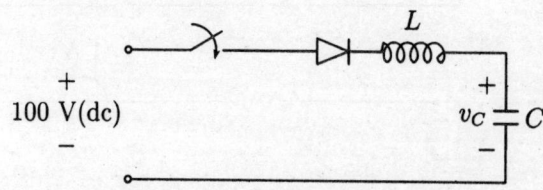

Hints : Ans (D). The steady-state voltage across the capacitor will be equal to the input voltage , that is 100 V.

26. In the circuit of given figure with $i_L(0)= 1$ A, $v_C(0) = 1$ V, The value of $v_R(0^+)$ is

   (a) 5 V
   (b) 2 V
   (c) 1 V
   (d) 0 V

   Hints :

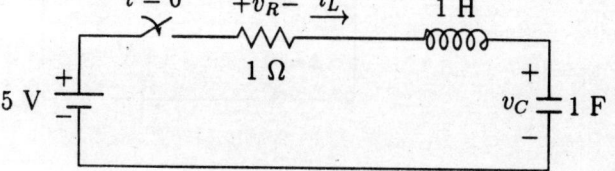

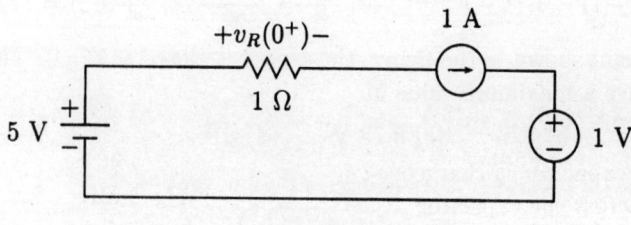

Circuit at $t = 0^+$

With reference to the above circuit at $t = 0^+$

$v_R(0^+) = (1\ \Omega)(1\ A) = 1$ V

27. In the circuit of given figure with $i_L(0)= 1$ A, $v_C(0) = 1$ V, the value of $v_L(0^+)$ is

   (a) 1 V
   (b) 2 V
   (c) 3 V
   (d) 0 V

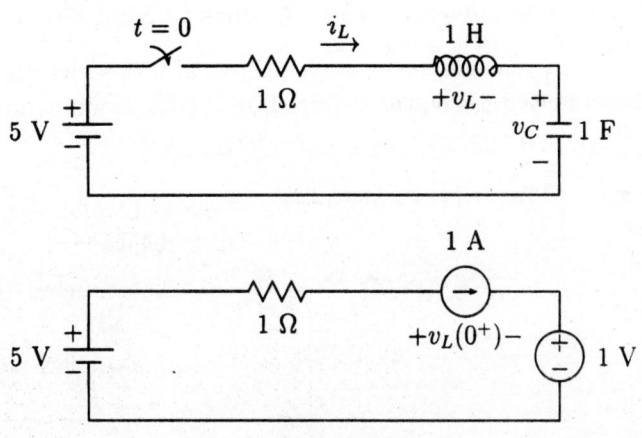

Circuit at t = 0⁺

Hints :

By KVL in the above circuit at $t = 0^+$

$-5 + 1 \times 1 + v_L(0^+) + 1 = \quad \Rightarrow v_L(0^+) = 3 \text{ V}$

28. In the circuit of given figure, the steady-state voltage across the capacitor v_C is

    (a) 0 V

    (b) 10 V

    (c) 10/3 V

    (d) 20/3 V

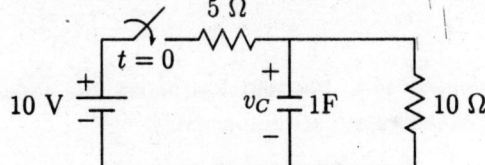

Hints :

$v_C(\infty) = \frac{10}{10+5}(10) = 20/3 \text{ V}$

29. A 1/2 F capacitor is charged to 10 V and then connected to 2 ohm resistor. The voltage across the capacitor at $t = 1$ sec. after the connection was made.

    (a) 0.362 V   (b) 3.60 V   (c) 3.68 V   (d) zero

    Hints : $v_C = v_C(0^+)e^{-t/RC} = 10e^{-t}$

    and $v_C(1sec) = 10e^{-1} = 3.68 \text{ V}$

30. In the circuit of given figure, the current i_L at $t = 0$ is

    (a) 0 A

    (b) 1 A

    (c) ∞

    (d) none of the above

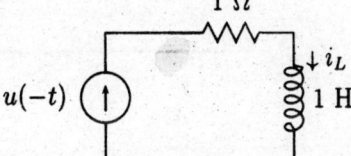

Hints : For t ¡ 0, steady-state, i_L is 1 A. Thus by continuity of inductor current, we obtain $i_L(0^+) = i_L(0^-) = 1$ A

31. In the circuit of given figure, the steady-state current i_1 and i_2 are respectively

    (a) 1/3 A and 2/3 A   (b) 2/3 A and 1/3 A

    (c) 1/2 A and 1/2 A   (d) 1 A and 1 A

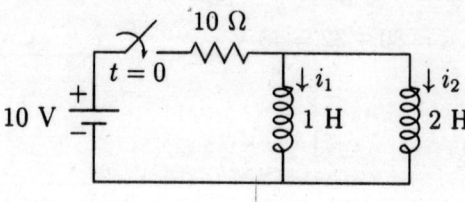

Hints :

In steady-state voltage across the inductors is zero, thus $i = 10/10 = 1$ A.

The current division formula is applicable to determine steady-state currents i_1 and i_2 as:

$$i_1 = \frac{L_2}{L_1+L_2}i = \frac{2}{1+2}(1) = 2/3 \text{ A}$$

$$i_2 = \frac{L_1}{L_1+L_2}i = \frac{1}{1+2}(1) = 1/3 \text{ A}$$

32. The voltage V_{C1}, V_{C2} and V_{C3} across the capacitors in the circuit of given figure under steady-state are respectively.

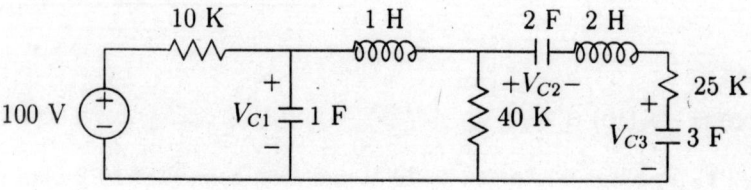

(A) 80 V, 32 V, 48 V   (B) 80 V, 48 V, 32 V

(C) 20 V, 8 V, 12 V   (D) 20 V, 12 V, 8 V

                                                    GATE-96(EC)

Hints :

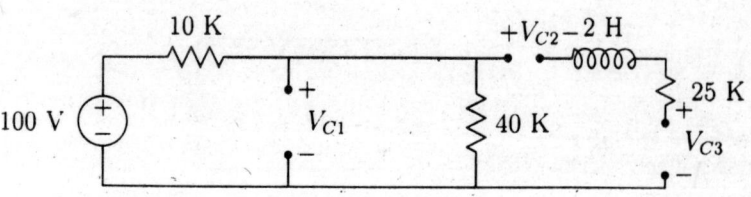

From the above figure, we have

$V_{C1} = \frac{40}{40+10}(100) = 80$ V

By KVL in the right most mesh gives

$V_{C2} + V_{C3} = 80$ and $\frac{V_{C2}}{V_{C3}} = \frac{C_3}{C_2} = 3/2$ $\Rightarrow V_{C2} = \frac{3}{2}V_{C3}$

Setting for V_{C2} gives $(3/2 + 1)V_{C3} = 80$ $\Rightarrow V_{C3} = 32$ V

and $V_{C2} = 80 - 32 = 48$ V

Second Order Circuits

33. A second order system is given by $\frac{d^2y}{dt^2} + 12\frac{dy}{dt} + 100y = 0$

The damped natural frequency in rad/sec is

(a) 100   (b) 10   (c) $\sqrt{44}$ (d) 8

IES- 97(EE)

Hints : $\omega_n = \sqrt{100} = 10$ and $2\zeta\omega_n = 12$ $\Rightarrow \zeta = 0.6$

Therefore the damped natural frequency is

$\omega_d = \omega_n\sqrt{1 - \zeta^2} = 10\sqrt{1 - (0.6)^2} = 8$

34. In the given circuit, the value of R that will give critical damping is

(a) 1 Ω

(b) 2 Ω

(c) 4 Ω

(d) 10 Ω

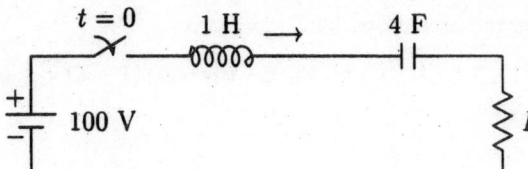

IES-96(EE)

Hints : $\zeta = \frac{R}{2}\sqrt{\frac{C}{L}}$ or, $1 = \frac{R}{2}\sqrt{\frac{4}{1}}$ or, $R = 1$ Ω

35. If a system is represented by the differential equation

$\frac{d^2y}{dt^2} + 6\frac{dy}{dt} + 9y = 0$ then the solution of y will be of the form

(a) $k_1e^{-t} + K_2e^{-9t}$ (b) $(k_1 + k_2t)e^{-3t}$

(c) $ke^{-3t}\sin(t + \phi)$ (d) $(k_1 + k_2t)e^{-3t}$

IES-96(EE)

Hints : The characteristic equation is $s^2 + 6s + 9 = 0$

and the roots are $s_{1,2} = -3, -3$. Thus solution form will be $(k_1 + k_2t)e^{-3t}$

36. Which one of these is not correct in case of series RLC circuit.

    (a) $\omega_n = 1/\sqrt{LC}$ (b) $R_c = 2\sqrt{L/C}$

    (c) $\zeta = \frac{1}{RC}\sqrt{L/C}$ (d) $\alpha = R/2L$

    Hints : Ans (c)

37. Series circuit containing R, L and C is excited by a step voltage across the capacitor exhibits oscillations. Damping coefficient (ratio) of this circuit is given by

    (a) $\zeta = \frac{R}{2\sqrt{LC}}$ (b) $\zeta = \frac{R}{LC}$

    (c) $\zeta = \frac{R}{2\sqrt{C/L}}$ (d) $\zeta = \frac{R}{2\sqrt{L/C}}$

                                                            IES-96(EE)

    Hint : Ans (d)

38. A DC voltage source is connected across a series RLC circuit. Under steady-state conditions, the applied voltage drops entirely across the

    (A) R only   (B) L only   (C) C only   (D) R and L combination

    Hint : Ans (C)

39. The damped natural frequency of oscillation in a series RLC circuit is

    (a) $\zeta\omega_n^2$ (b) $\omega_n\sqrt{(1-\zeta^2)}$ (c) $\omega_n^2\sqrt{(1-\zeta^2)}$ ·   (d) $\omega_n\sqrt{(1-\zeta^2)}$

    Hint : Ans (d)

40. The time constant of a series RLC circuit is

    (a) $\zeta\omega_n$ (b) ζ/ω_n (c) $1/\zeta\omega_n$ (d) $\omega_n\sqrt{(1-\zeta^2)}$

    Hint : Ans (c)

The Laplace transform

41. The Laplace transform of the function $i(t)$ is $I(s) = \frac{10s+4}{s(s+1)(s^2+4s+5)}$ its final value will be

    (a) 4/5   (b) 5/4   (c) 4   (d) 5.

                                                            IES-97(EE)

    Hints : $i(\infty) = \lim_{s\to 0} sI(s) = \lim_{s\to 0} \frac{10s+4}{(s+1)(s^2+4s+5)} = 4/5$

42. If $F(s) = \frac{s+1}{s(s+k)}$ and f(t) as $t \to \infty$ is 1/2, then the value of k is

    (a) 1/2   (b) 1   (c) 2   (d) ∞

                                                            IES-93(EE)

    Hints : $f(\infty) = \lim_{s\to 0} s\left[\frac{s+1}{s(s+k)}\right] = 1/2$ or, $1/k = 1/2$ or, $k = 2$

43. With symbols having usual meanings, the Laplace transform of $U(t - a)$ is

   (a) $1/s$ (b) $1/(s - a)$ (c) e^{-as}/s (d) e^{as}

IES-93(EE)

   Hint : Ans (c)

44. Given $F(s) = \frac{s+2}{s(s+1)}$, the initial and final values of f(t) will be respectively

   (a) 1, 2   (b) 2, 1   (c) 1, 1   (d) 2, 2

IES-94(EE)

   Hints : $f(0) = \lim_{s \to \infty} s\frac{s+2}{s(s+1)} = \lim_{s \to \infty} \frac{1+2/s}{1+1/s} = 1$

   and $f(\infty) = \lim_{s \to 0} s\frac{s+2}{s(s+1)} = 2$

45. If the $i(t) = \frac{1}{4}(1 - e^{-2t})u(t)$, where u(t) is a unit step voltage, then the complex frequencies associated with $i(t)$ would include

   (a) s = 0 and s = j2   (b) s = j2 and s = -j2

   (c) s = -j2 and s = -2   (d) s = 0 and s = -2

IES-98(EE)

   Hints : $\mathcal{L}[i(t)] = \frac{1}{4}\left(\frac{1}{s} + \frac{1}{s+2}\right) = \frac{1}{2}\frac{s+1}{s(s+2)}$

   The poles are at s = 0 and s = -2 and these are the complex frequencies of $i(t)$

46. Laplace transform of $\mathcal{L}\left[\frac{e^{-at} - e^{-bt}}{b-a}\right]$ is

   (a) $\frac{1}{(s+a)(s-b)}$ (b) $\frac{1}{(s-a)(s+b)}$

   (c) $\frac{1}{(s-a)(s-b)}$ (d) $\frac{1}{(s+a)(s+b)}$

IES-92(EE)

   Hints : $\mathcal{L}\left[\frac{e^{-at} - e^{-bt}}{b-a}\right] = \frac{1}{b-a}\left[\frac{1}{s+a} - \frac{1}{s+b}\right] = \frac{1}{(s+a)(s+b)}$

47. The Laplace transformation of f(t) is F(s). Given $F(s) = \frac{\omega}{s^2 + \omega^2}$, the final value of f(t) is

   (A) infinity   (B) zero   (C) one   (D) none of the above

GATE-95(EE)

   Hints : The $F(s)$ has complex poles at $j\omega$-axis, therefore the final value theorem can not be applied here. The final value is nothing but $\mathcal{L}^{-1}[F(s)] = \sin \omega t$. Thus answer is (D).

48. The Laplace transform of $\delta(t - nT)$ is

   (a) 1   (b) e^{-Ts} (c) 0   (d) e^{-nTs}

   Hints : Ans (d)

49. The value of of integral $\int_{-5}^{+6} e^{-2t}\delta(t - 1)dt$ is equal to

   (a) 0.1   (b) 0.2   (c) 0.1353   (d) zero

   Hints : $\int_{-5}^{+6} e^{-2t}\delta(t - 1)dt = e^{-2 \times 1} = 0.1353$

50. If $\mathcal{L}[f(t)] = \frac{2(s+1)}{s^2+2s+5}$, then $f(0)$ and $f(\infty)$ are given by

    (A) 0, 2 respectively   (B) 2, 0 respectively

    (C) 0, 1 respectively   (D) 2/5, 0 respectively

                                                           GATE-95(EC)

    Hints :

    $f(0) = \lim_{s\to\infty} sF(s) = \lim_{s\to\infty} \frac{2s(s+1)}{s^2+2s+5} = \lim_{s\to\infty} \frac{2(1+1/s)}{1+2/s+5/s^2} = 2$

    $f(\infty) = \lim_{s\to 0} sF(s) = \lim_{s\to 0} \frac{2s(s+1)}{s^2+2s+5} = 0$

51. The Laplace transform of $e^{\alpha t} \cos \alpha t$ is equal to

    (A) $\frac{s-\alpha}{(s-\alpha)^2+\alpha^2}$ (B) $\frac{s+\alpha}{(s-\alpha)^2+\alpha^2}$

    (B) $\frac{1}{(s-\alpha)^2}$ (D) None of the above

                                                           GATE-97(EC)

    Hints : Ans (A)

The Laplace Transforms of Signal Waveforms

52. A rectangular current pulse of duration T and magnitude has the Laplace transform

    (A) I/s (B) $I/s\, e^{-Ts}$ (C) $(I/s)e^{Ts}$ (D) $(I/s)[1 - e^{-Ts}]$

                                                           GATE-98(EE)

    Hint : Ans (D)

53. If a pulse voltage $v(t)$ of 4 V magnitude and 2 secs duration is applied to a pure inductor of 1 H, with zero initial current, the current (in A) drawn at t = 3 secs will be

    (a) zero   (b) 2   (c) 4   (d) 8

                                                           IES-94(EE)

    Hints : $v(t) = 4[u(t) - u(t-2)]$ thus $V(s) = \frac{4}{s}(1 - e^{-2s})$

    and $Z(s) = s$, then $I(s) = V(s)/Z(s) = \frac{4}{s^2}(1 - e^{-2s})$

    Therefore $i(t) = 4tu(t) - 4(t-2)u(t-2)$ and current at t = 3 secs is

    $i(3) = 4 \times 3 - 4(3-2) = 8$ A

54. The laplace transform of signal waveform shown in figure is

    (a) $\frac{1}{s^2}(e^{-2s} - e^{-3s}) + \frac{1}{s}(2e^{-2s} - 3e^{-3s})$

    (b) $\frac{1}{s^2}(e^{-2s} - e^{-3s}) - \frac{1}{s}(e^{-2s} - e^{-3s})$

    (c) $\frac{1}{s^2}(e^{-2s} - e^{-3s})$

    (d) none

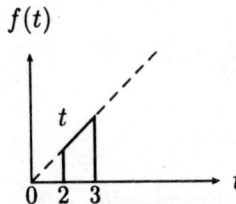

Hints : $f(t) = t[u(t-2) - u(t-3)] = (t-2+2)u(t-2) - (t-3+3)u(t-3) = (t-2)u(t-2) - (t-3)u(t-3) + 2u(t-2) - 3u(t-3)$

Thus $F(s) = e^{-2s}/s^2 - e^{-3s}/s^2 + 2e^{-2s}/s - 3e^{-3s}/s$ and answer is (a)

55. A pulse of unit amplitude and width a is applied to a series RL circuit as shown in figure. The current $i(t)$ as t tends to infinity will be

(a) zero

(b) 1 A

(c) a value between zero and one

(d) infinite.

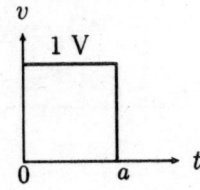

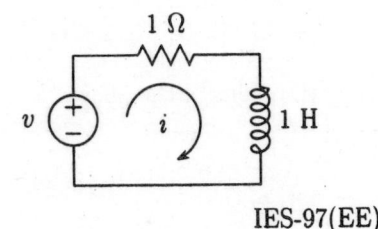

IES-97(EE)

Hint : Ans (a)

56. The Laplace transform of $v(t)$ as shown in the figure is

(a) $\frac{V}{s}e^{-s} - \frac{3V}{s}e^{-2s}$

(b) $\frac{2V}{s} - \frac{3V}{s}e^{-2s}$

(c) $\frac{2V}{s} + \frac{V}{s}e^{-s}$

(d) $\frac{2V}{s} - \frac{V}{s}e^{-s} - \frac{3V}{s}e^{-2s}$

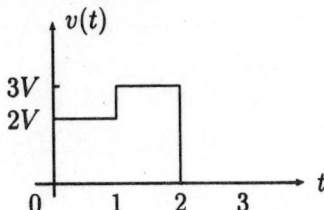

GATE-93(EC)

Hints :

$$v = 2V[u(t) - u(t-1)] + 3V[u(t-1) - u(t-2)]$$

$$V(s) = 2V\left[\frac{1}{s} - \frac{e^{-s}}{s}\right] + 3V\left[\frac{e^{-s}}{s} - \frac{e^{-2s}}{s}\right]$$

$$= \frac{2V}{s} + \frac{V}{s}e^{-s} - \frac{3V}{s}e^{-2s}$$

57. The equation of the given waveform is

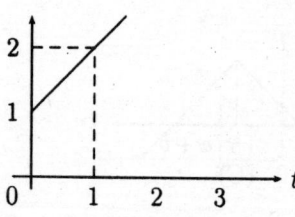

(A) $\frac{1}{2}tu(t)$ (B) (t-1)u(t)   (C) (t+1)u(t)   (D) None

Hint : Ans (C)

58. The Laplace transform of $v(t)$ shown in the figure is

(A) $\frac{3}{s+1}\left[e^{-s} - e^{-2s}\right]$

(B) $\frac{3}{s+1}\left[e^{-s} + e^{-2s}\right]$

(C) $\frac{3}{s+1}\left[e^{-(s+1)} - e^{-2(s+1)}\right]$

(D) None of the above

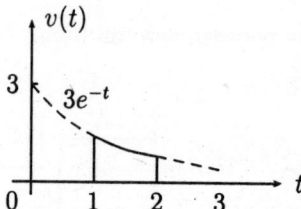

Hints :

$$
\begin{aligned}
v(t) &= 3e^{-t}[u(t-1) - u(t-2)] \\
&= 3e^{-(t-1+1)}u(t-1) - 3e^{-(t-2+2)}u(t-2) \\
V(s) &= \frac{e^{-s}}{s+1}(3e^{-1}) - \frac{e^{-2s}}{s+1}(3e^{-2}) \\
&= \frac{3}{s+1}\left[e^{-(s+1)} - e^{-(s+2)}\right]
\end{aligned}
$$

59. A signal is described by $s(t) = r(t-a) - r(t-b)$. a ¡ b where $r(t)$ is a unit ramp function starting at t = 0. The signal $s(t)$ is represented as

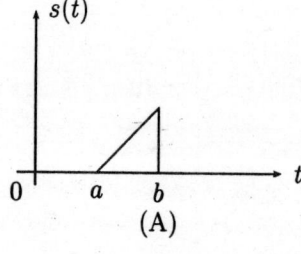

(A)

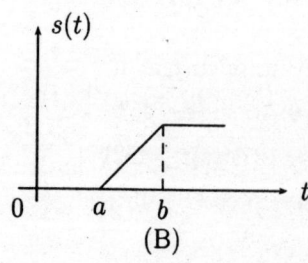

(B)

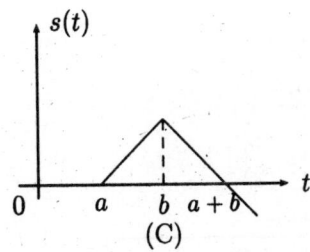

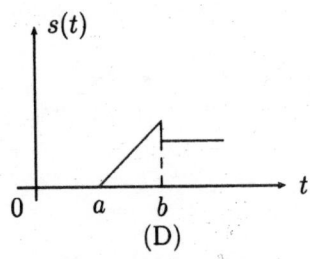

IES-94(EE)

Hint : Ans (B)

60. The Laplace transform of a unit ramp function starting at $t = a$, is

    (A) $\frac{1}{(s-a)^2}$ (B) $\frac{e^{-as}}{(s+a)^2}$ (C) $\frac{e^{-as}}{s^2}$ (D) $\frac{a}{s^2}$

GATE-94(EC)

Hint : Ans (C)

61. Consider the voltage waveform shown in the figure. The equation for $v(t)$ is

    (a) $u(t-1) + u(t-2) + u(t-3)$

    (b) $u(t-1) + 2u(t-2) + 3u(t-3)$

    (c) $u(t) + u(t-1) + u(t-2) + u(t-4)$

    (d) $u(t-1) + u(t-2) + u(t-3) - 3u(t-4)$

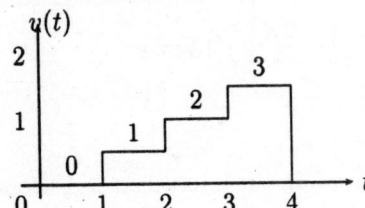

IES-95(EE)

Hints : $v(t) = [u(t-1) - u(t-2)] + 2[u(t-2) - u(t-3)] + 3[u(t-3) - u(t-4)] = u(t-1) + u(t-2) + u(t-3) - 3u(t-4)$

Systems With complex exponential inputs

62. The steady-state current $i(t)$ for the circuit of figure shown below, is

    (a) $i(t) = 0.707 \sin(t - 45^0)$

    (b) $i(t) = 0.707 \cos(t - 45^0)$

    (c) $i(t) = 0.707 \cos(t + 45^0)$

    (d) none

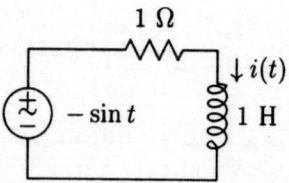

Hints : $i(t) = \frac{1}{1+p}(-\sin t)$ Thus the steady-state value is

$i_{ss}(t) = \text{Im}\left[\frac{-1}{1+j1}e^{j1t}\right] = \text{Im}\left[0.707\angle 135^0 e^{j1t}\right] = 0.707\sin(t + 135^0)$

$= 0.707\cos(t + 45^0)$

63. In the given circuit of figure, $i(t)$ under steady-state is

   (A) zero

   (B) 5

   (C) 7.07$\sin t$

   (D) 7.07$\sin(t - 45^0)$

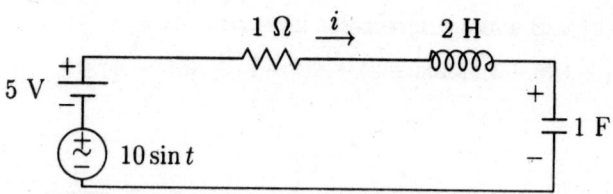

Hints : By superposition in the equivalent p-domain circuit, we have

$i(t) = i'(t) + i''(t)$

where $i'(t) = \frac{5}{1+2p+1/p} = \frac{5p}{2p^2+p+1}$

and $i''(t) = \frac{5p}{2p^2+p+1}(10\sin t)$

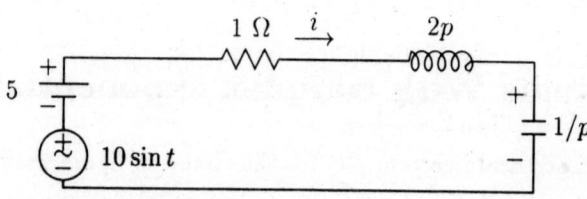

Steady-state of $i'(t)$, $i'_{ss} = \frac{5(0)}{2(0)^2+0+1}e^0 = 0$

and steady-state of $i''(t)$ is

$$i_{ss}'' = \text{Im}\left[\frac{j1}{2(j1)^2 + j1 + 1}(10e^{j1t})\right]$$
$$= \text{Im}\left[7.07\angle - 45^0 e^{j1t}\right]$$
$$= 7.07\sin(t - 45^0)$$

Thus steady-state of $i(t)$ is $0.707\sin(t - 45^0)$

64. Consider the following statements:

If the energy sources are connected to the unenergised systems as shown in fig. A and fig. B at t = 0, then

(1) $i_1(t)$ and $i_2(t)$ are equal

(2) the current in circuit of fig. A has one free frequency and two forced frequencies

(3) the circuit in fig. B has one free frequency and two forced frequencies

(4) the circuit in fig. A has two free frequencies and one forced frequency.

Of these statements

(a) 1, 2 and 3 are correct.

(b) 1, 3 and 4 are correct.

(c) 1 and 3 are correct.

(d) 4 alone is correct.

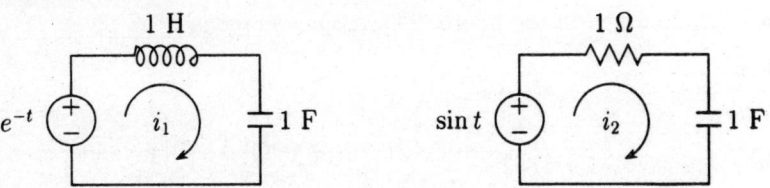

Hints : From fig. A $\quad (s + 1/s)I_1(s) = \frac{1}{s+1} \quad \Rightarrow I_1(s) = \frac{s}{(s+1)(s^2+1)}$

From fig. B $\quad (1 + 1/s)I_2(s) = \frac{1}{s^2+1} \quad \Rightarrow I_2(s) = \frac{s}{(s+1)(s^2+1)}$

Thus $i_1(t) = i_2(t)$. The fig. A has two free frequencies $(s + 1/s)$ and one forced frequency $(\mathcal{L}[e^{-t}] = \frac{1}{s+1})$ and fig. B has one free frequency $(1 + 1/s)$ and two forced frequencies $(\mathcal{L}[\sin t] = \frac{1}{s^2+1})$.

Sinusoidal Steady-state Analysis

65. Currents I_1, I_2 and I_3 meet a junction (node) in a circuit. All currents are marked as entering the node. If $I_1 = -6\sin\omega t$ mA and $I_2 = 8\cos\omega t$ mA, then I_3 will be

(A) $10\cos(\omega t + 36.87^0)$ mA   (b) $14\cos(\omega t + 36.87^0)$ mA

(C) $-14\cos(\omega t + 36.87^0)$ mA   (D) $-10\cos(\omega t + 36.87^0)$ mA

GATE-98(EE)

Hints : $I_1 = -6\sin\omega t = 6\cos(\omega t + 90^0)$ $\Rightarrow$ $6\angle 90^0$ and $I_2 = 8\cos\omega t$ $\Rightarrow$
$8\angle 0^0$ thus by KCL $I_1 + I_2 + I_3 = 0$ $\Rightarrow$ $-I_3 = 8\angle 0^0 + 6\angle 90^0 = 8 + j6 =$
$10\angle 36.87$ Therefore $I_3 = -10\cos(\omega t + 36.87)$ mA

66. The voltage phasor of a circuit is $10\angle 15^0$ V and current phasor is $2\angle -45^0$ A. The active and reactive power in the circuit are

    (A) 10 W and 17,32 Var   (B) 5 W and 8.66 Var

    (C) 20 W and 60 Var   (D) $20\sqrt{2}$ W and $10\sqrt{2}$ Var

GATE-98(EE)

Hints : Complex power $S = VI^* = (10\angle 15^0)(2\angle 45^0) = 20\angle 60^0 = 10 + j17.32$

Thus active power is 10 W and reactive power is 17.32 Var.

67. A water boiled at home is switched on to the ac mains supplying power at 230V/50 Hz. The frequency of instantaneous power consumed by the boiler is

    (A) 0 Hz   (B) 50 Hz   (C) 100 Hz   (D) 150 Hz

GATE-96(EE)

Hint: Ans (C)

68. The source in the circuit shown is a sinusoidal source. The voltage across various elements are marked in the figure. The input voltage is

    (a) 10 V

    (b) 5 V

    (c) 27 V

    (d) 24 V

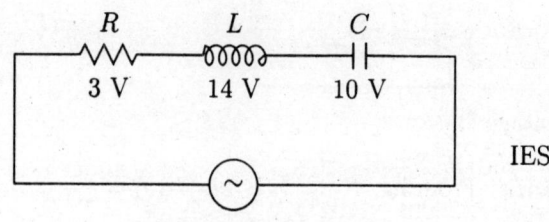

AC voltage source

IES-92(EE)

Hints : The total voltage across the reactive elements is (14 - 10) V = 4 V. Therefore the supply voltage will be $\sqrt{3^2 + 4^2} = 5$ V.

69. The current read by the ammeter in the ac circuit shown in the given figure is

    (a) 9 A

    (b) 5 A

    (c) 3 A

    (d) 1 A

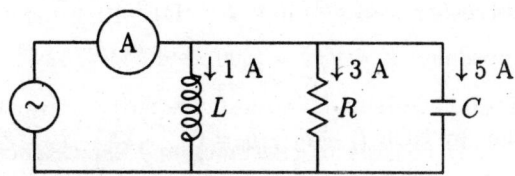

IES-95

Hints : Current read by ammeter $= \sqrt{3^2 + (5-1)^2} = 5$ A

70. Which of the following, are true for the circuit shown the figure?

1 $V_R = 1002$ V

2 $I = 2$ A

3 $L = 0.25$ H

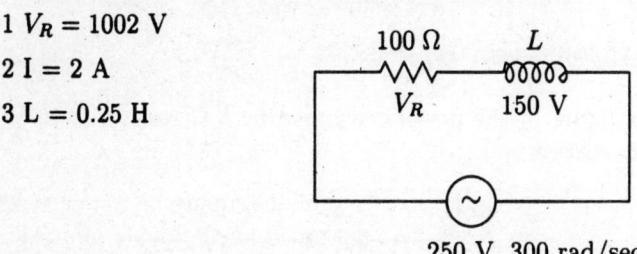

250 V, 300 rad/sec

Select the correct answer using the codes given below:

Codes : (a) 2 and 3　　(b) 1 and 2　　(c) 1 and 3　　(d) 1, 2, and 3

IES-96

Hints : $V_R = \sqrt{250^2 - 150^2} = 200$ V

$I = \frac{V_R}{R} = 200/100 = 2$ A

$(\omega L)I = 150 \quad \Rightarrow \quad 300L \times 2 = 150 \quad \Rightarrow \quad L = 1/4 = 0.25$ H

Thus Answer is (a)

71. A series circuit containing passive elements has the following current and applied voltage :

$v = 200 \sin(2000t + 50^0)$ V

$i = 4 \cos(2000t + 32.2^0)$ A

The circuit elements

(a) must be resistance and capacitance.

(b) must be resistance and inductance.

(c) must be inductance, capacitance and resistance.

(d) could be either resistance and capacitance or resistance, inductance and capacitance.

IES-96(EE)

Hints :Ans (d)

$v = 200\sin(2000t + 50^0) = 200\cos(2000t - 40^0)$ $\Rightarrow V = 200\angle - 40^0$

$i = 4\cos(2000t + 32.2)$ $\Rightarrow I = 4\angle 32.2^0$

Thus impedance $Z = V/I = (200\angle - 40^0)/(4\angle 32.2^0) = 50\angle - 53.2$

$= 29.95 - j40.04\ \Omega$

Thus Z may consists of resistance and capacitance (due to the negative sign of imaginary part) or combination of R, L and C in which $X_C > X_L$

72. A series circuit, with $R = 10\ \Omega$ and L = 20 mH, has current $i = 2\sin 500t$ A. The angle by which i lags v is

(a) 0^0 (b) 30^0 (c) 180^0 (d) 45^0

Hints : $Z = R + j\omega L = 10 + j20 \times 10^{-3} \times 500 = 10 + j10$

Thus $\theta = \tan^{-1}(10/10) = \tan^{-1}(1) = 45^0$

73. In the circuit of given figure, if the power consumed by 5 Ω resistor is 10 W, then the power factor of the circuit is

(a) 0.8

(b) 0.6

(c) 0.5

(d) zero

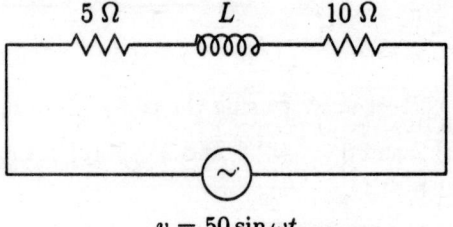

$v = 50\sin\omega t$

IES-96(EE)

Hints: $I^2 \times 5 = 10$ $\Rightarrow I = \sqrt{2}$ A

Total voltage across the resistor, $V_R = (5 + 10) \times \sqrt{2} = 15\sqrt{2}$ V

RMS voltage of the source $V = 50/\sqrt{2}$ V

Thus $\cos\phi = V_R/V = \frac{15\sqrt{2}}{50/\sqrt{2}} = 0.6$

74. When a resistor R is connected to a current source, it consumes a power of 18 W. When the same R is connected to a voltage source having the same magnitude as current source, the power absorbed by R is 4.5 W. The magnitude of current source and the value of R are

(A) $\sqrt{18}$ A and 1 Ω (B) 3 A and 2 Ω

(C) 1 A and 18 Ω (D) 6 A and 0.5 Ω

GATE-99(EE)

Hints : Let x be the magnitude of current and voltage each

Then $x^2 R = 18$ and $x^2/R = 4.5$ Thus $18 = x^2 \times \frac{x^2}{4.5}$ $\Rightarrow x^4 = 81$ $\Rightarrow x = 3$

and $R = 3^2/4.5 = 2\ \Omega$

75. In the circuit shown in the figure, $V_s = V_m \sin 2t$ and $Z_2 = 1 + j$, The value of C is chosen such the current is in phase with V_s. The value of C (in Farad) is

(a) 1/4

(b) $\frac{1}{2\sqrt{2}}$

(c) 2

(d) 4

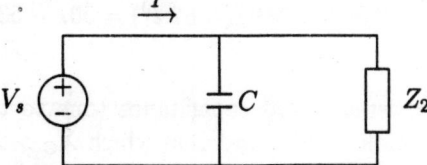

IES-93(EE)

Hints : $Z_2 = 1 + j1$ thus $Y_2 = 0.5 - j0.5$

Capacitive admittance $= j2C$ Thus total admittance seen by voltage source is

$Y = 0.5 + j(2C - 0.5)$. For getting current in phase with V_s, imaginary part of Y will be zero, i.e.; $2C - 0.5 = 0$ $\Rightarrow C = 1/4$ F

76. The phase angle of the current I with respect to the voltage V_1 in the circuit shown in the figure is

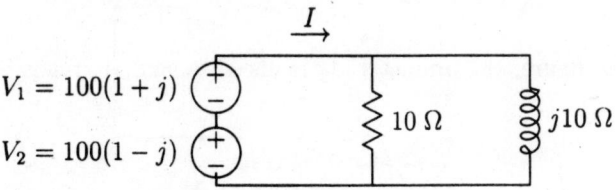

(a) 0^0 (b) 45^0 (c) -45^0 (d) -90^0

Hints : $I = \frac{100(1+j)+100(1-j)}{10+j10} = 10\sqrt{2}\angle - 45^0$

The polar form of V_1 is $100\sqrt{2}\angle45^0$. Thus phase angle of I with respect to V_1 is -90^0

77. The current waveform in a pure resistor of 10 Ω is shown in the figure, power dissipated in the resistor is

(a) 7.29 W

(b) 52.4 W

(c) 135 W

(d) 270 W

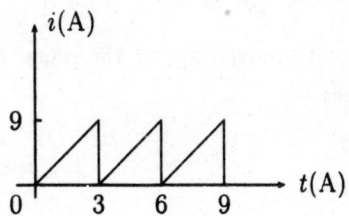

IES-94(EE)

Hints : $i_{rms} = \sqrt{\frac{1}{3}\int_0^3 (9t/3)^2 dt} = \sqrt{27}$

Thus $P = i_{rms}^2 R = 27 \times 10 = 270$ W

78. The rms value of the periodic waveform is

    (a) $\sqrt{3/2}A$

    (b) $\sqrt{2/3}A$

    (c) $\sqrt{1/3}A$

    (d) $\sqrt{2}A$

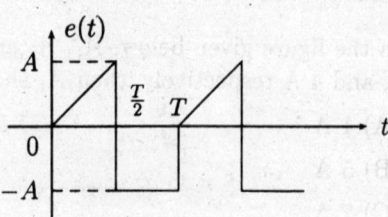

GATE-95(EE)

Hints :

$$e_{rms} = \sqrt{\frac{1}{T}\left[\int_0^{T/2}\left(\frac{2A}{T}t\right)^2 dt + \int_{T/2}^T (-A)^2 dt\right]}$$

$$= \sqrt{\frac{1}{T}\left[\frac{4A^2}{T^2}\left|\frac{t^3}{3}\right|_0^{T/2} + A^2|t|_{T/2}^T\right]} = \sqrt{2/3}A$$

79. In the circuit of given figure, the ammeter A_2 reads 12 A and A_3 reads 9 A. The reading of A_1 will be

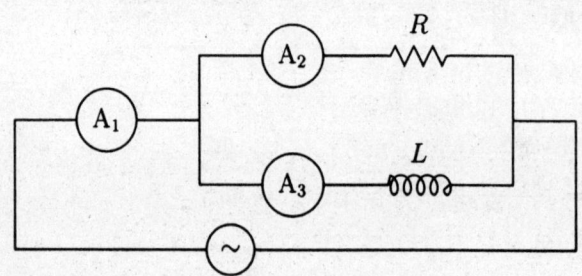

    (a) 5 A   (b) 21 A   (c) 3 A   (d) 15 A

80. In circuit of given figure, the phase angle of V_s with respect to V_L is

    (a) 45⁰

    (b) 30⁰

    (c) 60⁰

    (d) 90⁰

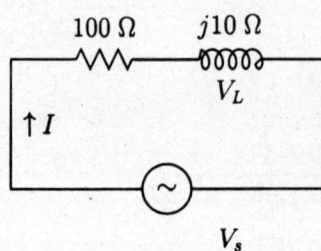

Hints : $\tan\theta = \frac{V_L}{V_R} = \frac{10I}{17.32I} = 0.577 \qquad \Rightarrow \theta = 30^0$

Thus $\theta' = 90^0 - 30^0 = 60^0$ where θ' is the angle between V_s and V_L.

81. In the figure given below, A_1, A_2 and A_3 are ideal ammeters. If A_2 and A_3 reads 3 A and 4 A respectively, then A_1 should read

   (A) 1 A

   (B) 5 A

   (C) 7 A

   (D) None of the above

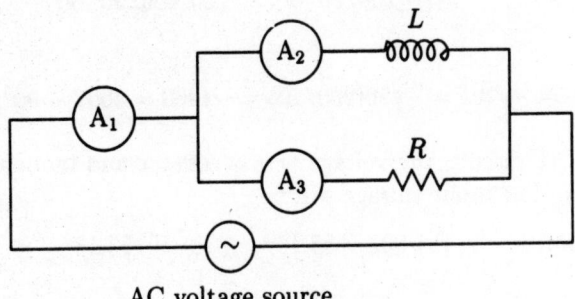

AC voltage source

GATE-96(EC)

Hints :

$A_1 = \sqrt{A_2^2 + A_3^2} = \sqrt{4^2 + 3^2} = 5$ A

82. The RMS value of a rectangular wave of period T, having a value of +V for a duration T_1 (¡ T) and -V for the duration, $T - T_1 = T_2$, equals

   (A) V (B) $\frac{T_1+T_2}{T}V$ (C) $\frac{V}{\sqrt{2}}$ (D) $\frac{T_1}{T_2}V$

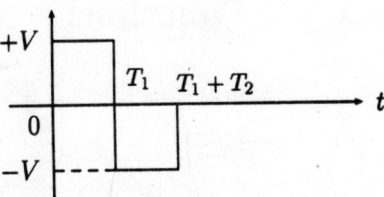

GATE-95(EC)

Hints :

RMS value $= \sqrt{\frac{1}{T}\left[\int_0^{T_1} V^2 dt + \int_{T_1}^{T_1+T_2} V^2 dt\right]} = V$

83. The current $i(t)$, through a 10 Ω resistor in series with an inductance, is given by

   $i(t) = 3 + 4\sin(100t + 45^0) + 4\sin(300t + 60^0)$ A

   RMS value of the current and the power dissipated in the circuit are :

(A) $\sqrt{41}$ A, 410 W respectively   (B) $\sqrt{35}$ A, 350 W respectively

(C) 5 A, 250 W respectively   (D) 11 A, 1210 W respectively

GATE-95(EC)

Hints : $i_{rms} = \sqrt{3^2 + (4/\sqrt{2})^2 + (4/\sqrt{2})^2} = 5$ A

And power $= i_{rms}^2 R = 5^2 \times 10 = 250$ W

84. A series LCR circuit consisting of R $= 10\ \Omega$, $|X_L| = 20\ \Omega$ and $|X_C| = 20\ \Omega$, is connected across a supply of 200 V rms. The rms value across the capacitor is

(A) $200\angle - 90^0$ V   (B) $200\angle 90^0$ V   (C) $400\angle 90^0$ V   (D) $400\angle - 90^0$ V

GATE-94(EC)

Hints : $V_C = -j20I = -j20(200/10) = -j400 = 400\angle - 90^0$ V

85. In a series RL circuit, the voltage across resistor and inductor are 30 V and 40 V respectively. The input voltage will be

(a) $70\angle - 53.13^0$ (b) $60\angle - 53.13^0$ (c) $50\angle 53.13^0$ (d) $40\angle 53.13^0$

Hints : Input voltage $= \sqrt{30^2 + 40^2}\angle (\tan^{-1} 40/30) = 50\angle 53.13^0$

86. In a series RC circuit, the voltage across resistor and inductor are 30 V and 40 V respectively. The input voltage will be

(a) $70\angle - 53.13^0$ (b) $60\angle - 53.13^0$ (c) $50\angle - 53.13^0$ (d) $40\angle 53.13^0$

Hints : Input voltage $= \sqrt{30^2 + 40^2}\angle (-\tan^{-1} 40/30) = 50\angle - 53.13^0$

Network Functions

87. The current transfer function $\alpha_{12} = I_1/I_2$ for the given circuit is

(a) $I_2/I_1 = \frac{1/sC}{R+1/sC}$

(b) $I_2/I_1 = \frac{R}{R+1/sC}$

(c) $I_2/I_1 = \frac{R/sC}{R+1/sC}$

(d) none

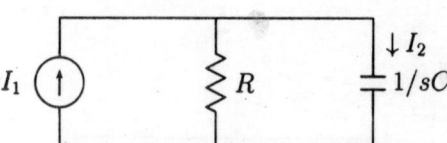

Hint : Ans (b)

88. The impedance $Z(s)$ of the one-port network shown in figure is given by

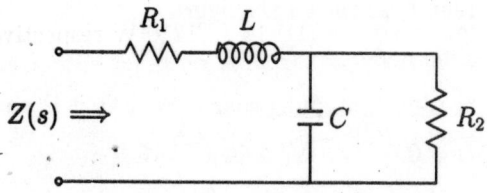

$$Z(s) \Rightarrow$$

(a) $$\frac{s^2 + sL/C + R_1/LC}{s + 1/LC}$$

(b) $$L\frac{s^2 + s\frac{L+R_1R_2C}{R_2LC} + \frac{R_1+R_2}{R_2LC}}{s + 1/R_2C}$$

(c) $$L\frac{s^2 + s\frac{L+C}{R_1R_2} + \frac{R_1}{LC}}{s + 1/R_1C}$$

(d) $$\frac{s^2 + s\frac{L+R_2C}{R_1LC} + \frac{R_1+R_2}{R_LC}}{s + 1/R_1C}$$

Hints :

$$Z(s) = (R_1 + sL) + (1/sC \| R_2) = R_1 + sL + \frac{R_2}{R_2Cs + 1}$$

$$= \frac{R_1R_2Cs + R_1 + R_2LCs^2 + sL + R_2}{R_2Cs + 1}$$

$$= \frac{R_2LCs^2 + s(R_1R_2C + L) + R_1 + R_2}{R_2Cs + 1}$$

$$= L\frac{s^2 + s\frac{L+R_1R_2C}{R_2LC} + \frac{R_1+R_2}{R_2LC}}{s + 1/R_2C}$$

89. The voltage transfer function G_{12} for the network shown in the given figure, is

(a) $G_{12} = \frac{1+\alpha}{3+2\alpha}$

(b) $G_{12} = \frac{1+\alpha}{2+3\alpha}$

(c) $G_{12} = \frac{2(1+\alpha)}{2+\alpha}$

(d) none

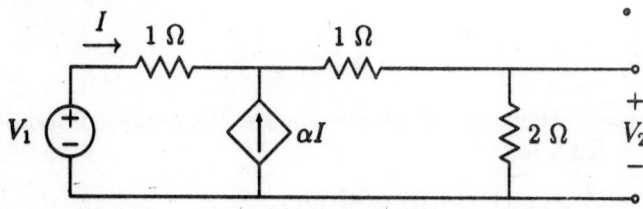

Hints : Label current I_2 as shown the figure

By KCL at the middle node $I_2 = (1 + \alpha)I$

By KVL around the loop containing source V_1, 1 Ω, 1 Ω and 2 Ω

$-V_1 + I + (1 + 2)I_2 = 0$ $\Rightarrow$ $V_1 = (4 + 3\alpha)I$

and $V_2 = 2I_2 = 2(1 + \alpha)I$ Thus $V_2/V_1 = \frac{2(1+\alpha)}{4+3\alpha}$

90. The open-circuit ratio $V_2(s)/V_1(s)$ of the network shown in the figure, is

    (a) $1 + 2s^2$ (b) $\frac{1}{1+2s^2}$

    (c) $1 + 2s$ (d) $\frac{1}{1+2s}$

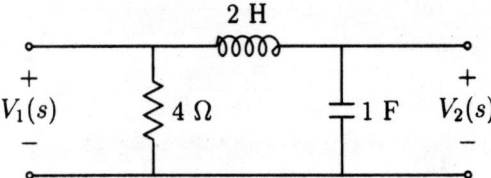

Hints : Using voltage division formula $V_2(s)/V_1(s) = \frac{1/s}{1/s+2s} = \frac{1}{1+2s^2}$

91. For the given circuit of figure, the transfer function is equal to

    (a) $\frac{1}{LCs^2+RLs+1}$

    (b) $\frac{1}{LCs^2+RCs+1}$

    (c) $\frac{1}{LCs^2+RCs+R}$

    (d) $\frac{1}{RCs^2+LCs+1}$

Hints : $E_0(s)/E_1(s) = \frac{1/sC}{R+sL+1/sC} = \frac{1}{LCs^2+RCs+1}$

92. For the compensated attenuator of the given figure, the impulse response under the condition $R_1C_1 = R_2C_2$ is

    (A) $\frac{R_2}{R_1+R_2}\left[1 - e^{-t/R_1C_1}\right]u(t)$ (B) $\frac{R_2}{R_1+R_2}u(t)$

    (C) $\frac{R_2}{R_1+R_2}u(t)$ (D) $\frac{R_2}{R_1+R_2}(1 - e^{-t/R_1C_1})u(t)$

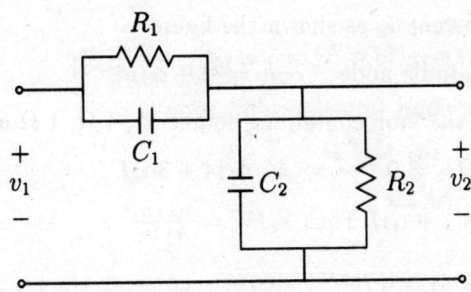

GATE-92(EC)

Hints : $Z_1 = R_1 \| (1/sC_1) = R_1/(R_1C_1s + 1)$ and $Z_2 = R_2/(R_2C_2s + 1)$

Thus $V_2/V_1 = \dfrac{Z_2}{Z_1+Z_2} = \dfrac{R_2(R_1C_1s+1)}{R_1R_2C_1s+R_1R_2C_2sR_1+R_2} = \dfrac{R_2}{R_1+R_2}$

And $v_2 = \dfrac{R_2}{R_1+R_2}u(t)$

93. For the network shown in the given figure, $Z(0) = 3$ ohm and $Z(\infty) = 2$ ohm. The values of R_1 and R_2 will respectively be

    (A) $2 \, \Omega, 1 \, \Omega$ (B) $1 \, \Omega, 2 \, \Omega$ (C) $3 \, \Omega, 2 \, \Omega$ (D) $2 \, \Omega, 3 \, \Omega$

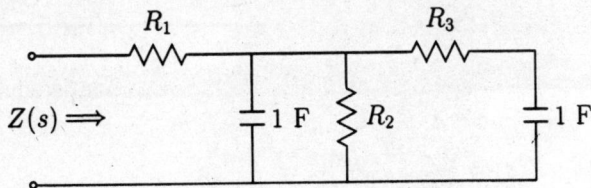

GATE-94(EC)

Hints : From the given network

$Z(0) = R_1 + R_2 = 3$ and $Z(\infty) = R_1 = 2$ Thus $R_1 = 2 \, \Omega$ and $R_2 = 1 \, \Omega$

94. The driving point impedance of the network shown in the figure is

    (a) $\dfrac{s^2+Rs}{1+Rs}$ (b) $\dfrac{Rs^2+s+R}{Rs+1}$

    (c) $\dfrac{s^2(R+1)+3}{1+Rs}$ (d) $\dfrac{1}{(1+Rs)^2}$

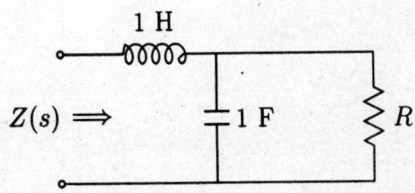

IES-93(EE)

Hints : $Z(s) = s + (1/s \| R) = s + \frac{R}{Rs+1} = \frac{Rs^2+s+1}{Rs+1}$

95. An LC -driving point impedance function is

(a) $\frac{s^3+s^2+s+1}{s^2+2s+5}$ (b) $\frac{s^4+s^2+1}{s^2+5}$

(c) $\frac{s^4+1}{s^3+2}$ (d) $\frac{2(s^2+1)}{s}$

IES-95(EE)

Hints : $Z(s) = sL + 1/sC = \frac{s^2LC+1}{sC}$ thus answer will be (d)

96. Consider the driving-point impedance function of a parallel circuit shown in the figure. When the pole of the function is made closer to the $j\omega$-axis, the memory of this circuit will

(a) increase   (b) decrease but not become zero

(c) remain unaffected   (d) become zero

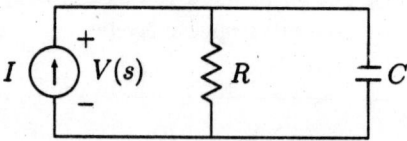

IES-93(EE)

Hints : $Z(s) = \frac{V(s)}{I(s)} = \frac{(1/sC)R}{R+1/sC} = \frac{1/RC}{s+1/RC}$

Thus for the given condition, the memory of the circuit will increase.

Two-port Parameters

97. Two two-port network α and β having ABCD parameters as :

$A_\alpha = 4 = D_\alpha;$ $B_\alpha = 5, C_\alpha = 3$ and

$A_\beta = 3 = D_\beta;$ $B_\beta = 4$ and $C_\beta = 2$

are connected in cascade in the order of α, β.

The equivalent 'A' parameter of the combination is

(a) 17   (b) 22   (c) 24   (d) 31

IES-94(EE)

Hints :
$$\begin{bmatrix} A & B \\ C & D \end{bmatrix} = \begin{bmatrix} A_\alpha & B_\alpha \\ C_\alpha & D_\alpha \end{bmatrix} \begin{bmatrix} A_\beta & B_\beta \\ C_\beta & D_\beta \end{bmatrix}$$
where $A = A_\alpha A_\beta + B_\alpha C_\beta = 4 \times 3 + 5 \times 2 = 22.$

98. Consider the following statements :

1. The two-port network shown below does NOT have an input impedance matrix representation

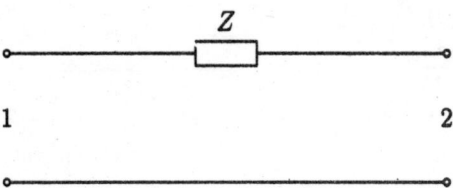

2. The following two-port network does NOT have an admittance matrix representation

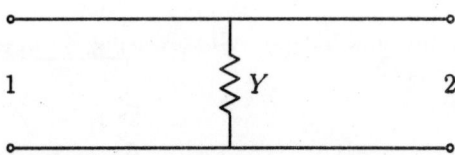

3. A two-port network is said to be reciprocal if it satisfies $z_{12} = z_{21}$ or, an equivalent relationship.

Of these statements

(a) 1 and 2 are correct

(b) 1, 2 and 3 are correct

(c) 1 and 3 are correct

(d) none is correct

IES-97(EE)

Hints : Ans (b)

99. Match list - I with list - II for the two-port network shown in the figure and select the correct answer using codes given below the list

List - I	List - II
A z_{11}	1 R
B z_{12}	2 $R + L$
C z_{21}	3 $R - Ls$
D z_{22}	4 $R + Ls$

Codes (a) A B C D

          1   2   1   4

      (b) A B C D

          2   1   1   3

      (c) A B C D

          1   1   1   4

      (d) A B C D

          2   1   3   4

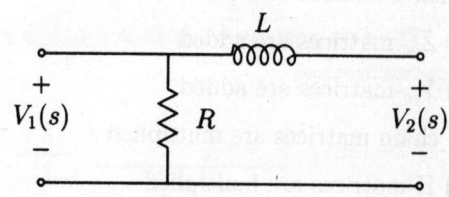

Hints : Expressing input and output voltages using KVL, assuming the standard directions of the currents

$V_1 = RI_1 + RI_2$

$V_2 = RI_1 + (R + Ls)I_2$ from them $z_{11} = Z_{12} = z_{21} = R$ and $z_{22} = R + Ls$. Thus answer is (c)

100. Two two-port networks with transmission parameters A_1, B_1, C_1, D_1 and A_2, B_2, C_2, D_2 respectively are cascaded. The transmission parameter matrix of the cascade network will be

(a) $\begin{bmatrix} A_1 & B_1 \\ rC_1 & D_1 \end{bmatrix} + \begin{bmatrix} A_2 & B_2 \\ C_2 & D_2 \end{bmatrix}$ (b) $\begin{bmatrix} A_1 & B_1 \\ C_1 & D_1 \end{bmatrix}\begin{bmatrix} A_2 & B_2 \\ C_2 & D_2 \end{bmatrix}$

(c) $\begin{bmatrix} A_1 A_2 & B_1 B_2 \\ C_1 C_2 & D_1 D_2 \end{bmatrix}$ (d) $\begin{bmatrix} A_1 A_2 + C_1 C_2 & A_1 A_2 - B_1 B_2 \\ C_1 A_2 - D_1 C_2 & C_1 C_2 + D_1 D_2 \end{bmatrix}$

Hint : Ans (b)

101. If a two-port network is passive, then we have with the usual notation, the following relationship

(A) $h_{12} = h_{22}$

(B) $h_{12} = -h_{21}$

(C) $h_{11} = h_{22}$

(D) $h_{11}h_{22} - h_{12}h_{21}$

GATE-94(EE)

Hint : Ans (B)

102. When a number of 2-port networks are connected in cascade, the individual

(a) Z_{oc} matrices are added

(b) Y_{sc}-matrices are added

(c) chain matrices are multiplied

(d) H-matrices are multiplied

IES-95(EE)

Hint : Ans (c)

103. A two-port network is defined by the relations

$$I_1 = 2V_1 + V_2$$

$$I_2 = 2V_1 + 3V_2$$

then z_{12} will be

(a) - 2 Ω (b) - 1 Ω (c) -1/2 Ω (d) -1/4 Ω

IES-95

Hints : $z_{12} = \frac{-y_{12}}{\Delta} = \frac{-1}{4} \, \Omega$

104. With the usual notations, a two-port resistive network satisfies the conditions

$A = D = \frac{3}{2}B = \frac{4}{3}C$ The z_{11} of the network is

(a) 5/3   (b) 4/3   (c) 2/3   (d) 1/3

IES-95(EE)

Hints : $z_{11} = \frac{V_1}{I_1}\big|_{I_2=0} = \frac{A}{C} = \frac{(4/3)C}{C} = 4/3$

105. The z and h parameters of the networks shown the figure will be

(a) $Z = \begin{bmatrix} 0 & 1 \\ -1 & 1 \end{bmatrix}$ $h = \begin{bmatrix} 1 & 1 \\ 1 & 1 \end{bmatrix}$

(b) $Z = \begin{bmatrix} 1 & 1 \\ 1 & 1 \end{bmatrix}$ $h = \begin{bmatrix} 0 & 1 \\ -1 & 1 \end{bmatrix}$

(c) $Z = \begin{bmatrix} 1 & 0 \\ -1 & 1 \end{bmatrix}$ $h = \begin{bmatrix} -1 & 1 \\ 1 & -1 \end{bmatrix}$

(d) $Z = \begin{bmatrix} 0 & 1 \\ 1 & 0 \end{bmatrix}$ $h = \begin{bmatrix} 0 & -1 \\ 2 & -1/2 \end{bmatrix}$

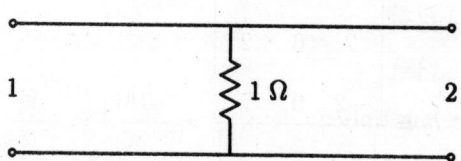

IES-96(EE)

Hint : Ans (b)

106. The y_{21} parameter of the network shown in the given figure will be

    (a) 1/6 Ω

    (b) -1/6 Ω

    (c) 1/3 Ω

    (d) -1/2 Ω

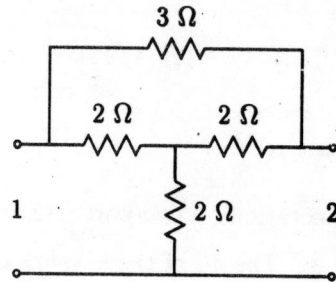

IES-96(EE)

Hints :

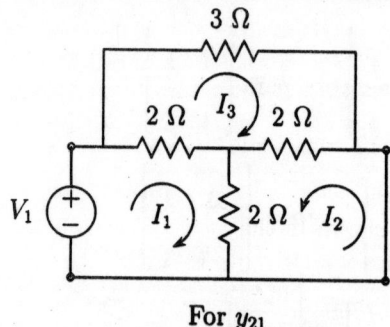

For y_{21}

Mesh equations are :

$$4I_1 + 2I_2 - 2I_3 = V_1$$
$$2I_1 + 4I_2 + 2I_3 = 0$$
$$-2I_1 + 2I_2 + 7I_3 = 0$$

$$I_2 = \frac{\begin{vmatrix} 4 & V_1 & -2 \\ 2 & 0 & 2 \\ -2 & 0 & 7 \end{vmatrix}}{\begin{vmatrix} 4 & 2 & -2 \\ 2 & 4 & 2 \\ -2 & 2 & 7 \end{vmatrix}} = \frac{-18V_1}{36} = -\frac{V_1}{2}$$

Thus $y_{21} = I_2/V_1 = -1/2$

107. The y_{22} parameter for the network shown in figure, is

(a) $Y_A + Y_B + Y_C$

(b) Y_B

(c) $Y_B + Y_C$

(d) none

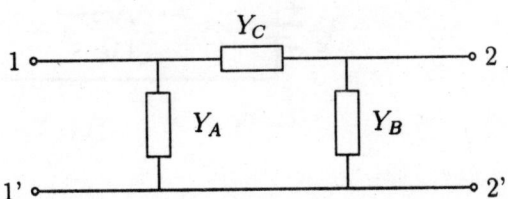

Hints : The π-network parameters and y parameters related as :

$Y_C = -y_{12}$ and $Y_B = y_{22} + y_{12}$ Thus $y_{22} = Y_B + Y_C$

108. A T-network is shown in the figure, the Y_{sc} matrix will be

(a) $\begin{bmatrix} 10/100 & 5/200 \\ 5/200 & 10/200 \end{bmatrix}$
(b) $\begin{bmatrix} 10/200 & -5/200 \\ -5/200 & 10/200 \end{bmatrix}$

(c) $\begin{bmatrix} 15/100 & -5/200 \\ -5/200 & 15/200 \end{bmatrix}$
(d) $\begin{bmatrix} 15/200 & 5/200 \\ 5/200 & 15/200 \end{bmatrix}$

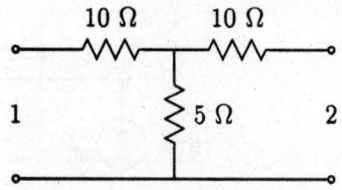

Hints : The z parameters can be determined directly from T-network as:

$z_{11} = 10 + 5 = 15\ \Omega$, $z_{12} = z_{21} = 5\ \Omega$, $z_{22} = 5 + 10 = 15\ \Omega$

The y parameters can obtain as:

$y_{11} = z_{22}/\Delta = 15/200$, $y_{12} = y_{21} = -5/200$, and $y_{22} = z_{11}/\Delta = 15/200$

Miscellaneous

109. In the network shown in figure, the effective resistance faced by the voltage source is

    (a) 4 Ω

    (b) 3 Ω

    (c) 2 Ω

    (d) 1 Ω

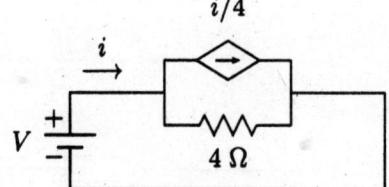

IES-98(EE)

Hints : By KCL $i = i/4 + V/4$ or $3i/4 = V/4$ or $V/i = 3\Omega$

110. The current, i_R for the circuit of given figure , is

    (a) 4 A

    (b) 2 A

    (c) 0 A

    (d) none

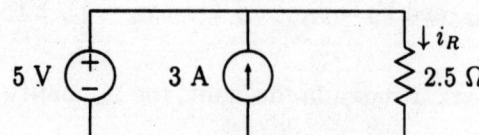

Hints : $i_R = \frac{5}{2.5} = 2$ A

111. The value of current i in the circuit of given figure is

    (a) 4 A

    (b) 0 A

(c) 2 A

(d) 1 A

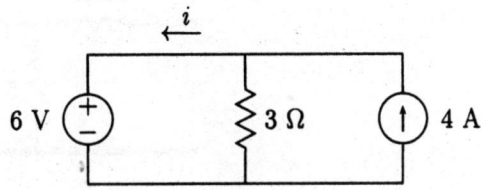

Hints : By KCL $\quad i + i_{3\Omega} = 4 \quad \Rightarrow i + \frac{6}{3} = 4 \quad \Rightarrow i = 2\,A$

112. The voltage v_R in the circuit of given figure is

(a) -6 V

(b) 6 V

(c) 5 V

(d) 1 V

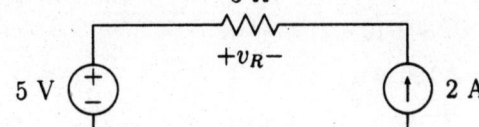

Hints :

The circulating current in the circuit will be equal to the value of the current source. Thus $v_R = 3(-2) = -6$ V

113. The voltage v in the circuit of given figure is

(a) 3 V

(b) 0 V

(c) 10 V

(d) 13 V

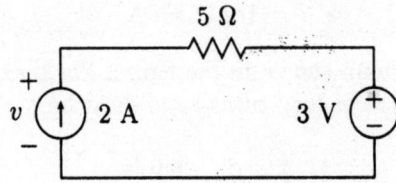

Hints : The circulating current is 2 A, then by KVL

$-v + 5 \times 2 + 3 = 0 \quad \Rightarrow v = 13$ V

114. A 10 V battery with an internal resistance of 1 Ω is connected across a non-linear load whose $v - i$ characteristic is given by

$7i = v^2 + 2v$

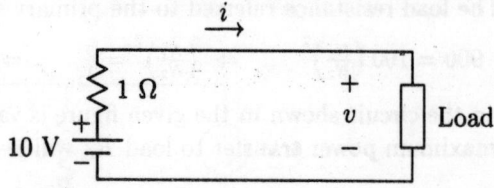

The current delivered by the battery is

(a) 2.5 A   (b) 5 A   (c) 6 A   (d) 7 A

IES-98(EE)

Hints : With reference to the figure

$v = 10 - i$ Thus

$7i = (10 - i)^2 + 2(10 - i)$

or, $i^2 - 29i + 120 = 0$

or, $i = \frac{29 \pm \sqrt{29^2 - 480}}{2}$

or, $i = 5, 24$ Thus $i = 5$ A

115. The V-I characteristic for the network shown in the given box is $V = 4I - 9$. If now a resistor $R = 2\,\Omega$ is connected across it, then the value of I will be

(a) -4.5 A   (b) -1.5 A   (c) 1.5 A   (d) 4.5 A

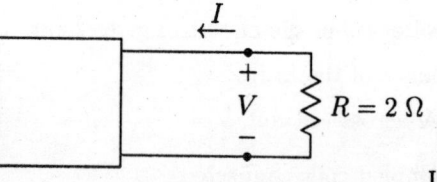

IES-95(EE)

Hints : $V = -2I$ Setting this value in the given relation, we have

$-2I = 4I - 9$ $\Rightarrow I = 3/2 = 1.5$ A

116. Consider the circuit shown in the figure. For maximum power transfer to the load, the primary to secondary turns ratio must be

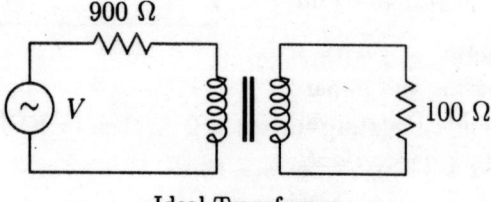

Ideal Transformer

(a) 9 : 1   (b) 3 : 1   (c) 1 : 3   (d) 1 : 9

IES-95(EE)

Hints : The load resistance referred to the primary should be equal to 900 ohm

Thus $\quad 900 = 100 \left(\frac{N_1}{N_2}\right)^2 \quad \Rightarrow \left(\frac{N_1}{N_2}\right)^2 = \frac{9}{1} \quad \Rightarrow \frac{N_1}{N_2} = \frac{3}{1}$

117. If the R_s in the circuit shown in the given figure is variable between 20 Ω and 80 Ω then the maximum power transfer to load R_L will be

(a) 15 W

(b) 13.33 W

(c) 6.67 W

(d) 2.4 W

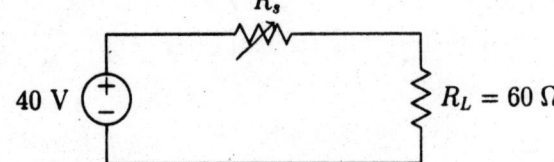

IES-95(EE)

Hints : For maximum power transfer $R_L = R_s = 60$ Ω

Thus $P_{max} = \left(\frac{40}{60+60}\right)^2 \times 60 = 6.67$ W

118. **Q 10.10** A coil having a resistance of 5 ohms and inductance of 0.1 H connected in series with a condenser of capacitance 50 μF. A constant alternating voltage of 200 volts is applied to the circuit. The voltage across the coil at resonance is

(a) 200 V   (b) 1788 V   (c) 1800 V   (d) 2000 V

IES-94(EE)

Hints : $\omega_0 = \frac{1}{\sqrt{LC}} = 2 \times 10^5$

Current at resonance, $I = \frac{200}{5} = 40$ A

Impedance of the coil, $Z = \sqrt{R^2 + \omega^2 L^2} = 45$ ohm

Voltage across the coil, $V = ZI = 1800$ V

119. Two coupled-coils connected in series have an equivalent inductance of 16 mH or 8 mH depending upon the interconnection. Then the mutual inductance M between the coils is

(a) 12 mH   (b) $8\sqrt{2}$ mH   (c) 4 mH   (d) 2 mH

IES-94(EE)

Hints : $L_1 + L_2 + 2M = 16$ and $L_1 + L_2 - 2M = 8$. On subtracting, we get $4M = 8 \quad \Rightarrow M = 2$ mH

120. Two inductive coil with self-inductances L_1 and L_2 are magnetically coupled in series opposing and in parallel aiding respectively. The mutual inductance between the coils is M. The equivalent inductances are respectively

(a) $L_1 + L_2 + 2M, \frac{L_1 L_2 - M^2}{L_1 + L_2 - 2M}$ (b)$L_1 + L_2 - 2M, \frac{L_1 L_2 - M^2}{L_1 + L_2 + 2M}$

(c) $L_1 + L_2 - 2M, \frac{L_1 L_2 - M^2}{L_1 + L_2 - 2M}$ (d) $L_1 + L_2 + 2M, \frac{L_1 L_2 - M^2}{L_1 + L_2 + 2M}$

IES-96(EE)

Hint : Ans (c)

121. A current shown in the given figure passes through a pure inductance of 3 mH. The instantaneous power in watts, during 0 ¡ t ¡ 2 ms is

    (a) 25000t

    (b) 50000t

    (c) 75000t

    (d) 100000t

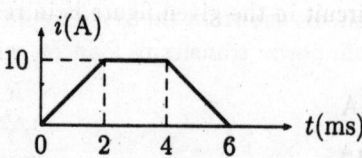

IES-96(EE)

Hints : $p = vi = L\frac{di}{dt}i = 3 \times 10^{-3} \times \frac{10}{2 \times 10^{-3}} \times \frac{10}{2 \times 10^{-3}}t = 75000t$

122. A series R-L-C circuit when excited by a 10 V sinusoidal voltage source of variable frequency, exhibits resonance at 100 Hz and has a 3 dB bandwidth of 5 Hz. The voltage across the inductor L at resonance is

    (A) 10 V   (B) $10\sqrt{2}$ V   (C) $10/\sqrt{2}$ (D) 200 V

GATE-99(EE)

Hints : $BW = \Delta f = \frac{f_0}{Q_0}$ or, $5 = 100/Q_0$ or, $Q_0 = 20$

Voltage across inductor $V_L = Q_0 V = 20 \times 10 = 200$ V

123. The current in the circuit of given figure, is

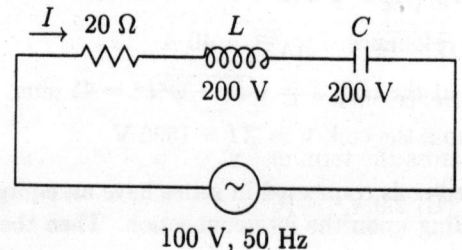

    (A) 5 A   (B) 10 A   (C) 15 A   (D) 25 A

GATE-99(EE)

Hints : $100 = \sqrt{V_R^2 + (200 - 200)^2}$

    $\Rightarrow V_R = 100$ V.

Thus $I = 100/20 = 5$ A

124. The voltage in the circuit of given figure is

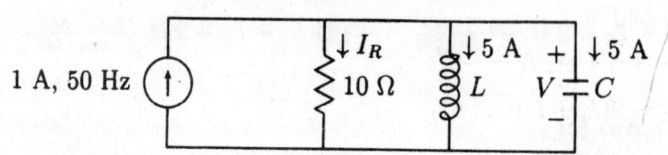

(a) $10/\sqrt{2}$ V (b) $10\sqrt{2}$ (c) 100 V (d) 10 V

Hints : $1 = \sqrt{I_R^2 + (5-5)^2}$ $\Rightarrow I_R = 1$ A. Thus $V = 10 \times 1 = 10$ V

125. The parallel RLC circuit in the given figure is in resonance . In this circuit

(A) $|I_R| < 1$ mA

(B) $|I_R + I_L| > 1$ mA

(C) $|I_R + I_C| < 1$ mA

(D) $|I_L + I_C| > 1$ mA

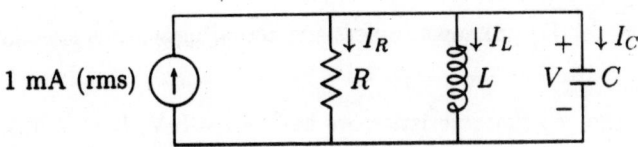

Hints :

At resonance, $|I_L| = |I_C|$, therefore

$|I_R| = 1$ mA

$|I_R + I_L| = \sqrt{I_R^2 + I_L^2} > 1$ mA

$|I_R + I_C| = \sqrt{I_R^2 + I_C^2} > 1$ mA

$|I_L + I_C| = 0$ Hence answer is (B)

126. The voltage across the terminals a and b in the given figure is

(A) 0.5 V (B) 2.0 V (C) 3.5 V (D) 4.0 V

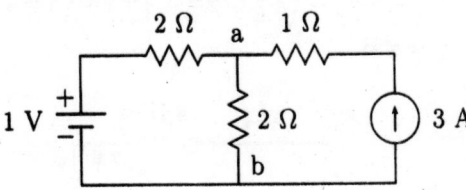

GATE-98(EE)

Hints : Using mesh analysis

$I_2 = 3$ A and

$4I_1 + 2 \times 3 = 1$ or, $I_1 = -5/4$ A

Thus $V_{ab} = 2(I_1 + I_2) = (-5/4 + 3) \times 2 = 3.5$ V

127. The v-i characteristic as seen from the terminal pair (A,B) of the network of given figure. If an inductance of value 6 mH is connected across the terminal pair (A,B), the time constant of the system will be

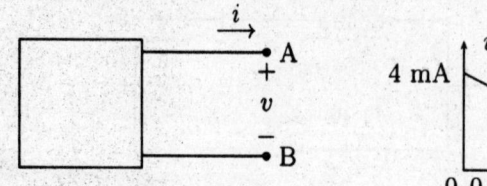 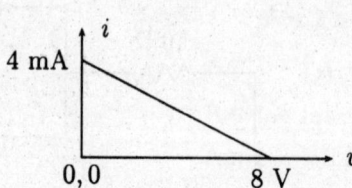

(A) 3 μsec   (B) 12 sec

(C) 32 sec   (D) unknown, unless the actual network is specified.

GATE-96(EE)

Hints : From v-i characteristics, we have $V_{oc} = 8$ V, $I_{sc} = 4$ mA

Then $R_{Th} = V_{oc}/I_{sc} = 2 \times 10^3\ \Omega$

Time constant $\tau = L/R_{Th} = (6 \times 10^{-3})/(2 \times 10^3) = 3\ \mu$sec

128. For the given circuit, the inductance seen by the terminal (A,B) is

(A) 12 H   (B) 4 H   (C) 8 H   (D) 16 H

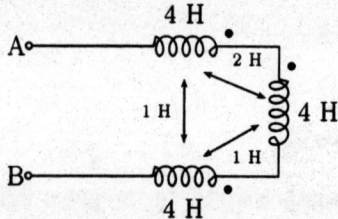

Hints : By KVL

$v = \left(4\frac{di}{dt} - 2\frac{di}{dt} - 1\frac{di}{dt}\right) + \left(4\frac{di}{dt} - 2\frac{di}{dt} + 1\frac{di}{dt}\right) + \left(4\frac{di}{dt} + 1\frac{di}{dt} - 1\frac{di}{dt}\right) = 8\frac{di}{dt}$

Therfore $L_{eq} = 8$ H

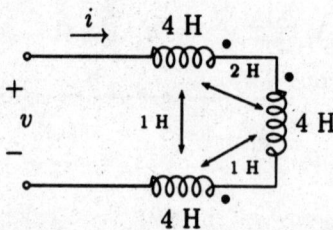

129. The following circuit resonates at

    (A) all frequencies

    (B) 0.5 rad/sec

    (C) 5 rad/sec

(D) 1 rad/sec

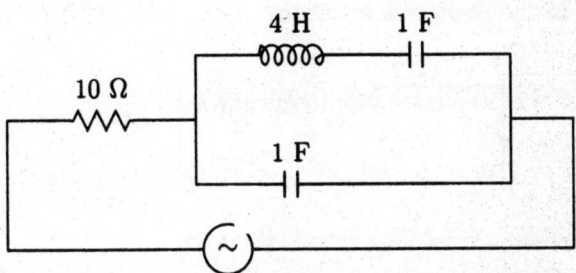

Hints :

$$Z = 10 + [(j4\omega + 1/j\omega)\|(1/j\omega)]$$
$$= 10 + \frac{(1/j\omega)(j4\omega + 1/j\omega)}{j4\omega + 1/j\omega + 1/j\omega}$$
$$= 10 + \frac{4 - 1/\omega^2}{j(4\omega - 2/\omega)}$$
$$= 10 - j\frac{4 - 1/\omega^2}{4\omega - 2/\omega}$$

For resonance $4 - \frac{1}{\omega^2} = 0$ $\Rightarrow \omega = 0.5$

130. A series RLC circuit has the following parameter values: R = 10 Ω, L = 0.01 H, C = 100 μF. The Q factor of the circuit at resonance is

(A) 0.5 (B) 1/3 (C) 1 (D) zero

Hints : $\omega_0 = \frac{1}{\sqrt{LC}} = \frac{1}{\sqrt{0.01 \times 100 \times 10^{-6}}} = 10^3$ rad/sec

$Q = \frac{\omega_0 L}{R} = \frac{10^3 \times 0.01}{10} = 1$

131. Two 2 H inductor coils are connected to a series and are and also magnetically coupled to each other, the coefficient of coupling being 0.1 . The inductance of the combination can be

(A) 0.4 H (B) 3.2 H (C) 4.0 H (D) 4.4 H

GATE-95(EC)

Hints :

$$L_{eq} = L_1 + L_2 \pm 2k\sqrt{L_1 L_2}$$
$$= 2 + 2 \pm 2 \times 0.1 \times \sqrt{2 \times 2}$$
$$= 4.4 \text{ H} \quad \text{or,} \quad 3.6 \text{ H}$$

132. For the series RLC circuit of given figure, the phasor diagram at a certain frequency is given. The operating frequency of the circuit is

(A) equal to the resonance frequency

(B) less than the resonance frequency

(C) greater than the resonance frequency

(D) not zero

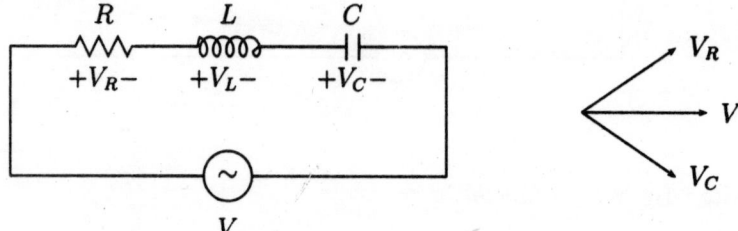

GATE-92(EC)

Hints: From the phasor diagram it is noted that V_R (hence I) leads the V, and this results in a capacitive circuit. Hence operating frequency is less than the resonant frequency.

133. In the circuit of given figure, the current i_D through the ideal diode (zero cutin voltage and forward resistance) equals

(A) 0 A

(B) 4 A

(C) 1 A

(D) None of the above

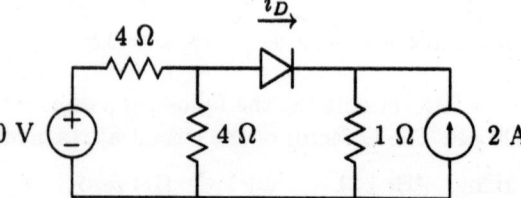

GATE-97(EC)

Hints : Find the Thevenin's equivalent across the diode terminals as :

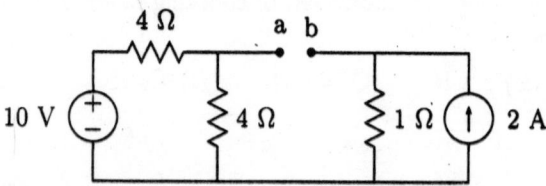

$V_{ab} = V_a - V_b = \frac{4}{4+4}(10) - (1 \times 2) = 5 - 2 = 3 \text{ V}$

$R_{Th} = (4||4) + 1 = 2 + 1 = 3 \, \Omega$

Hence the current through the diode will be

$$i_D = \frac{V_{ab}}{R_{Th}} = \frac{3}{3} = 1 \text{ A}$$

134. The voltage V in the circuit of given figure is equal to

    (A) 3 V   (B) -3 V   (C) 5 V   (D) None of the above

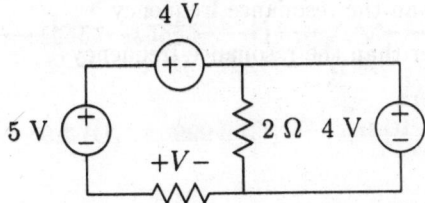

GATE-97(EC)

Hints : By KVL in the left mesh gives

$$4 + 4 - V - 5 \quad \Rightarrow V = 3 \text{ V}$$

135. The voltage V in the circuit of given figure is

    (A) 10 V   (B) 15 V   (C) 5 V   (D) None of the above

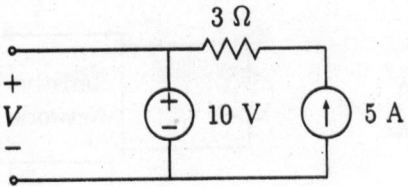

GATE-97(EC)

Hint : Ans (B)

136. If the secondary winding of the ideal transformer shown in the given figure has 40 turns, the number of turns in the primary winding for maximum power to the 2 Ω resistor will be

    (A) 20   (B) 40   (C) 80   (D) 160

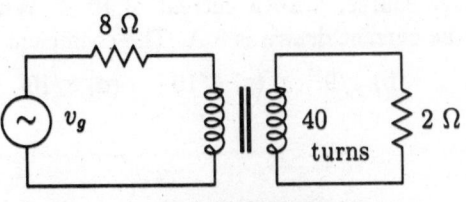

Ideal Transformer

GATE-93(EC)

Hints : For the maximum power transfer to the load

$$8 = \left(\frac{N_1}{40}\right)^2 \times 2 \quad \Rightarrow N_1/40 = 2 \quad \Rightarrow N_1 = 80$$

137. In the series circuit shown in figure given below. For the series resonance, the value of the coupling coefficient k will be

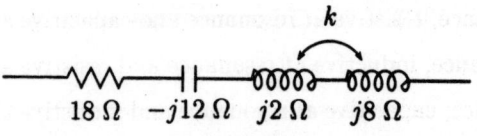

$$18 \,\Omega \quad -j12 \,\Omega \quad j2 \,\Omega \quad j8 \,\Omega$$

(A) 0.25   (B) 0.5   (C) 0.999   (D) 1.0

GATE-93(EE)

Hints : At resonance the reactive component of impedance will be zero, i.e.;

$$X = -12 + 2 + 8 + 2\omega M = 0 \quad \Rightarrow \omega M = 1$$

and $k = \frac{M}{\sqrt{L_1 L_2}} = \frac{\omega M}{\sqrt{(\omega L_1)(\omega L_2)}} = \frac{1}{\sqrt{2 \times 8}} = 1/4 = 0.25$

138. A dc circuit shown in the given figure has a voltage V, a current source I and several resistors. A particular resistor R dissipates a power of 4 watts when V alone is active. The same resistor R dissipates a power of 9 watts when I alone is active. The power dissipated by R when both sources are active will be

(A) 1 W

(B) 5 W

(C) 13 W

(D) 25 W

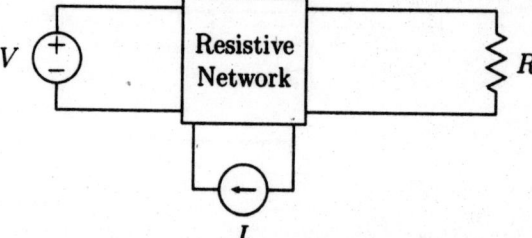

GATE-93(EC)

Hints : Given That $I'^2 R = 4$ and $I''^2 R = 9$ then

$\left(\frac{I'}{I''}\right)^2 = \frac{4}{9}$ or, $\frac{I'}{I''} = \frac{2}{3}$ or, $\frac{I'+I''}{I''} = \frac{2+3}{3} = \frac{5}{3}$

or, $\frac{(I'+I'')^2 R}{I''^2 R} = \frac{25}{9}$ or, $(I' + I'')^2 R = \frac{25}{9}(I''^2 R) = \frac{25}{9} \times 9 = 25$ W

139. Two identical coils of negligible resistance, when connected in series across a 50 Hz fixed voltage source, draw a current of 10 A. When the terminal of the coils are reversed, the current drawn is 8 A. The coefficient of coupling between two coils is

(a) 1/100   (b) 1/9   (c) 4/10   (d) 8/10

IES-98

Hints :

$\frac{V}{\omega(2L-2M)} = 10$ and $\frac{V}{\omega(2L+2M)} = 8$

Thus $\frac{L+M}{L-M} = \frac{10}{8} = \frac{5}{4}$ $\Rightarrow 4(L + M) = 5(L - M)$ $\Rightarrow 9M = L$ $\Rightarrow$

$M/L = 1/9$ Now $k = \frac{M}{\sqrt{L^2}} = M/L = 1/9$

140. As the frequency of a series RLC circuit is varied, the circuit behaves

(a) capacitive before resonance, resistive at resonance and inductive after resonance.

(b) inductive before resonance, resistive at resonance and capacitive after resonance.

(c) capacitive before resonance, inductive at resonance and resistive after resonance.

(d) resistive before resonance, capacitive at resonance and inductive after resonance.

Hint : Ans (a)

141. As the frequency of a parallel RLC circuit is varied, the circuit behaves

(a) capacitive before resonance, resistive at resonance and inductive after resonance.

(b) inductive before resonance, resistive at resonance and capacitive after resonance.

(c) capacitive before resonance, inductive at resonance and resistive after resonance.

(d) resistive before resonance, capacitive at resonance and inductive after resonance.

Hint : Ans (b)

142. The series RLC circuit in the given figure is in resonance. In this circuit

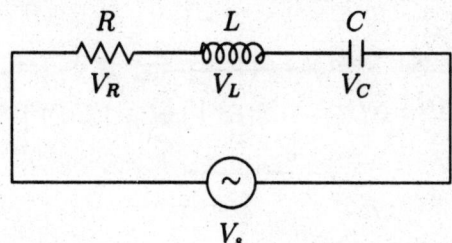

(a) $|V_s| = |V_R|$, $|V_s| < |V_L|$, $|V_s| < |V_C|$, and $|V_L| = |V_C|$

(b) $|V_s| = |V_R|$, $|V_s| > |V_L|$, $|V_s| > |V_C|$, and $|V_L| = |V_C|$

(c) $|V_s| > |V_R|$, $|V_s| > |V_L|$, $|V_s| > |V_C|$, and $|V_L| = |V_C|$

(d) none of the above

Hint : Ans (a)

143. The frequency at which the voltage across the capacitor is maximum in a series RLC circuit is

(a) $\frac{1}{2\pi}\sqrt{\frac{1}{LC} + \frac{R^2}{2L^2}}$ (b) $\frac{1}{2\pi}\sqrt{\frac{1}{LC} - \frac{R^2}{2L^2}}$

(c) $\frac{1}{2\pi}\sqrt{\frac{R^2}{2L^2} - \frac{1}{LC}}$ (d) None

Hints : For a series RLC circuit

$$I = \frac{V}{|Z|} = \frac{V}{\sqrt{R^2 + (\omega L - 1/\omega C)^2}}$$

$$|V_C| = X_C I = \frac{V}{\omega C \sqrt{R^2 + (\omega L - 1/\omega C)^2}}$$

The required frequency is determined from the equation $\frac{d|V_C|}{d\omega} = 0$, i.e;

$\frac{d|V_C|}{dt} = -\frac{V}{\omega^2 C}[R^2 + (\omega L - 1/\omega C)^2]^{-1/2} + \frac{V}{\omega C}(-1/2)[R^2 + (\omega L - 1/\omega C)^2]^{-3/2}(2)(\omega L - 1/\omega C)(L + 1/\omega^2 C) = 0$

$\Rightarrow \frac{1}{\omega} + \frac{(\omega L - 1/\omega C)(L + 1/\omega^2 C)}{R^2 + (\omega L - 1/\omega C)} = 0$

$\Rightarrow R^2 + 2\omega^2 L^2 - 2L/C = 0$

$\Rightarrow \omega = \omega_0 = \sqrt{\frac{1}{LC} - \frac{R^2}{2L^2}}$

$\Rightarrow f_0 = \frac{1}{2\pi}\sqrt{\frac{1}{LC} - \frac{R^2}{2L^2}}$

144. **Q 10.39** The frequency at which the voltage across the inductor is maximum in a series RLC circuit is

(a) $\frac{1}{2\pi\sqrt{LC + R^2 C^2/2}}$ (b) $\frac{1}{2\pi\sqrt{R^2 C^2/2 - LC}}$

(c) $\frac{1}{2\pi\sqrt{LC - R^2 C^2/2}}$ (d) None

Hints : For a series RLC circuit

$I = \frac{V}{|Z|} = \frac{V}{\sqrt{R^2 + (\omega L - 1/\omega C)^2}}$

$|V_L| = X_L I = \frac{\omega L V}{\sqrt{R^2 + (\omega L - 1/\omega C)^2}}$

The required frequency is determined from the equation $\frac{d|V_L|}{d\omega} = 0$, i.e;

$\frac{d|V_L|}{d\omega} = LV[R^2 + (\omega L - 1/\omega C)^2]^{-1/2} + \omega L V(-1/2)[R^2 + (\omega L - 1/\omega C)^2]^{-3/2}(2)(\omega L - 1/\omega C)(L + 1/\omega^2 C) = 0$

$\Rightarrow R^2 + 2/\omega^2 C^2 - 2L/C = 0$

$\Rightarrow 2/\omega^2 C^2 = 2L/C - R^2$

$\Rightarrow \omega = \omega_L = \frac{1}{\sqrt{LC - R^2 C^2/2}}$

$\Rightarrow \omega = \frac{1}{2\pi\sqrt{LC - R^2 C^2/2}}$

145. Q-factor of a series RL circuit excited by voltage source $V_m \sin \omega t$ is

(a) $\omega R/L$ (b) $\omega L/R$ (c) $\frac{1}{2}\frac{\omega L}{R}$ (d) none

Hints :

$$Q = 2\pi \frac{\text{Maximum energy stored per cycle}}{\text{Energy dissipated per cycle}}$$

$$= 2\pi \frac{(1/2)LI_m^2}{(I_m/\sqrt{2})^2 RT} = 2\pi \frac{(1/2)LI_m^2}{I_m^2 R/2f} = \frac{2\pi f L}{R} = \frac{\omega L}{R}$$

146. Q-factor of a series RC circuit excited by voltage source $V_m \sin \omega t$ is

(a) $\omega C/R$ (b) ωCR (c) $\frac{1}{\omega CR}$ (d) none

Hints :

$$Q = 2\pi \frac{\text{Maximum energy stored per cycle}}{\text{Energy dissipated per cycle}}$$

$$= 2\pi \frac{(1/2)CV_m^2}{(I_m/\sqrt{2})^2 RT} = 2\pi \frac{(1/2)C(\frac{1}{\omega C}I_m)^2}{I_m^2 R/2f} = \frac{1}{\omega CR}.$$

147. Q-factor of a parallel RC circuit excited by voltage source $V_m \sin\omega t$ is

    (a) $\omega C/R$ (b) ωCR (c) $1/\omega CR$ (d) none

    Hints :

    $$Q = 2\pi \frac{(1/2)CV_m^2}{V_m^2/2Rf} = \omega CR$$

148. The bandwidth of a series RLC circuit is

    (a) ω_0/Q_0 (b) Q_0/ω_0 (c) $Q_0\omega_0$ (d) none

    Hints :

    For a series RLC circuit

    $$I = \frac{V}{\sqrt{R^2 + (\omega L - 1/\omega C)^2}}$$

    We need to determine those frequencies where current has dropped to $\frac{1}{\sqrt{2}}$ times its resonance value. Thus

    $$\frac{V}{\sqrt{R^2 + (\omega L - 1/\omega C)^2}} = \frac{V}{\sqrt{2}R}$$

    $$\Rightarrow \sqrt{R^2 + (\omega L - 1/\omega C)^2} = \sqrt{2}R$$

    $$\Rightarrow \omega L - 1/\omega C = \pm R \qquad \Rightarrow \omega^2 \pm \frac{R}{L}\omega - \frac{1}{LC} = 0$$

    Roots are $\quad \omega_{1,2} = \pm R/2L \pm \sqrt{(R/2L)^2 + 1/LC} = \pm\zeta\omega_0 \pm \sqrt{\zeta^2\omega_0^2 + \omega_0^2}$

    $= \omega_0[\pm\zeta \pm \sqrt{\zeta^2 + 1}]$

    For the positive value of $\omega \quad \omega_{1,2} = \omega_0[\sqrt{\zeta^2 + 1} \pm \zeta]$

    Thus $\quad \omega_1 = \omega_0[\sqrt{\zeta^2 + 1} - \zeta]$ and $\quad \omega_2 = \omega_0[\sqrt{\zeta^2 + 1} + \zeta]$

    And bandwith is $\quad \omega_2 - \omega_1 = 2\zeta\omega_0 = R/L = \frac{\omega_0}{\omega_0 L/R} = \omega_0/Q_0$

149. For the circuit of given figure, the value of R for which the maximum power transfer to 100 ohm load takes place is

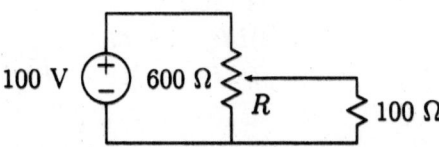

    (a) 100 ohm   (b) 600 ohm   (c) 600/7 ohm   (d) none

Hints : The power transfer to 100 ohm resistor will be maximum only when the voltage across (or current through) it is maximum. This happens when R becomes 600 ohm.

150. In series RLC circuit, the phase angles at half power frequencies are respectively

    (a) 45^0 at ω_1 and -45^0 at ω_2 (b) -45^0 at ω_1 and $+45^0$ at ω_2

    (c) 90^0 at ω_1 and -90^0 at ω_2 (d) -90^0 at ω_1 and $+90^0$ at ω_2

    Hints: With reference to the hints given for the question 148, at the half power frequencies

    $$\omega L - 1/\omega C = \pm R \quad \Rightarrow X = \pm R \text{ and the phase angle is}$$

    $$\phi = \tan^{-1}(\pm X/R) = \tan^{-1}(\pm 1) = \pm 45^0$$

    Since series RLC circuit is capacitive before resonance and inductive after resonance, thus answer is (a)

151. If ω_1 and ω_2 are the half power frequencies, then the resonant frequency is

    (a) $\omega_r = \frac{\omega_1 + \omega_2}{2}$ (b) $\omega_r = \omega_1/\omega_2$

    (c) $\omega_r = \frac{\omega_2 - \omega_1}{2}$ (d) $\omega_r = \sqrt{\omega_1 \omega_2}$

    Hints : With reference to the hints given for the question 148, we have

    $$\omega_1 = \omega_0[\sqrt{\zeta^2 + 1} - \zeta] \text{ and } \omega_2 = \omega_0[\sqrt{\zeta^2 + 1} + \zeta]$$

    Thus $\omega_1 \omega_2 = \omega_0^2 \quad \Rightarrow \omega_0 = \sqrt{\omega_1 \omega_2}$

    ****

Index